BIT-INTERLEAVED CODED MODULATION

BIT-INTERLEAVED CODED MODULATION

FUNDAMENTALS, ANALYSIS, AND DESIGN

Leszek Szczecinski

Institut National de la Recherche Scientifique (INRS)
University of Quebec
Montreal, Canada

Alex Alvarado

Department of Electronic & Electrical Engineering
University College London
London, United Kingdom

IEEE

WILEY

This edition first published 2015

Registered office
John Wiley & Sons Ltd, The Atrium, Southern Gate, Chichester, West Sussex, PO19 8SQ, United Kingdom

Library of Congress Cataloging-in-Publication Data

Szczecinski, Leszek.
Bit-interleaved coded modulation : fundamentals, analysis, and design / Leszek Szczecinski, Alex Alvarado.
pages cm
Includes bibliographical references and index.
978-0-470-68617-1 (hardback)
1. Bit-interleaved coded modulation. I. Alvarado, Alex. II. Title.
TK5103.256.S93 2015
621.384—dc23

2014022267

Typeset in 9/11pt TimesLTStd by Laserwords Private Limited, Chennai, India

Printed in Singapore by C.O.S. Printers Pte Ltd

1 2015

A ma femme, Karina, LS

A mi familia, Alex

Contents

About the Authors

Leszek Szczecinski is an Associate Professor at Institut National de la Recherche Scientifique (INRS-EMT), University of Quebec, Montreal, Canada, and an Adjunct Professor at the Electrical and Computer Engineering Department, McGill University, Montreal. In 1998–2001 he was an Assistant Professor at the Department of Electrical Engineering, University of Chile, and in 2009–2010 he was, as a Marie Curie Research Fellow, with CNRS, Laboratory of Signals and Systems, Gif-sur-Yvette, France. His research interests are in the area of communication theory, modulation and coding, ARQ, wireless communications, and digital signal processing.

Alex Alvarado is a Senior Research Associate at the Optical Networks Group, University College London, United Kingdom. In 2012–2014 he was a Marie Curie Intra-European Fellow at the University of Cambridge, United Kingdom, and during 2011–2012 he was a Newton International Fellow at the same institution. He has visited INRS-EMT on various occasions during 2007–2013. Dr. Alvarado's research interests are in the general areas of digital communications, coding, and information theory.

Preface

The objective of this book is to examine the main aspects of a particularly simple coded modulation scheme known as *bit-interleaved coded modulation* (BICM). BICM is based on bit-level operations, which allows for a straightforward combination of a binary encoder with a nonbinary modulation. Simplicity and flexibility are the reasons for the omnipresence of BICM in current communication systems and also what motivated us to look into BICM-based transceivers.

This work is intended for readers with a background in communication theory who are interested in theoretical aspects of BICM transmission. Various issues we have addressed in this work are motivated by questions we received from the audience during talks and lectures we have given over the years. We have tried to make the content tutorial in style hoping that the numerous examples we have included clarify and illustrate well the main results. The objective was to make the text accessible to graduate students and researchers working on coded modulation and BICM-related issues. To clearly delimitate the main results, we have separated the text—as much as possible—into definitions, lemmas, theorems, and corollaries.

From the beginning of this project, our objective was threefold. First, we wanted to clearly define all the building blocks in BICM transceivers. This is mostly done in Chapters 1–3, which contain definitions and some simple examples. When writing the text, we discovered that many of the "conventional" assumptions made in BICM, while often justified, were not necessarily as obvious as they seemed. We thus introduced elements of analysis to help explain these assumptions, often taken for granted.

The second objective is to present tools that allow us to analyze BICM receivers. This is the core of Chapters 4 and 6, which provide complementary views on the performance limits of BICM receivers, the former from an information-theoretic perspective and the latter from a communication-theoretic point of view. For completeness, in parallel with the analysis of BICM, we also present the analysis of the optimal (maximum-likelihood) decoder. Chapter 5 focuses on the probabilistic characterization of the so-called L-values (log-likelihood ratios), as these signals are the most distinguishable signature of BICM receivers. Chapters 4–6 can be read independently of each other.

The third objective was to indicate how the apparently simple BICM transceivers can be designed and optimized. This is done in Chapters 7–9, where we define and analyze the effects of mismatched L-values as well as focus on the design of the interleaver and code.

We opted in this book for a reference-free text, but to acknowledge the inspiration and contributions of many researchers, we have included at the end of each chapter a section with bibliographic notes. Owing to the large number of publications available in this area, we might have missed some previous works. We sincerely apologize if this is the case.

Acknowledgements

We would like to express our gratitude to those who, in one way or another, helped us to get through with this project. First of all, our sincere thanks go to those who motivated us to start and continue this work. In particular to Jacob Benesty (Institut National de la Recherche Scientifique, Canada) who provided inspiration to start this work and continuously encouraged us along the way. We also thank Lajos Hanzo (University of Southampton, UK) whose friendly advice pushed us into this adventure.

We would also like to thank our collaborators who participated in various aspects of this work, most often as coauthors of publications. In particular, we thank Rodolfo Feick (Universidad Técnica Federico Santa María, Chile), Erik Agrell, Fredrik Brännström, Arne Svensson, Alexandre Graell i Amat, Christian Häger and Mikhail Ivanov (all from Chalmers University of Technology, Sweden), Albert Guillèn i Fábregas and Alfonso Martinez (both from Universitat Pompeu Fabra, Spain), and Tobias Koch (Universidad Carlos III de Madrid, Spain). Many thanks to the alumni of Universidad Técnica Federico Santa María: Francisco Pellegrini, Christian Gonzalez, Marcos Bacic, Rolando Bettancourt, Andrés Cerón, Víctor Nuñez, as well as the alumni of Institut National de la Recherche Scientifique: Mustapha Benjillali and Thomas Chowdhury.

We are particularly grateful to Mikhail Ivanov, Christian Häger, Fredrik Brännström, Maël Le Treust (Centre National de la Recherche Scientifique, France), Martin Senst, and Jossy Sayir (University of Cambridge, UK) for providing feedback on preliminary versions of the manuscript. Thanks to Mr Aata El Hamss (Institut National de la Recherche Scientifique) for help with graphical presentation of some of the results.

We would also like to thank Wiley's staff: Mr Mark Hammond, Ms Suzan Barclay, and Ms Liz Wingett who were extremely patient with us during the project.

We acknowledge the support we received from the institutions we have had the chance to work at, including Institut National de la Recherche Scientifique (Canada), Chalmers University of Technology (Sweden), Laboratoire de Signaux et Systèmes of Centre National de la Recherche Scientifique (France), the University of Cambridge and University College London (UK).

Last but not least, our thanks go to different funding institutions which, over the years, financially supported our work. This includes the British Academy and The Royal Society, UK (under the Newton International Fellowship scheme), the European Community's Seventh's Framework Programme (under FP7/2007-2013 grants #271986 and #236068), the Swedish Research Council (under research grant #2006-5599), the Department of Signals and Systems, Chalmers University of Technology (under the Solveig and Karl G. Eliasson Memorial Fund), the European Commission (under projects NEWCOM++ #216715), the Comisión Nacional de Ciencia y Tecnología, Chile (under research grant PBCT-ACT-11/2004), the Natural Sciences and Engineering Research Council of Canada, and the Fonds québécois de la recherche sur la nature et les technologies, Quebec, Canada.

Acronyms

3GPP	third-generation partnership project
AG	asymptotic gain
AMC	adaptive modulation and coding
AWGN	additive white Gaussian noise
BCJR	Bahl–Cocke–Jelinek–Raviv
BEP	bit-error probability
BICM	bit-interleaved coded modulation
BICM-GMI	BICM generalized mutual information
BICM-ID	BICM with iterative demapping
BICO	binary-input continuous-output
BPSK	binary phase-shift keying
BRGC	binary reflected Gray code
BSGC	binary semi-Gray code
CBEDS	constellation bitwise Euclidean distance spectrum
CC	convolutional code
CDF	cumulative distribution function
CEDS	constellation Euclidean distance spectrum
CENC	convolutional encoder
CGF	cumulant-generating function
CM	coded modulation
CoM	consistent model
CoRe	constellation rearrangement
DS	distance spectrum
DVB	digital video broadcasting
ED	Euclidean distance
EDS	Euclidean distance spectrum
EP	edge profile
FBC	folded binary code
FHD	free Hamming distance
FLSA	full linear search algorithm
GCh	Gauss–Chebyshev
GDS	generalized distance spectrum
GH	Gauss–Hermite
GHW	generalized Hamming weight
GIDS	generalized input-dependent distance spectrum
GIODS	generalized input-output distance spectrum
GIWD	generalized input-dependent weight distribution

GMI	generalized mutual information
GWD	generalized weight distribution
HARQ	hybrid automatic repeat request
HD	Hamming distance
HW	Hamming weight
i.i.d.	independent and identically distributed
i.u.d.	independent and uniformly distributed
IDS	input-dependent distance spectrum
IEDS	input-dependent Euclidean distance spectrum
IWD	input-dependent weight distribution
l.h.s.	left-hand side
LDPC	low-density parity-check
LLR	logarithmic likelihood ratio
M-interleaver	multiple-input interleaver
MAP	maximum a posteriori
MED	minimum Euclidean distance
MFHD	maximum free Hamming distance
MFLSA	modified full linear search algorithm
MGF	moment-generating function
MI	mutual information
MIMO	multiple-input multiple-output
ML	maximum likelihood
MLC	multilevel coding
MSB	most significant bit
MSD	multistage decoding
MSP	modified set-partitioning
MUX	multiplexer
NBC	natural binary code
ODS	optimal distance spectrum
OFDM	orthogonal frequency-division multiplexing
PAM	pulse amplitude modulation
PCCENC	parallel concatenated convolutional encoder
PDF	probability density function
PDL	parallel decoding of the individual levels
PEP	pairwise-error probability
PMF	probability mass function
PSK	phase-shift keying
QAM	quadrature amplitude modulation
r.h.s.	right-hand side
S-interleaver	single-input interleaver
SED	squared Euclidean distance
SEP	symbol-error probability
SIR	signal-to-interference ratio
SL	Shannon limit
SNR	signal-to-noise ratio
SP	set-partitioning
SPA	saaddlepoint approximation
SSP	semi-set-partitioning
TC	turbo code
TCM	trellis-coded modulation
TENC	turbo encoder

TP	transition probability
TTCM	turbo trellis-coded modulation
u.d.	uniformly distributed
UEP	unequal error protection
WD	weight distribution
WEP	word-error probability
WLSF	weighted least-squares fit
ZcM	zero-crossing model

1

Introduction

1.1 Coded Modulation

The main challenge in the design of communication systems is to reliably transmit digital information (very often, bits generated by the source) over a medium which we call the *communication channel* or simply the *channel*. This is done by mapping a sequence of N_b bits $\underline{i} = [i[1], \ldots, i[N_b]]$ to a sequence of N_s *symbols* $\underline{\boldsymbol{x}} = [\boldsymbol{x}[1], \ldots, \boldsymbol{x}[N_s]]$. These symbols are then used to vary (modulate) parameters of the continuous-time waveforms (such as amplitude, phase, and/or frequency), which are sent over the channel every T_s seconds, i.e., at a symbol rate $R_s = 1/T_s$. The transmission rate of the system is thus equal to

$$\mathsf{R} \triangleq \frac{N_b}{N_s T_s} = R_s \frac{N_b}{N_s} \left[\frac{\text{bit}}{\text{s}}\right], \tag{1.1}$$

where the bandwidth occupied by the waveforms is directly proportional to the symbol rate R_s. Depending on the channel and the frequency used to carry the information, the waveforms may be electromagnetic, acoustic, optical, etc.

Throughout this book we will make abstraction of the actual waveforms and instead, consider a discrete-time model where the sequence of symbols is transmitted through the channel resulting in a sequence of received symbols $\underline{\boldsymbol{y}} = [\boldsymbol{y}[1], \ldots, \boldsymbol{y}[N_s]]$. In this discrete-time model, both the transmitted and received symbol at each time instant are N-dimensional column vectors. We also assume that linear modulation is used, that the transmitted waveforms satisfy the Nyquist condition, and that the channel is memoryless. Therefore, assuming perfect time/frequency synchronization, it is enough to model the relationship between the transmitted and received signals at time n, i.e.,

$$\boldsymbol{y}[n] = h[n]\boldsymbol{x}[n] + \boldsymbol{z}[n]. \tag{1.2}$$

In (1.2), we use $h[n]$ to model the channel attenuation (gain) and $\boldsymbol{z}[n]$ to model an unknown interfering signal (most often the noise). Using this model, we analyze the transmission rate (also known as *spectral efficiency*):

$$R \triangleq \frac{N_b}{N_s} \left[\frac{\text{bit}}{\text{symbol}}\right], \tag{1.3}$$

which is independent of R_s, thus allowing us to make abstraction of the bandwidth of the waveforms used for transmission. Clearly, R and R are related via

$$\mathsf{R} = R_s R \left[\frac{\text{bit}}{\text{s}}\right]. \tag{1.4}$$

The process of mapping the information bits $\underline{\boldsymbol{i}}$ to the symbols $\underline{\boldsymbol{x}}$ is known as *coding* and $\underline{\boldsymbol{x}}$ are called *codewords*. We will often relate to well-known results stemming from the works of Shannon [1, 2] which defined the fundamental limits for reliable communication over the channel. Modeling the transmitted and received symbols $\boldsymbol{x}[n]$ and $\boldsymbol{y}[n]$ as random vectors $\boldsymbol{X}$ and $\boldsymbol{Y}$ with distributions $p_{\boldsymbol{X}}(\boldsymbol{x})$ and $p_{\boldsymbol{Y}}(\boldsymbol{y})$, the rate R is upper bounded by the *mutual information* (MI) $I(\boldsymbol{X};\boldsymbol{Y})$. As long as $R < I(\boldsymbol{X};\boldsymbol{Y})$, the probability of decoding error (i.e., choosing the wrong information sequence) can be made arbitrarily small when N_{b} goes to infinity. The maximum achievable rate $\mathsf{C} = \max_{p_{\boldsymbol{X}}(\boldsymbol{x})} I(\boldsymbol{X};\boldsymbol{Y})$, called the *channel capacity*, is obtained by maximizing over the distribution of the symbols $\boldsymbol{X}$, and it represents the ultimate transmission rate for the channel. In the case when $\boldsymbol{z}[n]$ is modeled as a Gaussian vector, the *probability density function* (PDF) $p_{\boldsymbol{X}}(\boldsymbol{x})$ that maximizes the MI is also Gaussian, which is one of the most popular results establishing limits for a reliable transmission.

The achievability proof is typically based on random-coding arguments, where, to create the $2^{N_{\text{b}}}$ codewords which form the codebook $\mathcal{X}$, the symbols $\boldsymbol{x}[n]$ are generated from the distribution $p_{\boldsymbol{X}}(\boldsymbol{x})$. At the receiver's side, the decoder decides in favor of the most likely sequence from the codebook, i.e., it uses the *maximum likelihood* (ML) decoding rule:

$$\hat{\underline{\boldsymbol{x}}} = \underset{\underline{\boldsymbol{x}}\in\mathcal{X}}{\operatorname{argmax}}\{p_{\underline{\boldsymbol{Y}}|\underline{\boldsymbol{X}}}(\underline{\boldsymbol{y}}|\underline{\boldsymbol{x}})\}. \tag{1.5}$$

When N_{b} grows, however, the suggested encoding and decoding cannot be used as practical means for the construction of the coding scheme. First, because storing the $2^{N_{\text{b}}}$ codewords $\underline{\boldsymbol{x}}$ results in excessive memory requirements, and second, because an exhaustive enumeration over $2^{N_{\text{b}}}$ codewords in the set $\mathcal{X}$ in (1.5) would be prohibitively complex. In practice, the codebook $\mathcal{X}$ is not randomly generated, but instead, obtained using an appropriately defined deterministic algorithm taking information bits as input. The structure of the code should simplify the decoding but also the encoding, i.e., the transmitter can generate the codewords on the fly (the codewords do not need to be stored).

To make the encoding and decoding practical, some "structure" has to be imposed on the codewords. The first constraint typically imposed is that the transmitted symbols are taken from a discrete predefined set $\mathcal{S}$, called a *constellation*. Moreover, the structure of the code should also simplify the decoding, as only the codewords complying with the constraints of the imposed structure are to be considered.

Consider the simple case of $N = 1$ (i.e., $\boldsymbol{X}$ is in fact a scalar), when the constellation is given by $\mathcal{S} = \{-1, 1\}$, i.e., a 2-ary *pulse amplitude modulation* (PAM) (2PAM) constellation, and when the received symbol is corrupted by *additive white Gaussian noise* (AWGN), i.e., $\boldsymbol{Y} = \boldsymbol{X} + \boldsymbol{Z}$ where $\boldsymbol{Z}$ is a zero-mean Gaussian random variable with variance $\sigma_{\boldsymbol{Z}}^2$. We show in Fig. 1.1 the MI for this case and compare it with the capacity which relies on the optimization of the distribution of $\boldsymbol{X}$. This figure indicates that for $R < 0.5$ bit/symbol, both values are practically identical. This allows us to focus on the design of *binary* codes with rate $R \approx I(\boldsymbol{X};\boldsymbol{Y})$ that can operate reliably without bothering about the theoretical suboptimality of the chosen constellation. This has been in fact the focus of research for many years, resulting in various binary codes being developed and used in practice. Initially, *convolutional codes* (CCs) received quite a lot of attention, but more recently, the focus has been on "capacity-approaching" codes such as *turbo codes* (TCs) or *low-density parity-check* (LDPC) codes. Their performance is deemed to be very "close" to the limits defined by the MI, even though the decoding does not rely on the ML rule in (1.5).

One particular problem becomes evident when analyzing Fig 1.1: the MI curve for 2PAM saturates at $\log_2|\mathcal{S}| = 1$ bit/symbol, and thus, it is impossible to transmit at rates $R > 1$ bit/symbol. From (1.4) and $R \leq 1$ bit/symbol, we conclude that if we want to increase the transmission rate R, we need to increase the rate R_{s}, and therefore, the transmission bandwidth. This is usually called *bandwidth expansion* and might be unacceptable in many cases, including modern wireless communication systems with stringent constraints on the available frequency bands.

The solution to the problem of increasing the transmission rate R without bandwidth expansion is to use a *high-order* constellation $\mathcal{S}$ and move the upper bound on R from $R \leq 1$ bit/symbol to $R \leq \log_2|\mathcal{S}| = m$ bit/symbol. Combining coding with nonbinary modulation (i.e., high-order constellations)

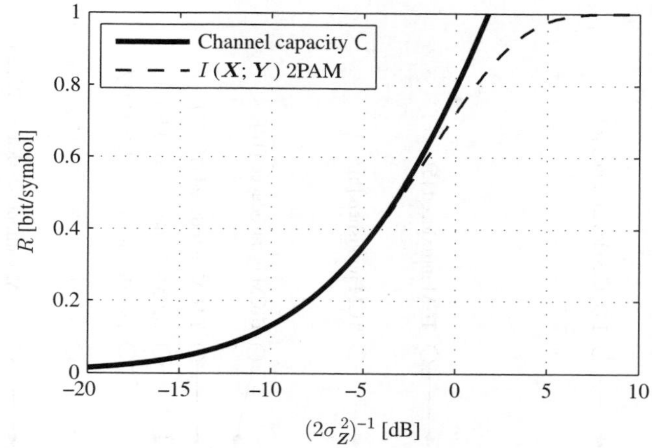

Figure 1.1 Channel capacity C and the MI for 2PAM

is often referred to as *coded modulation* (CM), to emphasize that not only coding but also the mapping from the code bits to the constellation symbols is important. The core problem in CM design is to choose the appropriate coding scheme that generates symbols from the constellation $\mathcal{S}$ and results in reliable transmission at rate $R > 1$ bit/symbol. On the practical side, we also need a CM which is easy to implement and which—because of the ever-increasing importance of wireless communications—allows us to adjust the coding rate R to the channel state, usually known as *adaptive modulation and coding* (AMC).

A well-known CM is the so-called *trellis-coded modulation* (TCM), where *convolutional encoders* (CENCs) are carefully combined with a high-order constellation $\mathcal{S}$. However, in the past decades, a new CM became prevalent and is the focus of this work: *bit-interleaved coded modulation* (BICM). The key component in BICM is a (suboptimal) two-step decoding process. First, *logarithmic likelihood ratios* (LLRs, also known as *L-values*) are calculated, and then a soft-input binary decoder is used. In the next section, we give a brief outline of the historical developments in the area of CM, with a particular emphasis on BICM.

1.2 The Road Toward BICM

The area of CM has been explored for many years. In what follows, we show some of the milestones in this area, which culminated with the introduction of BICM. In Fig. 1.2, we show a timeline of the CM developments, with emphasis on BICM.

The early works on CM in the 1970s include those by de Buda [27, 28], Massey [29], Miyakawa *et al.* [30], Anderson and Taylor [31], and Aulin [32]. The first breakthroughs for coding for $R > 1$ bit/symbol came with Ungerboeck and Csajka's TCM [5, 6, 33, 34] and Imai and Hirakawa's *multilevel coding* (MLC) [9, 35]. For a detailed historical overview of the early works on CM, we refer the reader to [36, Section 1.2] and [37, pp. 952–953]. Also, good summaries of the efforts made over the years to approach Shannon's limit can be found in [38, 39].

BICM's birth is attributed to the paper by Zehavi [10]. More particularly, Zehavi compared BICM based on CCs to TCM and showed that while BICM loses to TCM in terms of *Euclidean distance* (ED), it wins in terms of diversity. This fundamental observation spurred interest in BICM.

BICM was then analyzed in [17] by Caire *et al.*, where achievable rates for BICM were shown to be very close to those reached by CM, provided that a Gray labeling was used. Gray codes were then

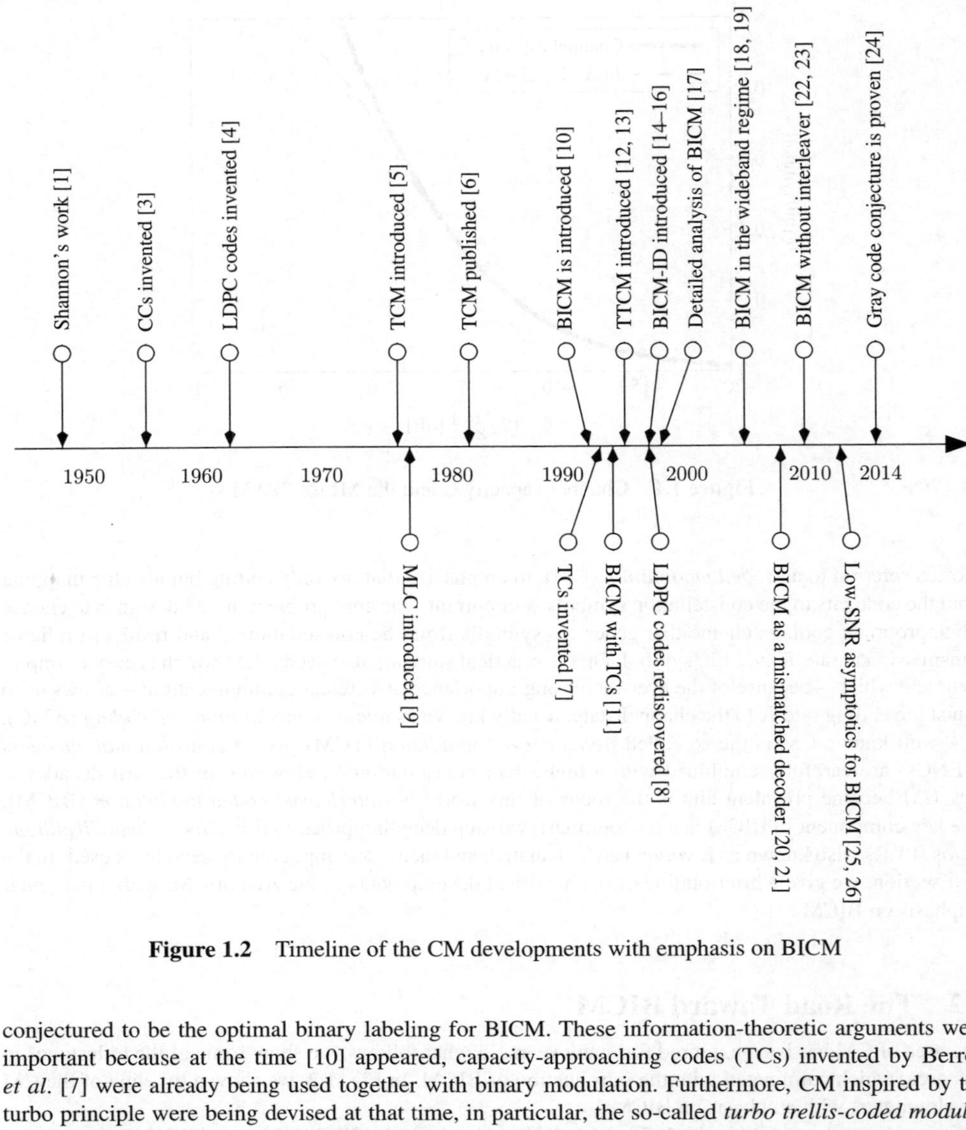

Figure 1.2 Timeline of the CM developments with emphasis on BICM

conjectured to be the optimal binary labeling for BICM. These information-theoretic arguments were important because at the time [10] appeared, capacity-approaching codes (TCs) invented by Berrou *et al.* [7] were already being used together with binary modulation. Furthermore, CM inspired by the turbo principle were being devised at that time, in particular, the so-called *turbo trellis-coded modulation* (TTCM) [12, 13, 40–43]. In TTCM, coding and modulation were tightly coupled, and thus, changing the targeted spectral efficiency typically required a change in the CM design.

BICM appeared then as an alternative to "capacity-approaching" CM such as TTCM, where, to eliminate the rigidity of such systems, a small decrease in terms of achievable rate was an acceptable price to pay. In 1994, the combination of a capacity-approaching TC using the BICM paradigm was proposed for the first time [11]. This combination is based on a fairly simple implementation and turned out to perform very well, as later shown, e.g., in [44].

In 2003, Le Goff [45] analyzed the relationship between achievable rates and the performance of practical coding schemes. It was argued that the gap between the CM and BICM rates translates into performance differences between TCM and BICM when both are based on "capacity-approaching" codes. However, these differences are in most cases so small that it may be difficult to relate them to the performance loss of different practical coding schemes whose "strengths" are difficult to compare. For example, the results in [45, Fig. 6], show that the TTCM of [13, 40] outperforms BICM for $R = 2\,\text{bit/symbol}$ (by a fraction of a decibel), however, for $R = 3\,\text{bit/symbol}$ BICM outperforms the TTCM of [42]. This comparison illustrates well the fact that using TCM based on a concatenated coding does not guarantee the "strength" of the resulting system. In fact, this example highlights the flexibility of BICM whose coding rate was chosen arbitrarily without any implementation hassle, in contrast to changing the rate for TCM, which usually requires a change of the encoding structure.

Recognizing the bit-to-symbol mapper as an encoder, BICM was later cast into the framework of serially concatenated codes in [14–16]. This led to the introduction of *bit-interleaved coded modulation with iterative demapping* (BICM-ID) studied, e.g., in [46–50]. From the very start, the key role of the binary labeling for BICM-ID was recognized and extensively studied in the literature; for details about this, we refer the reader to [47, 48, 51–56] and references therein.

An information-theoretic analysis of BICM[1] was always at the forefront of analytical developments. One of the most important information-theoretic papers in this area is [17], where a formal model to analyze the rates achievable in BICM was proposed. Later, Martinez *et al.* [20] recognized BICM as a mismatched decoder and the model of [17] was refined. It was also shown in [20] that in terms of achievable rates, the interleaver plays no role, and that the key element is the suboptimal (mismatched) decoder. This observation is closely linked to the results in [22, 23] where it is shown that when BICM with CCs is used in nonfading channels, a better error probability can be obtained if the interleaver is completely removed from the transceivers. BICM in the wideband regime was first studied in [18, 19, 57], and later generalized to multidimensional constellations and arbitrary input distributions in [25, 26]. The long-standing Gray code conjecture for BICM was recently proven for one-dimensional constellations in [24].

1.3 Why Everybody Loves BICM

Striving for a simple and flexible CM, we may need to trade off its performance. Indeed, this is what happens in practice. BICM is most often the answer to a simplicity and flexibility requirement, while the performance loss with respect to the potentially optimal/better CM are moderate, or at least acceptable. In our view, the main reasons that make BICM systems particularly appealing and explain their omnipresence in practical communication systems are the following.

1. **Flexibility**: Tributary to the bit-level operation, BICM provides a high design flexibility. Unlike other CM (e.g., TCM or TTCM) where owing to the well-structured relationship between the encoder and the modulator, the encoder and coding rate must be carefully matched to the constellation, BICM allows the designer to choose the constellation and the encoder independently. This is why BICM is usually considered to be a pragmatic approach to CM. This "unstructured" transmission characteristic of BICM yields a flexibility so high that it allows for a straightforward implementation of the so-called irregular modulation (where modulation varies within the transmitted block of symbols) [58] or irregular coding (where the coding/puncturing may vary within the block of bits) [47]. The flexibility is particularly important in wireless communications where the stream of data must be matched to the structures imposed by a multiuser communication context (such as slots and frames) when transmitting over variable channel conditions. The flexibility of BICM can be hardly beaten as all operations are done at the bit level. For example, to get the most out of a varying channel, puncturing, rate matching, bit padding, and other operations typical in wireless standards can be done with arbitrary (up to a bit level) granularity. Performing similar operations in other CM paradigms is less straightforward.

[1] Throughout this book, we use the acronym BICM to refer to noniterative BICM.

2. **Robustness**: As BICM uses one decoder, only one parameter (the rate) is to be adjusted, which makes AMC not only easy to implement but also more robust to changes in the channel variations. This may be contrasted with MLC, where multiple decoders are needed (which may or may not collaborate with each other) and their respective rates have to be jointly adjusted to the channel conditions. Although BICM and MLC may be ultimately fit into the same framework (see [59] for details), we will maintain their conventional separation and consider MLC as a different CM paradigm. However, many elements of the analysis we propose in this book are applicable also to MLC, particularly those related to the analysis of the reliability metrics for the code bits.
3. **Negligible Losses in Scalar Channels**: The appearance of "capacity-approaching" binary codes made the difference between optimal CM systems and BICM negligible for most practical purposes, when transmitting over scalar channels. The eventual differences of fractions of a decibel (if any) are not very relevant from a system-level perspective, and in most cases, can be considered a small price to pay compared to the advantages BICM systems offer.
4. **Increased Diversity**: When transmitting over fading channels, and thanks to the interleaving, the L-values used in the decision process are calculated from symbols affected by independent channel realizations, which increases the so-called *transmission diversity*. This particular observation (first made in [10]) spurred the initial interest in BICM. While this property is often evoked as the *raison d'être* of BICM, we note that communication systems such as *third-generation partnership project* (3GPP) or *long-term evolution* (LTE) are designed to operate with packetized transmission. As often the packets are much shorter than the channel coherence time, the time diversity is limited. In multicarrier transmission such as *orthogonal frequency-division multiplexing* (OFDM), the frequency diversity depends on the frequency selectivity of the channel and may be easier to exploit. Nevertheless, with the appearance of capacity-approaching codes such as TCs or LDPC codes, the diversity is less relevant than other parameters of the channel, such as the capacity, which these codes approach closely.

Nothing comes for free, and there are some drawbacks to using BICM:

1. **Suboptimality**: Owing to the bitwise processing, the BICM decoder is suboptimal. From an information-theoretic point of view, this translates into a decrease in achievable rates when compared to those obtained using an optimal decoder. These losses are small for transmission over scalar channels where a constellation with a Gray labeling is used. On the other hand, the losses become more important in *multiple-input multiple-output* (MIMO) channels, where the effective bit-to-symbol mapping is affected by the random channel variations. In these cases, the use of nonbinary codes might offer advantage over BICM. When using "weak" codes such as CCs, the suboptimality is observed also in nonfading scalar channels but may be remedied to some extent by appropriate modification (or even removal) of the interleaver.
2. **Long Interleavers/Delays**: To ensure independence of the L-values which the demapper calculates and passes to the channel decoder, relatively long interleavers might be required. This, in turn, causes a mandatory packet-based decoding, i.e., no symbol-after-symbol decoding can be performed. This fact, however, is less important with the presence of "capacity-approaching" codes, which rely on iterative decoding, and thus, inherently require packet-based processing.

1.4 Outline of the Contents

This book provides an exhaustive analysis of BICM, paying special attention to those properties that are of particular interest when compared to other CM paradigm. The book is therefore intended for those interested in theoretical aspects of CM transmission in general, and BICM in particular.

In order to give an overview of the chapters in this book, we consider the simple model of BICM transmission in Fig. 1.3. A block of binary data of length N_{b} is encoded by a binary *encoder* (ENC) which produces binary codewords of length N_{c}. The codewords are interleaved at bit level, and the resulting

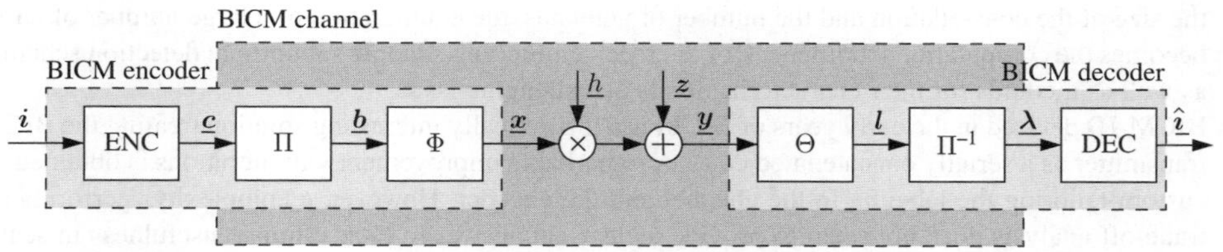

Figure 1.3 The BICM encoder is formed by a serial concatenation of a binary encoder (ENC), a bit-level interleaver (Π), and a memoryless mapper (Φ). The BICM decoder is based on a demapper (Θ) that computes L-values, a deinterleaver, and a *channel decoder* (DEC)

bitstream is separated into groups of m bits that are mapped to symbols of an M-ary constellation $\mathcal{S}$, where $M = 2^m$. The symbols are transmitted over the channel whose outcome is used at the receiver to calculate a posteriori probabilities for the transmitted bits, which are then used for decoding. Instead of probabilities, reliability metrics are most often expressed in the form of an LLR, also known as an *L-value*. These L-values are then deinterleaved and passed to the channel decoder.

- Chapter 2 introduces the basic definitions, concepts, and notational convention used throughout the book. CM transmission is discussed, as well as channel models, constellations, and binary labelings used in the book. A brief summary of CCs and TCs is also presented.
- Chapter 3 defines the L-values and their approximations/simplifications. The well-known *maximum a posteriori* (MAP) and ML decoding strategies are compared against the (suboptimal) BICM decoder shown in Fig. 1.3.
- Chapter 4 takes a closer look at the theoretical limits of CM with special attention paid to BICM. Achievable rates for BICM as a function of the constellation shape, the input distribution, and the binary labeling are studied.
- Chapter 5 characterizes the BICM channel in Fig. 1.3 as a binary-input real-output entity focusing on the probabilistic models of the L-values. PDFs of the L-values for one- and two-dimensional constellations are developed.
- Chapter 6 explains how to evaluate the performance of transmission without coding as well as different methods that can be used to approximate the performance of the encoder–decoder pair in BICM transmission.
- Chapter 7 deals with the problem of mismatched L-values. Suitable strategies for correcting the mismatched L-values are presented. Special attention is paid to the correction based on a linear scaling.
- Chapter 8 shows how to exploit *unequal error protection* (UEP) inherently present in BICM and caused by the binary labeling of the constellation. The performance evaluation tools developed in the previous chapters are refined and a design of the interleaver and the encoder is proposed, taking advantage of UEP.
- Chapter 9 studies the use of a BICM receiver for trellis codes in nonfading channels. This is motivated by results that show that for the AWGN channel and CENCs, the use of a trellis encoder (without an interleaver) and a BICM receiver has an improved performance when compared to standard BICM configurations. Equivalences between different encoders and the optimal selection of CCs are also analyzed.

There are some important issues related to BICM which are not addressed in this work. In what follows, we list some of the ones we feel deserve a more in-depth study.

1. BICM for MIMO (or more generally for high-dimensional constellations) has to rely on an efficient and most likely suboptimal calculation of the L-values. This is caused by the fact that by increasing

the size of the constellation and the number of antennas, the enumeration of a large number of terms becomes the computational bottleneck of the receiver. Finding suitable suboptimal detection schemes as well as methods for their characterization is a challenging issue.

2. BICM-ID devised in the early years of BICM is a theoretically interesting solution treating the BICM transmitter as a serially concatenated code. A remarkable improvement with iterations is obtained by custom-tailoring the labeling to the channel and the encoder. However, a complexity–performance trade-off analysis does not seem to provide a clear-cut answer to their ultimate usefulness in scalar channels. This is a promising solution when the channel introduces uncontrollable interference, as in the case of MIMO transmission.
3. Cooperative transmission is an increasingly popular subject and analyzing BICM transmission in the presence of multiple nodes using cooperative modulation schemes, network coding, and physical network coding is a promising venue to explore.

References

[1] Shannon, C. E. (1948) A mathematical theory of communications. *Bell Syst. Tech. J.*, **27**, 379–423 and 623–656.

[2] Shannon, C. E. (1949) Communication in the presence of noise. *Proc. IRE*, **37** (1), 10–21.

[3] Elias, P. (1955) Coding for noisy channels. *IRE Conv. Rec.*, **3**, 37–46.

[4] Gallager, R. (1962) Low-density parity-check codes. *IRE Trans. Inf. Theory*, **IT-8**, 21–28.

[5] Ungerboeck, G. and Csajka, I. (1976) On improving data-link performance by increasing channel alphabet and introducing sequence decoding. International Symposium on Information Theory (ISIT), June 1976, Ronneby, Sweden (book of abstracts).

[6] Ungerboeck, G. (1982) Channel coding with multilevel/phase signals. *IEEE Trans. Inf. Theory*, **28** (1), 55–67.

[7] Berrou, C., Glavieux, A., and Thitimajshima, P. (1993) Near Shannon limit error-correcting coding and decoding: Turbo codes. IEEE International Conference on Communications (ICC), May 1993, Geneva, Switzerland.

[8] MacKay, D. J. C. and Neal, R. M. (1997) Near Shannon limit performance of low density parity check codes. *Electron. Lett.*, **33** (6), 457–458.

[9] Imai, H. and Hirakawa, S. (1977) A new multilevel coding method using error-correcting codes. *IEEE Trans. Inf. Theory*, **IT-23** (3), 371–377.

[10] Zehavi, E. (1992) 8-PSK trellis codes for a Rayleigh channel. *IEEE Trans. Commun.*, **40** (3), 873–884.

[11] Le Goff, S., Glavieux, A., and Berrou, C. (1994) Turbo-codes and high spectral efficiency modulation. IEEE International Conference on Communications (ICC), May 1994, New Orleans, LA.

[12] Benedetto, S., Divsalar, D., Montorsi, G., and Pollara, F. (1995) Bandwidth efficient parallel concatenated coding schemes. *Electron. Lett.*, **31** (24), 2067–2069.

[13] Divsalar, D. and Pollara, F. (1995) On the design of turbo codes. TDA Progress Report 42-123, Jet Propulsion Laboratory, Pasadena, CA, pp. 99–121.

[14] Li, X. and Ritcey, J. A. (1997) Bit-interleaved coded modulation with iterative decoding. *IEEE Commun. Lett.*, **1** (6), 169–171.

[15] ten Brink, S., Speidel, J., and Yan, R.- H. (1998) Iterative demapping for QPSK modulation. *IEE Electron. Lett.*, **34** (15), 1459–1460.

[16] Benedetto, S., Montorsi, G., Divsalar, D., and Pollara, F. (1998) Soft-input soft-output modules for the construction and distributed iterative decoding of code networks. *Eur. Trans. Telecommun.*, **9** (2), 155–172.

[17] Caire, G., Taricco, G., and Biglieri, E. (1998) Bit-interleaved coded modulation. *IEEE Trans. Inf. Theory*, **44** (3), 927–946.

[18] Martinez, A., Guillén i Fàbregas, A., and Caire, G. (2008) Bit-interleaved coded modulation in the wideband regime. *IEEE Trans. Inf. Theory*, **54** (12), 5447–5455.

[19] Stierstorfer, C. and Fischer, R. F. H. (2008) Mappings for BICM in UWB scenarios. International ITG Conference on Source and Channel Coding (SCC), January 2008, Ulm, Germany.

[20] Martinez, A., Guillén i Fàbregas, A., Caire, G., and Willems, F. M. J. (2009) Bit-interleaved coded modulation revisited: a mismatched decoding perspective. *IEEE Trans. Inf. Theory*, **55** (6), 2756–2765.

[21] Guillén i Fàbregas, A., Martinez, A., and Caire, G. (2008) Bit-interleaved coded modulation. *Found. Trends Commun. Inf. Theory*, **5** (1–2), 1–153.

[22] Stierstorfer, C., Fischer, R. F. H., and Huber, J. B. (2010) Optimizing BICM with convolutional codes for transmission over the AWGN channel. International Zurich Seminar on Communications, March 2010, Zurich, Switzerland.
[23] Alvarado, A., Szczecinski, L., and Agrell, E. (2011) On BICM receivers for TCM transmission. *IEEE Trans. Commun.*, **59** (10), 2692–2702.
[24] Alvarado, A., Brännström, F., Agrell, E., and Koch, T. (2014) High-SNR asymptotics of mutual information for discrete constellations with applications to BICM. *IEEE Trans. Inf. Theory*, **60** (2), 1061–1076.
[25] Agrell, E. and Alvarado, A. (2011) Optimal alphabets and binary labelings for BICM at low SNR. *IEEE Trans. Inf. Theory*, **57** (10), 6650–6672.
[26] Agrell, E. and Alvarado, A. (2013) Signal shaping for BICM at low SNR. *IEEE Trans. Inf. Theory*, **59** (4), 2396–2410.
[27] de Buda, R. (1972) Fast FSK signals and their demodulation. *Can. Electron. Eng. J.*, **1**, 28–34.
[28] de Buda, R. (1972) Coherent demodulation of frequency-shift keying with low deviation ratio. *IEEE Trans. Commun.*, **COM-20**, 429–435.
[29] Massey, J. L. (1974) Coding and modulation in digital communications. International Zurich Seminar on Digital Communications, March 1974, Zurich, Switzerland.
[30] Miyakawa, H., Harashima, H., and Tanaka, Y. (1975) A new digital modulation scheme–Multi-mode binary CPFSK. 3rd International Conference Digital Satellite Communications, November 1975, Kyoto, Japan.
[31] Anderson, J. B. and Taylor, D. P. (1978) A bandwidth-efficient class of signal-space codes. *IEEE Trans. Inf. Theory*, **IT-24** (6), 703–712.
[32] Aulin, T. (1979) CPM–A power and bandwidth efficient digital constant envelope modulation scheme. PhD dissertation, Lund University, Lund, Sweden.
[33] Ungerboeck, G. (1987) Trellis-coded modulation with redundant signal sets Part I: Introduction. *IEEE Commun. Mag.*, **25** (2), 5–11.
[34] Ungerboeck, G. (1987) Trellis-coded modulation with redundant signal sets Part II: state of the art. *IEEE Commun. Mag.*, **25** (2), 12–21.
[35] Imai, H. and Hirakawa, S. (1977) Correction to 'A new multilevel coding method using error-correcting codes'. *IEEE Trans. Inf. Theory*, **IT-23** (6), 784.
[36] Anderson, J. B. and Svensson, A. (2003) *Coded Modulation Systems*, Springer.
[37] Lin, S. and Costello, D. J. Jr. (2004) *Error Control Coding*, 2nd edn, Prentice Hall, Englewood Cliffs, NJ.
[38] Costello, D. J. Jr. and Forney, G. D. Jr. (2007) Channel coding: the road to channel capacity. *Proc. IEEE*, **95** (6), 1150–1177.
[39] Forney, G. D. Jr. and Ungerboeck, G. (1998) Modulation and coding for linear Gaussian channels. *IEEE Trans. Inf. Theory*, **44** (6), 2384–2415 (Invited Paper).
[40] Benedetto, S. and Montorsi, G. (1996) Design of parallel concatenated convolutional codes. *IEEE Trans. Commun.*, **44** (5), 591–600.
[41] Benedetto, S., Divsalar, D., Montorsi, G., and Pollara, F. (1996) Parallel concatenated trellis coded modulation. IEEE International Conference on Communications (ICC), June 1996, Dallas, TX.
[42] Robertson, P. and Wörz, T. (1998) Bandwidth-efficient turbo trellis-coded modulation using punctured component codes. *IEEE J. Sel. Areas Commun.*, **16** (2), 206–218.
[43] Blackert, W. J. and Wilson, S. G. (1996) Turbo trellis coded modulation. Conference on Information Sciences and Systems (CISS), March 1996, Princeton, NJ.
[44] Abramovici, I. and Shamai, S. (1999) On turbo encoded BICM. *Ann. Telecommun.*, **54** (3–4), 225–234.
[45] Goff, S. Y. L. (2003) Signal constellations for bit-interleaved coded modulation. *IEEE Trans. Inf. Theory*, **49** (1), 307–313.
[46] Chindapol, A. and Ritcey, J. A. (2001) Design, analysis, and performance evaluation for BICM-ID with square QAM constellations in Rayleigh fading channels. *IEEE J. Sel. Areas Commun.*, **19** (5), 944–957.
[47] Tüchler, M. (2004) Design of serially concatenated systems depending on the block length. *IEEE Trans. Commun.*, **52** (2), 209–218.
[48] Schreckenbach, F., Görtz, N., Hagenauer, J. and Bauch, G. (2003) Optimization of symbol mappings for bit-interleaved coded modulation with iterative decoding. *IEEE Commun. Lett.*, **7** (12), 593–595.
[49] Szczecinski, L., Chafnaji, H., and Hermosilla, C. (2005) Modulation doping for iterative demapping of bit-interleaved coded modulation. *IEEE Commun. Lett.*, **9** (12), 1031–1033.
[50] Li, X., Chindapol, A., and Ritcey, J. A. (2002) Bit-interlaved coded modulation with iterative decoding and 8PSK signaling. *IEEE Trans. Commun.*, **50** (6), 1250–1257.

[51] Tan, J. and Stüber, G. L. (2002) Analysis and design of interleaver mappings for iteratively decoded BICM. IEEE International Conference on Communications (ICC), May 2002, New York City, NY.

[52] ten Brink, S. (2001) Convergence behaviour of iteratively decoded parallel concatenated codes. *IEEE Trans. Commun.*, **49** (10), 1727–1737.

[53] Zhao, L., Lampe, L., and Huber, J. (2003) Study of bit-interleaved coded space-time modulation with different labeling. IEEE Information Theory Workshop (ITW), March 2003, Paris, France.

[54] Clevorn, T., Godtmann, S., and Vary, P. (2006) Optimized mappings for iteratively decoded BICM on Rayleigh channels with IQ interleaving. IEEE Vehicular Technology Conference (VTC-Spring), May 2006, Melbourne, Australia.

[55] Tan, J. and Stüber, G. L. (2005) Analysis and design of symbol mappers for iteratively decoded BICM. *IEEE Trans. Wireless Commun.*, **4** (2), 662–672.

[56] Schreckenbach, F. (2007) Iterative decoding of bit-interleaved coded modulation. PhD dissertation, Technische Universität München, Munich, Germany.

[57] Alvarado, A., Agrell, E., Guillén i Fàbregas, A., and Martinez, A. (2010) Corrections to 'Bit-interleaved coded modulation in the wideband regime'. *IEEE Trans. Inf. Theory*, **56** (12), 6513.

[58] Schreckenbach, F. and Bauch, G. (2006) Bit-interleaved coded irregular modulation. *Eur. Trans. Telecommun.*, **17** (2), 269–282.

[59] Wachsmann, U., Fischer, R. F. H., and Huber, J. B. (1999) Multilevel codes: theoretical concepts and practical design rules. *IEEE Trans. Inf. Theory*, **45** (5), 1361–1391.

2

Preliminaries

In this chapter, we introduce the preliminaries for this book. In particular, we introduce all the building blocks of *coded modulation* (CM) and *bit-interleaved coded modulation* (BICM) transceivers. Although some chapters are more or less self-contained, reading this chapter before proceeding further is highly recommended. This chapter is organized as follows. We introduce the notation convention in Section 2.1 and in Section 2.2 we present a general model for the problem of binary transmission over a physical channel. Coded modulation systems are briefly analyzed in Section 2.3 and the channel models of interest are discussed in Section 2.4. In Section 2.5 constellations and binary labelings are discussed. Finally, in Section 2.6, we review some of the channel codes that will be used later in the book.

2.1 Notation Convention

We use lowercase letters x to denote a scalar, boldface letters $\boldsymbol{x}$ to denote a length-N column vector; both can be indexed as x_i and $\boldsymbol{x}_i = [x_{i,1}, \ldots, x_{i,N}]^{\mathrm{T}}$, where $[\cdot]^{\mathrm{T}}$ denotes transposition. A sequence of scalars is denoted by a row vector $\underline{x} = [x[1], \ldots, x[M]]$ and a sequence of N-dimensional vectors by

$$\underline{\boldsymbol{x}} = [\boldsymbol{x}[1], \ldots, \boldsymbol{x}[M]] = \begin{bmatrix} \underline{x}_1 \\ \vdots \\ \underline{x}_N \end{bmatrix}, \tag{2.1}$$

where $\underline{x}_k = [x_k[1], \ldots, x_k[M]], k = 1, \ldots, N$.

An $N \times M$ matrix is denoted as $\mathbf{X} = [\boldsymbol{x}_1, \ldots, \boldsymbol{x}_M]$ and its elements are denoted by $x_{i,j}$ where $i = 1, \ldots, N$ and $j = 1, \ldots, M$. The all-zeros and the all-ones column vectors are denoted by $\mathbf{1}$ and $\mathbf{0}$, respectively. The sequences of such vectors are denoted by $\underline{\mathbf{0}}$ and $\underline{\mathbf{1}}$, respectively. The identity matrix of size N is denoted by $\mathbf{I}_{[N]}$. The inner product of two vectors $\boldsymbol{x}$ and $\boldsymbol{y}$ is defined as $\langle \boldsymbol{x}_i, \boldsymbol{y}_j \rangle \triangleq \sum_{n=1}^{N} x_{i,n} y_{j,n}$, the Euclidean norm of a length-N vector $\boldsymbol{x}$ is defined as $\|\boldsymbol{x}\| \triangleq \sqrt{\langle \boldsymbol{x}, \boldsymbol{x} \rangle}$, and the L_1 (Manhattan) norm as $\|\boldsymbol{x}\|_1 \triangleq \sum_{n=1}^{N} |x_n|$. The Euclidean norm of the sequence $\underline{\boldsymbol{x}}$ in (2.1) is given by $\|\underline{\boldsymbol{x}}\| = \sqrt{\sum_{j=1}^{M} \|\boldsymbol{x}[j]\|^2}$.

Sets are denoted using calligraphic letters $\mathcal{C}$, where $\mathcal{C} \times \mathcal{C}'$ represents the Cartesian product between $\mathcal{C}$ and $\mathcal{C}'$. The exception to this notation is used for the number sets. The real numbers are denoted by $\mathbb{R}$, the nonnegative real numbers by $\mathbb{R}_+$, the complex numbers by $\mathbb{C}$, the natural numbers by $\mathbb{N} \triangleq \{0, 1, 2, \ldots\}$, the positive integers by $\mathbb{N}_+ \triangleq \{1, 2, 3, \ldots\}$, and the binary set by $\mathbb{B} \triangleq \{0, 1\}$. The set of real matrices of size $N \times M$ (or–equivalently–of length-M sequences of N-dimensional vectors) is denoted by $\mathbb{R}^{N \times M}$;

Bit-Interleaved Coded Modulation: Fundamentals, Analysis, and Design, First Edition.
Leszek Szczecinski and Alex Alvarado.

the same notation is used for matrices and vector sequences of other types, e.g., binary $\mathbb{B}^{N\times M}$ or natural $\mathbb{N}^{N\times M}$. The empty set is denoted by $\varnothing$. The negation of the bit b is denoted by $\overline{b}$.

An extension of the Cartesian product is the *ordered direct product*, which operates on vectors/matrices instead of sets. It is defined as

$$[\boldsymbol{a}_1,\ldots,\boldsymbol{a}_p]\otimes[\boldsymbol{b}_1,\ldots,\boldsymbol{b}_q]\triangleq[\boldsymbol{c}_1,\ldots,\boldsymbol{c}_{pq}], \tag{2.2}$$

where $\boldsymbol{c}_{q(i-1)+j}=[\boldsymbol{a}_i^{\mathrm{T}},\boldsymbol{b}_j^{\mathrm{T}}]^{\mathrm{T}}$ for $i=1,\ldots,p$ and $j=1,\ldots,q$.

The indicator function is defined as $\mathbb{I}_{[\nu]}=1$ when the statement ν is true and $\mathbb{I}_{[\nu]}=0$, otherwise. The natural logarithm is denoted by $\log(\cdot)$ and the base-2 logarithm by $\log_2(\cdot)$. The real part of a complex number z is denoted by $\Re[z]$ and its imaginary part by $\Im[z]$, thus $z=\Re[z]+\mathrm{j}\Im[z]$. The floor and ceiling functions are denoted by $\lfloor\cdot\rfloor$ and $\lceil\cdot\rceil$, respectively.

The *Hamming distance* (HD) between the binary vectors $\boldsymbol{b}$ and $\boldsymbol{c}$ is denoted by $d_{\mathrm{H}}(\boldsymbol{b},\boldsymbol{c})$, the *Hamming weight* (HW) of the vector $\boldsymbol{b}$ by $w_{\mathrm{H}}(\boldsymbol{b})$, which we generalize to the case of sequences of binary vectors $\underline{\boldsymbol{b}}\in\mathbb{B}^{m\times N}$ as

$$W_{\mathrm{H}}(\underline{\boldsymbol{b}})=[w_{\mathrm{H}}(\underline{b}_1),\ldots,w_{\mathrm{H}}(\underline{b}_m)]^{\mathrm{T}}, \tag{2.3}$$

i.e., the *generalized Hamming weight* (GHW) in (2.3) is the vector of HWs of each row of $\underline{\boldsymbol{b}}$. Similarly, the *total* HD between the sequences of binary vectors $\underline{\boldsymbol{b}},\underline{\boldsymbol{c}}\in\mathbb{B}^{m\times N}$ is denoted by $d_{\mathrm{H}}(\underline{\boldsymbol{b}},\underline{\boldsymbol{c}})=\sum_{n=1}^{N_{\mathrm{s}}}d_{\mathrm{H}}(\boldsymbol{b}[n],\boldsymbol{c}[n])$.

Random variables are denoted by capital letters Y, probabilities by $\Pr\{\cdot\}$, and the *cumulative distribution function* (CDF) of Y by $F_Y(y)=\Pr\{Y\leq y\}$. The *probability mass function* (PMF) of the random vector $\boldsymbol{Y}$ is denoted by $P_{\boldsymbol{Y}}(\boldsymbol{y})=\Pr\{\boldsymbol{Y}=\boldsymbol{y}\}$, the *probability density function* (PDF) of $\boldsymbol{Y}$ by $p_{\boldsymbol{Y}}(\boldsymbol{y})$. The joint PDF of the random vectors $\boldsymbol{X}$ and $\boldsymbol{Y}$ is denoted by $p_{\boldsymbol{X},\boldsymbol{Y}}(\boldsymbol{x},\boldsymbol{y})$ and the PDF of $\boldsymbol{Y}$ conditioned on $\boldsymbol{X}=\boldsymbol{x}$ by $p_{\boldsymbol{Y}|\boldsymbol{X}}(\boldsymbol{y}|\boldsymbol{x})$. The same notation applies to joint and conditional PMFs, i.e., $P_{\boldsymbol{X},\boldsymbol{Y}}(\boldsymbol{x},\boldsymbol{y})$ and $P_{\boldsymbol{Y}|\boldsymbol{X}}(\boldsymbol{y}|\boldsymbol{x})$. Note that the difference in notation between random vectors and matrices is subtle. The symbols for vectors are capital italic boldface while the symbols for matrices are not italic.

The expectation of an arbitrary function $f(\boldsymbol{X},\boldsymbol{Y})$ over the joint distribution of $\boldsymbol{X}$ and $\boldsymbol{Y}$ is denoted by $\mathbb{E}_{\boldsymbol{X},\boldsymbol{Y}}[f(\boldsymbol{X},\boldsymbol{Y})]$ and the expectation over the conditional distribution is denoted by $\mathbb{E}_{\boldsymbol{Y}|\boldsymbol{X}=\boldsymbol{x}}[f(\boldsymbol{x},\boldsymbol{Y})]$. The unconditional and conditional variance of a scalar function $f(\boldsymbol{X},\boldsymbol{Y})\in\mathbb{R}$ are respectively denoted as

$$\mathbb{V}\mathrm{ar}_{\boldsymbol{X},\boldsymbol{Y}}[f(\boldsymbol{X},\boldsymbol{Y})]\triangleq\mathbb{E}_{\boldsymbol{X},\boldsymbol{Y}}[f(\boldsymbol{X},\boldsymbol{Y})^2]-\mathbb{E}^2_{\boldsymbol{X},\boldsymbol{Y}}[f(\boldsymbol{X},\boldsymbol{Y})], \tag{2.4}$$

$$\mathbb{V}\mathrm{ar}_{\boldsymbol{Y}|\boldsymbol{X}=\boldsymbol{x}}[f(\boldsymbol{x},\boldsymbol{Y})]\triangleq\mathbb{E}_{\boldsymbol{Y}|\boldsymbol{X}=\boldsymbol{x}}[f(\boldsymbol{x},\boldsymbol{Y})^2]-\mathbb{E}^2_{\boldsymbol{Y},\boldsymbol{X}=\boldsymbol{x}}[f(\boldsymbol{x},\boldsymbol{Y})]. \tag{2.5}$$

The *moment-generating function* (MGF) of the random variable X is defined as

$$\mathsf{P}_X(s)\triangleq\mathbb{E}_X[\exp(sX)], \tag{2.6}$$

where $s\in\mathbb{C}$.

The notation $X\sim\mathcal{N}(\mu,\sigma^2)$ is used to denote a Gaussian random variable with mean value μ and variance σ^2; its PDF is given by

$$p_X(x)=\Psi(x,\mu,\sigma^2)\triangleq\frac{1}{\sqrt{2\pi\sigma^2}}\exp\left(-\frac{(x-\mu)^2}{2\sigma^2}\right). \tag{2.7}$$

The complementary CDF of $X\sim\mathcal{N}(0,1)$ is defined as

$$Q(x)\triangleq\Pr\{X>x\}=\frac{1}{\sqrt{2\pi}}\int_x^{\infty}\exp\left(-\frac{u^2}{2}\right)\mathrm{d}u. \tag{2.8}$$

We also use the complementary error function defined as

$$\mathrm{erfc}(x)\triangleq\frac{2}{\sqrt{\pi}}\int_x^{\infty}\exp(-u^2)\,\mathrm{d}u=2Q(\sqrt{2}x). \tag{2.9}$$

If $X_1 \sim \mathcal{N}(0,1)$ and $X_2 \sim \mathcal{N}(0,1)$, then $\boldsymbol{X} = [X_1, X_2]^{\mathrm{T}}$ is a normalized bivariate Gaussian variable with correlation $\rho = \mathbb{E}_{X_1,X_2}[X_1 X_2]$. Its PDF is given by

$$p_{X_1,X_2}(u,v) = \frac{1}{2\pi\sqrt{1-\rho^2}} \exp\left(-\frac{v^2 - 2vu\rho + u^2}{2(1-\rho^2)}\right), \tag{2.10}$$

and the complementary CDF by

$$\tilde{Q}(a,b,\rho) \triangleq \Pr\{X_1 > a \wedge X_2 > b\} \tag{2.11}$$

$$= \begin{cases} \int_a^\infty \int_b^\infty p_{X_1,X_2}(u,v)\,\mathrm{d}u\,\mathrm{d}v, & \text{if } |\rho| < 1 \\ Q(\max\{a,b\}), & \text{if } \rho = 1 \\ Q(a) - Q(-b), & \text{if } \rho = -1,\ a+b \leq 0 \\ 0, & \text{if } \rho = -1,\ a+b > 0 \end{cases}. \tag{2.12}$$

The convolution of two functions $f(x)$ and $g(x)$ is denoted by $f(x) * g(x)$ and the d-fold self-convolution of $f(x)$ is defined as

$$\{f(x)\}^{*1} \triangleq f(x), \tag{2.13}$$

$$\{f(x)\}^{*d} \triangleq \{f(x)\}^{*(d-1)} * f(x), \quad \text{if } d > 1. \tag{2.14}$$

For any function $f(x,y)$, we also define

$$[f(x,y)]_{x=a}^{x=b} \triangleq f(b,y) - f(a,y), \tag{2.15}$$

$$[f(x,y)]_{x=a,y=c}^{x=b,y=d} \triangleq f(b,d) - f(a,c). \tag{2.16}$$

For any $d > m$, with $d, m \in \mathbb{N}_+$ and $\boldsymbol{w} \in \{\mathbb{N}^m : \|\boldsymbol{w}\|_1 = d\}$, the multinomial coefficient defined as

$$\binom{d}{\boldsymbol{w}} \triangleq \frac{d!}{w_1! \cdot \ldots \cdot w_m!} \tag{2.17}$$

can be interpreted as the number of all different partitions of d elements into m sets, where the partition is defined by the number of elements w_k, $k = 1, \ldots, m$ in the kth set. In particular, we have

$$\left(\sum_{k=1}^{m} a_k\right)^d = \sum_{\substack{\boldsymbol{w}\in\mathbb{N}^m \\ \|\boldsymbol{w}\|_1 = d}} \binom{d}{\boldsymbol{w}} \prod_{k=1}^{m} a_k^{w_k}, \tag{2.18}$$

$$\sum_{\substack{\boldsymbol{w}\in\mathbb{N}^m \\ \|\boldsymbol{w}\|_1 = d}} \binom{d}{\boldsymbol{w}} = m^d. \tag{2.19}$$

2.2 Linear Modulation

In this book, we deal with transmission of binary information over a physical medium (channel) using linear modulation. At the transmitter side, the sequence of *independent and uniformly distributed* (i.u.d.) bits $\underline{i} = [i[1], \ldots, i[N_{\mathrm{b}}]]$ is mapped (encoded) into a sequence of symbols $\underline{x} = [x[1], \ldots, x[N_{\mathrm{s}}]]$ which is then used to alter the amplitude of a finite-energy real-valued waveform $v(t)$ which is sent over the channel. The whole transmitted waveform is then given by

$$s(t) = \sum_{n=0}^{N_{\mathrm{s}}-1} x[n] v(t - nT_{\mathrm{s}}), \tag{2.20}$$

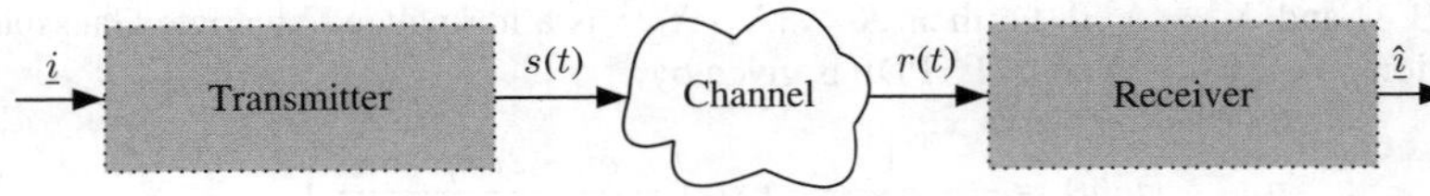

Figure 2.1 A simplified block diagram of a digital transmitter—receiver pair

where $R_s = 1/T_s$ is the signaling rate, which is proportional to the bandwidth occupied by the transmitted signals. The transmission rate is defined as

$$\mathsf{R} = \frac{N_b}{N_s T_s} = R R_s \left[\frac{\text{bit}}{\text{s}}\right], \tag{2.21}$$

where

$$R = \frac{N_b}{N_s} \left[\frac{\text{bit}}{\text{symbol}}\right]. \tag{2.22}$$

A general structure of such a system is schematically shown in Fig. 2.1.

The channel introduces scaling via the channel gain $h(t)$ and perturbations via the signal (noise) $z(t)$ which models the unknown thermal noise (Johnson–Nyquist noise) introduced by electronic components in the receiver. The signal observed at the receiver is given by

$$r(t) = h(t)s(t) + z(t). \tag{2.23}$$

The general problem in communications is to *guess* (estimate) the transmitted bits $\underline{i}$ using the channel observation $r(t)$.

We assume that $h(t)$ varies slowly compared to the duration of the waveform $v(t)$, that the noise $z(t)$ is a random white Gaussian process, and that the Nyquist condition for $v(t)$ is satisfied (i.e., for $r_v(t) = v(t) * v(-t)$, we have: $r_v(nT_s) = 1$ if $n = 0$ and $r_v(nT_s) = 0$ if $n \neq 0$). In this case, the receiver does not need to take into account the continuous-time signal $r(t)$, but rather, only its filtered (via the so-called matched filter $v(-t)$) and sampled version. Therefore, sampling the filtered output $y(t) = v(-t) * r(t)$ at time instants $0, T_s, 2T_s, \ldots$ provides us with sufficient statistics to estimate the transmitted sequence $\underline{x} = [x[1], x[2], \ldots, x[N_s]]$. Moreover, by considering N waveforms $v(t)$ that fulfill the Nyquist condition, we arrive at the N-dimensional discrete-time model

$$\boldsymbol{y}[n] = h[n]\boldsymbol{x}[n] + \boldsymbol{z}[n], \tag{2.24}$$

where $\boldsymbol{x}[n], \boldsymbol{y}[n], \boldsymbol{z}[n] \in \mathbb{R}^N$ and $h[n] \in \mathbb{R}$.

Using the model in (2.24), the channel and part of the transmitter and receiver in Fig. 2.1 are replaced by a discrete-time model, where the transmitted sequence of symbols is denoted by $\underline{\boldsymbol{x}} = [\boldsymbol{x}[1], \ldots, \boldsymbol{x}[N_s]]$ and $h[n] \in \mathbb{R}$ is the sequence of real channel attenuations (gains) varying independently of $\boldsymbol{x}[n]$ and assumed to be known (perfectly estimated) at the receiver. The transmitted sequence of symbols $\underline{\boldsymbol{x}} = [\boldsymbol{x}[1], \ldots, \boldsymbol{x}[N_s]]$ is corrupted by additive noise $\underline{\boldsymbol{z}} = [\boldsymbol{z}[1], \ldots, \boldsymbol{z}[N_s]]$, as shown in Fig. 2.2.

The main design challenge is now to find an encoder that maps the sequence of information bits $\underline{i}$ to a sequence of vectors $\underline{\boldsymbol{x}}$ so that reliable transmission of the information bits with limited decoding complexity is guaranteed. Typically, each element in $\underline{\boldsymbol{x}}$ belongs to a discrete constellation $\mathcal{S}$ of size $M = |\mathcal{S}|$. We assume $M = 2^m$, which leads us to consider the output of the encoder as M-ary symbols represented by length-m vectors of bits $\boldsymbol{b} = [b_1, \ldots, b_m]^{\mathrm{T}}$.

It is thus possible to consider that encoding is done in two steps: first a *binary encoder* (ENC) generates a sequence of N_c code bits $\underline{\boldsymbol{b}} = [\boldsymbol{b}[1], \ldots, \boldsymbol{b}[N_s]]$, and next, the code bits are mapped to constellation

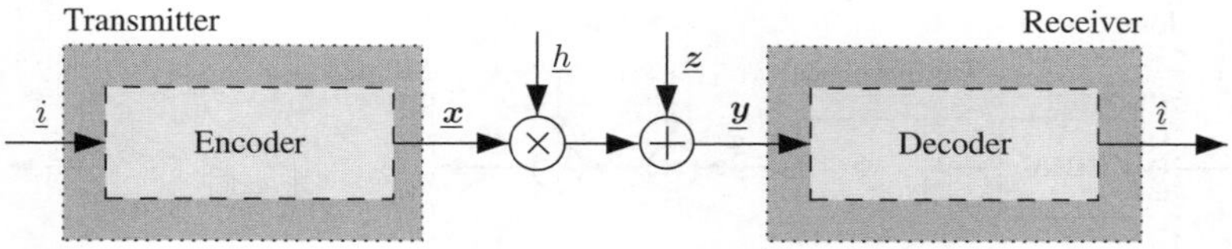

Figure 2.2 Simplified block diagram of the model in Fig. 2.1 where the continuous-time channel model is replaced by a discrete-time model

symbols. We will refer to the mapping between the information sequence $\underline{i}$ and the symbol sequence $\underline{x}$ as CM *encoding*. This operation is shown in Fig. 2.3, where the mapper Φ is a one-to-one mapping between the length-m binary vector $\boldsymbol{b}[n]$ and the constellation symbols $\boldsymbol{x}[n]$, i.e., $\Phi : \mathbb{B}^m \mapsto \mathcal{S}$. The modulation block is discussed in more detail in Section 2.5 and the encoding block in Section 2.6. At the receiver, the decoder will use *maximum likelihood* (ML) decoding, *maximum a posteriori* (MAP) decoding, or a BICM-type decoding; these are analyzed in Chapter 3.

The rate of the binary encoder (also known as the *code rate*) is given by

$$R_\mathrm{c} \triangleq \frac{N_\mathrm{b}}{N_\mathrm{c}} \left[\frac{\text{bit}}{\text{code bit}}\right], \tag{2.25}$$

where $N_\mathrm{c} = N_\mathrm{s} m$, and thus,

$$R = mR_\mathrm{c} \left[\frac{\text{bit}}{\text{symbol}}\right]. \tag{2.26}$$

The CM encoder in Fig. 2.2 maps each of the 2^{N_b} possible messages to the sequence $\underline{\boldsymbol{x}} = [\boldsymbol{x}[1], \ldots, \boldsymbol{x}[N_\mathrm{s}]]$. The code $\mathcal{X} \subset \mathcal{S}^{N_\mathrm{s}}$ with $|\mathcal{X}| = 2^{N_\mathrm{b}}$ is defined as the set of all codewords corresponding to the 2^{N_b} messages.[1] Similarly, we use $\mathcal{B} \subset \mathbb{B}^{N_\mathrm{c}}$ to denote the binary code, i.e., the set of binary codewords $\underline{\boldsymbol{b}}$. Having defined $\Phi(\cdot)$, the sets $\mathcal{X}$ and $\mathcal{B}$ are equivalent: the encoder is a one-to-one function that assigns each information message $\underline{i}$ to one of the 2^{N_b} possible symbol sequences $\underline{\boldsymbol{x}} \in \mathcal{X}$, or equivalently, to one of the corresponding bit sequences $\underline{\boldsymbol{b}} \in \mathcal{B}$. At the receiver's side, the decoder uses the vector of observations $\underline{\boldsymbol{y}}$ to generate an estimate of the information sequence $\hat{\underline{i}}$.

In the following section, we briefly review three popular ways of implementing the encoder–decoder pair in Fig. 2.2 or 2.3.

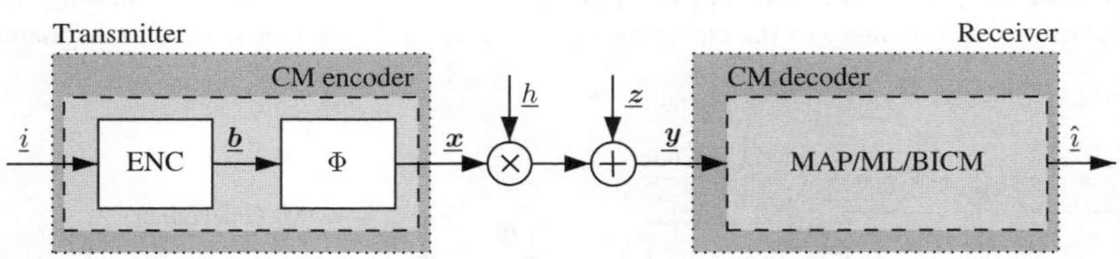

Figure 2.3 A simplified block diagram of the model in Fig. 2.2. The CM encoder is composed of a encoder (ENC) and a mapper (Φ). The CM decoder uses one of the rules defined in Chapter 3

[1] This is what is commonly done in practice and is general enough for the purposes of this book. We note, however, that the selection of $\underline{\boldsymbol{x}} \in \mathcal{S}^{N_\mathrm{s}}$ is, in fact, not entirely general as it assumes the use of the same constellation at each time instant.

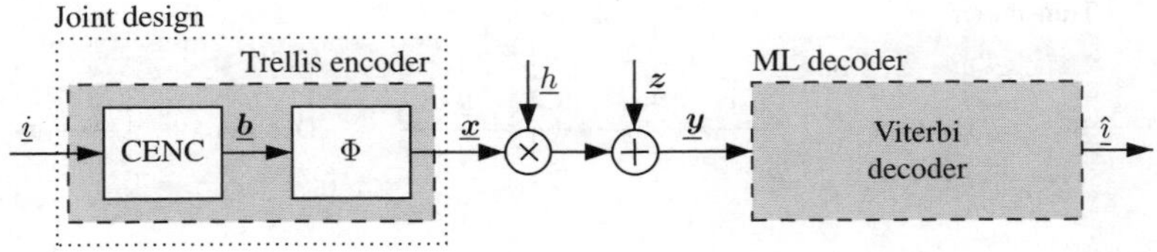

Figure 2.4 TCM: The encoder is formed by concatenating a CENC and a mapper Φ. The ML decoder may be implemented with low complexity using the Viterbi algorithm

2.3 Coded Modulation

We use the name *coded modulation* to refer to transceivers where $\log_2|\mathcal{S}| = m > 1$. In what follows we briefly describe popular ways of implementing such systems.

2.3.1 Trellis-Coded Modulation

Ungerboeck proposed *trellis-coded modulation* (TCM) in 1982 as a way of combining *convolutional encoders* (CENCs) (very popular at that time) with M-ary constellations. Targeting transmission over nonfading channels, TCM aimed at increasing the *Euclidean distance* (ED) between the transmitted codewords. In order to maximize the ED, the encoder and the mapper Φ are jointly designed; Ungerboeck proposed a design strategy based on the so-called labeling by *set-partitioning* (SP) and showed gains are attainable for various CENCs.

A typical TCM structure is shown in Fig. 2.4, where a CENC with rate $R_c = k_c/m$ is connected to an M-ary constellation. The resulting transmission rate is equal to $R = k_c$ [bit/symbol]. At the receiver, optimal decoding (i.e., finding the most likely codeword as in (1.5)) is implemented using the Viterbi algorithm. Throughout this book, we refer to the concatenation of a CENC and a memoryless mapper as a *trellis encoder*.

2.3.2 Multilevel Coding

Another approach to CM known as *multilevel coding* (MLC) was proposed by Imai and Hirakawa in 1977 and is shown in Fig. 2.5. The MLC encoder is based on the use of a *demultiplexer* (DEMUX) separating the information bits into m independent messages $\underline{i}_1, \ldots, \underline{i}_m$ which are then encoded by m parallel *binary* encoders. The outcomes of the encoders $[c_1[n], \ldots, c_m[n]]^{\mathrm{T}}$ are then passed to the mapper Φ.

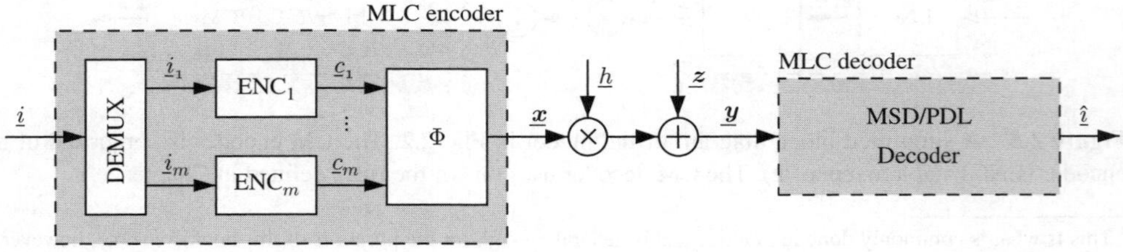

Figure 2.5 MLC: The encoder is formed by a demultiplexer, m parallel encoders, and a mapper. The decoder can be based on MSD or PDL

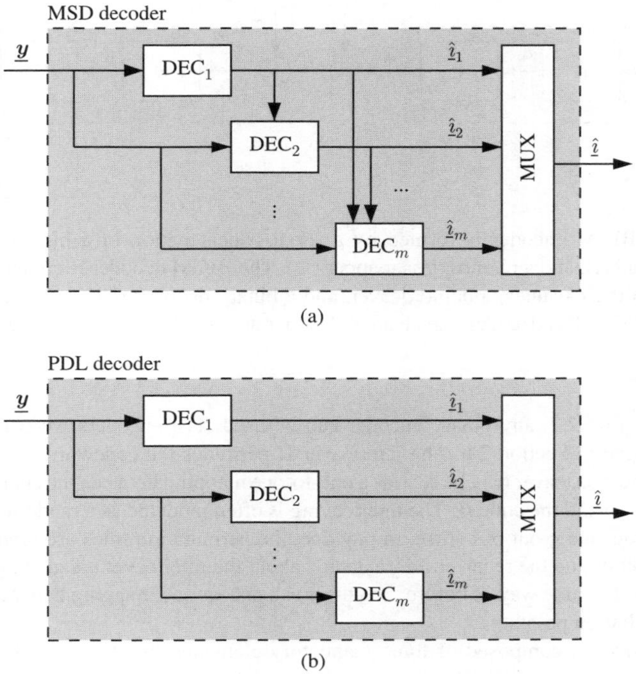

Figure 2.6 Decoding for MLC: (a) multistage decoding and (b) parallel decoding of the individual levels

The decoding of MLC can be done using the principle of successive interference cancelation, known as *multistage decoding* (MSD). First, the first sequence $\hat{\underline{\imath}}_1$ is decoded using the channel outcome according to the ML principle (see (1.5)), i.e., by using the conditional PDF $p_{\boldsymbol{Y}|C_1}(\boldsymbol{y}|c_1)$. Next, the decoding results $\hat{\underline{\imath}}_1$ (or equivalently $\hat{\underline{c}}_1$) are passed to the second decoder. Assuming error-free decoding, i.e., $\hat{\underline{c}}_1 = \underline{c}_1$, the second decoder can decode its own transmitted codeword using $p_{\boldsymbol{Y}|C_2,C_1}(\boldsymbol{y}|c_2, c_1)$. As shown in Fig. 2.6 (a), this continues to the last stage where all the bits (except the last one) are known.

Removing the interaction between the decoders and thus ignoring the relationships between the transmitted bitstreams $\underline{c}_1, \ldots, \underline{c}_m$ results in what is known as *parallel decoding of the individual levels* (PDL), also shown in Fig. 2.6 (b). Then, the decoders operate in parallel, yielding a shorter decoding time at the price of decoding suboptimality.

In MLC, the selection of the m rates of the codes is crucial. One way of doing this is based on information-theoretic arguments (which we will clarify in Section 4.4). One of the main issues when designing MLC is to take into account the decoding errors which propagate throughout the decoding process.

2.3.3 Bit-Interleaved Coded Modulation

BICM was introduced in 1992 by Zehavi and later analyzed from an information-theoretic point of view by Caire *et al.* A general BICM system model is shown in Fig. 2.7 where the CM encoder in Fig. 2.3 is realized as a serial concatenation of a binary *encoder* (ENC) of rate $R_c = N_b/N_c$, a bit-level interleaver Π, and a mapper Φ.

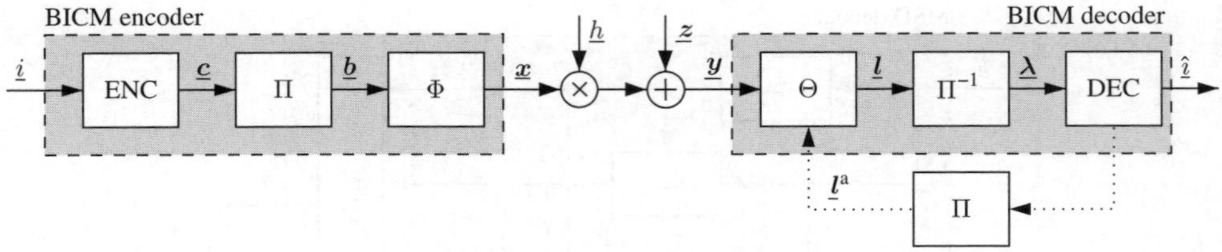

Figure 2.7 The BICM encoder is formed by a serial concatenation of a binary encoder (ENC), a bit-level interleaver (Π), and a memoryless mapper (Φ). The BICM decoder is composed of a demapper (Θ) that calculates the L-values, a deinterleaver, and a binary decoder (DEC). For a *BICM with iterative demapping* (BICM-ID) decoder, the channel decoder and the demapper exchange information in an iterative fashion

In the model in Fig. 2.7, any binary encoder can be used. The encoders we consider in this book are briefly introduced in Section 2.6. The interleaver Π permutes the codewords $\underline{\boldsymbol{c}}$ into the codewords $\underline{\boldsymbol{b}} = \Pi(\underline{\boldsymbol{c}})$. Thus, the interleaver may be seen as a one-to-one mapping between the channel code $\mathcal{C}$ (i.e., all possible sequences $\underline{\boldsymbol{c}}$) and the code $\mathcal{B}$. The interleaving is often modeled as a random permutation of the elements of the sequence $\underline{\boldsymbol{c}}$ but of course, in practice, the permutation rules are predefined and known by both the transmitter and the receiver. More details about the interleaver are given in Section 2.7. The mapper Φ operates the same way as before, i.e., there is a one-to-one mapping between length-m binary vectors and constellation points.

The BICM decoder is composed of three mandatory elements: the demapper Θ, the deinterleaver Π^{-1}, and the binary decoder DEC. The demapper Θ calculates the *logarithmic-likelihood ratio*s (LLRs, or L-values) for each code bit in $\underline{\boldsymbol{b}}$. These L-values are then deinterleaved by Π^{-1} and the resulting permuted sequence of L-values $\underline{\boldsymbol{\lambda}}$ is passed to the channel decoder DEC. The L-values are the basic signals exchanged in a BICM receiver. We analyze their properties in Chapters 3 and 5.

Recognizing the BICM encoder as a concatenation of a binary encoder and an M-ary mapper, the decoding can be done in an iterative manner. This is known as *bit-interleaved coded modulation with iterative decoding/demapping* (BICM-ID). In BICM-ID, the decoder and the demapper exchange information about the code bits using the so-called *extrinsic* L-values calculated alternatively by the demapper and the decoder. Soon after BICM-ID was introduced, binary labeling was recognized to play an important role in the design of BICM-ID transceivers and has been extensively studied in the literature.

2.4 Channel Models

Although in practice the codebook is obtained in a deterministic way, we model $\boldsymbol{x}[n]$ as N-dimensional independent random variables with PMF $P_{\boldsymbol{X}}(\boldsymbol{x}[n])$ (more details are given in Section 2.5.3), which is compatible with the assumption of the codebook $\mathcal{X}$ being randomly generated. Under these assumptions, and omitting the discrete-time index $[n]$, (2.24) can be modeled using random vectors

$$\boldsymbol{Y} = H\boldsymbol{X} + \boldsymbol{Z}, \tag{2.27}$$

where $\boldsymbol{Z} \in \mathbb{R}^N$ is an N-dimensional vector with *independent and identically distributed* (i.i.d.) Gaussian-distributed entries, so $\mathbb{E}_{\boldsymbol{Z}}[\boldsymbol{Z}\boldsymbol{Z}^{\mathrm{T}}] = \mathbf{I}_{[N]}\mathsf{N}_0/2$, where $\mathsf{N}_0/2$ is the power spectral density of the noise in (2.23).

The PDF of the channel output in (2.27), conditioned on the channel state $H = h$ and the transmitted symbol $\boldsymbol{X} = \boldsymbol{x}$, is given by

$$p_{\boldsymbol{Y}|\boldsymbol{X},H}(\boldsymbol{y}|\boldsymbol{x}, h) = \left(\frac{1}{\pi \mathsf{N}_0}\right)^{\frac{N}{2}} \exp\left(-\frac{1}{\mathsf{N}_0}\|\boldsymbol{y} - h\boldsymbol{x}\|^2\right). \tag{2.28}$$

The channel gain $H \in \mathbb{R}$ models variations in the propagation medium because of phenomena such as signal absorption and fading[2] and is captured by the PDF $p_H(h)$. Throughout this book, we assume that the receiver knows the channel realization perfectly, obtained using channel estimation techniques we do not cover.

Definition 2.1 (Instantaneous SNR) *For a given channel gain $H = h$, the instantaneous* signal-to-noise ratio *(SNR) is defined as the average received energy of the signal of interest over the average energy of the noise, i.e.,*

$$\mathsf{snr}(h) \triangleq h^2 \frac{\mathsf{E_s}}{\mathsf{N}_0} = \frac{h^2}{\mathsf{N}_0}, \tag{2.29}$$

where $\mathsf{E_s} = 1$ is the average symbol energy (see Section 2.5.1).

Example 2.2 (AWGN Channel) *When fading is not present, the channel is constant ($h = 1$), and thus, the instantaneous SNR is given by*

$$\mathsf{snr} = \mathsf{snr}(1) = \frac{1}{\mathsf{N}_0}. \tag{2.30}$$

To emphasize that the only distortion to the transmitted signals is due to the additive noise, this channel is also referred to as the additive white Gaussian noise (AWGN) channel. The conditional PDF (2.28) *is then expressed as*

$$p_{\boldsymbol{Y}|\boldsymbol{X}}(\boldsymbol{y}|\boldsymbol{x}) = \left(\frac{\mathsf{snr}}{\pi}\right)^{\frac{N}{2}} \exp\left(-\mathsf{snr}\,\|\boldsymbol{y} - \boldsymbol{x}\|^2\right). \tag{2.31}$$

If we use the so-called fast-fading channel model, the channel gain H is a random variable, so the instantaneous SNR in (2.29) is a random variable as well, i.e.,

$$\mathsf{SNR} = \mathsf{snr}(H) = \frac{H^2}{\mathsf{N}_0}. \tag{2.32}$$

Definition 2.3 (Average SNR) *The average SNR is defined through the expectation over the joint distribution of $\boldsymbol{X}$ and H, i.e.,*

$$\overline{\mathsf{snr}} \triangleq \frac{\mathbb{E}_{H,\boldsymbol{X}}[\|H\boldsymbol{X}\|^2]}{\mathsf{N}_0} = \frac{\mathbb{E}_H[H^2]\mathsf{E_s}}{\mathsf{N}_0}. \tag{2.33}$$

As the average SNR depends on the parameters of three random variables, it is customary to fix two of them and vary the third. For the numerical results presented in this book, we keep constant the transmitted signal energy and the variance of the channel gain, i.e., $\mathsf{E_s} = 1$ and $\mathbb{E}_H[H^2] = 1$.

[2] The modulated waveform arrives at the receiver via different paths. Each path contributes with a randomly varying phase adding up constructively or destructively, which produces the fading phenomenon.

Example 2.4 (Fast-Fading Channel) *We consider the case when H follows a Nakagami-m PDF, parameterized by the coefficient $\frac{1}{2} \leq \mathsf{m} < \infty$, i.e.,*

$$p_H(h, \mathsf{m}, \overline{\mathsf{snr}}) = \frac{2\mathsf{m}^{\mathsf{m}} h^{2\mathsf{m}-1}}{\Gamma(\mathsf{m})\, \overline{\mathsf{snr}}^{\mathsf{m}}} \exp\left(-\frac{\mathsf{m}}{\overline{\mathsf{snr}}} h^2\right), \tag{2.34}$$

where $\Gamma(v+1) \triangleq \int_0^\infty \lambda^v \mathrm{e}^{-\lambda}\, \mathrm{d}\lambda$ is the Gamma function. This PDF allows us to model a wide range of channels. For example, the Rayleigh fading channel is obtained when $\mathsf{m} = 1$, and the AWGN channel when $\mathsf{m} \to \infty$.

The instantaneous SNR then follows a Gamma distribution with PDF given by

$$p_{\mathsf{SNR}}(\mathsf{snr}, \mathsf{m}, \overline{\mathsf{snr}}) = \frac{\mathsf{m}^{\mathsf{m}} \mathsf{snr}^{\mathsf{m}-1}}{\Gamma(\mathsf{m}) \overline{\mathsf{snr}}^{\mathsf{m}}} \exp\left(-\frac{\mathsf{m}}{\overline{\mathsf{snr}}} \mathsf{snr}\right) \tag{2.35}$$

and it becomes an exponential function for $\mathsf{m} = 1$ (Rayleigh fading). The MGF of the random variable SNR is given by

$$\mathsf{P}_{\mathsf{SNR}}(s) = \left(\frac{\mathsf{m}}{\mathsf{m} - s \cdot \overline{\mathsf{snr}}}\right)^{\mathsf{m}}. \tag{2.36}$$

Each transmitted symbol conveys R information bits, and thus, the relation between the average symbol energy $\mathsf{E_s}$ and the average information bit energy $\mathsf{E_b}$ is given by $\mathsf{E_s} = R\mathsf{E_b}$. The average SNR can then be expressed as

$$\overline{\mathsf{snr}} = \mathbb{E}_H[H^2] \frac{\mathsf{E_s}}{\mathsf{N_0}} \tag{2.37}$$

$$= \frac{\mathsf{E_b}}{\mathsf{N_0}} R. \tag{2.38}$$

2.5 The Mapper

An important element of the BICM transmitter in Fig. 2.7 is the mapper Φ, which we show in Fig. 2.8. This mapper, using m bits as input, selects one of the symbols from the constellation to be transmitted. The mapper is then fully defined by the constellation $\mathcal{S}$ used for transmission and the way the bits are mapped to the constellation symbols. Although in a general case we can use many-to-one mappings (i.e., $M < 2^m$), our analysis will be restricted to bijective mapping operations $\Phi(\cdot)$, i.e., when $M = 2^m$.

2.5.1 *Constellation*

The constellation used for transmission is denoted by $\mathcal{S} \subset \mathbb{R}^N$, and the constellation symbols by $\boldsymbol{x}_i \in \mathcal{S}$, where $|\mathcal{S}| = M = 2^m$. The difference between two constellation symbols is defined as

$$\boldsymbol{d}_{i,j} \triangleq \boldsymbol{x}_i - \boldsymbol{x}_j. \tag{2.39}$$

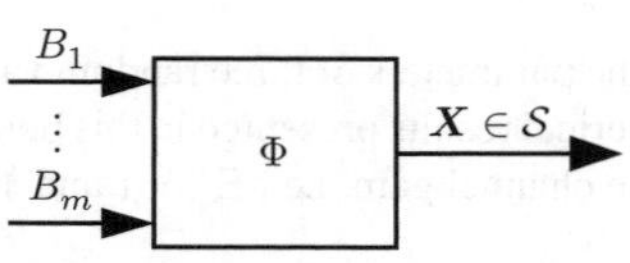

Figure 2.8 The role of the mapper Φ in Fig. 2.7 is to generate the symbols from the constellation $\mathcal{S}$ using m binary inputs $B_1, \ldots, B_m$

The *constellation Euclidean distance spectrum* (CEDS) is defined via a vector $\boldsymbol{d}_{\mathcal{S}} \triangleq [d_1, d_2, \ldots, d_D]^{\mathrm{T}}$. The CEDS contains all the D possible distances between pairs of constellation points in $\mathcal{S}$, where $d_1 < d_2 < \ldots < d_D$. The *minimum Euclidean distance* (MED) of the constellation is defined as the smallest element of the CEDS, i.e.,

$$d_{\mathcal{S}}^{\min} \triangleq \min_{\substack{i,j \in \mathcal{I} \\ i \neq j}} \|\boldsymbol{d}_{j,i}\| = d_1, \tag{2.40}$$

where from now on we use $\mathcal{I} \triangleq \{1, \ldots, M\}$ to enumerate all the constellation points in $\mathcal{S}$.

The input distribution is defined by the vector

$$\boldsymbol{p}_{\mathrm{s}} \triangleq [p_{\mathrm{s},1}, \ldots, p_{\mathrm{s},M}]^{\mathrm{T}} = [P_{\boldsymbol{X}}(\boldsymbol{x}_1), \ldots, P_{\boldsymbol{X}}(\boldsymbol{x}_M)]^{\mathrm{T}}. \tag{2.41}$$

For a uniform distribution, i.e., where all the symbols are equiprobable, we use the notation

$$\boldsymbol{p}_{\mathrm{s,u}} = \left[\frac{1}{M}, \ldots, \frac{1}{M}\right]^{\mathrm{T}}. \tag{2.42}$$

We assume that the constellation $\mathcal{S}$ is defined so that the transmitted symbols $\boldsymbol{x}$ have zero mean

$$\mathbb{E}_{\boldsymbol{X}}[\boldsymbol{X}] = \sum_{i \in \mathcal{I}} p_{\mathrm{s},i} \boldsymbol{x}_i = \boldsymbol{0}, \tag{2.43}$$

and, according to the convention we have adopted, the average symbol energy E_{s} is normalized as

$$\mathsf{E}_{\mathrm{s}} \triangleq \mathbb{E}_{\boldsymbol{X}}[\|\boldsymbol{X}\|^2] = \sum_{i \in \mathcal{I}} p_{\mathrm{s},i} \|\boldsymbol{x}_i\|^2 = 1. \tag{2.44}$$

We will often refer to the practically important constellations *pulse amplitude modulation* (PAM), *phase shift keying* (PSK), and *quadrature amplitude modulation* (QAM), defined as follows:

- An MPSK *constellation* is defined as

$$\mathcal{S}_{M\mathrm{PSK}} \triangleq \{[\cos((2i-1)\pi/M), \sin((2i-1)\pi/M)] : i \in \mathcal{I}\}, \tag{2.45}$$

where $d_{\mathcal{S}_{M\mathrm{PSK}}}^{\min} = 2\sin(\pi/M)$. In Fig. 2.9, three commonly used MPSK constellations are shown. For 8PSK, we also show the vector $\boldsymbol{d}_{4,6}$ in (2.39) as well as the MED (2.40). For $\boldsymbol{p}_{\mathrm{s}} = \boldsymbol{p}_{\mathrm{s,u}}$, (2.43) and (2.44) are satisfied.

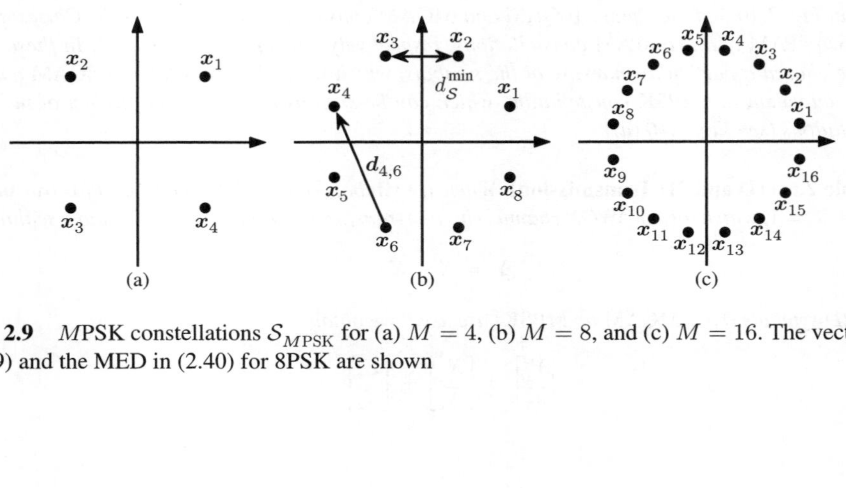

Figure 2.9 MPSK constellations $\mathcal{S}_{M\mathrm{PSK}}$ for (a) $M = 4$, (b) $M = 8$, and (c) $M = 16$. The vector $\boldsymbol{d}_{4,6}$ in (2.39) and the MED in (2.40) for 8PSK are shown

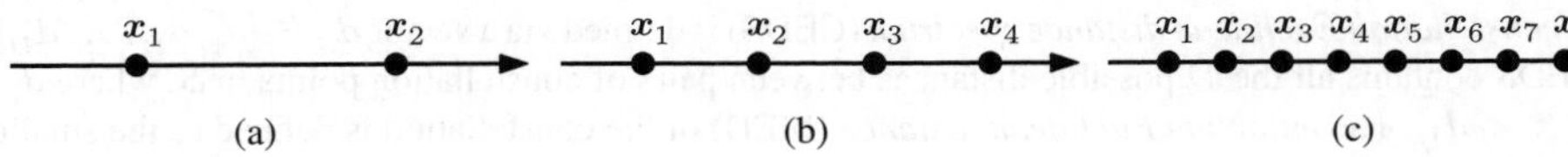

Figure 2.10 MPAM constellations $\mathcal{S}_{M\text{PAM}}$ for (a) $M = 2$, (b) $M = 4$, and (c) $M = 8$

- An MPAM *constellation* is defined as

$$\mathcal{S}_{M\text{PAM}} \triangleq \{\pm(M-1)\Delta, \pm(M-3)\Delta, \ldots, \pm 1\Delta\}, \tag{2.46}$$

where the CEDS is $\boldsymbol{d}_{\mathcal{S}_{M\text{PAM}}} = [2\Delta, 4\Delta, \ldots, (M-1)2\Delta]^{\text{T}}$ with $d^{\min}_{\mathcal{S}_{M\text{PAM}}} = 2\Delta$. For $\boldsymbol{p}_{\text{s}} = \boldsymbol{p}_{\text{s,u}}$,

$$\Delta = \sqrt{\frac{3}{M^2-1}}, \tag{2.47}$$

which guarantees that (2.43) and (2.44) are satisfied.
- A *rectangular* MQAM constellation is defined as the Cartesian product of two MPAM constellations, i.e.,

$$\mathcal{S}_{M\text{QAM}} \triangleq \mathcal{S}_{M_1\text{PAM}} \times \mathcal{S}_{M_2\text{PAM}}, \tag{2.48}$$

where $M_1 = 2^{m_1}$, $M_2 = 2^{m_2}$, and $|\mathcal{S}_{M\text{QAM}}| = M = 2^{m_1+m_2}$.
- A *square* MQAM is obtained as a particular case of a rectangular MQAM when $M_1 = M_2 = \sqrt{M}$. In this case, we have $\mathcal{S}_{M\text{QAM}} = \mathcal{S}^2_{\sqrt{M}\text{PAM}}$. Again, for a uniform input distribution $\boldsymbol{p}_{\text{s,u}}$, to guarantee (2.43) and (2.44), we will use the scaling

$$\Delta_{\text{QAM}} = \sqrt{\frac{3}{2(M-1)}}. \tag{2.49}$$

The MED is $d^{\min}_{\mathcal{S}_{M\text{QAM}}} = 2\Delta_{\text{QAM}}$.

One of the simplest constellations we will refer to is 2PAM, which is equivalent to 2PSK and is also known as *binary phase shift keying* (BPSK). Constellations with more than two constellation points are discussed in the following example.

Example 2.5 (PAM and QAM Constellations) *The* MPAM *constellations for* $M = 2, 4,$ and 8 *are shown in Fig. 2.10 and the square* 16QAM *and* 64QAM *constellations obtained as the Cartesian product of two* 4PAM *and two* 8PAM *constellations, respectively, are shown in Fig. 2.11. In these figures, we have chosen a particular indexing of the symbols, which we will discuss later. A* 4QAM *constellation is equivalent to a* 4PSK *constellation, which can be seen as the Cartesian product of two* 2PAM *constellations (see Fig. 2.10 (a)).*

Example 2.6 (1D and 2D Transmission) *When 1D symbols (e.g.,* MPAM*) are used for transmission, we have* $N = 1$*, which for the AWGN channel (*$h = 1$*) transforms the model in* (2.27) *into a scalar model*

$$Y = X + Z. \tag{2.50}$$

When 2D symbols (e.g., MQAM *or* MPSK*) are used, we obtain*

$$\begin{bmatrix} Y_1 \\ Y_2 \end{bmatrix} = \begin{bmatrix} X_1 \\ X_2 \end{bmatrix} + \begin{bmatrix} Z_1 \\ Z_2 \end{bmatrix}. \tag{2.51}$$

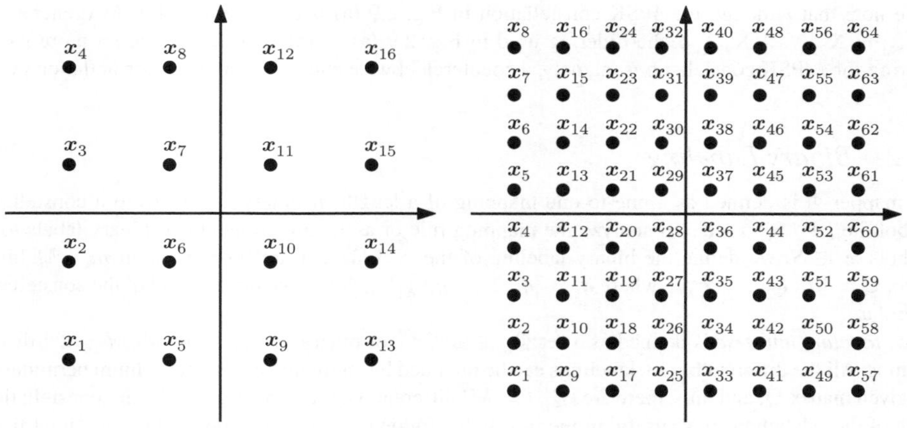

Figure 2.11 Square MQAM constellations $\mathcal{S}_{M\mathrm{QAM}}$ for (a) $M = 16$ and (b) $M = 64$. The 16QAM constellation is generated as the Cartesian product of two 4PAM constellations (shown in Fig. 2.10 (b)) and the 64QAM as the Cartesian product of two 8PAM constellations (shown in Fig. 2.10 (c))

We note that the model (2.51) *is often represented as a scalar model (like the one in* (2.50)*), but using complex numbers:*

$$Y_{\mathrm{c}} = h \cdot X_{\mathrm{c}} + Z_{\mathrm{c}}, \tag{2.52}$$

where $Y_{\mathrm{c}} = Y_1 + jY_2$, $X_{\mathrm{c}} = X_1 + jX_2$ *and* $Z_{\mathrm{c}} = Z_1 + jZ_2$, *and where* $j = \sqrt{-1}$.

Up to this point, we have considered the input constellation to be a set $\mathcal{S}$. As we will later see, it is useful to represent the constellation symbols taken from the constellation $\mathcal{S}$ in an *ordered* manner via the $N \times M$ matrix $\mathbf{X} = [\boldsymbol{x}_1, \boldsymbol{x}_2, \ldots, \boldsymbol{x}_M]$. Using this definition, the elements of $\mathbf{X}_{M\mathrm{PSK}}$ are $\boldsymbol{x}_i = [\cos((2i-1)\pi/M), \sin((2i-1)\pi/M)]^{\mathrm{T}}$ with $i \in \mathcal{I}$, and the elements of $\mathbf{X}_{M\mathrm{PAM}}$ are $\boldsymbol{x}_i = -(M - 2i + 1)\Delta$ with $i \in \mathcal{I}$. This notation will be useful in Section 2.5.2 where binary labelings are defined.

Another reason for using the notation $\mathbf{X}$ for a constellation is that with such a representation, 2D constellations generated via Cartesian products (such as rectangular MQAM constellations) can also be defined using the ordered direct product defined in (2.2). A *rectangular* MQAM constellation is then defined by the matrix $\mathbf{X}_{M\mathrm{QAM}}$ formed by the ordered direct product of two MPAM constellations $\mathbf{X}_{M_1\mathrm{PAM}}$ and $\mathbf{X}_{M_2\mathrm{PAM}}$, i.e.,

$$\mathbf{X}_{M\mathrm{QAM}} = \mathbf{X}_{M_1\mathrm{PAM}} \otimes \mathbf{X}_{M_2\mathrm{PAM}}. \tag{2.53}$$

Example 2.7 (16QAM **Constellation)** *The square* 16QAM *constellation in Example 2.5 can be represented as*

$$\mathbf{X}_{16\mathrm{QAM}} = \mathbf{X}_{4\mathrm{PAM}} \otimes \mathbf{X}_{4\mathrm{PAM}} \tag{2.54}$$

$$= [x_1, x_2, x_3, x_4] \otimes [x_1, x_2, x_3, x_4]. \tag{2.55}$$

The notation used in (2.55) (or more generally in (2.53)) will be used in Section 2.5.2 to formally define the binary labeling of 2D constellations.

We note that although the 4PSK constellation in Fig. 2.9 (a) is equivalent to 4QAM (generated as $\mathbf{X}_{4\text{QAM}} = \mathbf{X}_{2\text{PAM}} \otimes \mathbf{X}_{2\text{PAM}}$), the ordering used in Fig. 2.9 (a) is different, i.e., we use a more natural ordering for MPSK constellations, namely, a counterclockwise enumeration of the constellation points.

2.5.2 *Binary Labeling*

The mapper Φ is defined as a one-to-one mapping of a length-m binary codeword to a constellation symbol, i.e., $\Phi : \mathbb{B}^m \mapsto \mathcal{S}$. To analyze the mapping rule of assigning all length-m binary labels to the symbols $\boldsymbol{x}_i \in \mathcal{S}$, we define the binary labeling of the $N \times M$ constellation $\mathbf{X}$ as an $m \times M$ binary matrix $\mathbf{Q} = [\boldsymbol{q}_1, \boldsymbol{q}_2, \ldots, \boldsymbol{q}_M]$, where $\boldsymbol{q}_i = [q_{i,1}, \ldots, q_{i,m}]^{\text{T}} \in \mathbb{B}^m$ is the binary label of the constellation symbol $\boldsymbol{x}_i$.

The *labeling universe* is defined as a set $\mathcal{Q}_m \in \mathbb{B}^{m\times 2^m}$ of binary matrices $\mathbf{Q}$ with $M = 2^m$ distinct columns. All the different binary labelings can be obtained by studying all possible column permutations of a given matrix $\mathbf{Q}$, and thus, there are $|\mathcal{Q}_m| = M!$ different ways to label a given M-ary constellation.[3] Some of these labelings have useful properties. In the following, we define those which are of particular interest.

Definition 2.8 (Gray Labeling) *The binary labeling* $\mathbf{Q}$ *of the constellation* $\mathbf{X}$ *is said to be* Gray *if for all* i, j *such that* $\|\boldsymbol{d}_{j,i}\| = d_{\mathcal{S}}^{\min}$, $d_{\text{H}}(\boldsymbol{q}_i, \boldsymbol{q}_j) = 1$.

According to Definition 2.8, a constellation is Gray-labeled if all pairs of constellation points at MED have labels that differ only in one bit. As this definition depends on the constellation $\mathbf{X}$, a labeling construction is not always available; in fact, the existence of a Gray labeling cannot be guaranteed in general. However, it can be obtained for the constellations $\mathbf{X}_{M\text{PAM}}$ and $\mathbf{X}_{M\text{PSK}}$ we focus on.

While Gray labelings are not unique (for a given constellation $\mathbf{X}$, in general, there is more than one $\mathbf{Q}$ satisfying Definition 2.8), the most popular one is the *binary reflected Gray code* (BRGC), whose construction is given below. Throughout this book, unless stated otherwise, we use the BRGC, normally referred to in the literature as "Gray labeling", "Gray coding",[4] or "Gray mapping".

Definition 2.9 (Expansions) *To expand a labeling* $\mathbf{Q} = [\boldsymbol{q}_1, \ldots, \boldsymbol{q}_M]$, *where* $\boldsymbol{q}_i = [q_{i,1}, \ldots, q_{i,m}]^{\text{T}}$, *each binary codeword is repeated once to obtain a new matrix* $[\boldsymbol{q}_1, \boldsymbol{q}_1, \ldots, \boldsymbol{q}_M, \boldsymbol{q}_M]$, *and then one extra row* $[0, 1, 1, 0, 0, 1, 1, 0, \ldots, 0, 1, 1, 0]$ *of length* $2M$ *is appended.*

Definition 2.10 (Binary Reflected Gray Code) *The BRGC of order* $m \geq 1$, *denoted by* $\mathbf{Q}_{\text{BRGC}}$, *is generated by* $m - 1$ *recursive expansions of the trivial labeling* $[0, 1]$.

In Fig. 2.12, we show how the BRGC of order $m = 3$ is recursively constructed.

The importance of the BRGC is due to the fact that for all the constellations defined in Section 2.5, the BRGC (i) minimizes the *bit-error probability* (BEP) when the SNR tends to infinity and (ii) gives high achievable rates in BICM transmission for medium to high SNR. More details about this are given in Chapter 4.

Another labeling of particular interest is the *natural binary code* (NBC). The NBC is relevant for BICM because it maximizes its achievable rates for MPAM and MQAM constellations in the low SNR regime. For MPAM and MQAM constellations, the NBC is also one of the most natural ways of implementing Ungerboeck's SP labeling, and thus, it is very often used in the design of TCM systems.

Definition 2.11 (Natural Binary Code) *The NBC, denoted by* $\mathbf{Q}_{\text{NBC}}$, *is defined such that* $\boldsymbol{q}_i$ *is the base-2 representation of the integer* $i - 1$ *with* $i = 1, \ldots, M$ *and where* $\boldsymbol{q}_{i,1}$ *is the most significant bit (MSB).*

[3] The size of the labeling universe for M-ary constellations is very large, even for medium size constellations. For example, for $M = 16$, this number is approximately $2.1 \cdot 10^{13}$.

[4] The name "coding" is used for historical reasons and is unrelated to the encoder ENC we discuss in this book.

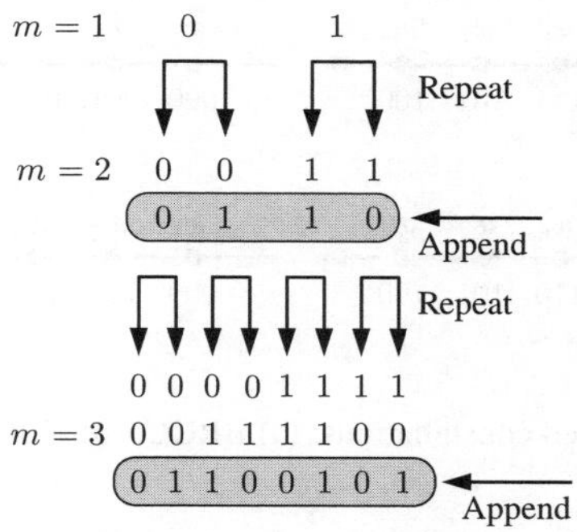

Figure 2.12 Recursive expansions of the labeling $[0, 1]$ used to generate the BRGC of order $m = 3$

If we denote the entries of $\mathbf{Q}_{\text{BRGC}}$ and $\mathbf{Q}_{\text{NBC}}$ by $b_{k,i}$ and $n_{k,i}$ for $k = 1, \dots, m$ and $i \in \mathcal{I}$, we find that $b_{1,i} = n_{1,i}$ and $b_{k,i} = n_{k-1,i} \oplus n_{k,i}$ for $k = 2, \dots, m$. Alternatively, we have $n_{k,i} = b_{1,i} \oplus \dots \oplus b_{k-1,i} \oplus b_{k,i}$ for $k = 1, \dots, m$, or, in matrix notation, $\mathbf{Q}_{\text{BRGC}} = \mathbf{T} \cdot \mathbf{Q}_{\text{NBC}}$ and $\mathbf{Q}_{\text{NBC}} = \mathbf{T}^{-1} \cdot \mathbf{Q}_{\text{BRGC}}$, where

$$\mathbf{T} = \begin{bmatrix} 1 & 0 & 0 & \dots & 0 & 0 & 0 \\ 1 & 1 & 0 & \dots & 0 & 0 & 0 \\ 0 & 1 & 1 & \dots & 0 & 0 & 0 \\ & \vdots & & \ddots & & \vdots & \\ 0 & 0 & 0 & \dots & 1 & 0 & 0 \\ 0 & 0 & 0 & \dots & 1 & 1 & 0 \\ 0 & 0 & 0 & \dots & 0 & 1 & 1 \end{bmatrix}, \quad \mathbf{T}^{-1} = \begin{bmatrix} 1 & 0 & 0 & \dots & 0 & 0 \\ 1 & 1 & 0 & \dots & 0 & 0 \\ 1 & 1 & 1 & \dots & 0 & 0 \\ \vdots & & & \ddots & & \vdots \\ 1 & 1 & 1 & \dots & 1 & 0 \\ 1 & 1 & 1 & \dots & 1 & 1 \end{bmatrix}. \tag{2.56}$$

Example 2.12 *The NBC and BRGC of order $m = 2$ are related via*

$$\mathbf{Q}_{\text{BRGC}} = \mathbf{T} \cdot \mathbf{Q}_{\text{NBC}} \tag{2.57}$$

$$\begin{bmatrix} 0 & 0 & 1 & 1 \\ 0 & 1 & 1 & 0 \end{bmatrix} = \begin{bmatrix} 1 & 0 \\ 1 & 1 \end{bmatrix} \begin{bmatrix} 0 & 0 & 1 & 1 \\ 0 & 1 & 0 & 1 \end{bmatrix}. \tag{2.58}$$

Example 2.13 *The NBC and BRGC of order $m = 3$ are*

$$\mathbf{Q}_{\text{NBC}} = \begin{bmatrix} 0 & 0 & 0 & 0 & \mathbf{1} & 1 & 1 & 1 \\ 0 & 0 & \mathbf{1} & 1 & 0 & 0 & 1 & 1 \\ 0 & \mathbf{1} & 0 & 1 & 0 & 1 & 0 & 1 \end{bmatrix}, \quad \mathbf{Q}_{\text{BRGC}} = \begin{bmatrix} 0 & 0 & 0 & 0 & \mathbf{1} & 1 & 1 & 1 \\ 0 & 0 & \mathbf{1} & 1 & 1 & 1 & 0 & 0 \\ 0 & \mathbf{1} & 1 & 0 & 0 & 1 & 1 & 0 \end{bmatrix}. \tag{2.59}$$

In (2.59), we highlighted in bold each first nonzero element of the kth row of the labeling matrix. These elements are called the *pivots* of $\mathbf{Q}$, and are used in the following definition.

Definition 2.14 (Reduced Echelon Matrix) *A matrix $\mathbf{Q} \in \mathbb{B}^{m \times M}$ is called a reduced echelon matrix if the following two conditions are fulfilled:*

1. *Every column with a pivot has all its other entries zero.*
2. *The pivot in row k is located to the right of the pivot in row $k+1$, for $k = 1, \dots, m-1$.*

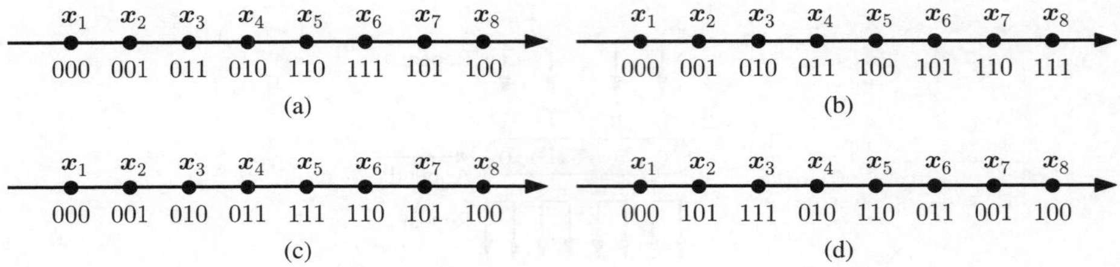

Figure 2.13 8PAM labeled constellations: (a) BRGC, (b) NBC, (c) FBC, and (d) BSGC

The matrix $\mathbf{Q}_{\text{NBC}}$ for $m = 3$ in Example 2.13 (or more generally $\mathbf{Q}_{\text{NBC}}$ for any m) is a reduced column echelon matrix while $\mathbf{Q}_{\text{BRGC}}$ is not because it does not fulfill the first condition in Definition 2.14.

Other binary labelings of interest include the *folded binary code* (FBC) and the *binary semi-Gray code* (BSGC). These labelings, together with that for the BRGC are shown in Fig. 2.13 for an 8PAM constellation.

Square and rectangular MQAM constellations are relevant in practice, and also very popular for BICM systems. In what follows, we describe one of the most common ways of labeling an MQAM constellation using the BRGC. Although we show it for the BRGC, the following way of labeling QAM constellations can also be used with the NBC, FBC, or BSGC. It is also important to notice at this point that to unequivocally define the labeling of a 2D constellation, the use of the ordered constellation points (via $\mathbf{X}$) is needed.

Definition 2.15 (BRGC for Rectangular MQAM Constellations) *A rectangular MQAM constellation defined by $\mathbf{X}_{M\text{QAM}} = \mathbf{X}_{M_1\text{PAM}} \otimes \mathbf{X}_{M_2\text{PAM}}$ with $M_1 = 2^{m_1}$ and $M_2 = 2^{m_2}$ is said to be labeled by the BRGC if its binary labeling $\mathbf{Q} \in \mathcal{Q}_{m_1+m_2}$ is formed by the ordered direct product of two BRGC labelings of order m_1 and m_2, i.e.,*

$$\mathbf{Q}_{\text{BRGC}} = \mathbf{Q}' \otimes \mathbf{Q}'', \tag{2.60}$$

where $\mathbf{Q}'$ is the BRGC of order m_1 and $\mathbf{Q}''$ is the BRGC of order m_2.

Again, a square MQAM is a particular case of Definition 2.15 when $M_1 = M_2 = \sqrt{M}$. One particularly appealing property of this labeling is that the first and second dimension of the constellation are decoupled in terms of the binary labeling. This is important because it reduces the complexity of the L-values' computation by the demapper.

Example 2.16 (BRGC for MQAM) *The BRGC for 4QAM ($M_1 = M_2 = \sqrt{M} = 2$) is given by the matrix*

$$\begin{aligned}\mathbf{Q}_{\text{BRGC}} &= \mathbf{Q}' \otimes \mathbf{Q}' \\ &= [0,1] \otimes [0,1] \\ &= \begin{bmatrix} 0 & 0 & 1 & 1 \\ 0 & 1 & 0 & 1 \end{bmatrix}\end{aligned} \tag{2.61}$$

and is shown in Fig. 2.14 (a). Similarly, for 16QAM ($M_1 = M_2 = \sqrt{M} = 4$) we obtain

$$\begin{aligned}\mathbf{Q}_{\text{BRGC}} &= \mathbf{Q}'' \otimes \mathbf{Q}'' \\ &= \begin{bmatrix} 0&0&0&0&0&0&0&0&1&1&1&1&1&1&1&1 \\ 0&0&0&0&1&1&1&1&1&1&1&1&0&0&0&0 \\ 0&0&1&1&0&0&1&1&0&0&1&1&0&0&1&1 \\ 0&1&1&0&0&1&1&0&0&1&1&0&0&1&1&0 \end{bmatrix},\end{aligned} \tag{2.62}$$

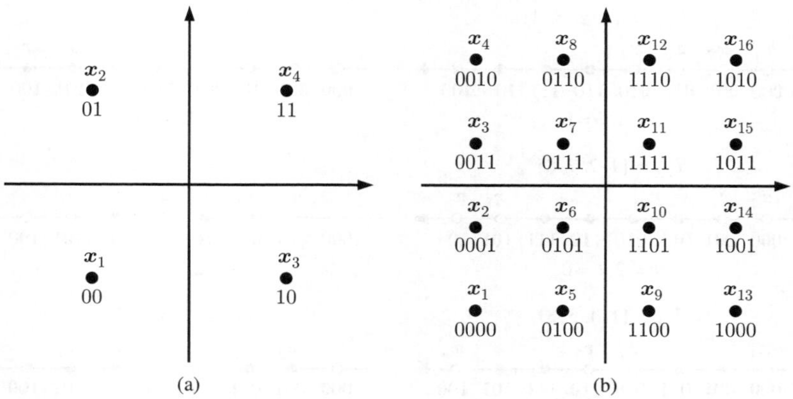

Figure 2.14 Constellations labeled by the BRGC: (a) 4QAM and (b) 16QAM

which is shown in Fig. 2.14 (b). From these figures, we see that, because of its construction, the BRGC labeling for 2D constellations can be decoupled into two dimensions, i.e., the first $m/2$ bits of the labeling are associated with the first dimension of the constellation and the last $m/2$ bits are associated with the second dimension.

It is important to note that $\mathbf{Q}_{\text{BRGC}}$ in (2.61) and (2.62) do not coincide with the definition of the BRGC in Definition 2.10. This is because each constituent labeling is a BRGC; however, the ordered direct product of two BRGC labelings is not necessarily a BRGC.

Example 2.17 (M16 **Labeling for BICM-ID)** *While the BRGC for MQAM is constructed via the ordered direct product of two BRGC labelings, not all labelings have this property. In the context of BICM-ID, e.g., this is a property that is in fact not desired. In Fig. 2.15, we show the so-called M16 labeling designed to yield low error floors in BICM-ID. Clearly, this labeling is not constructed as the ordered direct product of two labelings, and thus, it cannot be decoupled in each dimension.*

For our considerations, it is convenient to define the subconstellations

$$\mathcal{S}_{k,b} \triangleq \{\boldsymbol{x}_i \in \mathcal{S} : q_{i,k} = b\}, \quad k = 1, \ldots, m, \quad b \in \mathbb{B}. \tag{2.63}$$

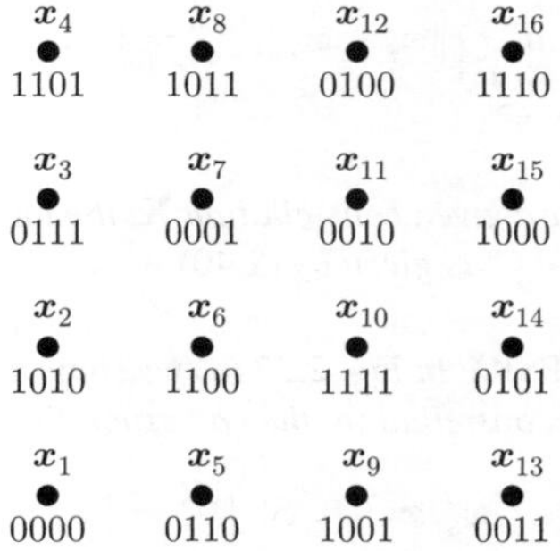

Figure 2.15 16QAM constellation with the M16 labeling

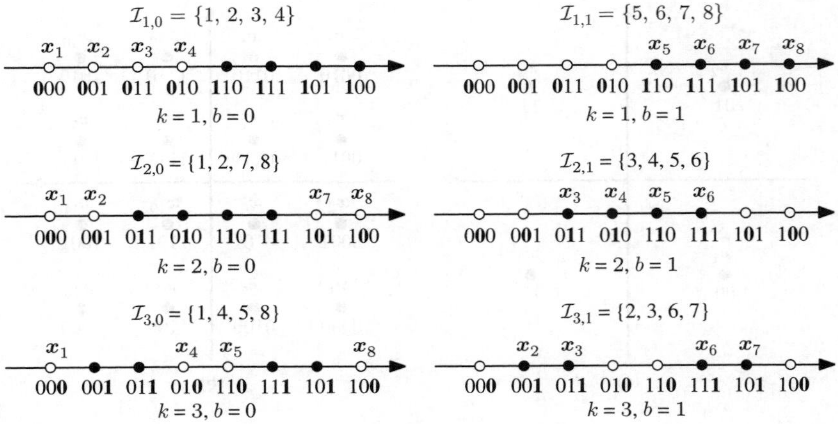

Figure 2.16 Subconstellations $\mathcal{S}_{k,b}$ for 8PAM labeled by the BRGC from Fig. 2.13 (a), where the values of $q_{i,k}$ for $k = 1, 2, 3$ are highlighted. The values of $\mathcal{I}_{k,b}$ are also shown

We also define $\mathcal{I}_{k,b} \subset \{1, \ldots, M\}$ where $|\mathcal{I}_{k,b}| = M/2$ as the set of indices to all the symbols in $\mathcal{S}_{k,b}$, i.e., $\mathcal{I}_{k,b} \triangleq \{i \in \mathcal{I} : q_{i,k} = b\}$.

Example 2.18 (Subconstellations $\mathcal{S}_{k,b}$ for 8PAM labeled by the BRGC) *In Fig. 2.16, we show $\mathcal{S}_{k,b}$ for 8PAM labeled by the BRGC. The set of indices $\mathcal{I}_{k,b}$ are*

$$\mathcal{I}_{1,0} = \{1, 2, 3, 4\}, \quad \mathcal{I}_{1,1} = \{5, 6, 7, 8\}, \tag{2.64}$$

$$\mathcal{I}_{2,0} = \{1, 2, 7, 8\}, \quad \mathcal{I}_{2,1} = \{3, 4, 5, 6\}, \tag{2.65}$$

$$\mathcal{I}_{3,0} = \{1, 4, 5, 8\}, \quad \mathcal{I}_{3,1} = \{2, 3, 6, 7\}. \tag{2.66}$$

We conclude this section by discussing the labeling by SP, initially proposed by Ungerboeck for TCM. To formally define this idea for a given constellation $\mathbf{X}$ and labeling $\mathbf{Q}$, we define the subsets $\mathcal{S}'_l([u_{m+1-l}, \ldots, u_m]) \subset \mathcal{S}$ for $l = 1, \ldots, m-1$, where

$$\mathcal{S}'_l([u_{m+1-l}, \ldots, u_m]) \triangleq \{\boldsymbol{x}_i \in \mathcal{S} : [q_{m+1-l,i}, \ldots, q_{m,i}] = [u_{m+1-l}, \ldots, u_m], i \in \mathcal{I}\}.$$

Additionally, we define the minimum intra-ED at level l as

$$d'_l \triangleq \min_{\substack{\boldsymbol{x}_i, \boldsymbol{x}_j \in \mathcal{S}'_l(\boldsymbol{u}^{\mathrm{T}}) \\ i \neq j, \boldsymbol{u}^{\mathrm{T}} \in \mathbb{B}^l}} \|\boldsymbol{x}_i - \boldsymbol{x}_j\|, \quad l = 1, \ldots, m-1. \tag{2.67}$$

Definition 2.19 (SP Labeling) *For a given constellation $\mathbf{X}$, the labeling $\mathbf{Q}$ is said to be an SP labeling if $d_{\mathcal{S}}^{\min} < d'_1 < \ldots < d'_{m-1}$, where $d_{\mathcal{S}}^{\min}$ is given by* (2.40).

Example 2.20 (SP Labeling for 8PSK) *In Fig. 2.17 (a) we show an 8PSK constellation labeled by the NBC (see Definition 2.11). For this constellation, the corresponding subsets of $\mathcal{S}$ are*

$$\mathcal{S}'_1([0]) = \{\boldsymbol{x}_1, \boldsymbol{x}_3, \boldsymbol{x}_5, \boldsymbol{x}_7\}, \quad \mathcal{S}'_1([1]) = \{\boldsymbol{x}_2, \boldsymbol{x}_4, \boldsymbol{x}_6, \boldsymbol{x}_8\}, \tag{2.68}$$

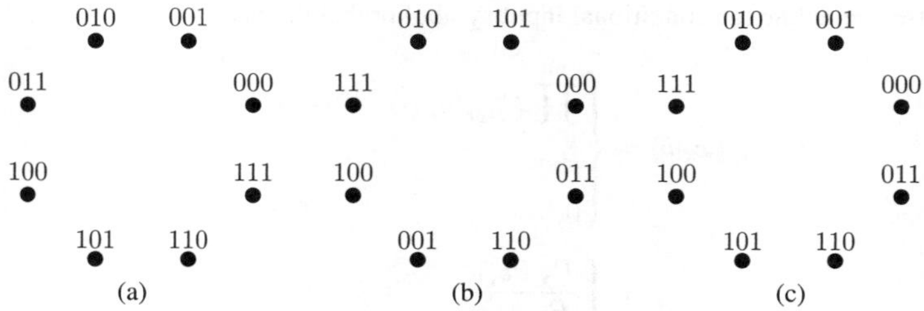

Figure 2.17 Three SP labelings: (a) NBC, (b) SSP, and (c) MSP

and

$$\mathcal{S}_2'([0,0]) = \{\boldsymbol{x}_1, \boldsymbol{x}_5\}, \quad \mathcal{S}_2'([0,1]) = \{\boldsymbol{x}_2, \boldsymbol{x}_6\}, \tag{2.69}$$

$$\mathcal{S}_2'([1,0]) = \{\boldsymbol{x}_3, \boldsymbol{x}_7\}, \quad \mathcal{S}_2'([1,1]) = \{\boldsymbol{x}_4, \boldsymbol{x}_8\}. \tag{2.70}$$

The corresponding intra-EDs are $d_1' = \sqrt{2}$ (e.g., obtained from $\|\boldsymbol{x}_3 - \boldsymbol{x}_1\|$) and $d_2' = 2$ (e.g., obtained from $\|\boldsymbol{x}_1 - \boldsymbol{x}_5\|$). As $d_{\mathcal{S}}^{\min} < d_1' < d_2'$, we conclude the NBC is an SP labeling for 8PSK*.*

For 8PSK *and in the context of BICM-ID, other labelings have been proposed. For example, the semi set-partitioning (SSP) labeling or the modified set-partitioning (MSP) labeling, shown in Fig. 2.17 (b) and (c), respectively. It can be shown that both of them are SP labelings according to Definition 2.19.*

2.5.3 Distribution and Shaping

Signal shaping refers to the use of nonequally spaced and/or nonequiprobable constellation symbols. The need for shaping may be understood using an information-theoretic argument: the *mutual information* (MI) $I(\boldsymbol{X};\boldsymbol{Y})$ can be increased by optimizing the distribution $\boldsymbol{p}_{\mathrm{s}}$ in (2.41) for a given constellation $\mathcal{S}$ (probabilistic shaping), by changing the constellation $\mathcal{S}$ for a given $\boldsymbol{p}_{\mathrm{s}}$ (geometrical shaping) or by optimizing both (mixed shaping). This argument relates to the random-coding principle, so it may be valid when capacity-approaching codes are used. In other cases (e.g., when *convolutional codes* (CCs) are used), different shaping principles might be devised because the MI is not necessarily a good indicator for the actual performance of the decoder.

We will assume that $B_1, B_2, \ldots, B_m$ are independent binary random variables. Sufficient conditions for this independence in the case of CENCs are given in Section 2.6.2. Moreover, even if in most cases the bits are i.u.d., nonuniform bit distributions can be imposed by an explicit injection of additional ones or zeros. This allows us to obtain $P_{B_k}(b) \neq 0.5$, and thus, the entire distribution is characterized by the vector

$$\boldsymbol{p}_{\mathrm{b}} \triangleq [p_{\mathrm{b},1}, \ldots, p_{\mathrm{b},m}]^{\mathrm{T}} = [P_{B_1}(0), \ldots, P_{B_m}(0)]^{\mathrm{T}}. \tag{2.71}$$

The probability of the transmitted symbols is given by

$$P_{\boldsymbol{X}}(\boldsymbol{x}_i) = \prod_{k=1}^{m} P_{B_k}(q_{i,k}), \tag{2.72}$$

or equivalently, by

$$p_{\mathrm{s},i} = \prod_{k=1}^{m} p_{\mathrm{b},k}^{\overline{q}_{i,k}} (1 - p_{\mathrm{b},k})^{q_{i,k}}. \tag{2.73}$$

For future use, we define the conditional input symbol probabilities as

$$P_{\mathbf{X}|B_k}(\boldsymbol{x}_i|b) = \begin{cases} \prod_{\substack{k'=1\\k'\neq k}}^{m} P_{B_{k'}}(q_{i,k'}), & \text{if } q_{i,k} = b \\ 0, & \text{if } q_{i,k} \neq b \end{cases} \tag{2.74}$$

$$= \begin{cases} \dfrac{P_{\mathbf{X}}(\boldsymbol{x}_i)}{P_{B_k}(b)}, & \text{if } i \in \mathcal{I}_{k,b} \\ 0, & \text{if } i \notin \mathcal{I}_{k,b} \end{cases}. \tag{2.75}$$

When the bits are *uniformly distributed* (u.d.) (i.e., $P_{B_k}(b) = 0.5$), we obtain a uniform input symbol distributions, i.e.,

$$P_{\mathbf{X}}(\boldsymbol{x}_i) = \frac{1}{M}, \quad \forall i \in \mathcal{I}, \tag{2.76}$$

and

$$P_{\mathbf{X}|B_k}(\boldsymbol{x}_i|b) = \frac{2}{M}, \quad \forall i \in \mathcal{I}_{k,b}. \tag{2.77}$$

2.6 Codes and Encoding

Throughout this book, we consider a rate $R_{\text{c}} = N_{\text{b}}/N_{\text{c}}$ binary encoder (see Fig. 2.18), where the information bits $\underline{i} = [i[1], \ldots, i[N_{\text{b}}]] \in \mathbb{B}^{N_{\text{b}}}$ are mapped onto n_{c} sequences of code bits $\underline{c}_q = [c_q[1], \ldots, c_q[N_{\text{q}}]]$, $q = 1, \ldots, n_{\text{c}}$, where

$$N_{\text{q}} = \frac{N_{\text{c}}}{n_{\text{c}}}. \tag{2.78}$$

Here, the index q in the sequences $\underline{c}_q$ represents a "class" of code bits sharing certain properties.[5] While this description is not entirely general,[6] it will allow us to highlight the encoder's properties and explain how these should be exploited during the design of BICM transceivers.

The encoder is a mapper between the information sequences $\underline{i} \in \mathbb{B}^{N_{\text{b}}}$ and the sequences $\underline{\boldsymbol{c}} \in \mathbb{B}^{n_{\text{c}} \times N_{\text{q}}}$, where $n_{\text{c}} N_{\text{q}} = N_{\text{c}}$. The ensemble of all codewords $\underline{\boldsymbol{c}}$ is called a *code*, which we denote by $\mathcal{C}$. In this book, we will consider *linear binary codes* where the relationship between the input and output bits can be defined via the generator matrix $\mathbf{G}_{\mathcal{C}} \in \mathbb{B}^{N_{\text{b}} \times N_{\text{c}}}$

$$\underline{c} = \underline{i}\, \mathbf{G}_{\mathcal{C}}, \tag{2.79}$$

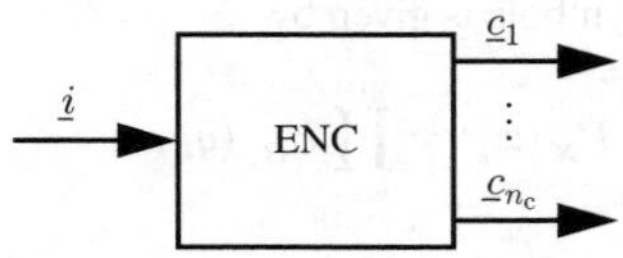

Figure 2.18 The encoder ENC in Fig. 2.7 has rate $R_{\text{c}} = N_{\text{b}}/N_{\text{c}}$ and n_{c} different "classes" of bits

[5] One such a class may be defined for the so-called systematic bits, i.e., when $\underline{c}_q = \underline{i}$.
[6] The number of bits in each class could be different.

where

$$\underline{c} = [\boldsymbol{c}[1]^{\mathrm{T}}, \boldsymbol{c}[2]^{\mathrm{T}}, \ldots, \boldsymbol{c}[N_{\mathrm{q}}]^{\mathrm{T}}] \tag{2.80}$$

gathers all the code bits, and all operations are in $\mathrm{GF}(2)$.

In order to analyze the performance of a receiver which takes coding into account, we need to analyze the decoding process. The latter depends, of course, on the structure of the code, and all the other elements in the BICM system. Nevertheless, we may start with the following simplified assertion: the probability of decoding a codeword $\hat{\underline{c}}$ that is different from the transmitted one $\underline{c}$ depends on the HD $d_{\mathrm{H}}(\hat{\underline{c}}, \underline{c})$. This motivates the following definition of the *distance spectrum* (DS) of a code.

Definition 2.21 (DS of the Code $\mathcal{C}$) *The DS of a code $\mathcal{C}$, denoted by $C_d^{\mathcal{C}}$, is defined as the average number of codewords at HD d, i.e.,*

$$C_d^{\mathcal{C}} \triangleq \frac{1}{|\mathcal{C}|}|\{(\underline{c}, \underline{c}') \in \mathcal{C}^2 : d_{\mathrm{H}}(\underline{c}, \underline{c}') = d\}|, \quad d = 1, \ldots, N_{\mathrm{c}}. \tag{2.81}$$

The next lemma particularizes Definition 2.21 to the case of linear codes.

Lemma 2.22 (DS of a Linear Code) *The DS of a linear code is given by the number of codewords with HW d*

$$C_d^{\mathcal{C}} = |\mathcal{C}(d)|, \quad d = 1, \ldots, N_{\mathrm{c}}, \tag{2.82}$$

where

$$\mathcal{C}(d) \triangleq \{\underline{c} \in \mathcal{C} : w_{\mathrm{H}}(\underline{c}) = d\}. \tag{2.83}$$

Proof: By using $d_{\mathrm{H}}(\underline{c}, \underline{c}') = w_{\mathrm{H}}(\underline{c} \oplus \underline{c}')$ in (2.81) and by noting that $\underline{c} \oplus \underline{c}' \in \mathcal{C}$ owing to (2.79). □

As the HD provides an idea about how well the codewords are "protected" against decoding errors, we may also be interested in analyzing this protection for each class of code bits. This motivates the following definition of the *generalized distance spectrum* (GDS).

Definition 2.23 (GDS of the Code $\mathcal{C}$) *The GDS of a linear code $\mathcal{C}$, denoted by $C_{\boldsymbol{w}}^{\mathcal{C}}$, is defined as the number of codewords with GHW equal to $\boldsymbol{w} \in \mathbb{N}^{n_{\mathrm{c}}}$, i.e.,*

$$C_{\boldsymbol{w}}^{\mathcal{C}} \triangleq |\{\underline{c} \in \mathcal{C} : W_{\mathrm{H}}(\underline{c}) = \boldsymbol{w}\}|, \quad \boldsymbol{w} \in \mathbb{N}^{n_{\mathrm{c}}}, \tag{2.84}$$

where $W_{\mathrm{H}}(\underline{c})$ is given by (2.3).

In this book, we consider two types of linear binary codes: CCs and TCs, where the codewords are obtained using a CENC and a *turbo encoder* (TENC), respectively.[7] CCs are an example of codes which can be decoded optimally with low decoding complexity and for which we can find the DS using well-known analytical approaches. TCs, on the other hand, are an example of capacity-approaching codes, where the corresponding analytical tools coincide in part with those used for the analysis of CCs.

The encoder calculates the codewords in a deterministic manner via (2.79). However, as the input bit sequence $\underline{i}$ is modeled as a sequence of i.u.d. binary random variables, the code bits $\underline{c}$ are also modeled

[7] In the case of CCs, the encoder is based on a convolution; in the case of *turbo code* (TC), the decoder employs an iterative (or *turbo*) principle. So, while there is nothing "convolutional" or "turbo" in the codes themselves, we keep both names for compatibility with the literature.

as random variables whose properties depend on the structure of $\mathbf{G}_{\mathcal{C}}$. To get insight into the probabilistic model of the code bits, we will need some simple relationships, which we define in the following.

Lemma 2.24 *The binary random variable* $A_k = \boldsymbol{C}\boldsymbol{g} = g_1C_1 \oplus \ldots \oplus g_kC_k$ *is u.d. for all* $\boldsymbol{g} = [g_1, \ldots, g_k]^{\mathrm{T}} \in \mathbb{B}^k$ *with* $w_{\mathrm{H}}(\boldsymbol{g}) > 0$, *if and only if the elements of the binary random vector* $\boldsymbol{C} = [C_1, \ldots, C_k] \in \mathbb{B}^k$ *are i.u.d.*

Proof: To prove the "if" part of the lemma, we proceed by induction. The proof for $k = 2$ follows from the fact that $A_2 = g_1C_1 \oplus g_2C_2$, and thus, $\Pr\{A_2 = 0\} = \Pr\{A_2 = 1\} = 1/2$ (it holds when $g_1 \neq 0$ or $g_2 \neq 0$). For $k + 1 \geq 2$ we assume that A_k is u.d. and we demonstrate that A_{k+1} is u.d. for any $\boldsymbol{g} \in \mathbb{B}^{k+1}$. As $A_{k+1} = A_k \oplus g_{k+1}C_{k+1}$, for $g_{k+1} = 0$ we obtain $A_{k+1} = A_k$, and thus, A_{k+1} is u.d. For $g_{k+1} = 1$, $A_{k+1} = A_k \oplus C_{k+1}$, where A_k and C_{k+1} are i.u.d. Because of the proof for $k = 2$, we conclude that their sum is u.d., which completes the "if" part of the proof.

The "only if" part is proven by contraposition: it is enough to demonstrate existence of $\boldsymbol{g}$ satisfying $w_{\mathrm{H}}(\boldsymbol{g}) > 0$ such that when $\boldsymbol{C}$ are not i.u.d. (i.e., there are two cases: $\boldsymbol{C}$ are not u.d. or are not independent) implies that A_k is not u.d. The first case is obvious: if there is one variable C_l which is not u.d., then we can use all-zero $\boldsymbol{g}$ except $g_l = 1$; this yields $A_k = C_l$ which is not u.d. In the second case, we assume that C_n and C_l are u.d. but not independent and defined by a joint PMF $P_{C_n,C_l}(c_n, c_l)$ such that $P_{C_n,C_l}(0, 1) + P_{C_n,C_l}(1, 0) \neq 1/2$. We take $\boldsymbol{g}$ with two nonzero elements $g_n = g_l = 1$, (i.e., $g_t = 0, t \notin \{n, l\}$) and obtain $\Pr\{A_k = 1\} = P_{C_n,C_l}(0, 1) + P_{C_n,C_l}(1, 0)$, which in turn proves that A_k is not u.d. □

The next theorem shows conditions on the matrix $\mathbf{G}$ so that the code bits are i.u.d.. We assume therein that the columns of $\mathbf{G} \in \mathbb{B}^{k\times l}$ are linearly independent, which means that for any $\boldsymbol{a}$ with $w_{\mathrm{H}}(\boldsymbol{a}) > 0$ we have $w_{\mathrm{H}}(\mathbf{G}\boldsymbol{a}) > 0$.

Theorem 2.25 *If the random variables gathered in the vector* $\boldsymbol{U} = [U_1, \ldots, U_k]$ *are i.u.d. and the columns of* $\mathbf{G} \in \mathbb{B}^{k\times l}$ *are linearly independent, the random variables in* $\boldsymbol{C} = \boldsymbol{U}\mathbf{G} = [C_1, \ldots, C_l]$ *are i.u.d. too.*

Proof: For an arbitrary nonzero $\boldsymbol{a} \in \mathbb{B}^l$, we define the binary random variable A as

$$A \triangleq \boldsymbol{C}\boldsymbol{a} \tag{2.85}$$

$$= \boldsymbol{U}\mathbf{G}\boldsymbol{a} \tag{2.86}$$

$$= \boldsymbol{U}\boldsymbol{v}. \tag{2.87}$$

If the columns of $\mathbf{G}$ are linearly independent, and thus, $w_{\mathrm{H}}(\mathbf{G}\boldsymbol{a}) = w_{\mathrm{H}}(\boldsymbol{v}) > 0$, using the "if" part of Lemma 2.24 and the i.u.d. condition on $\boldsymbol{U}$, we conclude that A in (2.87) is u.d. As A is also a linear combination of $\boldsymbol{C}$ (see (2.85)), using the "only if" part of Lemma 2.24, we conclude that $C_1, \ldots, C_l$ are i.u.d. □

If a matrix has fewer rows than columns, its columns cannot be linearly independent. Consequently, for any $R_{\mathrm{c}} = N_{\mathrm{b}}/N_{\mathrm{c}} < 1$, the generator matrix $\mathbf{G}_{\mathcal{C}}$ never satisfies this condition, and thus, the output bits $\underline{c}$ should be modeled as mutually dependent random variables. On the other hand, a subset of code bits can be independent if a subset of the columns of $\mathbf{G}_{\mathcal{C}}$ are linearly independent. In other words,

$$\tilde{\underline{c}} = \underline{i}\tilde{\mathbf{G}}_{\mathcal{C}}, \tag{2.88}$$

can be mutually independent provided that $\tilde{\mathbf{G}}_{\mathcal{C}}$ has linearly independent columns, where $\tilde{\mathbf{G}}_{\mathcal{C}}$ contains a subset of the columns of $\mathbf{G}_{\mathcal{C}}$. In Section 2.6.2, we will explain how to select the code bits to guarantee their independence for the particular case of CENCs.

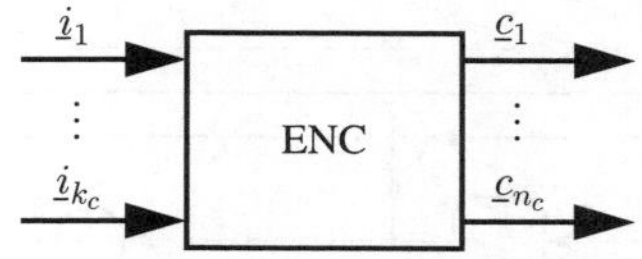

Figure 2.19 The encoder ENC in Fig. 2.7 can be a binary CENC with rate $R_c = k_c/n_c$

2.6.1 Convolutional Codes

We use the name *convolutional codes* to denote the codes obtained using CENCs. We focus on the so-called feedforward CENCs, where the sequence of input bits $\underline{i}$ is reorganized into k_c sequences $\underline{i}_q$, $q = 1, \ldots, k_c$ so that $\underline{i} = [\underline{i}_1, \ldots, \underline{i}_{k_c}]$. At each discrete-time instant n, the information bits $i_1[n], \ldots, i_{k_c}[n]$ are fed to the CENC shown in Fig. 2.19. The output of the CENC is fully determined by k_c shift registers and the way the input sequences are connected (through the registers) to its outputs (see Fig. 2.20 for an example). We denote the length of the pth shift register by ν_p, with $p = 1, \ldots, k_c$, the *overall constraint length* by $\nu = \sum_{p=1}^{k_c} \nu_p$, and the *number of states* by $S = 2^{\nu}$. The rate of such encoder is given by $R_c = k_c/n_c$.

A CENC is fully determined by the connection between the input and output bits, which is defined by the CENC *matrix*

$$\mathbf{G} = [\overline{\boldsymbol{g}}_1, \ldots, \overline{\boldsymbol{g}}_{n_c}] \triangleq \begin{bmatrix} \boldsymbol{g}_{1,1} & \boldsymbol{g}_{1,2} & \cdots & \boldsymbol{g}_{1,n_c} \\ \boldsymbol{g}_{2,1} & \boldsymbol{g}_{2,2} & \cdots & \boldsymbol{g}_{2,n_c} \\ \vdots & \vdots & \ddots & \vdots \\ \boldsymbol{g}_{k_c,1} & \boldsymbol{g}_{k_c,2} & \cdots & \boldsymbol{g}_{k_c,n_c} \end{bmatrix}, \quad (2.89)$$

where the binary vector $\boldsymbol{g}_{p,q} \triangleq [g_{p,q}[1], \ldots, g_{p,q}[\nu_p + 1]]^{\mathrm{T}} \in \mathbb{B}^{\nu_p+1}$ contains the convolution coefficients relating the input bits $i_p[n]$ and the output bits $c_q[n]$, i.e.,

$$\begin{aligned} c_q[n] &= \sum_{p=1}^{k_c} \sum_{l=0}^{\nu_p} i_p[n-l] g_{p,q}[l+1] \\ &= \boldsymbol{u}[n] \overline{\boldsymbol{g}}_q, \end{aligned} \quad (2.90)$$

where $\boldsymbol{u}[n] = [\boldsymbol{u}_1[n], \boldsymbol{u}_2[n], \ldots, \boldsymbol{u}_{k_c}[n]]$ and $\boldsymbol{u}_p[n] = [i_p[n], i_p[n-1], \ldots, i_p[n-\nu_p]]$, with $p = 1, 2, \ldots, k_c$.

It is customary to treat the vectors $\boldsymbol{g}_{p,q}$ in (2.89) as the binary representation of a number (with $g_{p,q}[1]$ being the MSB) given in decimal or (most often) in octal notation, as we do in the following example.

Example 2.26 (Rate $2/3$ **CENC)** *For the CENC in Fig. 2.20, we have* $R_c = 2/3$ *(*$k_c = 2$ *and* $n_c = 3$*),* $\nu_1 = 1$ *and* $\nu_2 = 3$*, and overall constraint length* $\nu = 4$*. Its encoder matrix is given by*

$$\mathbf{G} = \begin{bmatrix} 1 & 0 & 1 \\ 0 & 1 & 1 \\ \hline 0 & 1 & 1 \\ 0 & 1 & 0 \\ 0 & 1 & 1 \\ 0 & 1 & 1 \end{bmatrix}, \quad (2.91)$$

or in octal notation by

$$\mathbf{G} = \begin{bmatrix} 2 & 1 & 3 \\ 0 & 17 & 13 \end{bmatrix}. \quad (2.92)$$

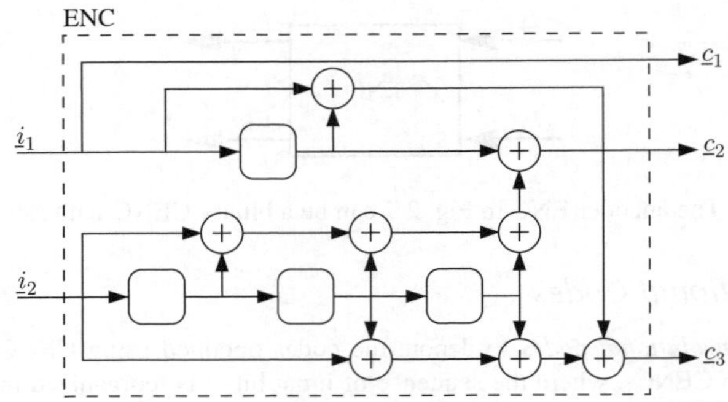

Figure 2.20 Rate $R_c = 2/3$ CENC with encoder matrix given by (2.92)

Usually, the design of CENCs relies on a criteria related to the performance of the codes for a particular transmission model. Quite often, as we will also do in Example 3.3, an implicit assumption of binary modulation appears. In such a case, the parameter of the CENC determining the performance is its *free Hamming distance* (FHD), which is defined as

$$d_{\text{free}} \triangleq \min_{\substack{d \in \mathbb{N}_+ \\ C_d^{\mathcal{C}} > 0}} d = \min_{\substack{\underline{c} \in \mathcal{C} \\ \underline{c} \neq \underline{0}}} w_{\text{H}}(\underline{c}). \tag{2.93}$$

In general, having defined an optimization criterion, "good" CENCs can be found by an exhaustive numerical search. For example, given the rate R_c and ν, *maximum free Hamming distance* (MFHD) CENCs can be found and tabulated by searching over the CENC *universe* $\mathcal{G}_{k_c,n_c,\nu}$, which is the set of all binary matrices $\mathbf{G} \in \mathbb{B}^{(\nu+k_c)\times n_c}$. Additional constraints are often imposed, e.g., catastrophic encoders[8] are avoided, or we may require the outputs $\boldsymbol{c}[n]$ to be u.d. (see Section 2.6.2). The search criterion can be further refined to take into account not only the MFHD but also the code's *weight distribution* (WD) $F_d^{\mathcal{C}}$ or its *input-dependent weight distribution* (IWD) $G_d^{\mathcal{C}}$.[9] This is the idea behind the so-called *optimal distance spectrum* (ODS) CENCs which consider distances $d_{\text{free}}, d_{\text{free}} + 1, d_{\text{free}} + 2, \ldots$ as well as the frequency of occurrence of events associated with each of these distances.

In Table 2.1, we show the ODS CENCs for $R_c = 1/2$ and $R_c = 1/3$ for different values of ν. In this table, we also show the FHD d_{free} of each encoder and the total input HW of sequences that generate codewords with HW d_{free}.

2.6.2 *Modeling the Code Bits in Convolutional Encoders*

Let $\boldsymbol{C}[n] = [C_1[1], \ldots, C_{n_c}[n]]^{\text{T}}$ be the binary random vector modeling the CENC's output bits $\boldsymbol{c}[n]$ at time n. From Theorem 2.25, we conclude that for CENCs defined by $\mathbf{G}$ with linearly independent columns and for i.u.d. information bits, the binary vectors $\boldsymbol{C}[n]$ are i.u.d., i.e., $P_{\boldsymbol{C}[n]}(\boldsymbol{c}_i) = \prod_{q=1}^{n_c} P_{C_q[n]}(c_{i,q}) = 2^{-n_c}$. This property will be used later to analyze the ML receiver for trellis encoders, where the code bits are directly connected to the mapper, and thus, the fact of having i.u.d. bits $\boldsymbol{C}[n] = \boldsymbol{B}[n]$ directly translates into i.u.d. symbols $\boldsymbol{X}[n] = \Phi(\boldsymbol{C}[n])$ as well.

[8] The encoder is catastrophic if there exists an infinite-weight input sequence $\underline{i}$ which generates a finite-weight codeword $\underline{\boldsymbol{c}}$. This property is undesirable because a small number of errors at the output can provoke an infinite number of errors in the detected input sequence.

[9] Formal definitions for the WD and IWD are given in Section 6.2.

Table 2.1 $R_c = 1/2$ and $R_c = 1/3$ ODS CENCs for different values of ν. The table shows the encoder $\mathbf{G}$, the FHD d_{free}, the first element of the WD $F_d^{\mathcal{C}}$, and the first element of the IWD $G_d^{\mathcal{C}}$

	$R_c = 1/2$				$R_c = 1/3$			
ν	$\mathbf{G}$	d_{free}	$F_{d_{\text{free}}}^{\mathcal{C}}$	$G_{d_{\text{free}}}^{\mathcal{C}}$	$\mathbf{G}$	d_{free}	$F_{d_{\text{free}}}^{\mathcal{C}}$	$G_{d_{\text{free}}}^{\mathcal{C}}$
2	[5, 7]	5	1	1	[5, 7, 7]	8	2	3
3	[15, 17]	6	1	2	[13, 15, 17]	10	3	6
4	[23, 35]	7	2	4	[23, 33, 37]	12	5	12
5	[53, 75]	8	1	2	[47, 53, 75]	13	1	1
6	[133, 171]	10	11	36	[133, 165, 171]	15	3	7
7	[247, 371]	10	1	2	[225, 331, 367]	16	1	1
8	[561, 753]	12	11	33	[575, 623, 727]	18	1	2
9	[1151, 1753]	12	1	2	[1233, 1375, 1671]	20	3	6

Let us have a look at the case of BICM. As the interleaver is present between the encoder and the mapper, the bits $b_1[n], \ldots, b_m[n]$ are the same as the bits $c_{q_1}[n-t_1], c_{q_2}[n-t_2], \ldots, c_{q_m}[n-t_m]$ taken from the output of the encoder at different time instants $n-t_1, \ldots, n-t_m$ and from different encoder's outputs $q_1, \ldots, q_m \in \{1, \ldots, n_c\}$. In what follows, we try to answer the following question: can we guarantee the independence of the bits $b_1[n], \ldots, b_m[n]$ in BICM? In particular, we will analyze the conventionally assumed independence of the bits in (2.72) for the particular case of CCs.

Theorem 2.27 *For any $q_1, \ldots, q_T \in \{1, \ldots, n_c\}$ and $t_1 \neq t_2 \neq \ldots \neq t_T$, T code bits $C_{q_1}[n-t_1]$, $C_{q_2}[n-t_2], \ldots, C_{q_T}[n-t_T]$ are i.u.d. for any CENC of rate $R_c = 1/n_c$ if its generator matrix $\mathbf{G}$ is such that for all q, $g_{1,q}[1] = 1$ or $g_{1,q}[\nu_1+1] = 1$.*

Proof: Without loss of generality, we assume $t_1 < t_2 < \ldots < t_T$. For any $q_1, \ldots, q_T \in \{1, \ldots, n_c\}$, we rewrite (2.90) as

$$\tilde{\boldsymbol{c}}[n] = \begin{bmatrix} c_{q_1}[n-t_1] \\ c_{q_2}[n-t_2] \\ \vdots \\ c_{q_T}[n-t_T] \end{bmatrix} = \tilde{\boldsymbol{u}}_1[n] \begin{bmatrix} g_{1,q_1}[1] & 0 & 0 & 0 \\ \vdots & \vdots & \vdots & \vdots \\ \vdots & 0 & \vdots & \vdots \\ \vdots & g_{1,q_2}[1] & \vdots & 0 \\ g_{1,q_1}[\nu_1+1] & \vdots & \vdots & g_{1,q_T}[1] \\ 0 & g_{1,q_2}[\nu_1+1] & \vdots & \vdots \\ \vdots & 0 & \vdots & \vdots \\ \vdots & \vdots & \vdots & \vdots \\ 0 & 0 & 0 & g_{1,q_T}[\nu_1+1] \end{bmatrix} \tag{2.94}$$

$$= \tilde{\boldsymbol{u}}_1[n]\tilde{\mathbf{G}}, \tag{2.95}$$

where $\tilde{\boldsymbol{u}}_1[n] = [i_1[n-t_1], i_1[n-t_1-1], \ldots, i_1[n-\nu_1-t_T]]$ accommodates the delayed input bits necessary to produce output with the maximum delay t_T. It is immediately seen that if $g_{1,q_1}[1] = g_{1,q_2}[1] = \ldots = g_{1,q_T}[1] = 1$, the columns of $\tilde{\mathbf{G}}$ are independent, and thus, from Theorem 2.25 and the i.u.d. assumption on the information bits, the bits $C_{q_1}[n-t_1], C_{q_2}[n-t_2], \ldots, C_{q_T}[n-t_T]$ are i.u.d. too. A similar argument is used to prove the case $g_{1,q_1}[\nu_1+1] = g_{1,q_2}[\nu_1+1] = \ldots = g_{1,q_T}[\nu_1+1] = 1$. □

Theorem 2.27 guarantees that, regardless of the memory ν, code bits taken at different time instants are i.u.d., even if the separation is only one time instant. The implication of this for BICM is that the input bits to the mapper will be i.u.d. if they correspond to code bits obtained from the encoder at different time instants. In other words, to guarantee the independence of the bits $b_1[n], \ldots, b_m[n]$ at the input of the mapper, it is enough to use m code bits taken at different time instants $n - t_1, n - t_2, \ldots, n - t_m$ for any $t_1 < t_2 < \ldots < t_m$. This condition is typically guaranteed by an appropriately designed interleaver or by the assumption of using a quasirandom interleaving, in which case, this condition is violated a relatively small number of times. For more details about this, see Section 2.7.

Example 2.28 (Code Bits for ODS CENCs) *Applying Theorem 2.25 to the encoders in Table 2.1, we easily verify that for $R_c = 1/2$, all matrices $\mathbf{G}$ have independent columns, and thus, they generate i.u.d. vectors $\boldsymbol{C}[n]$. The same happens for $R_c = 1/3$ except for the encoder with the matrix $\mathbf{G} = [5, 7, 7]$ which generates bits $c_2[n] = c_3[n]$ being dependent. We also note that the first and the last row of $\mathbf{G}$ contain no zeros, thus, according to Theorem 2.27, code bits at different time instants are i.u.d. If the interleaver guarantees that the bits at the input of the mapper $\boldsymbol{B}[n]$ are code bits taken at different time instants, we conclude that* (2.72) *holds for any ODS CENC in Table 2.1. Moreover, because of the i.u.d. assumption on the information bits, we also conclude that for any ODS CENC in Table 2.1, the symbol input distribution is given by* (2.76), *i.e., it is always uniform.*

For simplicity, Theorem 2.27 was given for $k_c = 1$. The following corollary generalizes Theorem 2.27 to any $k_c > 1$. Its proof is not included; however, it follows similar lines to the one used for Theorem 2.27.

Corollary 2.29 *For any $q_1, \ldots, q_T \in \{1, \ldots, n_c\}$ and $t_1 \neq t_2 \neq \ldots \neq t_T$, T code bits $C_{q_1}[n - t_1], C_{q_2}[n - t_2], \ldots, C_{q_T}[n - t_T]$ are i.u.d. for any CENC of rate $R_c = k_c/n_c$ if its generator matrix $\mathbf{G}$ is such that, $\forall q \in \{1, 2, \ldots, k_c\}$*

$$w_H([g_{1,q}[1], g_{2,q}[1], \ldots, g_{k_c,q}[1]]) > 0, \tag{2.96}$$

or

$$w_H([g_{1,q}[\nu_1 + 1], g_{2,q}[\nu_1 + 1], \ldots, g_{k_c,q}[\nu_1 + 1]]) > 0. \tag{2.97}$$

To clarify the conditions (2.96) and (2.97) in Corollary 2.29, consider the matrix of the CENC in Example 2.26 ($k_c = 2$) given by (2.91). In this case, we have $w_H([1, 0]) > 0$, $w_H([0, 1]) > 0$, and $w_H([1, 1]) > 0$, which corresponds to (2.96) for $q = 1, 2, 3$. The three relevant pairs of bits in this case are rows one and three in (2.91). Similarly, the condition in (2.97) gives $w_H([0, 0]) = 0$, $w_H([1, 1]) > 0$, and $w_H([1, 1]) > 0$, which corresponds to rows two and six in (2.91).

Example 2.30 (Dependent Code Bits) *To see when code bits are dependent, consider the ODS CENC with $\nu = 2$ and $R_c = 1/2$, i.e., $\mathbf{G} = [5, 7]$ in Table 2.1. To violate the conditions in Theorem 2.27, assume that the code bits are not taken at different time instants, i.e., $\tilde{\boldsymbol{c}}[n] = [c_1[n], c_2[n], c_2[n-1], c_1[n-2], c_2[n-2]]^T$, which corresponds to $T = 5$, $q_1 = q_4 = 1$, $q_2 = q_3 = q_5 = 2$, $t_1 = t_2 = 0$, $t_3 = 1$, and $t_4 = t_5 = 2$. In this case, the matrix we used in the proof of Theorem 2.27 has the following form*

$$\tilde{\mathbf{G}} = \begin{bmatrix} 1 & 1 & 0 & 0 & 0 \\ 1 & 0 & 1 & 0 & 0 \\ 1 & 1 & 0 & 1 & 1 \\ 0 & 0 & 1 & 1 & 0 \\ 0 & 0 & 0 & 1 & 1 \end{bmatrix}. \tag{2.98}$$

Although (2.98) *has equal number of columns and rows, its columns are linearly dependent. This can be seen by noting that the first column is equal to the sum of the last four columns. Thus,*

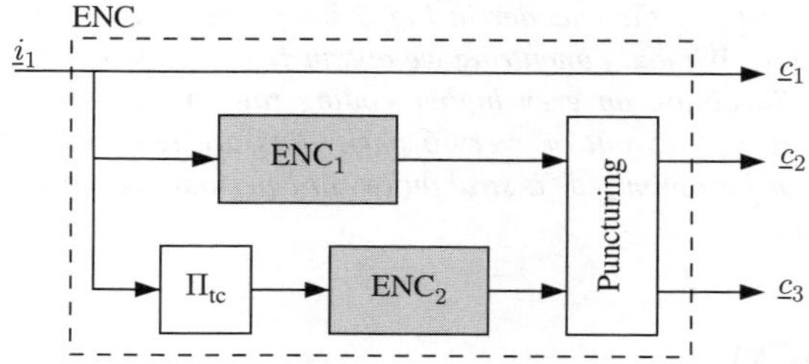

Figure 2.21 PCCENCs of rate $R_c = 1/3$. To increase the rate R_c, some of the parity bits are removed (punctured)

even for i.u.d. information bits, we cannot guarantee that the elements in $\tilde{\boldsymbol{C}}[n]$ are i.u.d. Moreover, it can be easily seen that the codewords $\tilde{\boldsymbol{C}}[n]$ are not observed with the same probability. In fact, some realizations of $\tilde{\boldsymbol{C}}[n]$ are observed with zero probability, e.g., those belonging to $\{[1,1,1,1,1]^{\mathrm{T}}, [0,0,1,1,1]^{\mathrm{T}}, [1,0,1,1,0]^{\mathrm{T}}\}$.

2.6.3 Turbo Codes

The TENC originally proposed in 1993 is based on two *parallel concatenated convolutional encoders* (PCCENCs) separated by an interleaver as we show in Fig. 2.21. The encoder produces a sequence of systematic bits $\underline{c}_1 = \underline{i}_1$, and two sequences of code (parity) bits $\underline{c}_2$ and $\underline{c}_3$. The latter is obtained by convolutionally encoding the interleaved sequence $\underline{i}_1$ (interleaved via Π_{tc}). At the receiver's side, the decoders of each CENC exchange information in an iterative manner. The iterations stop after a predefined number of iterations or when some stopping criterion is met. The encoder's interleaver is denoted by Π_{tc} to emphasize that it is different from the interleaver Π used in BICM transmitters between the mapper and the encoder.

Example 2.31 (PCCENCs with $\nu = 3$) *One of the simplest PCCENCs can be constructed by using two identical rate $1/2$ recursive systematic CENCs, each of them with overall constraint length $\nu = 3$, as shown in Fig. 2.22. The feedforward and feedback generator polynomials are 15 and 13, respectively.*

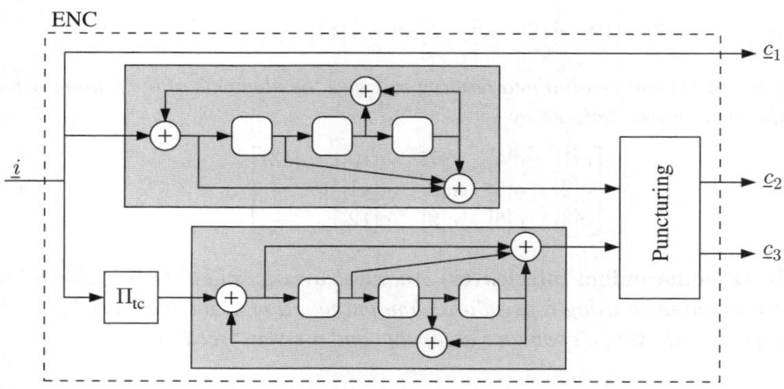

Figure 2.22 PCCENCs using two identical recursive systematic CENCs of rate $1/2$. The overall code rate is $R_c = 1/3$. To increase the rate, puncturing of nonsystematic bits is done most often but, in general, systematic bits can also be punctured

To increase the coding rate of the encoder in Fig. 2.22, puncturing can be applied, which is most often done to the parity bits. Without puncturing we obtain $R_c = \frac{1}{3}$, *but if we remove half of the parity bits we obtain* $R_c = \frac{1}{2}$. *To obtain an even higher coding rate, more parity bits could be removed. For example, to obtain* $R_c = \frac{3}{4}$, *5 out of every 6 parity bits are removed. While the corresponding "puncturing pattern" can be optimized, a straightforward periodic removal of the parity bits often provides satisfactory results.*

2.7 The Interleaver

The interleaver Π bijectively maps the binary codewords $\underline{\boldsymbol{c}} \in \mathbb{B}^{n_c \times N_q}$ into the codewords $\underline{\boldsymbol{b}} = \Pi(\underline{\boldsymbol{c}}; \boldsymbol{\pi}) \in \mathbb{B}^{m \times N_s}$. The interleaver is determined by the interleaving vector $\boldsymbol{\pi} = [\pi_1, \dots, \pi_{N_c}]$ which defines the mapping between $\underline{\boldsymbol{c}}$ and $\underline{\boldsymbol{b}}$ as follows:

$$\underline{b} = [c[\pi_1], c[\pi_2], \dots, c[\pi_{N_c}]], \tag{2.99}$$

where $\underline{c} = [\boldsymbol{c}[1]^T, \boldsymbol{c}[2]^T, \dots, \boldsymbol{c}[N_q]^T]$ (see (2.80)) and

$$\underline{b} = [\boldsymbol{b}[1]^T, \boldsymbol{b}[2]^T, \dots, \boldsymbol{b}[N_s]^T]. \tag{2.100}$$

The sequences $\underline{c}$ and $\underline{b}$ are obtained, respectively, by reading columnwise the elements of the sequences of binary vectors $\underline{\boldsymbol{c}}$ and $\underline{\boldsymbol{b}}$. Very often, we will skip the notation which includes the interleaving vector $\boldsymbol{\pi}$, i.e., we use $\underline{\boldsymbol{b}} = \Pi(\underline{\boldsymbol{c}})$.

We now define two simple and well-known interleavers.

Definition 2.32 (Rectangular Interleaver) *An interleaver is said to be rectangular with period T if it is defined via the following interleaving vector*

$$\boldsymbol{\pi} = [1, T+1, 2T+1, \dots, T'T+1, 2, T+2, \dots, T'T+2, \dots], \tag{2.101}$$

where $T' = \lceil \frac{N_c}{T} \rceil - 1$.

Example 2.33 (Rectangular Interleaver with Period 3) *For* $N_c = 13$ *and* $T = 3$, *we obtain the following rectangular interleaver*

$$\boldsymbol{\pi} = [1, 4, 7, 10, 13, 2, 5, 8, 11, 3, 6, 9, 12], \tag{2.102}$$

which gives the interleaved sequence

$$\underline{b} = [c[1], c[4], c[7], c[10], c[13], c[2], c[5], c[8], c[11], c[3], c[6], c[9], c[12]]. \tag{2.103}$$

We note that $\underline{\boldsymbol{b}}$ in (2.103) *can be obtained reading rowwise the elements of the following pseudo-matrix, and ignoring the void entires denoted by –*

$$\begin{bmatrix} c[1] & c[4] & c[7] & c[10] & c[13] \\ c[2] & c[5] & c[8] & c[11] & - \\ c[3] & c[6] & c[9] & c[12] & - \end{bmatrix}. \tag{2.104}$$

Definition 2.34 (Pseudorandom Interleaver) *An interleaver is said to be pseudorandom if the interleaving vector $\boldsymbol{\pi}$ is obtained using a pseudorandom permutation of the numbers* $1, \dots, N_c$, *where the permutation is generated using a computer algorithm and a given "seed"* ι.

As the interleaving transforms the code $\mathcal{C}$ into the code $\mathcal{B}$, considerations in the domain of the codewords $\underline{\boldsymbol{b}} \in \mathcal{B}$ may help clarify the relationships between modulation and coding. In what follows, we "reuse" the DS-related definitions from Section 2.6.

Definition 2.35 (DS of the Code $\mathcal{B}$) *The DS of a linear code $\mathcal{B}$, denoted by $C_d^{\mathcal{B}}$, is defined as the number of codewords with HW d,*

$$C_d^{\mathcal{B}} \triangleq |\mathcal{B}(d)|, \quad d = 1, \ldots, N_c, \tag{2.105}$$

where

$$\mathcal{B}(d) \triangleq \{\underline{\boldsymbol{b}} \in \mathcal{B} : w_H(\underline{\boldsymbol{b}}) = d\}. \tag{2.106}$$

Further, we adapt Definition 2.23 as follows.

Definition 2.36 (GDS of the Code $\mathcal{B}$) *The GDS of the code $\mathcal{B}$, denoted by $C_{\boldsymbol{w}}^{\mathcal{B}}$, is defined as the number of codewords with GHW $\boldsymbol{w} \in \mathbb{N}^m$, i.e.,*

$$C_{\boldsymbol{w}}^{\mathcal{B}} \triangleq |\mathcal{B}(\boldsymbol{w})|, \quad \boldsymbol{w} \in \mathbb{N}^m, \tag{2.107}$$

where

$$\mathcal{B}(\boldsymbol{w}) = \{\underline{\boldsymbol{b}} \in \mathcal{B} : W_H(\underline{\boldsymbol{b}}) = \boldsymbol{w}\}. \tag{2.108}$$

We note that the interleaving does not change the HW of the codewords, and thus, we have $C_d^{\mathcal{B}} = C_d^{\mathcal{C}}$. On the other hand, in general, $C_{\boldsymbol{w}}^{\mathcal{B}} \neq C_{\boldsymbol{w}}^{\mathcal{C}}$. In fact, the GHW $\boldsymbol{w}$ used in $C_{\boldsymbol{w}}^{\mathcal{B}}$ and in $C_{\boldsymbol{w}}^{\mathcal{C}}$ has, in general, different dimensions: $\boldsymbol{w} = [w_1, \ldots, w_{n_c}]$ for $C_{\boldsymbol{w}}^{\mathcal{C}}$ and $\boldsymbol{w} = [w_1, \ldots, w_m]$ for $C_{\boldsymbol{w}}^{\mathcal{B}}$.

At this point, it is also useful to define the *generalized input-output distance spectrum* (GIODS) of the code $\mathcal{B}$, which relates the input sequences $\underline{\boldsymbol{i}}$ and the corresponding codewords $\underline{\boldsymbol{b}}$.

Definition 2.37 (GIODS of the Code $\mathcal{B}$) *The GIODS of the code $\mathcal{B}$, denoted by $C_{v,\boldsymbol{w}}^{\mathcal{B}}$, is defined as the number of codewords with GHW $\boldsymbol{w} \in \mathbb{N}^m$, whose input sequence has HW v, i.e.,*

$$C_{v,\boldsymbol{w}}^{\mathcal{B}} \triangleq |\mathcal{B}(v, \boldsymbol{w})|, \quad v \in \mathbb{N}_+, \boldsymbol{w} \in \mathbb{N}^m, \tag{2.109}$$

where

$$\mathcal{B}(v, \boldsymbol{w}) = \{\underline{\boldsymbol{b}} \in \mathcal{B} : w_H(\underline{\boldsymbol{i}}_{\underline{\boldsymbol{b}}}) = v, W_H(\underline{\boldsymbol{b}}) = \boldsymbol{w}\} \tag{2.110}$$

and $\underline{\boldsymbol{i}}_{\underline{\boldsymbol{b}}} \in \mathbb{B}^{N_b}$ is the input sequence corresponding to the codeword $\underline{\boldsymbol{b}}$.

It is then convenient to introduce the *input-dependent distance spectrum* (IDS) and *generalized input-dependent distance spectrum* (GIDS) of the encoder.

Definition 2.38 (IDS and GIDS of the Encoder) *The IDS of the encoder is defined as*

$$B_d^{\mathcal{B}} \triangleq \sum_{\underline{\boldsymbol{b}} \in \mathcal{B}(d)} \frac{w_H(\underline{\boldsymbol{i}}_{\underline{\boldsymbol{b}}})}{N_b}, \tag{2.111}$$

while the respective GIDS is defined as

$$B_{\boldsymbol{w}}^{\mathcal{B}} \triangleq \sum_{\underline{\boldsymbol{b}} \in \mathcal{B}(\boldsymbol{w})} \frac{w_H(\underline{\boldsymbol{i}}_{\underline{\boldsymbol{b}}})}{N_b}. \tag{2.112}$$

As the IDS and GIDS defined above relate the codewords $\underline{\boldsymbol{b}}$ and the input sequences $\underline{\boldsymbol{i}}_{\underline{\boldsymbol{b}}}$, they depend on the encoding operation. They are thus different from the DS or GDS in Definitions 2.21 and 2.23 which are defined solely for the codewords, i.e., in abstraction of how the encoding is done.

The following relationships are obtained via marginalization of the GIODS

$$C^{\mathcal{B}}_{\boldsymbol{w}} = \sum_{v=1}^{N_{\mathrm{b}}} C^{\mathcal{B}}_{v,\boldsymbol{w}}, \tag{2.113}$$

$$B^{\mathcal{B}}_{\boldsymbol{w}} = \sum_{v=1}^{N_{\mathrm{b}}} C^{\mathcal{B}}_{v,\boldsymbol{w}} \frac{v}{N_{\mathrm{b}}}. \tag{2.114}$$

The codewords $\underline{\boldsymbol{b}}$ and $\underline{\boldsymbol{c}}$ are related to each other via the bijective operation of interleaving, which owes its presence to considerations similar to those we evoked in Section 2.6 to motivate the introduction of the DS. We repeat here the assertion we made that the "protection" against decoding errors can be related to the HW of the codewords (or the HD between pairs of codewords). Then, in order to simplify the decoder's task of distinguishing between the binary codewords, we postulate, not only to have a large distance between all pairs of codewords, but also to transmit every bit of the binary difference between the codewords, using different symbols $\boldsymbol{x}[n]$. In such a case, we decrease the probability that the respective observations $\boldsymbol{y}[n]$ are simultaneously affected by the channel in an adverse manner. These considerations motivate the following definition.

Definition 2.39 (Codeword Diversity) *For a given interleaving vector $\boldsymbol{\pi}$, the* diversity *of a codeword $\underline{\boldsymbol{c}}$ is defined as the number of labels $\boldsymbol{b}[n]$ in $\underline{\boldsymbol{b}} = \Pi(\underline{\boldsymbol{c}};\boldsymbol{\pi})$ with nonzero HWs, i.e.,*

$$\mathsf{D}(\underline{\boldsymbol{c}};\boldsymbol{\pi}) \triangleq \sum_{n=1}^{N_{\mathrm{s}}} \mathbb{I}_{[w_{\mathrm{H}}(\boldsymbol{b}[n])>0]}. \tag{2.115}$$

Clearly, the codeword diversity $\mathsf{D}(\underline{\boldsymbol{c}};\boldsymbol{\pi})$ is bounded as

$$\left\lceil \frac{w_{\mathrm{H}}(\underline{\boldsymbol{c}})}{m} \right\rceil \leq \mathsf{D}(\underline{\boldsymbol{c}};\boldsymbol{\pi}) \leq w_{\mathrm{H}}(\underline{\boldsymbol{c}}). \tag{2.116}$$

We thus say that the codeword $\underline{\boldsymbol{c}}$ achieves its *maximum diversity* if $\mathsf{D}(\underline{\boldsymbol{c}};\boldsymbol{\pi}) = w_{\mathrm{H}}(\underline{\boldsymbol{c}})$, which depends on how the interleaver allocates all the nonzero elements in $\underline{\boldsymbol{c}}$ to different labels $\boldsymbol{b}[n]$. In addition, if $w_{\mathrm{H}}(\underline{\boldsymbol{b}}) > N_{\mathrm{s}}$, the codeword cannot achieve its maximum diversity. From now on, we only consider the cases $w_{\mathrm{H}}(\underline{\boldsymbol{b}}) \leq N_{\mathrm{s}}$.

Definition 2.40 (Interleaver Diversity Efficiency) *For a given interleaving vector $\boldsymbol{\pi}$, the diversity efficiency of the interleaver Π is defined as the relative number of codewords in $\mathcal{C}(d)$ which attain their maximum diversity, i.e.,*

$$D^{\boldsymbol{\pi}}_{d} \triangleq \frac{1}{C^{\mathcal{C}}_{d}} |\mathcal{C}^{\mathsf{div}}(d)|, \tag{2.117}$$

where

$$\mathcal{C}^{\mathsf{div}}(d) \triangleq \{\underline{\boldsymbol{c}} \in \mathcal{C}(d) : \mathsf{D}(\underline{\boldsymbol{c}};\boldsymbol{\pi}) = d\} \tag{2.118}$$

is the set of codewords $\underline{\boldsymbol{c}}$ with HW d which attain their maximum diversity d.

The interleaver diversity efficiency is bounded as $0 \leq D^{\boldsymbol{\pi}}_{d} \leq 1$, where interleavers that achieve the upper bound are desirable. This is justified by the assertions we made above, namely that it is desirable to transmit bits in which the codewords differ using independent symbols. As a yardstick to compare the diversity efficiency of different interleavers, we define the random interleaver.

Definition 2.41 (Random Interleaver) *Let* $\{\boldsymbol{\pi}_i, i = 1, \ldots, N_c!\}$ *be the set of all distinct interleaver vectors. The interleaver is said to be random if the index i of the interleaving vector* $\boldsymbol{\pi}_i$ *is chosen randomly with uniform probability, i.e., the interleaving vector is modeled as a random variable* $\boldsymbol{\Pi}$ *with PMF*

$$\Pr\{\boldsymbol{\Pi} = \boldsymbol{\pi}_i\} = \frac{1}{N_c!}. \tag{2.119}$$

We note that the codewords $\mathcal{C}$ are generated in a deterministic manner, but for a random interleaver, the mapping between $\mathcal{C}$ and $\mathcal{B}$ is random. Therefore, both the codeword diversity $\mathsf{D}(\underline{\boldsymbol{c}}; \boldsymbol{\pi})$ and the interleaver diversity efficiency $D_d^{\boldsymbol{\pi}}$ are random. To evaluate them meaningfully, we then use expectations, i.e., we consider their average with respect to the distribution of the interleaving vector $\boldsymbol{\Pi}$ given in (2.119).[10]

Definition 2.42 (Average Interleaver Diversity Efficiency) *The average interleaver diversity efficiency of a random interleaver is defined as*

$$\overline{D}_d \triangleq \mathbb{E}_{\boldsymbol{\Pi}}[D_d^{\boldsymbol{\Pi}}]. \tag{2.120}$$

Theorem 2.43 *The average interleaver diversity efficiency can be expressed as*

$$\overline{D}_d = \prod_{l=1}^{d-1} \frac{N_c - lm}{N_c - l}. \tag{2.121}$$

Proof: As $|\mathcal{C}^{\mathsf{div}}(d)| = \sum_{\underline{\boldsymbol{c}} \in \mathcal{C}(d)} \mathbb{I}_{[\mathsf{D}(\underline{\boldsymbol{c}};\boldsymbol{\pi})=d]}$, we can write (2.120) using (2.119) and (2.117) as

$$\overline{D}_d = \frac{1}{N_c!} \sum_{i=1}^{N_c!} \frac{1}{C_d^{\mathcal{C}}} \sum_{\underline{\boldsymbol{c}} \in \mathcal{C}(d)} \mathbb{I}_{[\mathsf{D}(\underline{\boldsymbol{c}};\boldsymbol{\pi}_i)=d]} \tag{2.122}$$

$$= \frac{1}{N_c!} \frac{1}{C_d^{\mathcal{C}}} \sum_{\underline{\boldsymbol{c}} \in \mathcal{C}(d)} \sum_{i=1}^{N_c!} \mathbb{I}_{[\mathsf{D}(\underline{\boldsymbol{c}};\boldsymbol{\pi}_i)=d]}. \tag{2.123}$$

We note now that the enumeration over all the interleavers $\boldsymbol{\pi}_i$ in (2.123) yields all possible assignment of d bits to N_c positions. The set of resulting codeword diversities is the same for different codewords $\underline{\boldsymbol{c}} \in \mathcal{C}(d)$, and thus, the two innermost sums in (2.123) can be expressed as

$$\sum_{\underline{\boldsymbol{c}} \in \mathcal{C}(d)} \sum_{i=1}^{N_c!} \mathbb{I}_{[\mathsf{D}(\underline{\boldsymbol{c}};\boldsymbol{\pi}_i)=d]} = C_d^{\mathcal{C}} \sum_{i=1}^{N_c!} \mathbb{I}_{[\mathsf{D}(\underline{\boldsymbol{c}}';\boldsymbol{\pi}_i)=d]}, \tag{2.124}$$

where the equality holds for any $\underline{\boldsymbol{c}}' \in \mathcal{C}(d)$. Using (2.124) in (2.123) gives

$$\overline{D}_d = \frac{1}{N_c!} \sum_{i=1}^{N_c!} \mathbb{I}_{[\mathsf{D}(\underline{\boldsymbol{c}}';\boldsymbol{\pi}_i)=d]}, \quad \forall \underline{\boldsymbol{c}}' \in \mathcal{C}(d). \tag{2.125}$$

The last step in the proof is to find the number of permutations $\boldsymbol{\pi}_i$ which, when applied to an arbitrary codeword $\underline{\boldsymbol{c}}' \in \mathcal{C}(d)$, result in codewords $\underline{\boldsymbol{b}}$ with maximum diversity. To this end, we use the following procedure. We assign the first nonzero bit to any of the N_c "free" positions. Then, none of the m positions in the label $\boldsymbol{b}[n]$ taken by the first bit can be used by the second bit. To assign the second nonzero bit we

[10] Or more generally, with respect to any other PMF of $\boldsymbol{\Pi}$.

have only $N_c - m$ "free" positions. For the third nonzero bit there are $N_c - 2m$ "free" positions, and so on. Once all the d nonzero bits are assigned, the remaining $N_c - d$ zero bits can be assigned in $(N_c - d)!$ different ways. Taking the product of these possibilities in (2.125) gives

$$\overline{D}_d = \frac{N_c(N_c - m)\dots(N_c - (d-1)m)(N_c - d)!}{N_c!}, \tag{2.126}$$

which is the same as (2.121). This completes the proof. □

For large N_c, treating $\nu = 1/N_c$ as a continuous variable, we may apply a first-order Taylor-series approximation of $\overline{D}_d$ around $\nu = 0$, to obtain

$$1 - \overline{D}_d \approx \frac{m-1}{2}\frac{d(d-1)}{N_c}. \tag{2.127}$$

From now, we will refer to $1 - \overline{D}_d$ as the *average complementary interleaver diversity efficiency*.

In Fig. 2.23, we show the expression (2.121) as well as the approximation in (2.127). We observe that the average complementary interleaver diversity efficiency is strongly affected by an increase in d. For example, when $m = 2$, to keep the average complementary interleaver diversity efficiency at $1 - \overline{D}_d \leq 10^{-2}$, $N_c = 100$ is necessary for $d = 2$ but we need $N_c = 1000$ when $d = 5$. It is also interesting to note from Fig. 2.23 that a random interleaver does not guarantee an average interleaver diversity $\overline{D}_d = 1$, even for large N_c.

Another property of the code $\mathcal{B}$ we are interested in is related to the way the codewords' weights are distributed across the bit positions after interleaving, which we formalize in the following.

Definition 2.44 (Average GDS of the Code $\mathcal{B}$) *The average GDS of the code $\mathcal{B}$ (obtained by random interleaving) is defined as*

$$\overline{C}_{\boldsymbol{w}}^{\mathcal{B}} \triangleq \mathbb{E}_{\Pi}[C_{\boldsymbol{w}}^{\mathcal{B}}]. \tag{2.128}$$

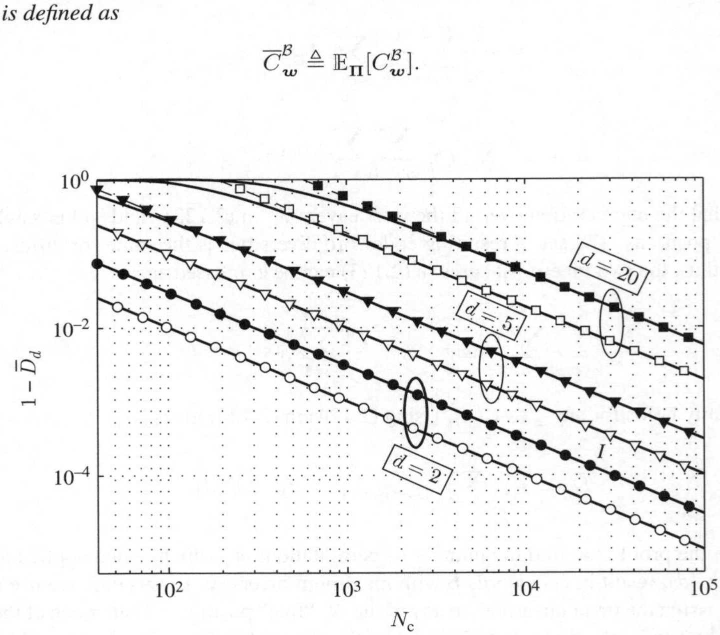

Figure 2.23 Average complementary interleaver diversity efficiency (solid lines) and its large-N_c approximation (2.127) (dashed lines) for $m = 2$ (hollow markers, e.g., ∘) and for $m = 4$ (filled markers, e.g., •)

Theorem 2.45 *For all d such that $C^{\mathcal{B}}_d \neq 0$, the average GDS of the code $\mathcal{B}$ in* (2.128) *satisfies*

$$\lim_{N_c\to\infty} \frac{\overline{C}^{\mathcal{B}}_{\boldsymbol{w}}}{C^{\mathcal{B}}_d} = \frac{1}{m^d}\binom{d}{\boldsymbol{w}}, \qquad \forall \boldsymbol{w} \in \mathbb{N}^m : d = \|\boldsymbol{w}\|_1. \tag{2.129}$$

Proof: Using (2.119) and Definition 2.36, we have that for any $\boldsymbol{w} \in \mathbb{N}^m : d = \|\boldsymbol{w}\|_1$, the average GDS of the code $\mathcal{B}$ in (2.128) can be expressed as

$$\overline{C}^{\mathcal{B}}_{\boldsymbol{w}} = \sum_{\underline{\boldsymbol{c}}\in\mathcal{C}(d)} \sum_{i=1}^{N_c!} \frac{\mathbb{I}_{[W_H(\Pi(\underline{\boldsymbol{c}};\boldsymbol{\pi}_i))=\boldsymbol{w}]}}{N_c!} \tag{2.130}$$

$$= \sum_{\underline{\boldsymbol{c}}\in\mathcal{C}(d)} \frac{\sum_{i=1}^{N_c!} \mathbb{I}_{[W_H(\Pi(\underline{\boldsymbol{c}};\boldsymbol{\pi}_i))=\boldsymbol{w}]}}{\sum_{i=1}^{N_c!}\sum_{\boldsymbol{v}:\|\boldsymbol{v}\|_1=d} \mathbb{I}_{[W_H(\Pi(\underline{\boldsymbol{c}};\boldsymbol{\pi}_i))=\boldsymbol{v}]}}, \tag{2.131}$$

where (2.131) follows from $\sum_{\boldsymbol{v}:\|\boldsymbol{v}\|_1=d} \mathbb{I}_{[W_H(\Pi(\underline{\boldsymbol{c}};\boldsymbol{\pi}_i))=\boldsymbol{v}]} = 1$, which holds for any $\underline{\boldsymbol{c}} \in \mathcal{C}(d)$ and any $\boldsymbol{\pi}_i$.

Using the same reasoning as in the proof of Theorem 2.43 (see (2.124)), we have

$$\sum_{i=1}^{N_c!} \mathbb{I}_{[W_H(\Pi(\underline{\boldsymbol{c}};\boldsymbol{\pi}_i))=\boldsymbol{v}]} = \sum_{i=1}^{N_c!} \mathbb{I}_{[W_H(\Pi(\underline{\boldsymbol{c}}';\boldsymbol{\pi}_i))=\boldsymbol{v}]}, \tag{2.132}$$

where the equality holds for any $\underline{\boldsymbol{c}}' \in \mathcal{C}(d)$ as long as $\|\boldsymbol{v}\|_1 = d$. Using (2.132) in (2.131), we obtain

$$\frac{\overline{C}^{\mathcal{B}}_{\boldsymbol{w}}}{|\mathcal{C}(d)|} = \frac{\sum_{i=1}^{N_c!} \mathbb{I}_{[W_H(\Pi(\underline{\boldsymbol{c}}';\boldsymbol{\pi}_i))=\boldsymbol{w}]}}{\sum_{\boldsymbol{v}:\|\boldsymbol{v}\|_1=d}\sum_{i=1}^{N_c!} \mathbb{I}_{[W_H(\Pi(\underline{\boldsymbol{c}}';\boldsymbol{\pi}_i))=\boldsymbol{v}]}}. \tag{2.133}$$

To calculate the *right-hand side* (r.h.s.) of (2.132) we note that we need to consider all the different assignments of d nonzero bits of the codeword $\underline{\boldsymbol{c}}'$ into m sets having $v_1, v_2, \ldots, v_m$ elements respectively; there are $\binom{d}{\boldsymbol{v}}$ such assignments. Moreover, the v_k bits in the kth set can be assigned into $N_s(N_s-1)\ldots(N_s-v_k+1) = \prod_{l=0}^{v_k-1}(N_s-l)$ different time positions. All remaining zeros can be assigned in $(N_c-d)!$ different ways. Thus the r.h.s. of (2.132) can be expressed as

$$\sum_{i=1}^{N_c!} \mathbb{I}_{[W_H(\Pi(\underline{\boldsymbol{c}}';\boldsymbol{\pi}_i))=\boldsymbol{v}]} = \binom{d}{\boldsymbol{v}} \prod_{k=1}^{m}\prod_{l=0}^{v_k-1}(N_s-l)(N_c-d)! \tag{2.134}$$

Then, using (2.134) in (2.133) we obtain

$$\frac{\overline{C}^{\mathcal{B}}_{\boldsymbol{w}}}{|\mathcal{C}(d)|} = \binom{d}{\boldsymbol{w}} \frac{\prod_{k=1}^{m}\prod_{l=0}^{w_k-1}(N_s-l)}{\sum_{\boldsymbol{v}:\|\boldsymbol{v}\|_1=d}\binom{d}{\boldsymbol{v}}\prod_{k=1}^{m}\prod_{l=0}^{v_k-1}(N_s-l)} \tag{2.135}$$

$$= \binom{d}{\boldsymbol{w}} \frac{\prod_{k=1}^{m}\prod_{l=0}^{w_k-1}(1-l/N_s)}{\sum_{\boldsymbol{v}:\|\boldsymbol{v}\|_1=d}\binom{d}{\boldsymbol{v}}\prod_{k=1}^{m}\prod_{l=0}^{v_k-1}(1-l/N_s)}. \tag{2.136}$$

We thus easily conclude that

$$\lim_{N_c\to\infty} \frac{\overline{C}^{\mathcal{B}}_{\boldsymbol{w}}}{|\mathcal{C}(d)|} = \lim_{N_c\to\infty} \frac{\overline{C}^{\mathcal{B}}_{\boldsymbol{w}}}{C^{\mathcal{C}}_d} \tag{2.137}$$

$$= \binom{d}{\boldsymbol{w}} \frac{1}{\sum_{\boldsymbol{v}:\|\boldsymbol{v}\|_1=d}\binom{d}{\boldsymbol{v}}} \tag{2.138}$$

$$= \binom{d}{\boldsymbol{w}} \frac{1}{m^d}, \tag{2.139}$$

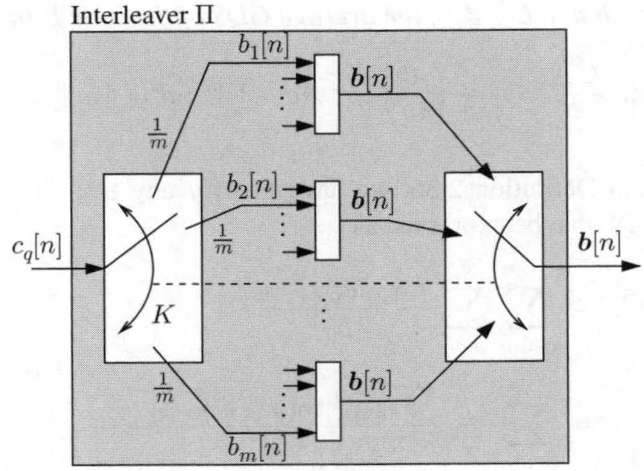

Figure 2.24 Model of the random interleaver: the bits $c_q[n]$ are mapped to the bit positions K within the label $\boldsymbol{b}[n]$, where K is a uniformly distributed random variable, i.e., $\Pr\{K = k\} = \frac{1}{m}, k = 1, \ldots, m$

where (2.137) follows from (2.82), (2.138) from (2.136), and (2.139) from (2.19). The proof is completed by noting that $C_d^{\mathcal{B}} = C_d^{\mathcal{C}}$. □

Up to now, we have considered two effects of the interleaving of the code $\mathcal{C}$ independently: the GDS in Definition 2.36 deals with the assignment of d nonzero bits from $\underline{\mathbf{c}}' \in \mathcal{C}(d)$ into different positions k at the modulator's input, while the interleaver diversity efficiency in Definition 2.40 deals with the assignment of these d bits into labels at different time instants n. We may also jointly consider these two "dimensions" of the interleaving. Namely, for all $\boldsymbol{w}$ with $\|\boldsymbol{w}\|_1 = d$, we have

$$\frac{\lim_{N_c\to\infty} |\{\underline{\mathbf{c}} \in \mathcal{C}^{\text{div}}(d) : \Pi(\underline{\mathbf{c}}) \in \mathcal{B}(\boldsymbol{w})\}|}{\overline{C}_{\boldsymbol{w}}^{\mathcal{B}}} = \overline{D}_d, \tag{2.140}$$

i.e., the ratio between the number of codewords in the set $\mathcal{B}(\boldsymbol{w})$ which achieve their maximum diversity and the number of codewords in the set $\mathcal{B}(\boldsymbol{w})$ depends solely on $d = \|\boldsymbol{w}\|_1$. The proof of (2.140) can be made along the lines of the proofs of Theorems 2.43 and 2.45.

Theorem 2.45 takes advantage of the enumeration over all possible interleavers $\boldsymbol{\pi}_i$, which makes the enumeration over all codewords unnecessary. As we can use any codeword $\underline{\mathbf{c}}' \in \mathcal{C}(d)$, the assignment of the bits $c_q[n]$ to the position k in the label $\boldsymbol{b}[n]$ becomes independent of the codeword $\underline{\mathbf{c}}$. Thus, we can simply treat the assigned position k as a random variable K. The assignment is not "biased" toward any position k, so K has a uniform distribution on the set $\{1, \ldots, m\}$, which leads to (2.139). Such a model is depicted in Fig. 2.24.

The concept of random interleavers leads to the random position-assignment model from Fig. 2.24. Of course, in practice, a fixed interleaving vector $\boldsymbol{\pi}$ is used. However, we will be able to apply the simple model depicted in Fig. 2.24 to analyze also a fixed interleaver, provided that it inherits some of the properties of the random interleaver. In such a case, we talk about *quasirandom interleavers* which we define in the following.

Definition 2.46 (Quasirandom Interleaver) *An interleaver* Π *is said to be quasirandom if for any* d *such that* $C_d^{\mathcal{B}} \neq 0$*, the following conditions hold true:*

1. The interleaver diversity efficiency tends to one

$$\lim_{N_c\to\infty} D_d^{\boldsymbol{\pi}} = 1. \tag{2.141}$$

2. *The ratio between the GDS and the DS tends to the ratio obtained for random interleavers in* (2.129), *i.e.,*

$$\lim_{N_c \to \infty} \frac{C_{\boldsymbol{w}}^{\mathcal{B}}}{C_d^{\mathcal{B}}} = \frac{\overline{C}_{\boldsymbol{w}}^{\mathcal{B}}}{C_d^{\mathcal{B}}} = \frac{1}{m^d} \binom{d}{\boldsymbol{w}}, \quad \forall \boldsymbol{w} \in \mathbb{N}^m : \|\boldsymbol{w}\|_1 = d. \tag{2.142}$$

3. *The relative number of codewords which achieve their maximum diversity within the set* $\mathcal{B}(\boldsymbol{w})$ *depends only on the HW of the codeword* $d = \|\boldsymbol{w}\|_1$, *i.e.,*

$$\frac{\lim_{N_c \to \infty} |\{\underline{\boldsymbol{c}} \in \mathcal{C}^{\mathsf{div}}(d) : \Pi(\underline{\boldsymbol{c}}) \in \mathcal{B}(\boldsymbol{w})\}|}{C_{\boldsymbol{w}}^{\mathcal{B}}} = D_d^{\boldsymbol{\pi}}. \tag{2.143}$$

As we refer to properties of the interleavers for $N_c \to \infty$, when discussing their effects, we have in mind a particular family of interleavers which defines how to obtain the interleaving vector $\boldsymbol{\pi}$ for each value of N_c. Therefore, strictly speaking, the definition above applies to a family of interleavers and not to a particular interleaver with interleaving vector $\boldsymbol{\pi}$.

The essential difference between the random interleaving and quasirandom interleaving lies in the analysis of the DS or GDS. In the case of random interleaving, instead of enumerating all possible codewords $\underline{\boldsymbol{c}}$, we fix one of them and average the spectrum over all possible realizations of the interleaver. In the case of a fixed (quasirandom) interleaving, we enumerate the codewords $\underline{\boldsymbol{b}}$ produced by interleaving all the codewords $\underline{\boldsymbol{c}}$ from $\mathcal{C}(d)$. The property of quasirandomness assumes that the enumeration over the codewords will produces the same average (spectrum) as the enumeration over the interleavers.

Of course, quasirandomness is a property which depends on the code and how the family of interleavers is defined. In the following example, we analyze a CC and two particular interleavers in the light of the conditions of quasirandomness.

Example 2.47 (Interleavers for CENCs) *Consider the CENC with* $\mathbf{G} = [5, 7]$ *in Table 2.1 where we use the so-called trellis termination, i.e., the sequence of* N_b *information bits is padded with* $\nu = 2$ *dummy zeros (which forces the encoder to terminate in the zero state). We assume* $m = 2$ *and consider two different interleavers: the rectangular interleaver with period* $T = \lfloor\sqrt{N_c}\rfloor$ *in Definition 2.32, and the pseudorandom interleaver in Definition 2.34 based on a pseudorandom seed* $\iota = 1$.[11]

Finding the values of the interleaver diversity efficiency may be difficult for large values of N_c, *so we restrict our analysis to the leading terms of the DS of the CC for* $d \geq d_{\text{free}} = 5$. *The first four nonzero values of the DS* $C_d^{\mathcal{C}}$ *can be shown to be* $C_5^{\mathcal{C}} = N_b$, $C_6^{\mathcal{C}} = 2N_b - 3$, $C_7^{\mathcal{C}} = 4N_b - 12$, $C_8^{\mathcal{C}} = 8N_b - 36$.

The complementary interleaver diversity efficiency $1 - D_d^{\boldsymbol{\pi}}$ *is shown in Fig. 2.25 for various values of* $N_c \geq 20$, *and indeed decreases with* N_c *for the pseudorandom interleaver. Moreover, the rectangular interleaver, even with moderate values of* N_c, *guarantees all codewords will reach their maximum interleaver diversity efficiency. In particular, for all values of* $N_c \geq 20$, *we have* $D_5^{\boldsymbol{\pi}} = 1$, *and thus,* $1 - D_5^{\boldsymbol{\pi}}$ *is not shown on the plot. Thus, both interleavers can be considered quasirandom with respect to the first condition of Definition 2.46.*

Let us verify now the second condition in Definition 2.46. As we set $m = 2$, *there are only two positions in the label* $\mathbf{b}[n]$, *and thus, we only need to consider* $\boldsymbol{w} = [w_1, d - w_1]$. *Then* $\binom{d}{\boldsymbol{w}} = \frac{d!}{w_1!(d-w_1)!}$. *We show in Fig. 2.26 how the ratio for the random interleaver* $\frac{\overline{C}_{\boldsymbol{w}}^{\mathcal{B}}}{C_d^{\mathcal{B}}} = \frac{1}{2^d}\binom{d}{\boldsymbol{w}}$ *compares with* $\frac{C_{\boldsymbol{w}}^{\mathcal{B}}}{C_d^{\mathcal{B}}}$ *for various values of* N_c. *A formal comparison might be carried out; however, from Fig. 2.26, it is rather clear that the rectangular interleaver does not satisfy the second condition of quasi-randomness in Definition 2.46. This means, in particular, that we cannot apply the random model shown in Fig. 2.24.*

We show similar results in Fig. 2.27 for the case of the pseudorandom interleaver. In this case, the behavior is significantly different. We can see that the second condition of Definition 2.46 is approximatively satisfied. For $N_c = 40$ *and* $N_c = 200$, *the empirically obtained ratio does not match very well the*

[11] We used the pseudorandom permutation defined in MATLAB.

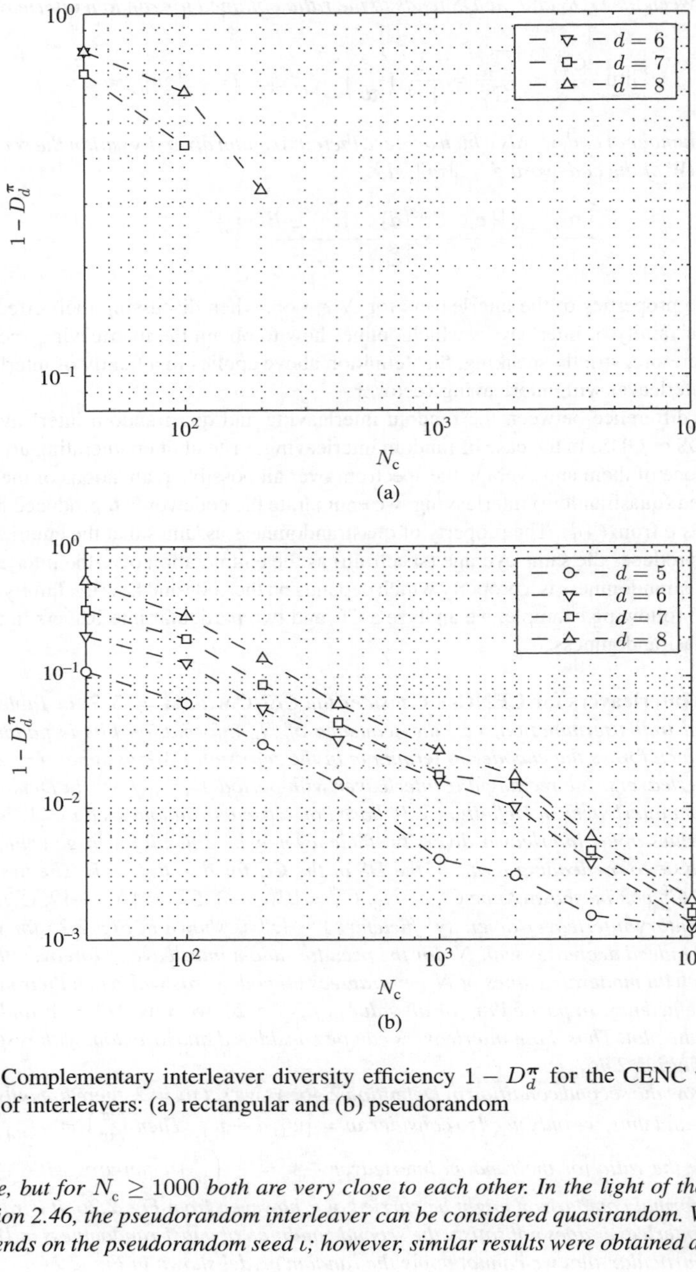

Figure 2.25 Complementary interleaver diversity efficiency $1 - D_d^\pi$ for the CENC with $\mathbf{G} = [5, 7]$ and two types of interleavers: (a) rectangular and (b) pseudorandom

theoretical one, but for $N_c \geq 1000$ both are very close to each other. In the light of the second condition of Definition 2.46, the pseudorandom interleaver can be considered quasirandom. We note that this condition depends on the pseudorandom seed ι; however, similar results were obtained after changing ι.

2.8 Bibliographical Notes

TCM was originally proposed at ISIT 1976 [1] and then developed in [2–4]. TCM quickly became a very popular research topic and improved TCM paradigms were soon proposed: rotationally invariant

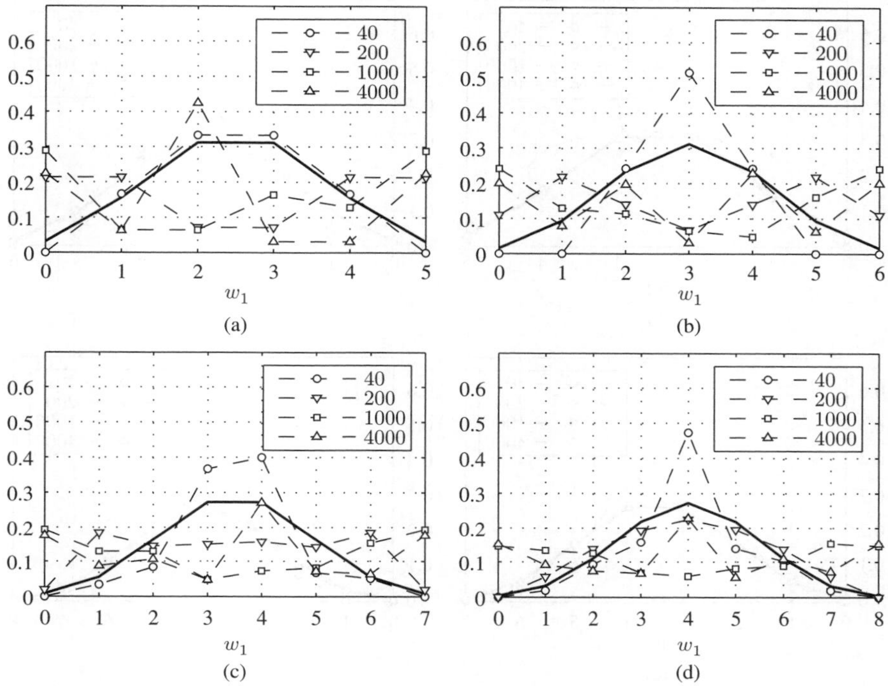

Figure 2.26 Values of $\frac{C^{\mathcal{B}}_{\boldsymbol{w}}}{C^{\mathcal{B}}_{d}}$ for $\boldsymbol{w} = [w_1, d - w_1]$ when using a rectangular interleaver for different values of N_c: (a) $d = 5$, (b) $d = 6$, (c) $d = 7$, and (d) $d = 8$. The thick solid lines are the distributions of the random interleaver $\frac{1}{m^d}\binom{d}{\boldsymbol{w}}$

TCM [5, 6], multidimensional TCM [3, 7–9], TCM based on cosets and lattices [10, 11], TCM with nonequally spaced symbols [12–15], etc. TCM went also quickly from research to practice; it was introduced in the modem standards in the early 1990s (V.32 [16] and V.32bis [17]) increasing the transmission rates up to 14.4 kbps. TCM is a well-studied topic and extensive information about it can be found, e.g., in [18, 19, Section 8.12], [20, Chapter 4], [21, Section 8.2], [22, Chapter 14], [23, Chapter 18]. TCM for fading channels is studied in [20, Chapter 5].

MLC was proposed by Imai and Hirakawa in [24, 25]. MLC with MSD as well as the design rules for selecting the m rates of the encoders were analyzed in detail in [26, 27]. MLC for fading channels, which includes bit interleavers in each level, has been proposed in [28], and MLC using capacity-approaching (turbo) codes was proposed in [29].

BICM was introduced in [30] and later analyzed from an information-theoretic point of view in [31, 32]. BICM-ID was introduced in [33–35] where BICM was recognized as a serial concatenation of encoders (the encoder and the mapper) and further studied in [36–40]. For relevant references about the topics related to BICM treated in this book, we refer the reader to the end of each chapter.

The discrete-time AWGN model and the detection of signals in a continuous-time AWGN channel is a well-studied topic in the literature, see, e.g., [19, Chapter 3], [41, Chapter 2], [42, Chapter 5], [43, Section 2.5], [44, Chapters 26 and 28]. For more details about models for fading channels, we refer the reader to [19, Chapter 13] or to [42, Chapter 3]. In particular, more details on the Nakagami fading distribution [45] are given in [42, Section 3.2.2].

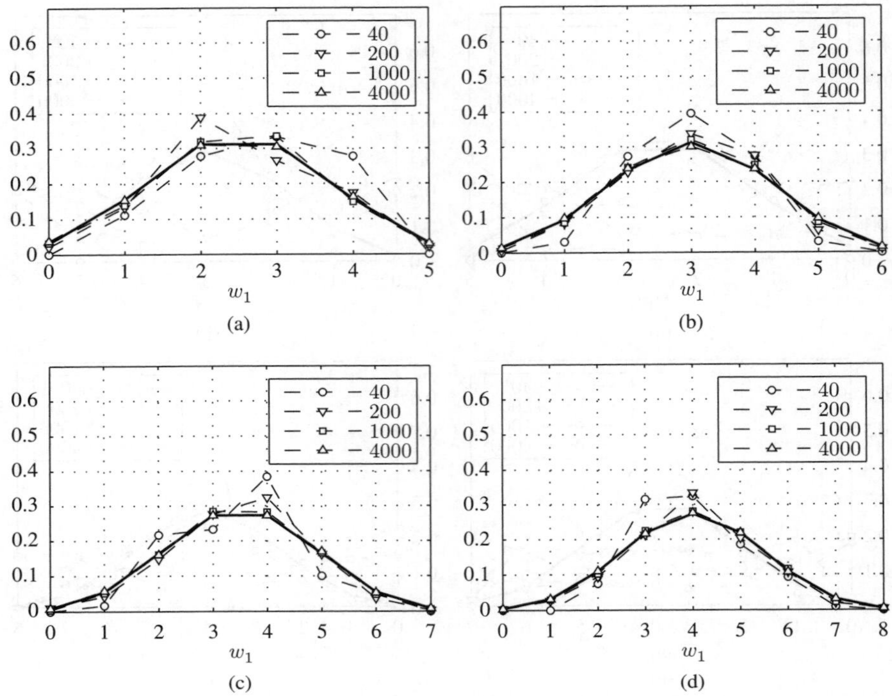

Figure 2.27 Values of $\frac{C^{\mathcal{B}}_{\boldsymbol{w}}}{C^{\mathcal{B}}_{d}}$ for $\boldsymbol{w} = [w_1, d - w_1]$ when using a pseudorandom interleaver for different values of N_c: (a) $d = 5$, (b) $d = 6$, (c) $d = 7$, and (d) $d = 8$. The thick solid lines are the distributions of the random interleaver $\frac{1}{m^d}\binom{d}{\boldsymbol{w}}$

The BRGC was introduced in [46] and studied for uncoded transmission in [47, 48] where its asymptotic optimality for PAM, PSK, and QAM constellations was proved. For more details about Gray labelings, we refer the reader also to [49]. The expansion used to define the BRGC was introduced in [47]. An alternative construction that can be used is based on *reflections*, which is detailed in [47, Section IV]. The FBC was analyzed in [50] for uncoded transmission and the BSGC was recently introduced in [51].

For 8PSK and in the context of BICM-ID, other labelings have been proposed; see [52] and references therein. For example, the SSP labeling was proposed in [40, Fig. 2 (c)], (later found via an algorithmic search in [38, Fig. 2 (a)], and called M8), or the MSP labeling [53, Fig. 2 (b)]. These two labelings are shown in Fig. 2.17 (b) and (c). The M16 labeling used in BICM-ID for 16QAM was first proposed in [38, Fig. 2 (b)].

The design of bit labelings for improving the performance of BICM-ID has been studied in [37, 38, 54–59] and references therein. Most of the works consider one-to-one mappers (i.e., a bijective mapping from labels to constellation symbols); however, when signal shaping is considered, a many-to-one mapping ($M < 2^m$) may be useful, as shown in [59, Section 6.2].

The channel capacity defined in 1948 by Shannon [60] spurred a great deal of activity and approaching Shannon's limit became one of the most important problems among researchers for about 45 years. The complexity of encoding/decoding was always an issue but these limits have been continuously pushed by the development of integrated circuits. CENCs provided a low-complexity encoding strategy and the Viterbi algorithm [61] gave a clever and relatively low-complexity decoding method. Introduced by Elias [62] in 1955, CENCs were studied extensively and their detailed description can be found in popular textbooks, e.g., [23, Chapter 11], [63, Chapter 5], [64]. The name for CENCs with ODS was coined by Frenger *et al.* in [65], however, most of the newly reported spectra in [65] had already been presented in [66, Tables III–V], [67, Tables II–IV], as later clarified in [68].

TCs were invented by Berrou *et al.* [69] and surprised the coding community with their performance maintaining relatively simple encoding and decoding via iterative processing. Since then, they have been analyzed in detail in many works and textbooks, e.g., [63, Chapter 8.2; 23, Chapter 16]; TCs are also used in many communication standards such as 3G and 4G telephony [70, Section 16.5.3], *digital video broadcasting* (DVB) standards [71], as well as in deep space communications [72, 73].

Most often, the BICM literature assumes that random interleaving with infinite length is used [31], which leads to the simple random multiplexing model we have shown in Fig. 2.24. The formulas describing the diversity efficiency of the finite-length interleaver can be found in [32, Chapter 4.3].

References

[1] Ungerboeck, G. and Csajka, I. (1976) On improving data-link performance by increasing channel alphabet and introducing sequence decoding. International Symposium on Information Theory (ISIT), June 1976, Ronneby, Sweden (book of abstracts).

[2] Ungerboeck, G. (1982) Channel coding with multilevel/phase signals. *IEEE Trans. Inf. Theory*, **28** (1), 55–67.

[3] Ungerboeck, G. (1987) Trellis-coded modulation with redundant signal sets Part I: introduction. *IEEE Commun. Mag.*, **25** (2), 5–11.

[4] Ungerboeck, G. (1987) Trellis-coded modulation with redundant signal sets Part II: state of the art. *IEEE Commun. Mag.*, **25** (2), 12–21.

[5] Wei, L.-F. (1984) Rotationally invariant convolutional channel coding with expanded signal space—Part I: 180°. *IEEE J. Sel. Areas Commun.*, **SAC-2** (5), 659–671.

[6] Wei, L.-F. (1984) Rotationally invariant convolutional channel coding with expanded signal space—Part II: Nonlinear codes. *IEEE J. Sel. Areas Commun.*, **SAC-2** (5), 672–686.

[7] Wei, L.-F. (1987) Trellis-coded modulation with multidimensional constellations. *IEEE Trans. Inf. Theory*, **IT-33** (4), 483–501.

[8] Gersho, A. and Lawrence, V. B. (1984) Multidimensional signal constellations for voiceband data transmission. *IEEE J. Sel. Areas Commun.*, **SAC-2** (5), 687–702.

[9] Forney, G. D. Jr., Gallager, R., Lang, G. R., Longstaff, F. M., and Qureshi, S. U. (1984) Efficient modulation for band-limited channels. *IEEE J. Sel. Areas Commun.*, **SAC-2** (5), 632–647.

[10] Calderbank, A. R. and Sloane, N. J. A. (1987) New trellis codes based on lattices and cosets. *IEEE Trans. Inf. Theory*, **IT-33** (2), 177–195.

[11] Forney, G. D. Jr. (1988) Coset codes I: introduction and geometrical classification. *IEEE Trans. Inf. Theory*, **34** (6), 1123–1151 (invited paper).

[12] Calderbank, R. and Mazo, J. E. (1984) A new description of trellis codes. *IEEE Trans. Inf. Theory*, **IT-30** (6), 784–791.

[13] van der Vleuten, R. J. and Weber, J. H. (1996) Optimized signal constellations of trellis-coded modulation on AWGN channels. *IEEE Trans. Commun.*, **44** (6), 646–648.

[14] Simon, M. K. and Divsalar, D. (1985) Combined trellis coding with asymmetric MPSK modulation. Jet Propulsion Laboratory, Pasadena, CA, JPL Pub. 85-24, May 1985.

[15] Divsalar, D., Simon, M., and Yuen, J. (1987) Trellis coding with asymmetric modulations. *IEEE Trans. Commun.*, **35** (2), 130–3141.

[16] ITU (1993) A family of 2-wire, duplex modems operating at data signalling rates of up to 9600 bit/s for use on the general switched telephone network and on leased telephone-type circuits. ITU-T Recommendation V.32, Tech. Rep., International Telecommunication Union (ITU).

[17] ITU (1991) A duplex modem operating at data signalling rates of up to 14 400 bit/s for use on the general switched telephone network and on leased point-to-point 2-wire telephone-type circuits. ITU-T Recommendation V.32 bis, Tech. Rep., International Telecommunication Union (ITU).

[18] Biglieri, E., Divsalar, D., McLane, P. J., and Simon, M. K. (1992) *Introduction to Trellis-Coded Modulation with Applications*, Prentice Hall.

[19] Proakis, J. G. and Salehi, M. (2008) *Digital Communications*, 5th edn, McGraw-Hill.

[20] Jamali, S. H. and Le-Ngoc, T. (1994) *Coded-Modulation Techniques for Fading Channels*, Kluwer Academic Publishers.

[21] Burr, A. (2001) *Modulation and Coding for Wireless Communications*, Prentice Hall.

[22] Wicker, S. B. (1995) *Error Control Systems for Digital Communication and Storage*, Prentice Hall.

[23] Lin, S. and Costello, D. J. Jr. (2004) *Error Control Coding*, 2nd edn, Prentice Hall, Englewood Cliffs, NJ.

[24] Imai, H. and Hirakawa, S. (1977) A new multilevel coding method using error-correcting codes. *IEEE Trans. Inf. Theory*, **IT-23** (3), 371–377.

[25] Imai, H. and Hirakawa, S. (1977) Correction to 'A new multilevel coding method using error-correcting codes'. *IEEE Trans. Inf. Theory*, **IT-23** (6), 784.

[26] Wachsmann, U., Fischer, R. F. H., and Huber, J. B. (1999) Multilevel codes: theoretical concepts and practical design rules. *IEEE Trans. Inf. Theory*, **45** (5), 1361–1391.

[27] Beygi, L., Agrell, E., Karlsson, M., and Makki, B. (2010) A novel rate allocation method for multilevel coded modulation. IEEE International Symposium on Information Theory (ISIT), June 2010, Austin, TX.

[28] Kofman, Y., Zehavi, E., and Shamai, S. (1994) Performance analysis of a multilevel coded modulation system. *IEEE Trans. Commun.*, **42** (2/3/4), 299–312.

[29] Wachsmann, U. and Huber, J. B. (1995) Power and bandwidth efficient digital communication using turbo codes in multilevel codes. *Eur. Trans. Telecommun.*, **6** (5), 557–567.

[30] Zehavi, E. (1992) 8-PSK trellis codes for a Rayleigh channel. *IEEE Trans. Commun.*, **40** (3), 873–884.

[31] Caire, G., Taricco, G., and Biglieri, E. (1998) Bit-interleaved coded modulation. *IEEE Trans. Inf. Theory*, **44** (3), 927–946.

[32] Guillén i Fàbregas, A., Martinez, A., and Caire, G. (2008) Bit-interleaved coded modulation. *Found. Trends Commun. Inf. Theory*, **5** (1–2), 1–153.

[33] Li, X. and Ritcey, J. A. (1997) Bit-interleaved coded modulation with iterative decoding. *IEEE Commun. Lett.*, **1** (6), 169–171.

[34] ten Brink, S., Speidel, J., and Yan, R.-H. (1998) Iterative demapping for QPSK modulation. *IEE Electron. Lett.*, **34** (15), 1459–1460.

[35] Benedetto, S., Montorsi, G., Divsalar, D., and Pollara, F. (1998) Soft-input soft-output modules for the construction and distributed iterative decoding of code networks. *Eur. Trans. Telecommun.*, **9** (2), 155–172.

[36] Chindapol, A. and Ritcey, J. A. (2001) Design, analysis, and performance evaluation for BICM-ID with square QAM constellations in Rayleigh fading channels. *IEEE J. Sel. Areas Commun.*, **19** (5), 944–957.

[37] Tüchler, M. (2004) Design of serially concatenated systems depending on the block length. *IEEE Trans. Commun.*, **52** (2), 209–218.

[38] Schreckenbach, F., Görtz, N., Hagenauer, J., and Bauch, G. (2003) Optimization of symbol mappings for bit-interleaved coded modulation with iterative decoding. *IEEE Commun. Lett.*, **7** (12), 593–595.

[39] Szczecinski, L., Chafnaji, H., and Hermosilla, C. (2005) Modulation doping for iterative demapping of bit-interleaved coded modulation. *IEEE Commun. Lett.*, **9** (12), 1031–1033.

[40] Li, X., Chindapol, A., and Ritcey, J. A. (2002) Bit-interlaved coded modulation with iterative decoding and 8PSK signaling. *IEEE Trans. Commun.*, **50** (6), 1250–1257.

[41] Viterbi, A. J. and Omura, J. K. (1979) *Principles of Digital Communications and Coding*, McGraw-Hill.

[42] Goldsmith, A. (2005) *Wireless Communications*, Cambridge University Press, New York.

[43] Anderson, J. B. (2005) *Digital Transmission Engineering*, 2nd edn, John Wiley & Sons, Inc.

[44] Lapidoth, A. (2009) *A Foundation in Digital Communication*, Cambridge University Press.

[45] Nakagami, M. (1960) The m-distribution, a general formula of intensity distribution of rapid fading, in *Statistical Methods in Radio Wave Propagation* (ed. W.G. Hoffman), Pergamon, Oxford.

[46] Gray, F. (1953) Pulse code communications. US patent 2 632 058.

[47] Agrell, E., Lassing, J., Ström, E. G., and Ottosson, T. (2004) On the optimality of the binary reflected Gray code. *IEEE Trans. Inf. Theory*, **50** (12), 3170–3182.

[48] Agrell, E., Lassing, J., Ström, E. G., and Ottosson, T. (2007) Gray coding for multilevel constellations in Gaussian noise. *IEEE Trans. Inf. Theory*, **53** (1), 224–235.

[49] Savage, C. (1997) A survey of combinatorial Gray codes. *SIAM Rev.*, **39** (4), 605–629.

[50] Lassing, J., Ström, E. G., Agrell, E., and Ottosson, T. (2003) Unequal bit-error protection in coherent M-ary PSK. IEEE Vehicular Technology Conference (VTC-Fall), October 2003, Orlando, FL.

[51] Agrell, E. and Alvarado, A. (2011) Optimal alphabets and binary labelings for BICM at low SNR. *IEEE Trans. Inf. Theory*, **57** (10), 6650–6672.

[52] Brännström, F. and Rasmussen, L. K. (2009) Classification of unique mappings for 8PSK based on bit-wise distance spectra. *IEEE Trans. Inf. Theory*, **55** (3), 1131–1145.

[53] Tran, N. H. and Nguyen, H. H. (2006) Signal mappings of 8-ary constellations for bit interleaved coded modulation with iterative decoding. *IEEE Trans. Broadcast.*, **52** (1), 92–99.
[54] Tan, J. and Stüber, G. L. (2002) Analysis and design of interleaver mappings for iteratively decoded BICM. IEEE International Conference on Communications (ICC), May 2002, New York City, NY.
[55] ten Brink, S. (2001) Convergence behaviour of iteratively decoded parallel concatenated codes. *IEEE Trans. Commun.*, **49** (10), 1727–1737.
[56] Zhao, L., Lampe, L., and Huber, J. (2003) Study of bit-interleaved coded space-time modulation with different labeling. IEEE Information Theory Workshop (ITW), March 2003, Paris, France.
[57] Clevorn, T., Godtmann, S., and Vary, P. (2006) Optimized mappings for iteratively decoded BICM on Rayleigh channels with IQ interleaving. IEEE Vehicular Technology Conference (VTC-Spring), May 2006, Melbourne, Australia.
[58] Tan, J. and Stüber, G. L. (2005) Analysis and design of symbol mappers for iteratively decoded BICM. *IEEE Trans. Wireless Commun.*, **4** (2), 662–672.
[59] Schreckenbach, F. Iterative decoding of bit-interleaved coded modulation. PhD dissertation, Technische Universität München, Munich, Germany, 2007.
[60] Shannon, C. E. (1948) A mathematical theory of communications. *Bell Syst. Tech. J.*, **27**, 379–423 and 623–656.
[61] Viterbi, A. J. (1967) Error bounds for convolutional codes and an asymptotically optimum decoding algorithm. *IEEE Trans. Inf. Theory*, **13** (2), 260–269.
[62] Elias, P. (1955) Coding for noisy channels. *IRE Conv. Rec.*, **3**, 37–46.
[63] Morelos-Zaragoza, R. H. (2002) *The Art of Error Correcting Coding*, 2nd edn, John Wiley & Sons.
[64] Johannesson, R. and Zigangirov, K. S. (1999) *Fundamentals of Convolutional Coding*, 1st edn, IEEE Press.
[65] Frenger, P., Orten, P., and Ottosson, T. (1999) Convolutional codes with optimum distance spectrum. *IEEE Trans. Commun.*, **3** (11), 317–319.
[66] Chang, J.-J., Hwang, D.-J., and Lin, M.-C. (1997) Some extended results on the search for good convolutional codes. *IEEE Trans. Inf. Theory*, **43** (6), 1682–1697.
[67] Bocharova, I. E. and Kudryashov, B. D. (1997) Rational rate punctured convolutional codes for soft-decision Viterbi decoding. *IEEE Trans. Inf. Theory*, **43** (4), 1305–1313.
[68] Frenger, P. K., Orten, P., and Ottosson, T. (2001) Comments and additions to recent papers on new convolutional codes. *IEEE Trans. Inf. Theory*, **47** (3), 1199–1201.
[69] Berrou, C., Glavieux, A., and Thitimajshima, P. (1993) Near Shannon limit error-correcting coding and decoding: Turbo codes. IEEE International Conference on Communications (ICC), May 1993, Geneva, Switzerland.
[70] Dahlman, E., Parkvall, S., Sköld, J., and Beming, P. (2008) *3G Evolution: HSPA and LTE for Mobile Broadband*, 2nd edn, Academic Press.
[71] ETSI (2009) Digital video broadcasting (DVB); Frame structure channel coding and modulation for a second generation digital terrestrial television broadcasting system (DVB-T2). Technical Report ETSI EN 301 790 V1.5.1 (2009-05), ETSI.
[72] Divsalar, D. and Pollara, F. (1995) Turbo codes for deep-space communications. TDA Progress Report 42-120, Jet Propulsion Laboratory, Pasadena, CA, pp. 29–39.
[73] Divsalar, D. and Pollara, F. (1995) On the design of turbo codes. TDA Progress Report 42-123, Jet Propulsion Laboratory, Pasadena, CA, pp. 99–121.

3

Decoding

As explained in Chapter 2, the information bits that were mapped onto the codewords need to be correctly recovered at the receiver. This is the very essence of data transmission, where the role of the receiver is to "decode" the information bits contained in the received signals.

This chapter studies different decoding strategies and is organized as follows. The optimal *maximum a posteriori* (MAP) and *maximum likelihood* (ML) decoding strategies are introduced in Section 3.1 where we also introduce the L-values which become a key element used throughout this book. The *bit-interleaved coded modulation* (BICM) decoder is defined in Section 3.2. In Section 3.3, some important properties of L-values are discussed and we conclude with a discussion about hard-decision decoding in Section 3.4.

3.1 Optimal Decoding

In this section, we review well-known decoding rules. They will be compared in Section 3.2 to the BICM decoding rule, which will be based on the L-values which we will define shortly.

Definition 3.1 (MAP Decoding) *The MAP decoding rule is defined as*

$$\hat{\underline{\boldsymbol{x}}}^{\text{map}} \triangleq \underset{\underline{\boldsymbol{x}}\in\mathcal{X}}{\text{argmax}}\{P_{\underline{\boldsymbol{X}}|\underline{\boldsymbol{Y}}}(\underline{\boldsymbol{x}}|\underline{\boldsymbol{y}})\} \tag{3.1}$$

$$= \underset{\underline{\boldsymbol{x}}\in\mathcal{X}}{\text{argmax}}\{\log P_{\underline{\boldsymbol{X}}|\underline{\boldsymbol{Y}}}(\underline{\boldsymbol{x}}|\underline{\boldsymbol{y}})\}, \tag{3.2}$$

where $\mathcal{X}$ is the set of symbol sequences (codewords) defined in Section 2.2.

The logarithm is used in (3.2) to remove an exponential function which will ultimately appear in the expressions when a Gaussian channel is considered. As the logarithm is a monotonically increasing function, it does not change the maximization result in (3.1).

From Bayes' rule we obtain

$$P_{\underline{\boldsymbol{X}}|\underline{\boldsymbol{Y}}}(\underline{\boldsymbol{x}}|\underline{\boldsymbol{y}}) = \frac{P_{\underline{\boldsymbol{X}}}(\underline{\boldsymbol{x}})p_{\underline{\boldsymbol{Y}}|\underline{\boldsymbol{X}}}(\underline{\boldsymbol{y}}|\underline{\boldsymbol{x}})}{p_{\underline{\boldsymbol{Y}}}(\underline{\boldsymbol{y}})}$$

$$= \frac{P_{\underline{\boldsymbol{X}}}(\underline{\boldsymbol{x}})}{p_{\underline{\boldsymbol{Y}}}(\underline{\boldsymbol{y}})}\prod_{n=1}^{N_\text{s}} p_{\boldsymbol{Y}|\boldsymbol{X}}(\boldsymbol{y}[n]|\boldsymbol{x}[n]), \tag{3.3}$$

Bit-Interleaved Coded Modulation: Fundamentals, Analysis, and Design, First Edition.
Leszek Szczecinski and Alex Alvarado.

where the factorization into the product of N_s conditional channel transition probabilities is possible thanks to the assumption of transmission over a memoryless channel. Using the fact that $p_{\underline{\boldsymbol{Y}}}(\underline{\boldsymbol{y}})$ in (3.3) does not depend on $\underline{\boldsymbol{x}}$, we can express (3.2) as

$$\hat{\underline{\boldsymbol{x}}}^{\mathrm{map}} = \operatorname*{argmax}_{\underline{\boldsymbol{x}}\in\mathcal{X}} \left\{ \log P_{\underline{\boldsymbol{X}}}(\underline{\boldsymbol{x}}) + \sum_{n=1}^{N_s} \log \left(p_{\boldsymbol{Y}|\boldsymbol{X}}(\boldsymbol{y}[n]|\boldsymbol{x}[n]) \right) \right\}. \tag{3.4}$$

Throughout this book, the information bits are assumed to be *independent and uniformly distributed* (i.u.d.). This, together with the one-to-one mapping performed by the *coded modulation* (CM) encoder (see Section 2.2 for more details), means that the codewords $\underline{\boldsymbol{x}}$ are equiprobable. This leads to the definition of the ML decoding rule.

Definition 3.2 (ML Decoding) *The ML decoding rule is defined as*

$$\hat{\underline{\boldsymbol{x}}}^{\mathrm{ml}} \triangleq \operatorname*{argmax}_{\underline{\boldsymbol{x}}\in\mathcal{X}} \{ \log p_{\underline{\boldsymbol{Y}}|\underline{\boldsymbol{X}}}(\underline{\boldsymbol{y}}|\underline{\boldsymbol{x}}) \} \tag{3.5}$$

$$= \operatorname*{argmax}_{\underline{\boldsymbol{x}}\in\mathcal{X}} \left\{ \sum_{n=1}^{N_s} \log p_{\boldsymbol{Y}|\boldsymbol{X}}(\boldsymbol{y}[n]|\boldsymbol{x}[n]) \right\}. \tag{3.6}$$

As the mapping between code bits and symbols is bijective and memoryless (at a symbol level), we can write the ML rule in (3.6) at a *bit level* as

$$\hat{\underline{\boldsymbol{b}}}^{\mathrm{ml}} = \operatorname*{argmax}_{\underline{\boldsymbol{b}}\in\mathcal{B}} \left\{ \sum_{n=1}^{N_s} \log p_{\boldsymbol{Y}|\boldsymbol{B}}(\boldsymbol{y}[n]|\boldsymbol{b}[n]) \right\}. \tag{3.7}$$

For the *additive white Gaussian noise* (AWGN) channel with conditional channel transition probabilities given in (2.31), the optimal decoding rule in (3.7) can be expressed as

$$\hat{\underline{\boldsymbol{b}}}^{\mathrm{ml}} = \operatorname*{argmin}_{\underline{\boldsymbol{b}}\in\mathcal{B}} \left\{ \sum_{n=1}^{N_s} \mathsf{snr}\|\boldsymbol{y}[n] - \Phi(\boldsymbol{b}[n])\|^2 \right\}, \tag{3.8}$$

where $\Phi(\boldsymbol{b}[n]) = \boldsymbol{x}[n]$.

As the optimal operation of the decoder relies on exhaustive enumeration of the codewords via operators $\operatorname{argmax}_{\underline{\boldsymbol{x}}\in\mathcal{X}}$ or $\operatorname{argmax}_{\underline{\boldsymbol{b}}\in\mathcal{B}}$, the expressions we have shown are entirely general, and do not depend on the structure of the encoder. Moreover, we can express the ML decoding rule in the domain of the codewords generated by the binary encoder, i.e.,

$$\hat{\underline{\boldsymbol{c}}}^{\mathrm{ml}} = \operatorname*{argmax}_{\underline{\boldsymbol{c}}\in\mathcal{C}} \{ \log p_{\underline{\boldsymbol{Y}}|\underline{\boldsymbol{C}}}(\underline{\boldsymbol{y}}|\underline{\boldsymbol{c}}) \}, \tag{3.9}$$

where $\mathcal{C}$ is defined in Section 2.3.3.

Example 3.3 (Decoding for Binary Modulation) *Assume that bits taken from the output of the encoder* $c[n], n = 1, \ldots, N_c$ *are interleaved into the bits* $b_k[n], k = 1, \ldots, m, n = 1, \ldots, N_s$*, where* $N_c = mN_s$*. We consider a particular constellation where each bit* $b_k[n]$ *is mapped to a symbol from a binary constellation* $\mathcal{S} = \{-1, 1\}$ *using the rule* $x_k[n] = \Phi(b_k[n]) = 2b_k[n] - 1$*. This means that we treat* m *1D modulations as one modulation in a space of* $N = m$ *dimensions. At each time* n*, the symbols* $x_k[n], k = 1, \ldots, m$ *are transmitted, yielding observations* $y_k[n] = x_k[n] + z_k[n]$*, which using a vectorial notation can be expressed as* $\boldsymbol{y}[n] = \boldsymbol{x}[n] + \mathbf{z}[n]$*,* *where* $\boldsymbol{x}[n] = \Phi(\boldsymbol{b}[n]) = [\Phi(b_1[n]), \ldots, \Phi(b_m[n])]$. *Then* (3.8) *becomes*

$$\hat{\underline{\boldsymbol{b}}}^{\mathrm{ml}} = \underset{\underline{\boldsymbol{b}}\in\mathcal{B}}{\operatorname{argmax}} \left\{ \sum_{n=1}^{N_{\mathrm{s}}} \log p_{\boldsymbol{Y}|\boldsymbol{B}}(\boldsymbol{y}[n]|\boldsymbol{b}[n]) \right\} \tag{3.10}$$

$$= \underset{\underline{\boldsymbol{b}}\in\mathcal{B}}{\operatorname{argmax}} \left\{ \sum_{n=1}^{N_{\mathrm{s}}} \log p_{\boldsymbol{Y}|\boldsymbol{X}}(\boldsymbol{y}[n]|\Phi(\boldsymbol{b}[n])) \right\} \tag{3.11}$$

$$= \underset{\underline{\boldsymbol{b}}\in\mathcal{B}}{\operatorname{argmax}} \left\{ \sum_{n=1}^{N_{\mathrm{s}}} \sum_{k=1}^{m} \log p_{Y_k|X_k}(y_k[n]|\Phi(b_k[n])) \right\} \tag{3.12}$$

$$= \underset{\underline{\boldsymbol{b}}\in\mathcal{B}}{\operatorname{argmax}} \left\{ \sum_{n=1}^{N_{\mathrm{s}}} \sum_{k=1}^{m} \log p_{Y_k|B_k}(y_k[n]|b_k[n]) \right\}, \tag{3.13}$$

where (3.12) *follows from taking into account the fact that the bits* $b_k[n]$ *are transmitted over independent channels.*

Adding a constant term to the maximized function does not change the optimization results, so we can subtract the term $\log p_{Y_k|B_k}(y_k[n]|0)$, *which gives*

$$\hat{\underline{\boldsymbol{b}}}^{\mathrm{ml}} = \underset{\underline{\boldsymbol{b}}\in\mathcal{B}}{\operatorname{argmax}} \left\{ \sum_{n=1}^{N_{\mathrm{s}}} \sum_{k=1}^{m} \log p_{Y_k|B_k}(y_k[n]|b_k[n]) - \log p_{Y_k|B_k}(y_k[n]|0) \right\} \tag{3.14}$$

$$= \underset{\underline{\boldsymbol{b}}\in\mathcal{B}}{\operatorname{argmax}} \left\{ \sum_{n=1}^{N_{\mathrm{s}}} \sum_{k=1}^{m} \log \frac{p_{Y_k|B_k}(y_k[n]|b_k[n])}{p_{Y_k|B_k}(y_k[n]|0)} \right\} \tag{3.15}$$

$$= \underset{\underline{\boldsymbol{b}}\in\mathcal{B}}{\operatorname{argmax}} \left\{ \sum_{n=1}^{N_{\mathrm{s}}} \sum_{k=1}^{m} l_k[n] b_k[n] \right\}, \tag{3.16}$$

where to pass from (3.15) *to* (3.16) *we used the simple relationships*

$$b_k[n] \log \frac{p_{Y_k|B_k}(y_k[n]|1)}{p_{Y_k|B_k}(y_k[n]|0)} = \log \frac{p_{Y_k|B_k}(y_k[n]|b_k[n])}{p_{Y_k|B_k}(y_k[n]|0)} \tag{3.17}$$

and defined

$$l_k[n] = \log \frac{p_{Y_k|B_k}(y_k[n]|1)}{p_{Y_k|B_k}(y_k[n]|0)}. \tag{3.18}$$

Further, from (2.31) *we can easily calculate* $l_k[n] = 4\,\mathsf{snr}\, y_k[n]$.

As (3.16) *is equivalent to* (3.13), $l_k[n]$ *in* (3.18) *captures all the information about the likelihoods* $p_{\boldsymbol{Y}|\boldsymbol{B}}(\boldsymbol{y}[n]|\boldsymbol{b}[n])$ *which is relevant for the decoding. Moreover, the interleaving between the code bits* $\underline{c}$ *and the bits* $\underline{\boldsymbol{b}}$ *is bijective (i.e.,* $\underline{c} = \Pi^{-1}(\underline{\boldsymbol{b}})$*) thus, we can deinterleave the metrics* $\underline{\boldsymbol{l}} \triangleq [\boldsymbol{l}[1], \ldots, \boldsymbol{l}[N_{\mathrm{s}}]]$ *(where* $\boldsymbol{l}[n] = [l_1[n], \ldots, l_m[n]]^{\mathrm{T}}$*) into the corresponding metrics as* $\underline{\lambda} = \Pi^{-1}(\underline{\boldsymbol{l}})$, *and then rewrite* (3.16) *as*

$$\hat{\underline{c}}^{\mathrm{ml}} = \underset{\underline{c}\in\mathcal{C}}{\operatorname{argmax}} \left\{ \sum_{n=1}^{N_{\mathrm{c}}} c[n]\lambda[n] \right\}. \tag{3.19}$$

The decoding solution in (3.19) is indeed appealing: if the decoder is fed with the deinterleaved metrics $\lambda[n]$, it remains unaware of the form of the mapper $\Phi(\cdot)$, the channel model, noise distribution, demapper, and so on. In other words, the metrics $\lambda[n]$ act as an "interface" between the channel outcomes $\boldsymbol{y}[n]$ and the decoder. We call these metrics *L-values* and they are a core element in BICM decoding and bit-level information exchange. The underlying assumption for optimality of the processing based on the L-values is that the observations associated with each transmitted bit are independent (allowing us to pass from (3.11) to (3.12)). Of course, this assumption does not hold in general, nevertheless, the possibility of bit-level operation, and *binary decoding* in particular, may outweigh the eventual suboptimality of the resulting processing.

We note that most of the encoders/decoders used in practice are designed and analyzed using the simple transmission model of Example 3.3. In this way, the design of the encoder/decoder is done in abstraction of the observation $\boldsymbol{y}[n]$, which allows the designer to use well-studied and well-understood off-the-shelf binary encoders/decoders. The price to pay for the resulting design simplicity is that, when the underlying assumptions of the channel are not correct, the design is suboptimal. The optimization of the encoders taking into account the underlying channel for various transmission parameters are at the heart of Chapter 8.

3.2 BICM Decoder

As we have seen in Example 3.3, the decoding may be carried out optimally using L-values. This is the motivation for the following definition of the BICM decoder.

Definition 3.4 (BICM Decoder) *The BICM decoder shown in Fig. 2.7 uses the following* binary *decoding rule*

$$\hat{\underline{\boldsymbol{b}}}^{\text{bi}} \triangleq \underset{\underline{\boldsymbol{b}}\in\mathcal{B}}{\operatorname{argmax}} \left\{ \sum_{n=1}^{N_\text{s}} \sum_{k=1}^{m} b_k[n] l_k[n] \right\}, \tag{3.20}$$

where

$$l_k[n] \triangleq \log \frac{p_{\boldsymbol{Y}|B_k}(\boldsymbol{y}[n]|1)}{p_{\boldsymbol{Y}|B_k}(\boldsymbol{y}[n]|0)}. \tag{3.21}$$

Alternatively, (3.20) *can be equivalently formulated in the domain of code bits* $\underline{\boldsymbol{c}}$ *and deinterleaved L-values, i.e.,*

$$\hat{\underline{\boldsymbol{c}}}^{\text{bi}} \triangleq \underset{\underline{\boldsymbol{c}}\in\mathcal{C}}{\operatorname{argmax}} \left\{ \sum_{n=1}^{N_\text{c}} \sum_{q=1}^{n_\text{c}} c_q[n] \lambda_q[n] \right\}. \tag{3.22}$$

Using (3.17), the BICM decoding rule in (3.20) can be expressed as

$$\hat{\underline{\boldsymbol{b}}}^{\text{bi}} = \underset{\underline{\boldsymbol{b}}\in\mathcal{B}}{\operatorname{argmax}} \left\{ \sum_{n=1}^{N_\text{s}} \log \prod_{k=1}^{m} p_{\boldsymbol{Y}|B_k}(\boldsymbol{y}[n]|b_k[n]) \right\}, \tag{3.23}$$

which, compared to the optimal ML rule in (3.7), allows us to conclude that the BICM decoder uses the following approximation:

$$p_{\boldsymbol{Y}|\boldsymbol{B}}(\boldsymbol{y}[n]|\boldsymbol{b}[n]) \approx \prod_{k=1}^{m} p_{\boldsymbol{Y}|B_k}(\boldsymbol{y}[n]|b_k[n]). \tag{3.24}$$

As (3.24) shows, the decoding metric in BICM is, in general, not proportional to $P_{\boldsymbol{Y}|\boldsymbol{B}}(\boldsymbol{y}[n]|\boldsymbol{b}[n])$, and thus, the BICM decoder can be seen as a *mismatched* version of the ML decoder in (3.7). Furthermore, (3.24) shows that the BICM decoder assumes that $\boldsymbol{y}[n]$ is not affected by the bits $b_1[n], \ldots, b_{k-1}[n], b_{k+1}[n], \ldots, b_m[n]$. The approximation becomes the identity if the bits $b_k[n]$ are independently mapped into each of the dimensions of the constellation $\mathcal{S} \in \mathbb{R}^N$ as $x_k[n] = \Phi(b_k[n])$, i.e., we require $N \geq m$, cf. Example 3.3.

The principal motivation for using the BICM decoder in Definition 3.4 is due to the bit-level operation (i.e., the binary decoding). While, as we said, this decoding rule is ML-optimal only in particular cases, cf. Example 3.3, the possibility of using any binary decoder most often outweighs the eventual decoding mismatch/suboptimality.

We note also that we might have defined the BICM decoder via likelihoods as in (3.23); however, defining it in terms of L-values as in (3.20) does not entail any loss and is simpler to implement because (i) we avoid dealing with very small numbers appearing when likelihoods are multiplied and (ii) we move the formulation to the logarithmic domain, so that the decoder uses additions instead of multiplications, which are simpler to implement.

Example 3.5 (BICM Decoding in Quaternary Modulation) *For quaternary modulation ($m = 2$), the BICM decoding rule in* (3.23) *is*

$$\hat{\underline{\boldsymbol{b}}}^{\text{bi}} = \underset{\underline{\boldsymbol{b}} \in \mathcal{B}}{\operatorname{argmax}} \left\{ \sum_{n=1}^{N_s} \log \left(p_{\boldsymbol{Y}|B_1}(\boldsymbol{y}[n]|b_1[n]) \cdot p_{\boldsymbol{Y}|B_2}(\boldsymbol{y}[n]|b_2[n]) \right) \right\}. \tag{3.25}$$

For equally likely bits B_1 and B_2 and for a given time instant n (which we omit for notation simplicity), we express the argument of the logarithm in (3.25) *as*

$$p_{\boldsymbol{Y}|B_1}(\boldsymbol{y}|b_1)p_{\boldsymbol{Y}|B_2}(\boldsymbol{y}|b_2) = K_1 P_{B_1|\boldsymbol{Y}}(b_1|\boldsymbol{y}) P_{B_2|\boldsymbol{Y}}(b_2|\boldsymbol{y}), \tag{3.26}$$

where K_1 is a real constant. Furthermore, we have

$$\begin{aligned} P_{B_1|\boldsymbol{Y}}(b_1|\boldsymbol{y}) P_{B_2|\boldsymbol{Y}}(b_2|\boldsymbol{y}) &= P_{B_1|\boldsymbol{Y}}(b_1|\boldsymbol{y}) \left[P_{B_2|\boldsymbol{Y},B_1}(b_2|\boldsymbol{y}, b_1) + P_{B_2|\boldsymbol{Y},B_1}(b_2|\boldsymbol{y}, \bar{b}_1) \right] \\ &= P_{B_1,B_2|\boldsymbol{Y}}(b_1, b_2|\boldsymbol{y}) + P_{B_1|\boldsymbol{Y}}(b_1|\boldsymbol{y}) P_{B_2|\boldsymbol{Y},B_1}(b_2|\boldsymbol{y}, \bar{b}_1) \\ &= P_{\boldsymbol{B}|\boldsymbol{Y}}(\boldsymbol{b}|\boldsymbol{y}) + P_{B_1|\boldsymbol{Y}}(b_1|\boldsymbol{y}) P_{B_2|\boldsymbol{Y},B_1}(b_2|\boldsymbol{y}, \bar{b}_1). \end{aligned} \tag{3.27}$$

Combining (3.26) *and* (3.27)*, we obtain*

$$p_{\boldsymbol{Y}|B_1}(\boldsymbol{y}|b_1)p_{\boldsymbol{Y}|B_2}(\boldsymbol{y}|b_2) = K_2 p_{\boldsymbol{Y}|\boldsymbol{B}}(\boldsymbol{y}|\boldsymbol{b}) + K_1 P_{B_1|\boldsymbol{Y}}(b_1|\boldsymbol{y}) P_{B_2|\boldsymbol{Y},B_1}(b_2|\boldsymbol{y}, \bar{b}_1), \tag{3.28}$$

where K_2 is a real constant. The first term in the right-hand side *(r.h.s.) of* (3.28)[1] *corresponds in fact to the correct metric in* (3.7). *The metric in* (3.28), *however, includes a second term, which accounts for the suboptimality expressed in* (3.24).

As we will see, the expression in (3.20) is essential to analyze the performance of the BICM decoder. It will be used in Chapter 6 to study its error probability and in Chapter 7 to study the correction of suboptimal L-values. We will also use it in Chapter 8 to study *unequal error protection* (UEP).

[1] Disregarding the factor K_2 which is irrelevant for decoding.

3.3 L-values

In BICM, the L-values are calculated at the receiver and are used to convey information about the a posteriori probability of the transmitted bits. These signals are then used by the BICM decoder as we explain in Section 3.2.

The *logarithmic likelihood ratio* (LLR) or L-value for the kth bit in the nth transmitted symbol is given by

$$l_k[n] = \log \frac{p_{\boldsymbol{Y}|B_k}(\boldsymbol{y}[n]|1)}{p_{\boldsymbol{Y}|B_k}(\boldsymbol{y}[n]|0)} \tag{3.29}$$

$$= l_k^{\text{apo}}[n] - l_k^{\text{a}}, \tag{3.30}$$

where the *a posteriori* and *a priori* L-values are, respectively, defined as

$$l_k^{\text{apo}}[n] \triangleq \log \frac{P_{B_k|\boldsymbol{Y}}(1|\boldsymbol{y}[n])}{P_{B_k|\boldsymbol{Y}}(0|\boldsymbol{y}[n])}, \tag{3.31}$$

$$l_k^{\text{a}} \triangleq \log \frac{P_{B_k}(1)}{P_{B_k}(0)}. \tag{3.32}$$

We note that unlike $l_k[n]$ and $l_k^{\text{apo}}[n]$, l_k^{a} in (3.32) does not include a time index $[n]$. This is justified by the model imposed on the bits B_k, namely, that $B_k[1], \ldots, B_k[N_{\text{s}}]$ are *independent and identically distributed* (i.i.d.) random variables (but possibly distributed differently across the bit positions), and thus, $l_k^{\text{a}} \triangleq l_k^{\text{a}}[1] = \ldots = l_k^{\text{a}}[N_{\text{s}}]$. We mention that there are, nevertheless, cases when a priori information may vary with n. This occurs in *bit-interleaved coded modulation with iterative demapping* (BICM-ID), where the L-value calculated by the demapper may use a priori L-values obtained from the decoder. In those cases, we will use $l_k^{\text{a}}[n]$.

It is well accepted to use the name LLR to refer to $l_k[n]$, $l_k^{\text{apo}}[n]$ and l_k^{a}. However, strictly speaking, only $l_k[n]$ should be called an LLR. The other two quantities ($l_k^{\text{apo}}[n]$ and l_k^{a}) are a logarithmic ratio of a posteriori probabilities and a logarithmic ratio of a priori probabilities, respectively, and not *likelihood* ratios. That is why using the name "L-value" to denote $l_k[n]$, $l_k^{\text{apo}}[n]$ and l_k^{a} is convenient and avoids confusion. Furthermore, the L-value in (3.30) is sometimes called an *extrinsic* L-value to emphasize that it is calculated from the channel outcome without information about the a priori distribution of the bit the L-value is being calculated for.

To alleviate the notation, from now on we remove the time index n, i.e., the dependence of the L-values on n will remain implicit unless explicitly mentioned. Using (3.29) and the fact that $P_{B_k|\boldsymbol{Y}}(0|\boldsymbol{y}) + P_{B_k|\boldsymbol{Y}}(1|\boldsymbol{y}) = 1$, we obtain the following relation between the a posteriori probabilities of B_k and its L-value l_k^{apo}

$$P_{B_k|\boldsymbol{Y}}(b|\boldsymbol{y}) = \frac{\exp(b \cdot l_k^{\text{apo}})}{1 + \exp(l_k^{\text{apo}})}, \tag{3.33}$$

$$\log P_{B_k|\boldsymbol{Y}}(b|\boldsymbol{y}) = b \cdot l_k^{\text{apo}} + C. \tag{3.34}$$

An analogous relationship links the a priori L-value l_k^{a} and the a priori probabilities, i.e.,

$$P_{B_k}(b) = \frac{\exp(b \cdot l_k^{\text{a}})}{1 + \exp(l_k^{\text{a}})}, \tag{3.35}$$

$$\log P_{B_k}(b) = b \cdot l_k^{\text{a}} + C'. \tag{3.36}$$

Combining (3.33), (3.35), and (3.30) we obtain

$$p_{\boldsymbol{Y}|B_k}(\boldsymbol{y}|b) = \exp(b \cdot l_k)\frac{1+\exp(l_k^{\mathrm{a}})}{1+\exp(l_k^{\mathrm{apo}})}p_{\boldsymbol{Y}}(\boldsymbol{y}), \tag{3.37}$$

$$\log p_{\boldsymbol{Y}|B_k}(\boldsymbol{y}|b) = b \cdot l_k + C''. \tag{3.38}$$

The constants C, C', and C'' in the above equations are independent of b. Therefore, moving operations to the logarithmic domain highlights the relationship between the L-values and corresponding probabilities and/or the likelihood; this relationship, up to a proportionality factor, is in the same form for a priori, a posteriori, and extrinsic L-values.

We note that the L-values in (3.29) may alternatively be defined as

$$l_k = \log\frac{p_{\boldsymbol{Y}|B_k}(\boldsymbol{y}|0)}{p_{\boldsymbol{Y}|B_k}(\boldsymbol{y}|1)}, \tag{3.39}$$

which could have some advantages in practical implementations. For example, in a popular complement-to-two binary format, the *most significant bit* (MSB) carries the sign, i.e., when an MSB is equal to zero (0), it means that a number is positive, and when an MSB is equal to one (1), it means that the number is negative. Then, if (3.39) is used, the transmitted bit obtained via hard decisions can be recovered directly from the MSB. This, of course, has no impact on any theoretical consideration, and thus, we will always use the definition (3.29).

Example 3.6 (Uncoded MAP Estimate Based on L-values) *The relationship between the a posteriori probabilities and the L-values in* (3.33) *is shown in Fig. 3.1 where we see that for* $l_k^{\mathrm{apo}} \geq 0$ *we have* $P_{B_k|\boldsymbol{Y}}(1|\boldsymbol{y}) \geq P_{B_k|\boldsymbol{Y}}(0|\boldsymbol{y})$ *and for* $l_k^{\mathrm{apo}} < 0$ *we obtain* $P_{B_k|\boldsymbol{Y}}(1|\boldsymbol{y}) < P_{B_k|\boldsymbol{Y}}(0|\boldsymbol{y})$*. Thus, the sign of* l_k^{apo} *defines the MAP estimate of the transmitted bit:*

$$\hat{b}_k = \begin{cases} 1, & \text{if } l_k^{\mathrm{apo}} \geq 0 \\ 0, & \text{if } l_k^{\mathrm{apo}} < 0 \end{cases}, \tag{3.40}$$

which corresponds to the so-called "hard decision" on the bit B_k.

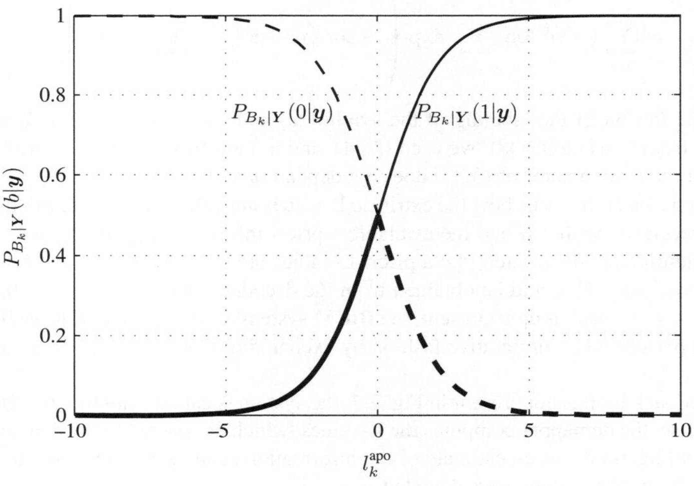

Figure 3.1 A posteriori probabilities $P_{B_k|\boldsymbol{Y}}(b|\boldsymbol{y})$ as a function of the L-value l_k^{apo}. The thick-line portion of the plot corresponds to the function (3.46)

The probability of error of the hard decision given an observation $\boldsymbol{Y} = \boldsymbol{y}$ is given by

$$P_{\mathrm{b},k}(\boldsymbol{y}) \triangleq \Pr\{\hat{B}_k \neq B_k | \boldsymbol{Y} = \boldsymbol{y}\} \tag{3.41}$$

$$= P_{\hat{B}_k|\boldsymbol{Y},B_k}(0|\boldsymbol{y},1)P_{B_k|\boldsymbol{Y}}(1|\boldsymbol{y}) + P_{\hat{B}_k|\boldsymbol{Y},B_k}(1|\boldsymbol{y},0)P_{B_k|\boldsymbol{Y}}(0|\boldsymbol{y}) \tag{3.42}$$

$$= \mathbb{I}_{[l_k^{\mathrm{apo}}<0]}P_{B_k|\boldsymbol{Y}}(1|\boldsymbol{y}) + \mathbb{I}_{[l_k^{\mathrm{apo}}\geq 0]}P_{B_k|\boldsymbol{Y}}(0|\boldsymbol{y}) \tag{3.43}$$

$$= \mathbb{I}_{[l_k^{\mathrm{apo}}<0]}\frac{\exp(l_k^{\mathrm{apo}})}{1+\exp(l_k^{\mathrm{apo}})} + \mathbb{I}_{[l_k^{\mathrm{apo}}\geq 0]}\frac{1}{1+\exp(l_k^{\mathrm{apo}})} \tag{3.44}$$

$$= \mathbb{I}_{[l_k^{\mathrm{apo}}<0]}\frac{1}{1+\exp(-l_k^{\mathrm{apo}})} + \mathbb{I}_{[l_k^{\mathrm{apo}}\geq 0]}\frac{1}{1+\exp(l_k^{\mathrm{apo}})} \tag{3.45}$$

$$= \frac{1}{1+\exp(|l_k^{\mathrm{apo}}|)}, \tag{3.46}$$

where (3.42) *follows from using the law of total probability,* (3.43) *from* (3.40) *and* (3.44) *from* (3.33).

According to (3.46)*, the probability of error $P_{\mathrm{b},k}(\boldsymbol{y})$ decreases when the magnitude of the L-value increases, and thus, the magnitude of the L-value represents the "reliability" of the hard decision. The function* (3.46) *is shown as a thick (solid or dashed) line in Fig. 3.1.*

Assuming that the bits B_k are independent[2] (see (2.72)), the L-values in (3.29) for the AWGN channel with conditional *probability density function* (PDF) given by (2.31) can be calculated as

$$l_k = \log \frac{\sum_{i\in\mathcal{I}_{k,1}} p_{\boldsymbol{Y}|\boldsymbol{X}}(\boldsymbol{y}|\boldsymbol{x}_i)P_{\boldsymbol{X}|B_k}(\boldsymbol{x}_i|1)}{\sum_{i\in\mathcal{I}_{k,0}} p_{\boldsymbol{Y}|\boldsymbol{X}}(\boldsymbol{y}|\boldsymbol{x}_i)P_{\boldsymbol{X}|B_k}(\boldsymbol{x}_i|0)} \tag{3.47}$$

$$= \log\left(\frac{\sum\limits_{i\in\mathcal{I}_{k,1}} \exp\left(-\,\mathsf{snr}\,\|\boldsymbol{y}-\boldsymbol{x}_i\|^2\right)\prod\limits_{\substack{k'=1\\k'\neq k}}^{m} P_{B_{k'}}(q_{i,k'})}{\sum\limits_{i\in\mathcal{I}_{k,0}} \exp\left(-\,\mathsf{snr}\,\|\boldsymbol{y}-\boldsymbol{x}_i\|^2\right)\prod\limits_{\substack{k'=1\\k'\neq k}}^{m} P_{B_{k'}}(q_{i,k'})}\right) \tag{3.48}$$

$$= \sum_{b\in\mathbb{B}}(-1)^{\bar{b}}\log\sum_{i\in\mathcal{I}_{k,b}}\exp\left(-\,\mathsf{snr}\,\|\boldsymbol{y}-\boldsymbol{x}_i\|^2 + \sum_{\substack{k'=1\\k'\neq k}}^{m} q_{i,k'}l_{k'}^{\mathrm{a}}\right), \tag{3.49}$$

where $q_{i,k}$ is the kth bit of the labeling of the symbol $\boldsymbol{x}_i$ and $l_{k'}^{\mathrm{a}}$ is the a priori information given by (3.32). To pass from (3.47) to (3.48), we used (2.74), and to pass from (3.48) to (3.49), we used (3.33) and the fact that the denominator of (3.33) does not depend on b.

The expression (3.49) shows us how the extrinsic L-values are calculated for the kth bit in the symbol, from both the received symbol $\boldsymbol{y}$ and the available a priori information $l_{k'}^{\mathrm{a}}$ for $k' \neq k$. When the code bits B_k are nonuniformly distributed, the a priori L-values are obtained from the a priori distributions of the bits. Alternatively, $l_{k'}^{\mathrm{a}}$ could be obtained from the decoder, which is shown by the dotted lines in Fig. 2.7. When this feedback loop is present, the BICM system is referred to as BICM-ID. In BICM-ID, the decoding is performed in an iterative fashion by exchanging L-values between the decoder and the demapper.

When the feedback loop is not present in Fig. 2.7, the system is called a noniterative BICM, or simply BICM. In this case, the demapper computes the L-values, which are passed to the deinterleaver and then to the decoder, which produces an estimate of the information sequence $\hat{\underline{i}}$. When the bits are uniformly distributed, the extrinsic L-values are calculated as

$$l_k = \sum_{b\in\mathbb{B}}(-1)^{\bar{b}}\log\sum_{i\in\mathcal{I}_{k,b}}\exp\left(-\mathsf{snr}\,\|\boldsymbol{y}-\boldsymbol{x}_i\|^2\right). \tag{3.50}$$

[2] We discussed the independence of the bits in Section 2.6.1.

This expression shows how the demapper Θ computes the L-values, which depends on the received signal $\boldsymbol{y}$, the constellation and its binary labeling, and the SNR.

Having clarified the importance of the L-values for the decoding, we will now have a closer look at their properties and models. A detailed probabilistic modeling of the L-values is covered in Chapter 5.

3.3.1 PDF of L-values for Binary Modulation

The L-value l_k in (3.50) is a function of the received symbol $\boldsymbol{y}$. For a given transmitted symbol $\boldsymbol{X} = \boldsymbol{x}_i$, the received symbol $\boldsymbol{y}$ is a random variable $\boldsymbol{Y} = \boldsymbol{x}_i + \boldsymbol{Z}$, and thus, the L-value is also a random variable L_k. In this section, we show how to compute the PDF of the L-values for an arbitrary binary modulation. These developments are an introduction to more general results we present in Chapter 5.

For notation simplicity, we use $\theta_k(\boldsymbol{y})$ to denote the functional relationship between the L-value l_k and the (random) received symbol $\boldsymbol{y}$, i.e., $\theta_k(\boldsymbol{y}) = l_k$. Even though binary modulation may be analyzed in one dimension, we continue to use a vectorial notation to be consistent with the notation used in Chapter 5, where we develop PDFs of 2D constellations.

We consider binary modulation over the AWGN channel (i.e., $M = 2$ and $h = 1$) where the constellation $\mathcal{S} = \{\boldsymbol{x}_1, \boldsymbol{x}_2\}$ consists of two elements, and where the labeling is $\mathbf{Q} = [0, 1]$, i.e., $\boldsymbol{x}_1 = \Phi(0)$ and $\boldsymbol{x}_2 = \Phi(1)$. In this case, (3.50) becomes

$$l_1 = \theta_1(\boldsymbol{y}) = \log \frac{\exp\left(-\ \mathsf{snr}\, \|\boldsymbol{y} - \boldsymbol{x}_2\|^2\right)}{\exp\left(-\ \mathsf{snr}\, \|\boldsymbol{y} - \boldsymbol{x}_1\|^2\right)} \tag{3.51}$$

$$= \mathsf{snr}\left(\|\boldsymbol{y} - \boldsymbol{x}_1\|^2 - \|\boldsymbol{y} - \boldsymbol{x}_2\|^2\right) \tag{3.52}$$

$$= 2\ \mathsf{snr}\left(-\ \langle \boldsymbol{x}_1, \boldsymbol{y}\rangle + \langle \boldsymbol{x}_2, \boldsymbol{y}\rangle + \frac{\|\boldsymbol{x}_1\|^2 - \|\boldsymbol{x}_2\|^2}{2}\right). \tag{3.53}$$

By introducing[3]

$$\xi_{j,i}(\boldsymbol{y}) \triangleq \boldsymbol{d}_{j,i}^{\mathrm{T}}(\boldsymbol{y} - \overline{\boldsymbol{x}}_{j,i}), \tag{3.54}$$

$$\overline{\boldsymbol{x}}_{j,i} \triangleq \frac{1}{2}(\boldsymbol{x}_i + \boldsymbol{x}_j), \tag{3.55}$$

where $\boldsymbol{d}_{j,i}$ is given by (2.39), we obtain from (3.53)

$$\theta_1(\boldsymbol{y}) = 2\ \mathsf{snr}\ \xi_{2,1}(\boldsymbol{y}). \tag{3.56}$$

Given a transmitted symbol $\boldsymbol{X} = \boldsymbol{x}_i$, $\boldsymbol{Y} = \boldsymbol{x}_i + \boldsymbol{Z}$, and thus, (3.56) can be expressed as

$$\theta_1(\boldsymbol{Y}) = \theta_1(\boldsymbol{x}_i + \boldsymbol{Z}) \tag{3.57}$$

$$= 2\ \mathsf{snr}\ \boldsymbol{d}_{2,1}^{\mathrm{T}}\left(\boldsymbol{x}_i + \boldsymbol{Z} - 0.5(\boldsymbol{x}_1 + \boldsymbol{x}_2)\right) \tag{3.58}$$

$$= 2\ \mathsf{snr}\left(\frac{\boldsymbol{d}_{2,1}^{\mathrm{T}}(2\boldsymbol{x}_i - \boldsymbol{x}_1 - \boldsymbol{x}_2)}{2} + \boldsymbol{d}_{2,1}^{T}\boldsymbol{Z}\right) \tag{3.59}$$

$$= 2\ \mathsf{snr}\left((-1)^i \frac{\|\boldsymbol{d}_{2,1}\|^2}{2} + \boldsymbol{d}_{2,1}^{\boldsymbol{T}}\boldsymbol{Z}\right). \tag{3.60}$$

[3] It is worth mentioning at this point that the expressions in (3.54) and (3.55) are in fact quite important, as they will be used in different chapters of this book.

As for binary modulation we have $(-1)^i = (-1)^{\overline{b}}$, we can express (3.60) conditioned on the transmitted bit $B_1 = b$ as

$$\theta_1(\boldsymbol{Y}) = 2\ \mathsf{snr}\left((-1)^{\overline{b}}\frac{\|\boldsymbol{d}_{2,1}\|^2}{2} + \mathbf{d}_{2,1}^{\mathrm{T}}\boldsymbol{Z}\right) \tag{3.61}$$

$$= (-1)^{\overline{b}}\mathsf{snr}\,\|\boldsymbol{d}_{2,1}\|^2 + 2\ \mathsf{snr}\boldsymbol{d}_{2,1}^{\mathrm{T}}\boldsymbol{Z}. \tag{3.62}$$

To calculate the PDF of the L-value $L_k = \theta_k(\boldsymbol{Y})$ conditioned on the transmitted bit, we recognize (3.62) as a linear transformation of the Gaussian vector $\boldsymbol{Z} = [Z_1, Z_2, \ldots, Z_N]^{\mathrm{T}}$, where Z_n are i.i.d. $Z_n \sim \mathcal{N}(0, \mathsf{N}_0/2)$. So we conclude that L_1 is a Gaussian random variable, $L_1 \sim \mathcal{N}\left((-1)^{\overline{b}}\mu_L, \sigma_L^2\right)$.

$$p_{L_1|B_1}(l|b) = \Psi(l, (-1)^{\overline{b}}\mu_L, \sigma_L^2), \tag{3.63}$$

where

$$\mu_L = \mathsf{snr}\,\|\boldsymbol{d}_{2,1}\|^2, \qquad \sigma_L^2 = 2\mu_L. \tag{3.64}$$

The results in (3.63) and (3.64) show that transmitting any N-dimensional binary modulation over the AWGN channel yields L-values whose PDF is Gaussian, with the parameters fully determined by the *squared Euclidean distance* (SED) between the two constellation points. The following example shows the results for the case $N = 1$, i.e., 2PAM (*binary phase shift keying* (BPSK)).

Example 3.7 (PDF of L-values for 2PAM) *Using 2PAM we have $\boldsymbol{x}_1 = -1$ and $\boldsymbol{x}_2 = 1$, and $\boldsymbol{d}_{2,1} = 2$. The linear transformation (3.56) is then given by*

$$l_1 = 4\ \mathsf{snr}\ y, \tag{3.65}$$

which is shown in Fig. 3.2. Note that this is the L-value we also found in Example 3.3. As $N = 1$, we temporarily remove the vectorial notation and use Y and y instead of $\boldsymbol{Y}$ and $\boldsymbol{y}$.

Using (3.63), we find

$$p_{L_1|B_1}(l|b) = \Psi(l, (-1)^{\overline{b}}4\ \mathsf{snr}, 8\ \mathsf{snr}). \tag{3.66}$$

In Fig. 3.2, we also plot the PDF $p_{Y|\mathbf{X}}(y|\boldsymbol{x}_2)$ and the PDF of the L-values (3.63) conditioned on $B_1 = 1$.

The result in (3.63) shows that the PDF of the L-values for binary modulation is Gaussian and this form of the PDF is often used for modeling of the L-values in different contexts. In reality, the PDF is not Gaussian in the case of multilevel modulations, as we will show in Chapter 5.

3.3.2 Fundamental Properties of the L-values

In what follows, we discuss two fundamental properties of the PDF of the L-values which will become useful in the following chapters.

Definition 3.8 (Consistency) *A PDF $p_{L|B}(l|b)$ with $b \in \mathbb{B}$ is said to be consistent if for any $l \in \mathbb{R}$*

$$p_{L|B}(l|1) = p_{L|B}(l|0)\mathrm{e}^{l}. \tag{3.67}$$

As we will see later, consistency is an inherent property of the L-values if they are calculated via (3.29).

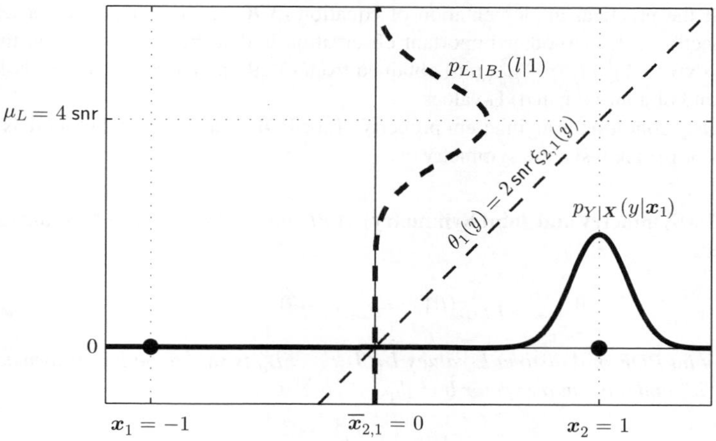

Figure 3.2 The L-values for binary modulation are known to be Gaussian distributed as they are a linear transformation of a Gaussian random variable

Example 3.9 (Consistency Condition for 2PAM) *The consistency condition in (3.67) for the L-values whose PDF is given by (3.63) can be expressed as*

$$\log \frac{p_{L_1|B_1}(l|1)}{p_{L_1|B_1}(l|0)} = \log \frac{\exp\left(-\frac{(l-\mu_L)^2}{2\sigma_L^2}\right)}{\exp\left(-\frac{(l+\mu_L)^2}{2\sigma_L^2}\right)} = \frac{2\mu_L}{\sigma_L^2} l = l, \tag{3.68}$$

and thus, the consistency condition (3.67) is satisfied for $l \in \mathbb{R}$.

Theorem 3.10 *The conditional PDF $p_{L_k|B_k}(l|b)$ is consistent if l_k is calculated via (3.29).*

Proof: As $l_k = \theta_k(\boldsymbol{y})$, the joint PDF of $\boldsymbol{Y}$ and L_k is given by

$$p_{\boldsymbol{Y},L_k|B_k}(\boldsymbol{y}, l_k|b) = p_{\boldsymbol{Y}|B_k}(\boldsymbol{y}|b)\mathbb{I}_{[l_k=\theta_k(\boldsymbol{y})]}. \tag{3.69}$$

To obtain $p_{L_k|B_k}(l|b)$, we marginalize (3.69) over $\boldsymbol{Y}$

$$\begin{aligned} p_{L_k|B_k}(l_k|b) &= \int p_{\boldsymbol{Y},L_k|B_k}(\boldsymbol{y}, l_k|b)\,\mathrm{d}\boldsymbol{y} \\ &= \int_{\boldsymbol{y}:\theta_k(\boldsymbol{y})=l_k} p_{\boldsymbol{Y}|B_k}(\boldsymbol{y}|b)\,\mathrm{d}\boldsymbol{y}. \end{aligned} \tag{3.70}$$

Further, using (3.29) we have $p_{\boldsymbol{Y}|B_k}(\boldsymbol{y}|b) = \exp((-1)^{\overline{b}} \cdot l_k)p_{\boldsymbol{Y}|B_k}(\boldsymbol{y}|\overline{b})$, and thus, (3.70) becomes

$$p_{L_k|B_k}(l_k|b) = \int_{\boldsymbol{y}:\theta_k(\boldsymbol{y})=l_k} \exp((-1)^{\overline{b}} \cdot l_k)p_{\boldsymbol{Y}|B_k}(\boldsymbol{y}|\overline{b})\,\mathrm{d}\boldsymbol{y} \tag{3.71}$$

$$= \exp((-1)^{\overline{b}} \cdot l_k)\int_{\boldsymbol{y}:\theta_k(\boldsymbol{y})=l_k} p_{\boldsymbol{Y}|B_k}(\boldsymbol{y}|\overline{b})\,\mathrm{d}\boldsymbol{y} \tag{3.72}$$

$$= \exp((-1)^{\overline{b}} \cdot l_k)p_{L_k|B_k}(l_k|\overline{b}), \tag{3.73}$$

where, to pass from (3.72) to (3.73), we used (3.70). □

We note that the practical implementation of equation (3.70) is, in general, rather tedious, as will be shown in Section 5.1.2. Another important observation is that the proof relies on the relationship $p_{\boldsymbol{Y}|B_k}(\boldsymbol{y}|b) = \exp((-1)^{\bar{b}} \cdot l_k) p_{\boldsymbol{Y}|B_k}(\boldsymbol{y}|\bar{b})$ obtained from (3.29), and thus, it does not hold for max-log (or any other kind of approximated) L-values.

The consistency condition is an inherent property of the PDF of an L-value, but there is another property important for the analysis: the symmetry.

Definition 3.11 (Symmetry and Joint Symmetry) *A PDF $p_{L|B}(l|b)$ with $b \in \mathbb{B}$ is said to be symmetric if for any $l \in \mathbb{R}$*

$$p_{L|B}(l|1) = p_{L|B}(-l|0). \tag{3.74}$$

Moreover, the joint PDF of d distinct L-values $L_1, L_2, \ldots, L_d$ is said to be jointly symmetric if for any $[l_1, \ldots, l_d]^{\mathrm{T}} \in \mathbb{R}^d$ and any binary vector $\boldsymbol{b} = [b_1, \ldots, b_d]^{\mathrm{T}} \in \mathbb{B}^d$,

$$\begin{aligned} &p_{L_1,\ldots,L_d|B_1,\ldots,B_d}(l_1, \ldots, l_d|b_1, \ldots, b_d) \\ &\quad = p_{L_1,\ldots,L_d|B_1,\ldots,B_d}((-1)^{b_1} l_1, \ldots, (-1)^{b_d} l_d|0, \ldots, 0). \end{aligned} \tag{3.75}$$

Clearly, a necessary condition for a PDF to satisfy the symmetry condition is that its support is symmetric with respect to zero.

Example 3.12 (Symmetry Condition for 2PAM) *The symmetry condition in* (3.74) *for the PDF in* (3.63) *is fulfilled, i.e.,*

$$\begin{aligned} p_{L_1|B_1}(-l|0) &= \frac{1}{\sqrt{2\pi\sigma_L^2}} \exp\left(-\frac{(-l+\mu_L)^2}{2\sigma_L^2}\right) \\ &= \frac{1}{\sqrt{2\pi\sigma_L^2}} \exp\left(-\frac{(l-\mu_L)^2}{2\sigma_L^2}\right) \\ &= p_{L_1|B_1}(l|1). \end{aligned} \tag{3.76}$$

Intuitively, this can be understood from the symmetry of the input distribution $\boldsymbol{x}_1 = -\boldsymbol{x}_2$, i.e., $p_{\boldsymbol{Y}|\boldsymbol{X}}(\boldsymbol{y}|\boldsymbol{x}_1) = p_{\boldsymbol{Y}|\boldsymbol{X}}(-\boldsymbol{y}|\boldsymbol{x}_2)$, and by the fact that the L-values in this case are a linear transformation of $\boldsymbol{Y}$.

The symmetry condition in Definition 3.11 can be seen as a "natural" extension of the symmetry of the 2PAM case in Example 3.12. As we will see, this condition is not always satisfied; however, it it is convenient to enforce it as then, the analysis of the BICM receiver is simplified. To this end, we assume that before being passed to the mapper, the bits b_k are scrambled, as shown in Fig. 3.3, i.e., the bits at the input of the mapper are a result of a binary *exclusive OR* (XOR) operation

$$b'_k[n] = s_k[n] \oplus b_k[n] \tag{3.77}$$

with bits s_k generated in a pseudorandom manner. We model these scrambled bits as i.i.d. random variables S_k with $P_{S_k}(1) = P_{S_k}(0) = \frac{1}{2}$.

We assume that both the transmitter and the receiver know the scrambling sequence $\underline{\boldsymbol{s}}$, as schematically shown in Fig. 3.3. Thus, the "scrambling" and the "descrambling" may be considered, respectively, as part of the mapper and the demapper, and consequently, are not shown in the BICM block diagrams. In fact, the scrambling we described is often included in practical communication systems.

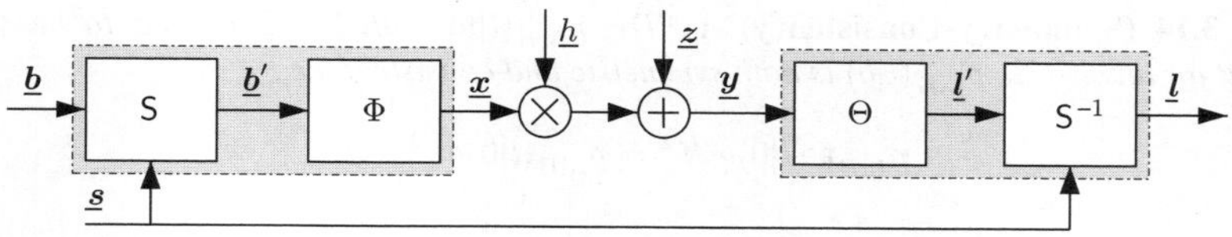

Figure 3.3 The sequence of pseudorandom bits $\underline{s}$ is known by the scrambler S and the descrambler S^{-1}. The dashed-dotted boxes indicate that these two processing units may be considered as a part of the mapper and demapper

The L-values at the output of the descrambler can be expressed as

$$l_k = \log \frac{p_{\boldsymbol{Y}|B_k}(\boldsymbol{y}|1)}{p_{\boldsymbol{Y}|B_k}(\boldsymbol{y}|0)} \tag{3.78}$$

$$= \log \frac{p_{\boldsymbol{Y}|B'_k}(\boldsymbol{y}|s_k \oplus 1)}{p_{\boldsymbol{Y}|B'_k}(\boldsymbol{y}|s_k \oplus 0)} \tag{3.79}$$

$$= (-1)^{s_k} \log \frac{p_{\boldsymbol{Y}|B'_k}(\boldsymbol{y}|1)}{p_{\boldsymbol{Y}|B'_k}(\boldsymbol{y}|0)}, \tag{3.80}$$

which gives

$$l_k = (-1)^{s_k} l'_k, \tag{3.81}$$

where l'_k is the L-value obtained for the bits b'_k. From (3.81), we see that the operation of the descrambler simply corresponds to changing the sign of the L-value l'_k according to the scrambling sequence.

Theorem 3.13 *For any PDF $p_{L'_k|B'_k}(l|b)$, the PDF of random variables L_k modeling the L-values l_k after descrambling in Fig. 3.3 is symmetric.*

Proof: We first note that because of (3.77) and (3.81), the following holds

$$p_{L_k|B_k,S_k}(l|b,s) = p_{L'_k|B'_k}((-1)^s l|b \oplus s), \tag{3.82}$$

so the PDF of the L-value L_k is obtained as

$$p_{L_k|B_k}(l|b) = \sum_{s \in \mathbb{B}} p_{L_k|B_k,S_k}(l|b,s) P_{S_k}(s) \tag{3.83}$$

$$= \frac{1}{2}[p_{L'_k|B'_k}(l|b) + p_{L'_k|B'_k}(-l|\bar{b})], \tag{3.84}$$

where (3.83) follows from marginalizing over S_k and (3.84) from (3.82). Then, $p_{L_k|B_k}(l|b)$ must fulfill (3.81) as a direct consequence of (3.84). □

Thus, thanks to the scrambler, the PDF of the L-values l_k in Fig. 3.3 are symmetric regardless of the form of the PDF $p_{L'_k|B'_k}(l|b)$. In analogy to the above development, we conclude that the joint symmetry condition (3.75) is also satisfied if the scrambler is used.

Definition 3.14 (Symmetry-Consistency) *A PDF $p_{L|B}(l|b)$ with $b \in \mathbb{B}$ is said to be symmetric-consistent if for any $l \in \mathbb{R}$, $p_{L|B}(l|b)$ is both symmetric and consistent, i.e.,*

$$p_{L|B}(-l|0)\mathrm{e}^{-l/2} = p_{L|B}(l|0)\mathrm{e}^{l/2} \tag{3.85}$$

and

$$p_{L|B}(l|1)\mathrm{e}^{-l/2} = p_{L|B}(-l|1)\mathrm{e}^{l/2}. \tag{3.86}$$

The symmetry-consistency condition (also referred to as the exponential symmetric condition) in Definition 3.14 will be used in Chapters 6 and 7. Clearly, the PDF of the L-values for 2PAM fulfills this condition.

Example 3.15 (Consistency Condition with Mismatched SNR) *Consider a receiver for* 2PAM, *where the signal-to-noise ratio (SNR) is not perfectly known, i.e., the receiver uses* $\hat{\mathsf{snr}}$ *instead of the true value* snr. *In this case,* (3.50) *is*

$$\tilde{l}_1 = \tilde{\theta}_1(\boldsymbol{y}) = \log \frac{\exp(-\hat{\mathsf{snr}}\|\boldsymbol{y} - \boldsymbol{x}_2\|^2)}{\exp(-\hat{\mathsf{snr}}\|\boldsymbol{y} - \boldsymbol{x}_1\|^2)}, \tag{3.87}$$

which can be interpreted as a mismatch in the channel transition probability, i.e., the demapper computes the L-values using an incorrect $p_{\boldsymbol{Y}|\boldsymbol{X}}(\boldsymbol{y}|\boldsymbol{x})$. *Following a procedure analogous to* (3.52)–(3.62), *we obtain*

$$\tilde{L}_1 = \tilde{\theta}_1(\boldsymbol{Y}) = (-1)^{\bar{b}}\hat{\mathsf{snr}}\|\boldsymbol{d}_{2,1}\|^2 + 2\,\hat{\mathsf{snr}}\,\boldsymbol{d}_{2,1}^{\mathrm{T}}\boldsymbol{Z}. \tag{3.88}$$

In this case, the L-value is still a Gaussian random variable as in (3.63); *however, the mean value and the variance are given by*

$$\mu_{\tilde{L}} = (-1)^{\bar{b}} 4\,\hat{\mathsf{snr}}, \qquad \sigma_{\tilde{L}}^2 = 8\frac{\hat{\mathsf{snr}}^2}{\mathsf{snr}}. \tag{3.89}$$

Then, the PDF is defined by

$$p_{\tilde{L}_1|B_1}(l|b) = \Psi\left(l, (-1)^{\bar{b}} 4\,\hat{\mathsf{snr}}, 8\frac{\hat{\mathsf{snr}}^2}{\mathsf{snr}}\right). \tag{3.90}$$

In the case of a perfect SNR estimation ($\hat{\mathsf{snr}} = \mathsf{snr}$), (3.90) *becomes* (3.66).

The symmetry condition in this case still checks; however, the consistency condition can be shown to be

$$\log \frac{p_{\tilde{L}_1|B_1}(l|1)}{p_{\tilde{L}_1|B_1}(l|0)} = l\frac{\mathsf{snr}}{\hat{\mathsf{snr}}}, \tag{3.91}$$

and thus, is satisfied only if $\hat{\mathsf{snr}} = \mathsf{snr}$, *i.e., when the SNR is perfectly known.*

3.3.3 Approximations

The computation of l_k in (3.50) involves computing ($M = 2^m$ times) the function $\exp(\cdot)$. To deal with the exponential increase of the complexity as m grows, approximations are often used. One of the most common approaches stems from the factorization of the logarithm of a sum of exponentials via the so-called Jacobian logarithm, i.e.,

$$\log(\exp(x_1) + \exp(x_2)) = \max{}^*\{x_1, x_2\}, \tag{3.92}$$

where

$$\max{}^*\{x_1, x_2\} \triangleq \max\{x_1, x_2\} + g(|x_1 - x_2|), \tag{3.93}$$

with

$$g(z) = \log(1 + \exp(-z)), \quad z \geq 0. \tag{3.94}$$

The function $g(z)$ in (3.94) is shown in Fig. 3.4.

The "max-star" operation in (3.93) can be applied recursively to more than two arguments, i.e.,

$$\log \sum_{i=1}^{I} \exp(x_i) = \max{}^*\{x_1, \max{}^*\{x_2, \max{}^*\{x_3, \ldots\}\}\}. \tag{3.95}$$

The advantage of using (3.95) is that the function $g(z)$ in (3.94) fulfills $0 < g(z) \leq \log 2$, and thus, can be efficiently implemented via a lookup table or via an analytical approximation based on piecewise linear functions. This is interesting from an implementation point of view because the function exp() does not have to be explicitly evoked.

Example 3.16 (Approximating $\max^*\{x_1, x_2\}$**)** *To approximate* $\max^*\{x_1, x_2\}$ *we replace* (3.94) *by a piecewise linear function, i.e.,* $g(z) \approx g_1(z)$ *where*

$$g_1(z) \triangleq (b_1 - a_1 z)\mathbb{I}_{[a_1 z < b_1]}, \quad z \geq 0, \tag{3.96}$$

and where a_1 *and* b_1 *are appropriately chosen. As an example, the function* $g_1(z)$ *with* $a_1 = 0.2$ *and* $b_1 = 0.6$ *is shown in Fig. 3.4. A second, and even simpler alternative, is to approximate* $g(z)$ *via the windowed zero-order linear function* $g(z) \approx g_0(z)$ *where*

$$g_0(z) \triangleq b_2 \mathbb{I}_{[z < a_2]}, \quad z \geq 0. \tag{3.97}$$

In Fig. 3.4, we show $g_1(z)$ *in* (3.97) *with* $b_2 = 0.5 \cdot \log(2)$ *and* $a_2 = 2$. *We have chosen these parameters arbitrarily but in general they should be optimized. Defining the optimality criterion is not necessarily trivial, however. The issue of dealing with suboptimally calculated L-values is addressed in more detail in Chapter 7.*

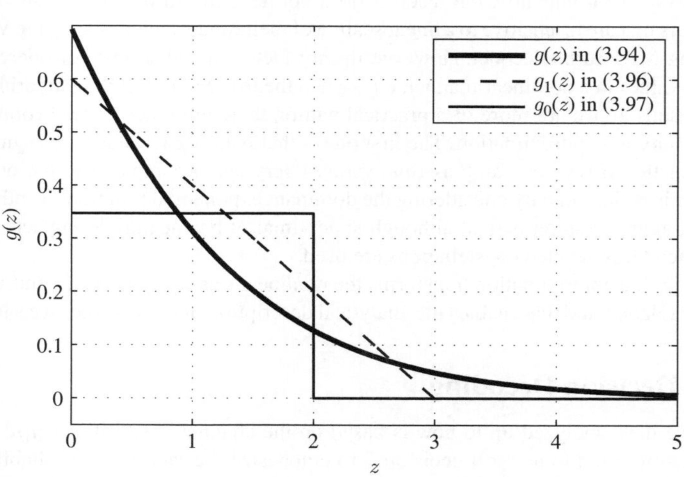

Figure 3.4 The function $g(z)$ in (3.94) and the two approximations in Example 3.16

Example 3.16 shows two simple ways of approximating $\max^*\{x_1, x_2\}$ in (3.93), and in general, other simplifications can be devised. Arguably, the simplest approximation is obtained ignoring $g(z)$, i.e., setting $g(z) = 0$ which used in (3.95) yields the popular *max-log approximation*

$$\log \sum_{i=1}^{I} \exp(x_i) \approx \max_{i} \{x_i\}. \tag{3.98}$$

By using (3.98) in (3.50), we obtain the following approximation for the L-values

$$l_k = \mathsf{snr} \sum_{b \in \mathbb{B}} (-1)^b \min_{i \in \mathcal{I}_{k,b}} \|\boldsymbol{y} - \boldsymbol{x}_i\|^2, \tag{3.99}$$

which are usually called *max-log L-values*.

Although the max-log L-values in (3.99) are suboptimal, they are frequently used in practice. One reason for their popularity is that some arithmetic operations are removed, and thus, the complexity of the L-values calculation is reduced. This is relevant from an implementation point of view and it explains why the max-log approach was recommended by the *third-generation partnership project* (3GPP) working groups. Most of the analysis and design of BICM transceivers in this book will be based on max-log L-values, which we will simply call L-values, despite their suboptimal calculation.

It is worthwhile mentioning that the max-log approximation given by (3.98) as well as the max-star approximation in (3.93) can also be used in the decoding algorithms of *turbo codes* (TCs), i.e., in the soft-input soft-output blocks described in Section 2.6.3. When MAP decoding is implemented in the *logarithmic domain* (Log-MAP) using the *Bahl–Cocke–Jelinek–Raviv* (BCJR) algorithm, computations involving logarithms of sum of terms appear. Consequently, the max-star and max-log approximations can be used to reduce the decoding complexity, which yields the Max-Log-MAP algorithm. Different approaches for simplified metric calculation have been investigated in the literature, for both the decoding algorithm and the metric calculation in the demapper. Some of the issues related to optimizing the simplified metrics' calculation in the demapper will be addressed in Chapter 7.

Another interesting feature of the max-log approximation will become apparent on the basis of the discussion in Section 3.2. As we showed in (3.20), the decisions made by a BICM decoder are based on calculating linear combinations of the L-values. Then, a linear scaling of all the L-values by a common factor is irrelevant to the decoding decision, so we may remove the factor snr in (3.99) without affecting the decoding results. Consequently, the receiver does not need to estimate the noise variance N_0. The family of decoders that are insensitive to a linear scaling of their inputs includes, e.g., the Viterbi algorithm and the Max-Log-MAP turbo decoder. However, the snr factor should still be considered if the decoder processes the L-values in a nonlinear manner, e.g., when the true MAP decoding algorithm is used.

While the reasons above are more of a practical nature, there are other more theoretical reasons for considering the max-log approximation. The first one is that max-log L-values are asymptotically tight, i.e., when one of the SEDs $\|\boldsymbol{y} - \boldsymbol{x}_i\|^2$ is (on average) very small compared to the others (i.e., when $\mathsf{snr} \to \infty$), there is no loss in only considering the dominant exponential function. Another reason favoring the max-log approximation is that, although suboptimal, it has negligible impact on the receiver's performance when Gray-labeled constellations are used.

Finally, the max-log approximation transforms the nonlinear relation between l_k and $\boldsymbol{y}$ in (3.50) into a piecewise linear relation and this enables the analytical description of the L-values we show in Chapter 5.

3.4 Hard-Decision Decoding

The decoding we have analyzed up to now is based on the channel observations $\boldsymbol{y}[n]$ or the L-values $l_k[n]$. This is also referred to as "soft decisions" to emphasize the fact that the reliability of the MAP estimate of the transmitted bits may be deduced from their amplitude (see Example 3.6).

By ignoring the reliability, we may still detect the transmitted symbols/bits before decoding. This is mostly considered for historical reasons, but still may be motivated by the simplicity of the resulting processing. The idea is to transform the hard decisions on $\boldsymbol{y}[n]$ or $\boldsymbol{l}[n]$ into MAP estimates of the symbols/bits. Nowadays, such a "hard-decision" decoding is rather a benchmark for the widely used soft-decision decoding we are mostly interested in.

In hard-decision symbol detection, we ignore coding, i.e., we take $N_s = 1$. From (3.11), assuming equiprobable symbols and omitting the time instant n, we obtain

$$\hat{\boldsymbol{x}} = \underset{\boldsymbol{x}\in\mathcal{X}}{\operatorname{argmin}}\{\|\boldsymbol{y} - \boldsymbol{x}\|^2\}, \tag{3.100}$$

which finds the closest (in terms of *Euclidean distance* (ED)) symbol in the constellation to the received vector $\boldsymbol{y}$, where the set of all $\boldsymbol{y}$ producing the decision $\hat{\boldsymbol{x}} = \boldsymbol{x}_j$ is called the *Voronoi region* of the symbol $\boldsymbol{x}_j$, $\mathcal{Y}_j = \{\boldsymbol{y} : \hat{\boldsymbol{x}} = \boldsymbol{x}_j\}$ (discussed in greater detail in Chapter 6).

Once hard decisions are made via (3.100), we obtain the sequence of symbol estimates $\underline{\hat{\boldsymbol{x}}} = [\hat{\boldsymbol{x}}[1], \ldots, \hat{\boldsymbol{x}}[N_s]]$, which can be seen as a discretization of the received signal $\underline{\boldsymbol{y}}$. The discretized sequence of received signals can still be used for decoding the transmitted sequence in an ML sense. To this end, the obtained estimates $\underline{\hat{\boldsymbol{x}}}$ should be considered the observations and modeled as random variables (because they are functions of the random variable $\boldsymbol{Y}$) with the conditional *probability mass function* (PMF)

$$P_{\hat{\boldsymbol{X}}|\boldsymbol{X}}(\boldsymbol{x}_j|\boldsymbol{x}_i) = \Pr\{\boldsymbol{Y} \in \mathcal{Y}_j | \boldsymbol{X} = \boldsymbol{x}_i\}. \tag{3.101}$$

The ML sequence decoding using the hard decisions $\underline{\hat{\boldsymbol{x}}}$ is then obtained using (3.101) in (3.6), i.e.,

$$\underline{\hat{\boldsymbol{x}}} = \underset{\underline{\boldsymbol{x}}\in\mathcal{X}}{\operatorname{argmax}}\left\{\sum_{n=1}^{N_s} \log P_{\hat{\boldsymbol{X}}|\boldsymbol{X}}(\hat{\boldsymbol{x}}[n]|\boldsymbol{x}[n])\right\}. \tag{3.102}$$

While we can use the hard decisions to perform ML sequence decoding, this remains a suboptimal approach due to the loss of information in the discretization of the signals $\boldsymbol{y}[n]$ to M points (when going from $\underline{\boldsymbol{y}}$ to $\underline{\hat{\boldsymbol{x}}}$ via (3.100)).

In the same way that we considered hard-decision symbol detection, we may consider the BICM decoder in (3.20) when hard decisions on the transmitted *bits* are made according to (3.40). Formulating the problem in the optimal framework of Section 3.1, the ML sequence decoding is given by

$$\underline{\hat{\boldsymbol{b}}} = \underset{\underline{\boldsymbol{b}}\in\mathcal{B}}{\operatorname{argmax}}\left\{\sum_{n=1}^{N_s} \log P_{\hat{\boldsymbol{B}}|\boldsymbol{B}}(\hat{\boldsymbol{b}}[n]|\boldsymbol{b}[n])\right\}. \tag{3.103}$$

The decision rule in (3.103) uses the "observations" $\hat{\boldsymbol{b}}[n] = [\hat{b}_1[n], \ldots, \hat{b}_m[n]]^T$ in an optimal way, but is not equivalent to the sequence decoding (3.102) because $\hat{\boldsymbol{x}} \neq \Phi(\hat{\boldsymbol{b}})$. As we will see, the equivalence occurs only when making a decision from the max-log L-values.

Theorem 3.17 *Decoding* (3.103) *using hard decisions* $\hat{b}_k$ *obtained from max-log L-values is equivalent to the hard-decision symbol decoder* (3.102).

Proof: Because the sets $\mathcal{B}$ and $\mathcal{X}$ are bijectively mapped, it is enough to show that $\hat{\boldsymbol{x}} = \Phi(\hat{\boldsymbol{b}})$ as then $P_{\hat{\boldsymbol{B}}|\boldsymbol{B}}(\hat{\boldsymbol{b}}[n]|\boldsymbol{b}[n]) = P_{\hat{\boldsymbol{X}}|\boldsymbol{X}}(\hat{\boldsymbol{x}}[n]|\boldsymbol{x}[n])$, and thus, (3.102) is equal to (3.103).

The decision on the bits using L-values can be written as

$$\hat{b}_k = \underset{b\in\mathbb{B}}{\operatorname{argmax}}\{bl_k\} \tag{3.104}$$

$$= \begin{cases} 1, & \text{if } \min_{\boldsymbol{x}\in\mathcal{X}_{k,0}} \|\boldsymbol{y} - \boldsymbol{x}\|^2 \geq \min_{\boldsymbol{x}\in\mathcal{X}_{k,1}} \|\boldsymbol{y} - \boldsymbol{x}\|^2 \\ 0, & \text{if } \min_{\boldsymbol{x}\in\mathcal{X}_{k,0}} \|\boldsymbol{y} - \boldsymbol{x}\|^2 < \min_{\boldsymbol{x}\in\mathcal{X}_{k,1}} \|\boldsymbol{y} - \boldsymbol{x}\|^2 \end{cases} \tag{3.105}$$

$$= \underset{b\in\mathbb{B}}{\operatorname{argmin}}\left\{\min_{\boldsymbol{x}\in\mathcal{S}_{k,b}} \|\boldsymbol{y} - \boldsymbol{x}\|^2\right\}. \tag{3.106}$$

As $\min_{b\in\mathbb{B}}\{\min_{\boldsymbol{x}\in\mathcal{S}_{k,b}}\|\boldsymbol{y}-\boldsymbol{x}\|^2\} = \min_{\boldsymbol{x}\in\mathcal{X}}\|\boldsymbol{y}-\boldsymbol{x}\|^2$ for any mapper $\Phi(\cdot)$, the symbol found by the BICM decoder based on max-log metrics in (3.106) will always be the closest $\boldsymbol{x} \in \mathcal{X}$ to $\boldsymbol{y}$ in terms of ED, regardless of bit position k. This is the same rule used by the symbol-by-symbol detector in (3.100), which completes the proof. □

The decision rule in (3.103) uses the observations $\hat{\boldsymbol{b}}[n]$ in an ML sense, and thus, is optimal. On the other hand, if we want to use a BICM decoder (i.e., L-values passed to the deinterleaver followed by decoding), we need to transform the hard decisions on the bits $\hat{\boldsymbol{b}}[n]$ into "hard-decision" L-values. To this end, we use (3.29) with $\hat{\boldsymbol{b}}[n]$ as the observations, i.e.,

$$l_k^{\mathrm{hd}}[n] = \log \frac{P_{\hat{B}_k|B_k}(\hat{b}_k[n]|1)}{P_{\hat{B}_k|B_k}(\hat{b}_k[n]|0)}. \tag{3.107}$$

The hard-decision L-values in (3.107) are a binary representation of the reliability of the hard decisions made on the bits, and can be expressed as

$$l_k^{\mathrm{hd}}[n] = \begin{cases} \log \dfrac{1-P_{\mathrm{b},k}}{P_{\mathrm{b},k}}, & \text{if } \hat{b}_k[n] = 1 \\ \log \dfrac{P_{\mathrm{b},k}}{1-P_{\mathrm{b},k}}, & \text{if } \hat{b}_k[n] = 0 \end{cases} \tag{3.108}$$

$$= (-1)^{1-\hat{b}_k[n]} \log \frac{1-P_{\mathrm{b},k}}{P_{\mathrm{b},k}}, \tag{3.109}$$

where the *bit-error probability* (BEP) for the kth bit is defined as

$$P_{\mathrm{b},k} \triangleq \Pr\{\hat{B}_k \neq B_k\}. \tag{3.110}$$

The BEP in (3.110) is an unconditional BEP (i.e., not the same as the conditional BEP in (3.41)) which can be calculated using methods described in Section 6.1.1.

3.5 Bibliographical Notes

BICM was introduced in [1], where it was shown that despite the suboptimality of the BICM decoder, BICM transmission provides diversity gains in fading channels. This spurred the interest in the BICM decoding principle which was later studied in detail in [2–4].

The symmetry-consistency condition of the L-values appeared in [5] and was later used in [6], for example. The symmetry condition for the channel and the decoding algorithm was studied in [7]. The symmetry and consistency conditions were also described in [8, Section III]. The idea of scrambling the bits to "symmetrize" the channel was already postulated in [2] and it is nowadays a conventional assumption when studying the performance of BICM.

The use of the max-star/max-log approximation has been widely used from the 1990s for both the decoding algorithm and the metrics calculation in the demapper (see, e.g., [9–15] and references therein). This simplification, proposed already in the original BICM paper [1] has also been proposed by the 3GPP working groups [16]. The small losses caused by using the max-log approximation with Gray-labeled constellations has been reported in [17, Fig. 9; 18].

Recognizing the BICM decoder as a mismatched decoder [19–21] is due to [22] where the *generalized mutual information* (GMI) was shown to be an achievable rate for BICM (discussed in greater detail in Section 4.3). Mismatched metrics for BICM have also been studied in [23, 24]. The equivalence between the bitwise hard decisions based on max-log L-values and the symbolwise decisions was briefly mentioned in [25, Section IV-A] and proved in [26, Section II-C] (see also [27]).

References

[1] Zehavi, E. (1992) 8-PSK trellis codes for a Rayleigh channel. *IEEE Trans. Commun.*, **40** (3), 873–884.

[2] Caire, G., Taricco, G., and Biglieri, E. (1998) Bit-interleaved coded modulation. *IEEE Trans. Inf. Theory*, **44** (3), 927–946.

[3] Wachsmann, U., Fischer, R. F. H., and Huber, J. B. (1999) Multilevel codes: theoretical concepts and practical design rules. *IEEE Trans. Inf. Theory*, **45** (5), 1361–1391.

[4] Guillén i Fàbregas, A., Martinez, A., and Caire, G. (2008) Bit-interleaved coded modulation. *Found. Trends Commun. Inf. Theory*, **5** (1–2), 1–153.

[5] Hoeher, P., Land, I., and Sorger, U. (2000) Log-likelihood values and Monte Carlo simulation—some fundamental results. In Proceedings International Symposium on Turbo Codes and Related Topics, September 2000, Brest, France.

[6] ten Brink, S. (2001) Convergence behaviour of iteratively decoded parallel concatenated codes. *IEEE Trans. Commun.*, **49** (10), 1727–1737.

[7] Richardson, T. J. and Urbanke, R. L. (2001) The capacity of low-density-parity-check codes under message-passing decoding. *IEEE Trans. Inf. Theory*, **47** (2), 599–618.

[8] Tüchler, M. (2004) Design of serially concatenated systems depending on the block length. *IEEE Trans. Commun.*, **52** (2), 209–218.

[9] Viterbi, A. J. (1998) An intuitive justification and a simplified implementation of the MAP decoder for convolutional codes. *IEEE J. Sel. Areas Commun.*, **16** (2), 260–264.

[10] Robertson, P., Villebrun, E., and Hoeher, P. (1995) A comparison of optimal and sub-optimal MAP decoding algorithms operating in the log domain. IEEE International Conference on Communications (ICC), June 1995, Seattle, WA.

[11] Pietrobon, S. S. (1995) Implementation and performance of a serial MAP decoder for use in an iterative turbo decoder. IEEE International Symposium on Information Theory (ISIT), September 1995, Whistler, BC, Canada.

[12] Benedetto, S., Divsalar, D., Montorsi, G., and Pollara, F. (1995) Soft-output decoding algorithms in iterative decoding of turbo codes. TDA Progress Report 42-87, Jet Propulsion Laboratory, Pasadena, CA, pp. 63–121.

[13] Gross, W. and Gulak, P. (1998) Simplified MAP algorithm suitable for implementation of turbo decoders. *Electron. Lett.*, **34** (16), 1577–1578.

[14] Valenti, M. and Sun, J. (2001) The UMTS turbo code and an efficient decoder implementation suitable for software defined radios. *Int. J. Wirel. Inf. Netw.*, **8** (4), 203–216.

[15] Hu, X.-Y., Eleftheriou, E., Arnold, D.-M., and Dholakia, A. (2001) Efficient implementations of the sum-product algorithm for decoding LDPC codes. IEEE Global Telecommunications Conference (GLOBECOM), November 2001, San Antonio, TX.

[16] Ericsson, Motorola, and Nokia (2000) Link evaluation methods for high speed downlink packet access (HSDPA). TSG-RAN Working Group 1 Meeting #15, TSGR1#15(00)1093, Tech. Rep., August 2000.

[17] Classon, B., Blankenship, K., and Desai, V. (2002) Channel coding for 4G systems with adaptive modulation and coding. *IEEE Wirel. Commun. Mag.*, **9** (2), 8–13.

[18] Alvarado, A., Carrasco, H., and Feick, R. (2006) On adaptive BICM with finite block-length and simplified metrics calculation. IEEE Vehicular Technology Conference (VTC-Fall), September 2006, Montreal, QC, Canada.

[19] Ganti, A., Lapidoth, A., and Telatar, I. E. (2000) Mismatched decoding revisited: general alphabets, channels with memory, and the wide-band limit. *IEEE Trans. Inf. Theory*, **46** (7), 2315–2328.

[20] Kaplan, G. and Shamai, S. (Shitz) (1993) Information rates and error exponents of compound channels with application to antipodal signaling in a fading environment. *Archiv fuer Elektronik und Uebertragungstechnik*, **47** (4), 228–239.

[21] Merhav, N., Kaplan, G., Lapidoh, A., and Shamai (Shitz), S. (1994) On information rates for mismatched decoders. *IEEE Trans. Inf. Theory*, **40** (6), 1953–1967.

[22] Martinez, A., Guillén i Fàbregas, A., Caire, G., and Willems, F. M. J. (2009) Bit-interleaved coded modulation revisited: A mismatched decoding perspective. *IEEE Trans. Inf. Theory*, **55** (6), 2756–2765.

[23] Nguyen, T. and Lampe, L. (2011) Bit-interleaved coded modulation with mismatched decoding metrics. *IEEE Trans. Commun.*, **59** (2), 437–447.

[24] Nguyen, T. and Lampe, L. (2011) Mismatched bit-interleaved coded noncoherent orthogonal modulation. *IEEE Commun. Lett.*, **5**, 563–565.

[25] Fertl, P., Jaldén, J., and Matz, G. (2012) Performance assessment of MIMO-BICM demodulators based on mutual information. *IEEE Trans. Signal Process.*, **60** (3), 2764–2772.

[26] Ivanov, M., Brännström, F., Alvarado, A., and Agrell, E. (2012) General BER expression for one-dimensional constellations. IEEE Global Telecommunications Conference (GLOBECOM), December 2012, Anaheim, CA

[27] Ivanov, M., Brännström, F., Alvarado, A., and Agrell, E. (2013) On the exact BER of bit-wise demodulators for one-dimensional constellations. *IEEE Trans. Commun.*, **61** (4), 1450–1459.

4

Information-Theoretic Elements

Information theory provides us with fundamental limits on the transmission rates supported by a channel. In this chapter, we analyze these limits for the *coded modulation* (CM) schemes presented in Chapter 2, paying special attention to *bit-interleaved coded modulation* (BICM).

This chapter is structured as follows. In Section 4.1, we introduce the concepts of *mutual information* (MI) and channel capacity; in Section 4.2, we study the maximum transmission rates for general CM systems; and in Section 4.3, for BICM systems. We review their relation and we analyze how they are affected by the choice of the (discrete) constellation. We pay special attention to the selection of the binary labeling and use of probabilistic shaping in BICM. We conclude the chapter by showing in Section 4.5 ready-to-use numerical quadrature formulas to efficiently compute MIs.

4.1 Mutual Information and Channel Capacity

We analyze the transmission model defined by (2.27), i.e., $\boldsymbol{Y} = H\boldsymbol{X} + \boldsymbol{Z}$, where $\boldsymbol{Z}$ is an N-dimensional zero-mean Gaussian vector with covariance matrix $\mathbb{E}_{\boldsymbol{Z}}[\boldsymbol{Z}\boldsymbol{Z}^{\mathrm{T}}] = \mathrm{I}_{[N]}\mathrm{N}_0/2$. In this chapter, we are mostly interested in discrete constellations. However, we will initially assume that the input symbols $\boldsymbol{X}$ are modeled as *independent and identically distributed* (i.i.d.) continuous random variables characterized by their *probability density function* (PDF) $p_{\boldsymbol{X}}$. Considering constellations with continuous support ($\mathcal{S} = \mathbb{R}^N$) provides us with an upper limit on the performance of discrete constellations. As mentioned before, we also assume that the channel state $H = h$ is known at the receiver (through perfect channel estimation) and is not available at the transmitter. The assumption of the transmitter not knowing the channel state is important when analyzing the channel capacity (defined below) for fading channels.

The MI between the random vectors $\boldsymbol{X}$ and $\boldsymbol{Y}$ conditioned on the channel state $H = h$ is denoted by $I(\boldsymbol{X};\boldsymbol{Y})$ and given by[1]

$$I(\boldsymbol{X};\boldsymbol{Y}) \triangleq \mathbb{E}_{\boldsymbol{X},\boldsymbol{Y}|H=h}\left[\log_2 \frac{p_{\boldsymbol{X},\boldsymbol{Y}|H}(\boldsymbol{X},\boldsymbol{Y}|h)}{p_{\boldsymbol{Y}|H}(\boldsymbol{Y}|h)p_{\boldsymbol{X}}(\boldsymbol{X})}\right] \tag{4.1}$$

$$= \int_{\mathbb{R}^N} p_{\boldsymbol{X}}(\boldsymbol{x}) \int_{\mathbb{R}^N} p_{\boldsymbol{Y}|\boldsymbol{X},H}(\boldsymbol{y}|\boldsymbol{x},h) \log_2 \frac{p_{\boldsymbol{Y}|\boldsymbol{X},H}(\boldsymbol{y}|\boldsymbol{x},h)}{p_{\boldsymbol{Y}|H}(\boldsymbol{y}|h)}\,\mathrm{d}\boldsymbol{y}\,\mathrm{d}\boldsymbol{x}, \tag{4.2}$$

where $p_{\boldsymbol{Y}|\boldsymbol{X},H}(\boldsymbol{y}|\boldsymbol{x},h)$ is given by (2.28).

[1] Throughout this book, we use base-2 logarithms, so MIs are expressed in binary units (or bits).

Bit-Interleaved Coded Modulation: Fundamentals, Analysis, and Design, First Edition.
Leszek Szczecinski and Alex Alvarado.

For future use, we also define the conditional MIs

$$I(\boldsymbol{X};\boldsymbol{Y}|H) = \mathbb{E}_{\boldsymbol{X},\boldsymbol{Y},H}\left[\log_2 \frac{p_{\boldsymbol{Y}|\boldsymbol{X},H}(\boldsymbol{Y}|\boldsymbol{X},H)}{p_{\boldsymbol{Y}|H}(\boldsymbol{Y}|H)}\right], \tag{4.3}$$

$$I(\boldsymbol{X};\boldsymbol{Y}|H,B_k) = \mathbb{E}_{\boldsymbol{X},\boldsymbol{Y},H,B_k}\left[\log_2 \frac{p_{\boldsymbol{Y}|\boldsymbol{X},H,B_k}(\boldsymbol{Y}|\boldsymbol{X},H,B_k)}{p_{\boldsymbol{Y}|H,B_k}(\boldsymbol{Y}|H,B_k)}\right]. \tag{4.4}$$

Although $I(\boldsymbol{X};\boldsymbol{Y})$ is the most common notation for MI found in the literature (which we also used in Chapter 1), throughout this chapter, we also use an alternative notation

$$\mathsf{I}(\mathsf{snr}, p_{\boldsymbol{X}}) \triangleq I(\boldsymbol{X};\boldsymbol{Y}), \tag{4.5}$$

which shows that conditioning on $H = h$ in (4.1) is equivalent to conditioning on the instantaneous *signal-to-noise ratio* (SNR); this notation also emphasizes the dependence of the MI on the input PDF $p_{\boldsymbol{X}}$. This notation allows us to express the MI for a fast fading channel (see Section 2.4) by averaging the MI in (4.5) over the SNR, i.e.,

$$\begin{aligned}\bar{\mathsf{I}}(\overline{\mathsf{snr}}, p_{\boldsymbol{X}}) &= I(\boldsymbol{X};\boldsymbol{Y}|H)\\ &= \mathbb{E}_{\mathsf{SNR}}[\mathsf{I}(\mathsf{SNR}, p_{\boldsymbol{X}})],\end{aligned} \tag{4.6}$$

where $\mathsf{I}(\mathsf{SNR}, p_{\boldsymbol{X}})$ is given by (4.5). Throughout this chapter, we use the notation $\mathsf{I}(\mathsf{snr}, p_{\boldsymbol{X}})$ and $\bar{\mathsf{I}}(\overline{\mathsf{snr}}, p_{\boldsymbol{X}})$ to denote MIs for the *additive white Gaussian noise* (AWGN) and fading channels, respectively.

The MIs above have units of bit/symbol (or equivalently bit/channel use) and they define the maximum transmission rates that can be reliably[2] used when the codewords $\underline{\boldsymbol{x}}$ are symbols generated randomly according to the continuous distribution PDF $p_{\boldsymbol{X}}$. More precisely, the converse of Shannon's channel coding theorem states that it is not possible to transmit information reliably above the MI, i.e.,

$$R \le \bar{\mathsf{I}}(\overline{\mathsf{snr}}, p_{\boldsymbol{X}}). \tag{4.7}$$

The *channel capacity* of a continuous-input continuous-output memoryless channel under average power constraint is defined as the maximum MI, i.e.,

$$\bar{\mathsf{C}}(\overline{\mathsf{snr}}) \triangleq \max_{p_{\boldsymbol{X}}:\mathsf{E}_\mathsf{s}\le 1} \bar{\mathsf{I}}(\overline{\mathsf{snr}}, p_{\boldsymbol{X}}), \tag{4.8}$$

where the optimization is carried out over all input distributions that satisfy the average energy constraint $\mathsf{E}_\mathsf{s} = \mathbb{E}_{\boldsymbol{X}}[\|\boldsymbol{X}\|^2] \le 1$. Furthermore, we note that $\boldsymbol{X}$ is independent of H because we assumed that the transmitter does not know the channel. Because of this, we do not consider the case when the transmitter, knowing the channel, adjusts the signal's energy to the channel state.

Example 4.1 (Capacity of the AWGN Channel) *For the AWGN channel in* (2.27), *the noise is independent in each dimension, so the transmission of* $\boldsymbol{X}$ *can be considered as N transmissions over independent AWGN channels, i.e.,* $Y_n = X_n + Z_n$, *where* Z_n *are i.i.d. zero-mean Gaussian random variables with variance* $\mathsf{N}_0/2$, *with* $n = 1, \ldots, N$. *Then, it can be shown that the maximum in* (4.8) *is obtained when* $p_{X_n}(x) = \Psi(x, 0, \mathsf{E}_\mathsf{s}/N)$ *for* $n = 1, \ldots, N$, *yielding*

$$\mathsf{C}(\mathsf{snr}) = \frac{N}{2}\log_2\left(1 + \frac{2}{N}\mathsf{snr}\right). \tag{4.9}$$

[2] That is, with vanishing probability of decoding error when $N_\mathrm{s} \to \infty$.

Example 4.2 (Capacity of Fading Channels) *Since the Gaussian PDF that yields* $\mathsf{C}(\mathsf{snr})$ *in* (4.9) *is independent of* snr, *it also maximizes the MI in fading channels. The capacity of a fading channel is then given by*

$$\overline{\mathsf{C}}(\overline{\mathsf{snr}}) = \mathbb{E}_{\mathsf{SNR}}[\mathsf{C}(\mathsf{SNR})] \tag{4.10}$$

$$= \frac{N}{2}\int_0^\infty p_{\mathsf{SNR}}(\mathsf{snr})\log_2\left(1+\frac{2}{N}\mathsf{snr}\right)\mathrm{d}\mathsf{snr}. \tag{4.11}$$

For Rayleigh fading, using (2.35) *with* $\mathsf{m}=1$ *in* (4.11), *we obtain*

$$\overline{\mathsf{C}}(\overline{\mathsf{snr}}) = \frac{N}{2}\frac{1}{\overline{\mathsf{snr}}}\int_0^\infty \exp\left(-\frac{\mathsf{snr}}{\overline{\mathsf{snr}}}\right)\log_2\left(1+\frac{2}{N}\mathsf{snr}\right)\mathrm{d}\mathsf{snr}, \tag{4.12}$$

which solved via integration by parts yields

$$\overline{\mathsf{C}}(\overline{\mathsf{snr}}) = \frac{N}{2}\log_2(\mathrm{e})\exp\left(\frac{N}{2\overline{\mathsf{snr}}}\right)\mathrm{E}_1\left(\frac{N}{2\overline{\mathsf{snr}}}\right), \tag{4.13}$$

where the exponential integral $\mathrm{E}_1(x) \triangleq \int_1^\infty t^{-1}\mathrm{e}^{-tx}\,\mathrm{d}t$ *is available as a numerical routine in popular computational software.*

MI curves are typically plotted versus SNR, however, to better appreciate their behavior at low SNR, plotting MI versus the average information bit energy-to-noise ratio $\mathsf{E_b}/\mathsf{N_0}$ is preferred. To do this, first note that (4.7) can be expressed using (2.38) as

$$R \leq \mathsf{I}\left(R\frac{\mathsf{E_b}}{\mathsf{N_0}}, p_{\boldsymbol{X}}\right). \tag{4.14}$$

The notation $\mathsf{I}(\mathsf{snr}, p_{\boldsymbol{X}})$ emphasizes that the MI is a function of both snr and $p_{\boldsymbol{X}}$, which will be useful throughout this chapter. In what follows, however, we discuss $\mathsf{I}(\mathsf{snr}, p_{\boldsymbol{X}})$ as a function of snr for a given input distribution, i.e., $\mathsf{I}(\mathsf{snr}, p_{\boldsymbol{X}}) : \mathbb{R}_+ \mapsto \mathbb{R}_+$.

The MI is an increasing function of the SNR, and thus, it has an inverse, which we denote by $\mathsf{I}^{-1}(\cdot, p_{\boldsymbol{X}})$. By using this inverse function (for any given $p_{\boldsymbol{X}}$) in both sides of (4.14), we obtain

$$\mathsf{I}^{-1}(R, p_{\boldsymbol{X}}) \leq R\frac{\mathsf{E_b}}{\mathsf{N_0}}. \tag{4.15}$$

Furthermore, by rearranging the terms, we obtain

$$\frac{\mathsf{E_b}}{\mathsf{N_0}} \geq f_{\mathsf{I}}(R, p_{\boldsymbol{X}}) \triangleq \frac{\mathsf{I}^{-1}(R, p_{\boldsymbol{X}})}{R}, \tag{4.16}$$

which shows that $\mathsf{E_b}/\mathsf{N_0}$ is bounded from below by $f_{\mathsf{I}}(R, p_{\boldsymbol{X}})$, which is a function of the rate R.

In other words, the function $f_{\mathsf{I}}(R, p_{\boldsymbol{X}})$ in (4.16) gives, for a given input distribution $p_{\boldsymbol{X}}$, a lower bound on the $\mathsf{E_b}/\mathsf{N_0}$ needed for reliable transmission at rate R. For example, for the AWGN channel in Example 4.1, and by using (4.9) in (4.16), we obtain

$$f_{\mathsf{C}}(R) = \frac{N}{2R}(2^{2R/N} - 1). \tag{4.17}$$

In this case, the function $f_{\mathsf{C}}(R)$ depends solely on R (and not on $p_{\boldsymbol{X}}$, as in (4.16)), which follows because $\mathsf{C}(\mathsf{snr})$ in (4.9) depends only on snr.

The expressions in (4.17) allow us to find a lower bound on $\mathsf{E_b}/\mathsf{N_0}$ when $\mathsf{snr} \to 0$, or equivalently, when $R \to 0^+$, i.e., in the low-SNR regime. Using (4.17), we obtain

$$\lim_{R\to 0^+} f_{\mathsf{C}}(R) = -1.59 \text{ dB}, \tag{4.18}$$

which we refer to as the *Shannon limit* (*SL*). The bound in (4.18) corresponds to the minimum average information bit energy-to-noise ratio $\mathsf{E_b}/\mathsf{N_0}$ needed to reliably transmit information when $\mathsf{snr} \to 0$.

For notation simplicity, in (4.14)–(4.16), we consider nonfading channels. It is important to note, however, that because $\mathbb{E}_H[H^2] = 1$, exactly the same expressions apply to fading channels, i.e., when $\mathsf{I}(\mathsf{snr}, p_{\boldsymbol{X}})$ is replaced by $\bar{\mathsf{I}}(\overline{\mathsf{snr}}, p_{\boldsymbol{X}})$.

4.2 Coded Modulation

In this section, we focus on the practically relevant case of discrete constellations, and thus, we restrict our analysis to *probability mass functions* (PMFs) $p_{\boldsymbol{X}}$ with M nonzero mass points, i.e., $P_{\boldsymbol{X}}(\boldsymbol{x}_i) > 0$, $\forall i \in \mathcal{I}$.

4.2.1 CM Mutual Information

We define the *coded modulation mutual information* (CM-MI) as the MI between the input and the output of the channel when a discrete constellation is used for transmission. As mentioned in Section 2.5.1, in this case, the transmitted symbols are fully determined by the PMF $p_{\boldsymbol{X}}$, and thus, we use the matrix $\mathbf{X}$ to denote the support of the PMF (the constellation $\mathcal{S}$) and the vector $\boldsymbol{p}_{\mathrm{s}}$ to denote the probabilities associated with the symbols (the input distribution). The CM-MI can be expressed using (4.2), where the first integral is replaced by a sum and the PDF $p_{\boldsymbol{X}}$ by the PMF $P_{\boldsymbol{X}}$, i.e.,

$$\bar{\mathsf{I}}^{\mathrm{cm}}(\overline{\mathsf{snr}}, \mathbf{X}, \boldsymbol{p}_{\mathrm{s}}) \triangleq I(\boldsymbol{X}; \boldsymbol{Y}|H) \tag{4.19}$$

$$= \mathbb{E}_H\left[\sum_{i\in\mathcal{I}} p_{\mathrm{s},i} \int_{\mathbb{R}^N} p_{\boldsymbol{Y}|\boldsymbol{X},H}(\boldsymbol{y}|\boldsymbol{x}_i, H) \log_2 \frac{p_{\boldsymbol{Y}|\boldsymbol{X},H}(\boldsymbol{y}|\boldsymbol{x}_i, H)}{p_{\boldsymbol{Y}|H}(\boldsymbol{y}|H)}\, \mathrm{d}\boldsymbol{y}\right] \tag{4.20}$$

$$= \sum_{i\in\mathcal{I}} p_{\mathrm{s},i} \int_0^\infty p_H(h) \int_{\mathbb{R}^N} p_{\boldsymbol{Y}|\boldsymbol{X},H}(\boldsymbol{y}|\boldsymbol{x}_i, h) \log_2 \frac{p_{\boldsymbol{Y}|\boldsymbol{X},H}(\boldsymbol{y}|\boldsymbol{x}_i, h)}{p_{\boldsymbol{Y}|H}(\boldsymbol{y}|h)}\, \mathrm{d}\boldsymbol{y}\, \mathrm{d}h, \tag{4.21}$$

where the notation $\bar{\mathsf{I}}^{\mathrm{cm}}(\overline{\mathsf{snr}}, \mathbf{X}, \boldsymbol{p}_{\mathrm{s}})$ emphasizes the dependence of the MI on the input PMF $P_{\boldsymbol{X}}$ (via $\mathbf{X}$ and $\boldsymbol{p}_{\mathrm{s}}$, see (2.41)).

For the AWGN channel with channel transition probability (2.31), the CM-MI in (4.21) can be expressed as

$$\mathsf{I}^{\mathrm{cm}}(\mathsf{snr}, \mathbf{X}, \boldsymbol{p}_{\mathrm{s}}) = I(\boldsymbol{X}; \boldsymbol{Y}) \tag{4.22}$$

$$= \int_{\mathbb{R}^N} \left(\frac{\mathsf{snr}}{\pi}\right)^{\frac{N}{2}} \sum_{i\in\mathcal{I}} p_{\mathrm{s},i} \exp(-\mathsf{snr}\|\boldsymbol{y} - \boldsymbol{x}_i\|^2) \cdot \log_2 \frac{\exp(-\mathsf{snr}\|\boldsymbol{y} - \boldsymbol{x}_i\|^2)}{\sum_{j\in\mathcal{I}} p_{\mathrm{s},j} \exp(-\mathsf{snr}\|\boldsymbol{y} - \boldsymbol{x}_j\|^2)}\, \mathrm{d}\boldsymbol{y}. \tag{4.23}$$

For a uniform input distribution $\boldsymbol{p}_{\mathrm{s}} = \boldsymbol{p}_{\mathrm{s,u}}$ in (2.42), the above expression particularizes to

$$\mathsf{I}^{\mathrm{cm}}(\mathsf{snr}, \mathbf{X}, \boldsymbol{p}_{\mathrm{s,u}}) = m - \frac{1}{M}\int_{\mathbb{R}^N} \left(\frac{\mathsf{snr}}{\pi}\right)^{\frac{N}{2}} \sum_{i\in\mathcal{I}} \exp(-\mathsf{snr}\|\boldsymbol{y} - \boldsymbol{x}_i\|^2) \cdot \log_2 \frac{\sum_{j\in\mathcal{I}} \exp(-\mathsf{snr}\|\boldsymbol{y} - \boldsymbol{x}_j\|^2)}{\exp(-\mathsf{snr}\|\boldsymbol{y} - \boldsymbol{x}_i\|^2)}\, \mathrm{d}\boldsymbol{y}. \tag{4.24}$$

Example 4.3 (CM-MI for M**PSK)** *In Fig. 4.1 (a), the CM-MI given by* (4.24) *is shown for the* M*PSK constellations defined in Section 2.5.1. We also show in this figure the AWGN capacity* $\mathsf{C}(\mathsf{snr})$ *in* (4.9) *(where we use* $N = 2$ *because* M*PSK constellations have two dimensions). As a consequence of using a discrete constellation, for high SNR, the MI curves flatten out at* m[bit/symbol]. *On the other hand, in the low-SNR regime, any* M*PSK constellation yields MIs very close to the AWGN capacity. To observe this effect better, we show in Fig. 4.1 (b) the function* $f_{\mathsf{I}^{\mathrm{cm}}}(R, \mathbf{X}_{M\mathrm{PSK}}, \boldsymbol{p}_{\mathrm{s,u}})$ *from* (4.16)*, where the vertical axis should be regarded as the argument of this function. We can now clearly observe that all the CM-MI curves achieve the SL* -1.59 dB *given by* (4.18).

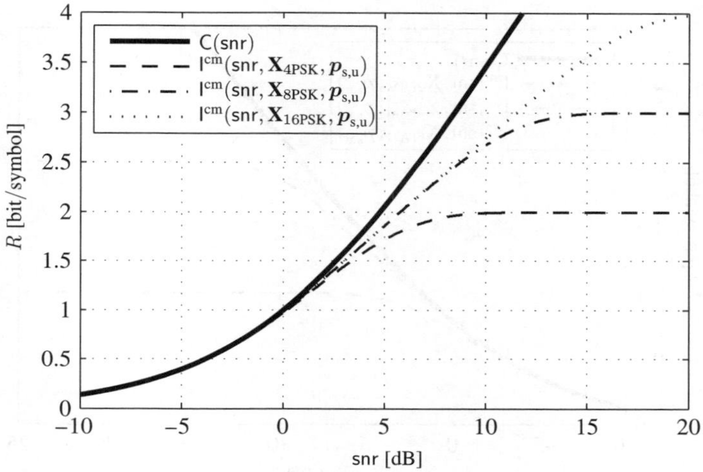

(a)

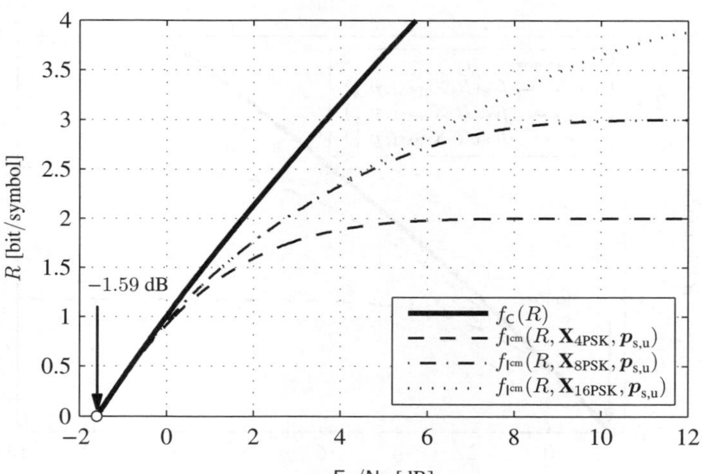

(b)

Figure 4.1 MPSK constellations and AWGN channel: (a) CM-MI and (b) $f_{\mathsf{I}^{\mathrm{cm}}}(R, \cdot, \boldsymbol{p}_{\mathrm{s,u}})$ in (4.16). The AWGN capacity $\mathsf{C}(\mathsf{snr})$ and the corresponding $f_{\mathsf{C}}(R)$ are shown for reference

Example 4.4 (CM-MI for M**PAM)** *The results obtained for the case of the* M*PAM constellations in Fig. 2.10 are shown in Fig. 4.2. On the basis of these results, similar conclusions to the ones regarding* M*PSK constellations can be drawn, namely, these results show that for a given value of SNR, the CM-MI increases with* M*, and thus, to maximize the transmission rates, the largest available constellation should be used. While theoretically possible, using large constellations in the low SNR regime would be impractical because of time/phase synchronization problems. However, we will not consider these effects here. As we will see in the next section, this is not the case in BICM.*

Example 4.5 (CM-MI for M**QAM)** *The CM-MI as a function of the SNR for* M*QAM constellations are shown in Fig. 4.3. Because of the construction of* M*QAM constellations, the obtained results*

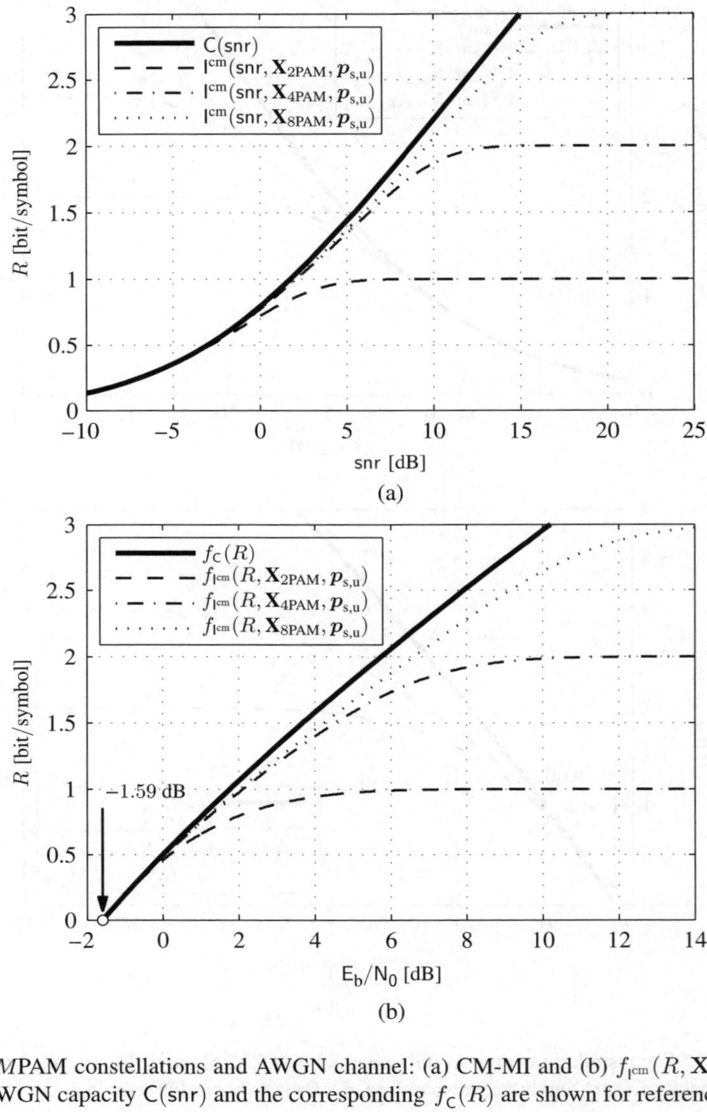

Figure 4.2 MPAM constellations and AWGN channel: (a) CM-MI and (b) $f_{\mathsf{I}^{\mathrm{cm}}}(R, \mathbf{X}_{M\mathrm{PAM}}, \boldsymbol{p}_{\mathrm{s,u}})$ in (4.16). The AWGN capacity $\mathsf{C}(\mathsf{snr})$ and the corresponding $f_{\mathsf{C}}(R)$ are shown for reference

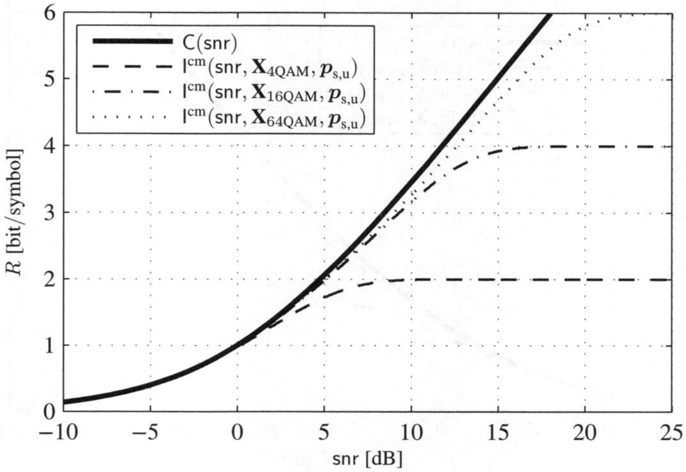

Figure 4.3 CM-MI for 4PSK (4QAM) and for the MQAM constellations in Fig. 2.11 ($M = 16, 64$) over the AWGN channel. The AWGN capacity $\mathsf{C}(\mathsf{snr})$ in (4.9) is also shown

are equivalent to those obtained for MPAM *except for an SNR scaling difference (a consequence of splitting the energy* E_s *over two dimensions of the constellation), i.e.,* $\mathsf{I}^{\mathrm{cm}}(\mathsf{snr}, \mathbf{X}_{\sqrt{M}\mathrm{PAM}}, \boldsymbol{p}_{\mathrm{s,u}}) = 0.5 \cdot \mathsf{I}^{\mathrm{cm}}(2\mathsf{snr}, \mathbf{X}_{M\mathrm{QAM}}, \boldsymbol{p}_{\mathrm{s,u}})$.

Example 4.6 (CM-MI for MPAM in Fading Channels) *In Fig. 4.4, we show the results obtained for* M*PAM constellations in a Rayleigh fading channel. When compared to the results in Fig. 4.2, we see that fading introduces a penalty in terms of MI (note that both figures are shown for the same range of SNR). Nevertheless, all the constellations achieve the SL as the rate tends to zero.*

4.2.2 CM Capacity

The CM-MI $\mathsf{I}^{\mathrm{cm}}(\mathsf{snr}, \mathbf{X}, \boldsymbol{p}_\mathrm{s})$ corresponds to the maximum transmission rate when the codewords' symbols $\boldsymbol{x}[n]$ are taken from the constellation $\boldsymbol{X}$ following the PMF $P_{\boldsymbol{X}}$. In such cases, the role of the receiver is to find the transmitted codewords using $p_{\boldsymbol{Y}|\boldsymbol{X},H}(\boldsymbol{y}|\boldsymbol{x}, h)$ by applying the *maximum likelihood* (ML) decoding rule we defined in Chapter 3. In practice, the CM encoder must be designed having the decoding complexity in mind. To ease the implementation of the ML decoder, particular encoding structures are adopted. This is the idea behind *trellis-coded modulation* (TCM) (see Fig. 2.4), where the *convolutional encoder* (CENC) generates symbols $\boldsymbol{b}[n]$ which are mapped directly to constellation symbols $\boldsymbol{x}[n]$. In this case, the code can be represented using a trellis structure, which means that the Viterbi algorithm can be used to implement the ML decoding rule, and thus, the decoding complexity is manageable.

The most popular CM schemes are based on uniformly distributed constellation points. However, using a uniform input distribution is not mandatory, and thus, one could think of using an arbitrary PMF (and/or a nonequally spaced constellation) so that the CM-MI is increased. To formalize this, and in analogy with (4.8), we define the *CM capacity* for a given constellation size M as

$$\mathsf{C}^{\mathrm{cm}}(\mathsf{snr}) \triangleq \max_{\mathbf{X}, \boldsymbol{p}_\mathrm{s}:\mathsf{E}_\mathrm{s}=1} \mathsf{I}^{\mathrm{cm}}(\mathsf{snr}, \mathbf{X}, \boldsymbol{p}_\mathrm{s}), \tag{4.25}$$

where the optimization over $\mathbf{X}$ and $\boldsymbol{p}_\mathrm{s}$ is equivalent to the optimization over the PMF $P_{\boldsymbol{X}}(\boldsymbol{x})$.

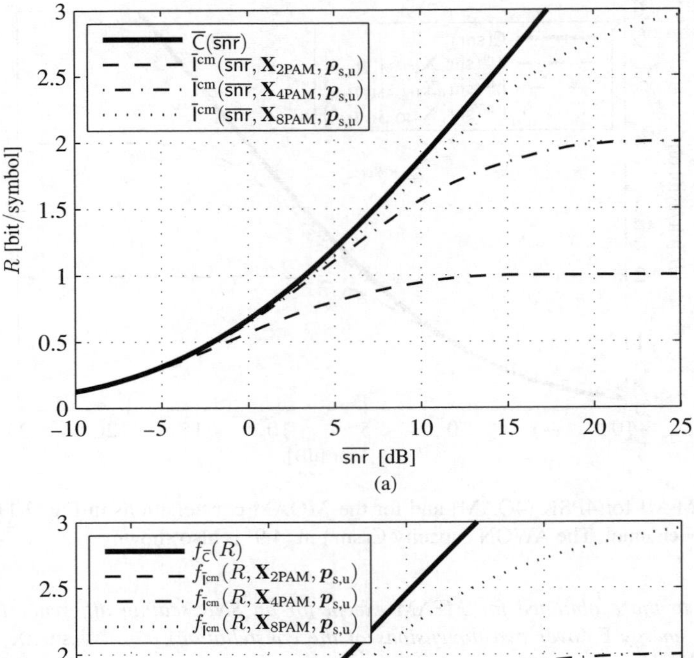

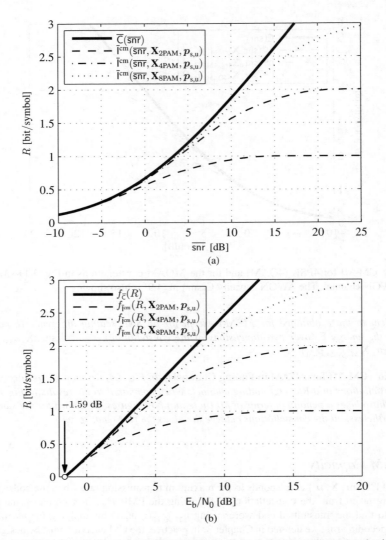

Figure 4.4 *M*PAM constellations and Rayleigh fading channel: (a) CM-MI and (b) $f_{\overline{\mathsf{I}}^{\mathrm{cm}}}(R, \mathbf{X}_{M\mathrm{PAM}}, \boldsymbol{p}_{\mathrm{s,u}})$ in (4.16). The average capacity $\overline{\mathsf{C}}(\overline{\mathsf{snr}})$ and the corresponding $f_{\overline{\mathsf{C}}}(R)$ are shown for reference

The optimization problem in (4.25) is done under constraint $\mathsf{E_s} = \sum_{j\in\mathcal{I}} p_{\mathrm{s},j}\|\boldsymbol{x}_j\|^2 = 1$, where $\mathsf{E_s}$ depends on both $\mathbf{X}$ and $\boldsymbol{p}_{\mathrm{s}}$. In principle, a constraint $\mathsf{E_s} \leq 1$ could be imposed; however, for the channel in (2.27) we consider here, an increase in $\mathsf{E_s}$ always results in a higher MI and thus, the constraint is always active.

Again, the optimization result (4.25) should be interpreted as the maximum number of bits per symbol that can be reliably transmitted using a fully optimized M-point constellation, i.e., when for each SNR value, the constellation and the input distribution are selected in order to maximize the CM-MI. This is usually referred to as *signal shaping*.

The CM capacity for fading channels is defined as

$$\overline{\mathsf{C}}^{\mathrm{cm}}(\overline{\mathsf{snr}}) \triangleq \max_{\mathbf{X},\, \boldsymbol{p}_{\mathrm{s}}:\mathsf{E_s}=1} \overline{\mathsf{I}}^{\mathrm{cm}}(\overline{\mathsf{snr}}, \mathbf{X}, \boldsymbol{p}_{\mathrm{s}}). \tag{4.26}$$

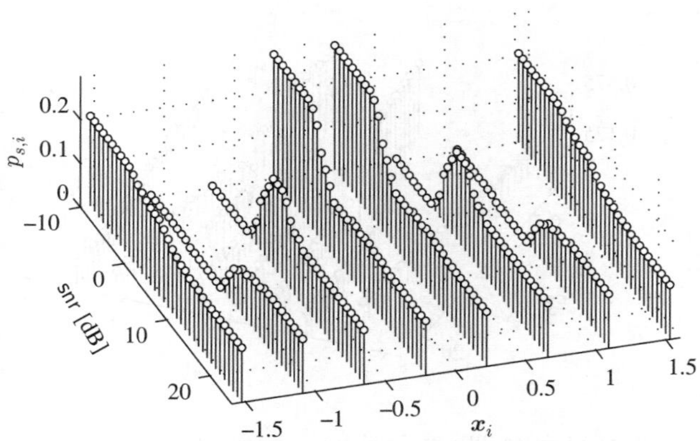

Figure 4.5 Solution of problem (4.27): optimal PMF for an equidistant 8PAM constellation in the AWGN channel

We note that the optimal constellations $\mathbf{X}$ and $\boldsymbol{p}_{\text{s}}$, i.e., those solving (4.25) or (4.26), are not the same for the AWGN and fading channels. This is different from the case of continuous distributions $p_{\boldsymbol{X}}(\boldsymbol{x})$, where the Gaussian distribution is optimal for each value of snr over the AWGN channel and thus, the same distribution yields the capacity of fading channels, cf. Example 4.2.

The joint optimization over both $\mathbf{X}$ and $\boldsymbol{p}_{\text{s}}$ is a difficult problem, and thus, one might prefer to solve simpler ones:

$$\mathsf{C}^{\text{cm}}(\mathsf{snr}, \mathbf{X}) = \max_{\boldsymbol{p}_{\text{s}}:\mathsf{E}_{\text{s}}=1} \mathsf{I}^{\text{cm}}(\mathsf{snr}, \mathbf{X}, \boldsymbol{p}_{\text{s}}), \tag{4.27}$$

$$\mathsf{C}^{\text{cm}}(\mathsf{snr}, \boldsymbol{p}_{\text{s}}) = \max_{\mathbf{X}:\mathsf{E}_{\text{s}}=1} \mathsf{I}^{\text{cm}}(\mathsf{snr}, \mathbf{X}, \boldsymbol{p}_{\text{s}}). \tag{4.28}$$

The expression (4.28) is typically called *geometrical shaping* as only the constellation symbols (geometry) are being optimized, while the problem (4.27) is called *probabilistic shaping* as the probabilities of the constellation symbols are optimized.

Finally, as a pragmatic compromise between the fully optimized signal shaping of (4.25) and the geometric/probabilistic solutions of (4.28)–(4.27), we may optimize the distribution $\boldsymbol{p}_{\text{s}}$ while maintaining the "structure" of the constellation, i.e., allowing for a scaling of a predefined constellation with a factor $\eta > 0$

$$\mathsf{C}^{\text{cm}}(\mathsf{snr}, \hat{\eta}\mathbf{X}) = \max_{\boldsymbol{p}_{\text{s}},\eta:\mathsf{E}_{\text{s}}=1} \mathsf{I}^{\text{cm}}(\mathsf{snr}, \eta\mathbf{X}, \boldsymbol{p}_{\text{s}}), \tag{4.29}$$

where $\hat{\eta}$ is the optimal value of η.

Example 4.7 (Probabilistic Shaping for 8PAM**)** *Let us consider first an* MPAM *constellation defined in* (2.46). *The distance between adjacent points* 2Δ *is defined in* (2.47) *so as to guarantee* $\mathsf{E}_{\text{s}} = 1$ *under uniformly distributed constellation symbols (i.e., for* $\boldsymbol{p}_{\text{s}} = \boldsymbol{p}_{\text{s,u}}$*). Having fixed* $\mathbf{X}$, *we solve* (4.27), *which is relatively simple because* $\mathsf{I}^{\text{cm}}(\mathsf{snr}, \mathbf{X}, \boldsymbol{p}_{\text{s}})$ *is concave in* $\boldsymbol{p}_{\text{s}}$. *The results of this optimization for* $M = 8$ *are shown in Fig. 4.5.*

Example 4.8 (Geometrical Shaping for $M = 8$**)** *In Fig. 4.6 we show the optimal input distribution found by solving* (4.28) *for different SNR values. For low SNR values the optimal input distribution degenerates to a* 4*-ary constellation, i.e., the constellation symbols overlap, which is equivalent to changing their probability. This may be seen as a rudimentary probabilistic shaping where the probabilities are multiples of* $\frac{1}{M}$. *On the other hand, for* $\mathsf{snr} \to \infty$ *the symbols are uniformly distributed: for high SNR we*

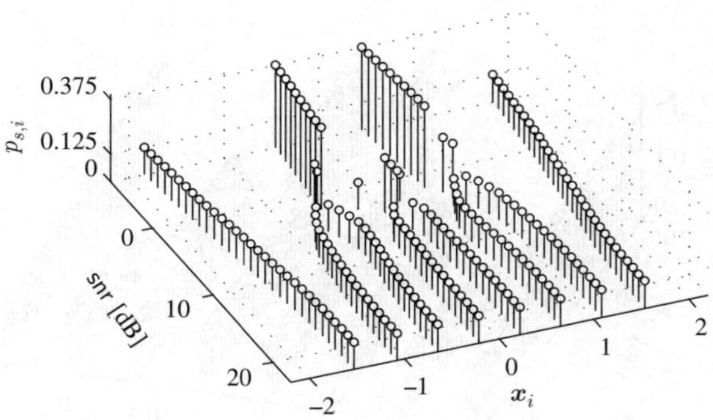

Figure 4.6 Solution of problem (4.28): optimal input distribution for an 8-ary constellation in the AWGN channel

may neglect the noise, and thus, the MI can be approximated by the entropy of $\boldsymbol{X}$*, which is maximized by a uniform distribution.*

Example 4.9 (Mixed Shaping for 8PAM**)** *The solution to the "mixed" shaping problem in* (4.29) *is shown in Fig. 4.7. The mixed optimization* (4.29) *illustrates well how the optimal distribution changes: for medium SNR it mimics a "Gaussian-like" distribution and for high SNR it is uniform again.*

We conclude this section by quantifying the gains obtained by changing the input distribution. We show in Fig. 4.8 (a)[3] the CM-MI $\mathsf{I}^{\mathrm{cm}}(\mathsf{snr}, \mathbf{X}_{8\mathrm{PAM}}, \boldsymbol{p}_{\mathrm{s,u}})$ (8PAM with a uniform input distribution), as well as $\mathsf{C}^{\mathrm{cm}}(\mathsf{snr}, \mathbf{X}_{8\mathrm{PAM}})$ in (4.27), $\mathsf{C}^{\mathrm{cm}}(\mathsf{snr}, \boldsymbol{p}_{\mathrm{s,u}})$ in (4.28), and $\mathsf{C}^{\mathrm{cm}}(\mathsf{snr}, \hat{\eta}\mathbf{X}_{8\mathrm{PAM}})$ in (4.29). These results show that the gains offered by probabilistic and geometric shaping are quite small; however, when the mixed optimization in (4.29) is done (i.e., when the probability and the gain η are jointly optimized), the gap to the AWGN capacity is closed for any $R \leq 2$ bit/symbol. To clearly observe this effect, we show in Fig. 4.8 (b) the MI gains offered (with respect to $\mathsf{I}^{\mathrm{cm}}(\mathsf{snr}, \mathbf{X}_{8\mathrm{PAM}}, \boldsymbol{p}_{\mathrm{s,u}})$) by the two approaches as well

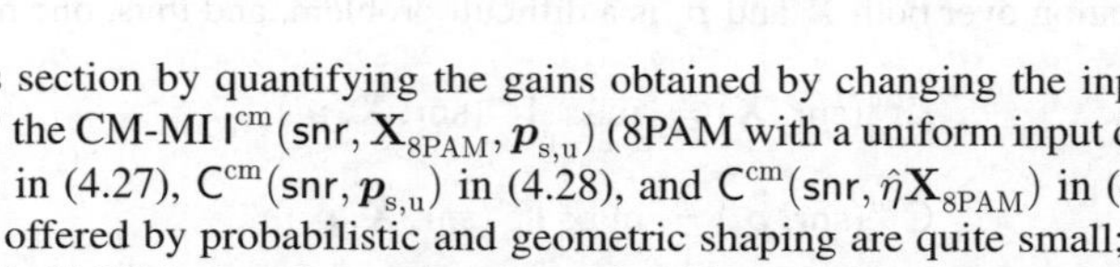

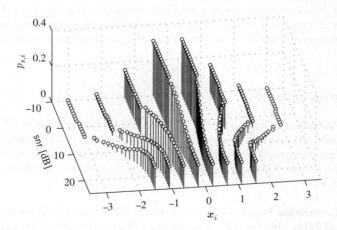

Figure 4.7 Solution of problem (4.29): optimal PMF for an equidistant 8PAM constellation in the AWGN channel

[3] To better appreciate the gains, the curves are plotted versus $\mathsf{E_b/N_0}$.

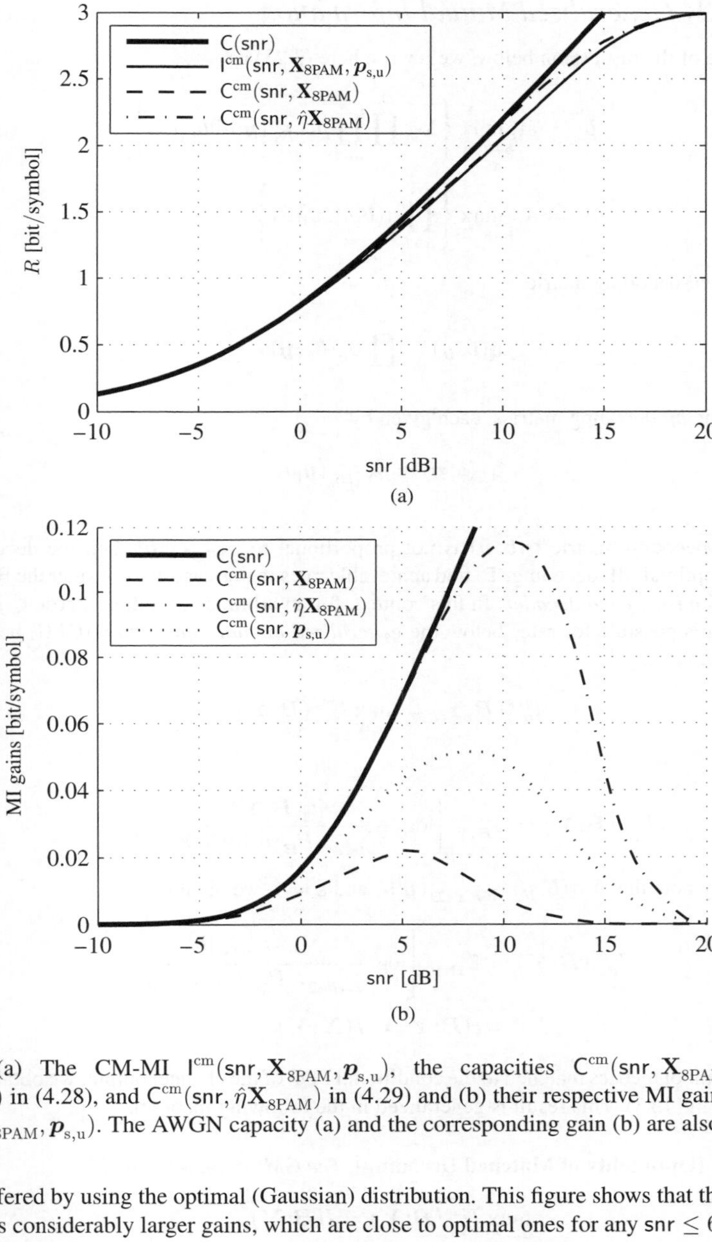

Figure 4.8 (a) The CM-MI $\mathsf{I}^{\mathrm{cm}}(\mathsf{snr}, \mathbf{X}_{8\mathrm{PAM}}, \boldsymbol{p}_{\mathrm{s,u}})$, the capacities $\mathsf{C}^{\mathrm{cm}}(\mathsf{snr}, \mathbf{X}_{8\mathrm{PAM}})$ in (4.27), $\mathsf{C}^{\mathrm{cm}}(\mathsf{snr}, \boldsymbol{p}_{\mathrm{s,u}})$ in (4.28), and $\mathsf{C}^{\mathrm{cm}}(\mathsf{snr}, \hat{\eta}\mathbf{X}_{8\mathrm{PAM}})$ in (4.29) and (b) their respective MI gains with respect to $\mathsf{I}^{\mathrm{cm}}(\mathsf{snr}, \mathbf{X}_{8\mathrm{PAM}}, \boldsymbol{p}_{\mathrm{s,u}})$. The AWGN capacity (a) and the corresponding gain (b) are also shown

as the gain offered by using the optimal (Gaussian) distribution. This figure shows that the optimization in (4.29) gives considerably larger gains, which are close to optimal ones for any $\mathsf{snr} \leq 6$ dB.

4.3 Bit-Interleaved Coded Modulation

In this section, we are interested in finding the rates that can be reliably used by the BICM transceivers shown in Fig. 2.7. We start by defining achievable rates for BICM transceivers with arbitrary input distributions, and we then move to study the problem of optimizing the system's parameters to increase these rates.

4.3.1 BICM Generalized Mutual Information

For the purpose of the discussion below, we rewrite here (3.23) as

$$\hat{\underline{\boldsymbol{b}}}^{\mathrm{bi}} = \underset{\underline{\boldsymbol{b}} \in \mathcal{B}}{\operatorname{argmax}} \left\{ \log \prod_{n=1}^{N_\mathrm{s}} \prod_{k=1}^{m} p_{\boldsymbol{Y}|B_k}(\boldsymbol{y}[n]|b_k[n]) \right\} \tag{4.30}$$

$$= \underset{\underline{\boldsymbol{b}} \in \mathcal{B}}{\operatorname{argmax}} \left\{ \prod_{n=1}^{N_\mathrm{s}} \mathbb{q}(\boldsymbol{b}[n], \boldsymbol{y}[n]) \right\}, \tag{4.31}$$

where the *symbol*-decoding metric

$$\mathbb{q}(\boldsymbol{b}, \boldsymbol{y}) = \prod_{k=1}^{m} \mathbb{q}_k(b_k, \boldsymbol{y}) \tag{4.32}$$

is defined via the *bit* decoding metrics, each given by

$$\mathbb{q}_k(b, \boldsymbol{y}) = p_{\boldsymbol{Y}|B_k}(\boldsymbol{y}|b). \tag{4.33}$$

The symbol-decoding metric $\mathbb{q}(\boldsymbol{b}, \boldsymbol{y})$ is not proportional to $p_{\boldsymbol{Y}|\boldsymbol{B}}(\boldsymbol{y}|\boldsymbol{b})$, i.e., the decoder does not implement the optimal ML decoding. To find achievable rates in this case, we consider the BICM decoder as the so-called *mismatched decoder*. In this context, for an arbitrary decoding metric $\mathbb{q}(\boldsymbol{b}, \boldsymbol{y})$, reliable communication is possible for rates below the *generalized mutual information* (GMI) between $\boldsymbol{B}$ and $\boldsymbol{Y}$, which is defined as

$$I_{\mathbb{q}}^{\mathrm{gmi}}(\boldsymbol{B}; \boldsymbol{Y}) \triangleq \max_{s \geq 0} I_{\mathbb{q},s}^{\mathrm{gmi}}(\boldsymbol{B}; \boldsymbol{Y}), \tag{4.34}$$

where

$$I_{\mathbb{q},s}^{\mathrm{gmi}}(\boldsymbol{B}; \boldsymbol{Y}) \triangleq \mathbb{E}_{\boldsymbol{B},\boldsymbol{Y}} \left[\log_2 \frac{[\mathbb{q}(\boldsymbol{B}, \boldsymbol{Y})]^s}{\sum_{\boldsymbol{b} \in \mathbb{B}^m} P_{\boldsymbol{B}}(\boldsymbol{b}) [\mathbb{q}(\boldsymbol{b}, \boldsymbol{Y})]^s} \right]. \tag{4.35}$$

We immediately note that if $\mathbb{q}(\boldsymbol{b}, \boldsymbol{y}) \propto p_{\boldsymbol{Y}|\boldsymbol{B}}(\boldsymbol{y}|\boldsymbol{b})$ and $s = 1$, we obtain

$$I_{\mathbb{q},s}^{\mathrm{gmi}}(\boldsymbol{B}; \boldsymbol{Y}) = \mathbb{E}_{\boldsymbol{B},\boldsymbol{Y}} \left[\log_2 \frac{p_{\boldsymbol{Y}|\boldsymbol{B}}(\boldsymbol{y}|\boldsymbol{b})}{\sum_{\boldsymbol{b} \in \mathbb{B}^m} P_{\boldsymbol{B}}(\boldsymbol{b}) p_{\boldsymbol{Y}|\boldsymbol{B}}(\boldsymbol{y}|\boldsymbol{b})} \right] \tag{4.36}$$

$$= I(\boldsymbol{B}; \boldsymbol{Y}) = I(\boldsymbol{X}; \boldsymbol{Y}), \tag{4.37}$$

that is, using symbol metrics matched to the conditional PDF of the channel output, we obtain the CM-MI $\mathsf{I}^{\mathrm{cm}}(\mathsf{snr}, \mathbf{X}, \boldsymbol{p}_\mathrm{s})$ in (4.23). This result is generalized in the following theorem.

Theorem 4.10 (Optimality of Matched Decoding) *The GMI in* (4.34) *fulfills*

$$\max_{\mathbb{q}, s \geq 0} I_{\mathbb{q},s}^{\mathrm{gmi}}(\boldsymbol{B}; \boldsymbol{Y}) = I(\boldsymbol{B}; \boldsymbol{Y}), \tag{4.38}$$

which is achieved by $\mathbb{q}(\boldsymbol{b}, \boldsymbol{y}) \propto p_{\boldsymbol{Y}|\boldsymbol{B}}(\boldsymbol{y}|\boldsymbol{b})$ *and* $s = 1$.

Proof: Using the definition of the MI and (4.35), we can write

$$I(\boldsymbol{B};\boldsymbol{Y}) - I^{\text{gmi}}_{\mathbb{q},s}(\boldsymbol{B};\boldsymbol{Y}) = \mathbb{E}_{\boldsymbol{B},\boldsymbol{Y}}\left[\log_2 \frac{P_{\boldsymbol{B}|\boldsymbol{Y}}(\boldsymbol{B}|\boldsymbol{Y})}{P_{\boldsymbol{B}}(\boldsymbol{B})}\right] - \mathbb{E}_{\boldsymbol{B},\boldsymbol{Y}}\left[\log_2 \frac{[\mathbb{q}(\boldsymbol{B},\boldsymbol{Y})]^s}{\sum_{\boldsymbol{b}\in\mathbb{B}^m} P_{\boldsymbol{B}}(\boldsymbol{b})[\mathbb{q}(\boldsymbol{b},\boldsymbol{Y})]^s}\right] \tag{4.39}$$

$$= \mathbb{E}_{\boldsymbol{B},\boldsymbol{Y}}\left[\log_2 \frac{P_{\boldsymbol{B}|\boldsymbol{Y}}(\boldsymbol{B}|\boldsymbol{Y})}{P_{\boldsymbol{B}}(\boldsymbol{B})[\mathbb{q}(\boldsymbol{b},\boldsymbol{Y})]^s / \sum_{\boldsymbol{b}\in\mathbb{B}^m} P_{\boldsymbol{B}}(\boldsymbol{b})[\mathbb{q}(\boldsymbol{b},\boldsymbol{Y})]^s}\right] \tag{4.40}$$

$$= \mathbb{E}_{\boldsymbol{Y}}[t(\boldsymbol{Y})], \tag{4.41}$$

where

$$t(\boldsymbol{y}) \triangleq \mathbb{E}_{\boldsymbol{B}|\boldsymbol{Y}=\boldsymbol{y}}\left[\log_2 \frac{P_{\boldsymbol{B}|\boldsymbol{Y}}(\boldsymbol{B}|\boldsymbol{y})}{P_{\boldsymbol{B}}(\boldsymbol{B})[\overline{\mathbb{q}}(\boldsymbol{B},\boldsymbol{y})]^s}\right], \tag{4.42}$$

and where

$$\overline{\mathbb{q}}(\boldsymbol{b},\boldsymbol{y}) \triangleq \mathbb{q}(\boldsymbol{b},\boldsymbol{y}) \cdot \Upsilon(\boldsymbol{y}), \tag{4.43}$$

$$\frac{1}{[\Upsilon(\boldsymbol{y})]^s} \triangleq \sum_{\boldsymbol{b}\in\mathbb{B}^m} P_{\boldsymbol{B}}(\boldsymbol{b})[\mathbb{q}(\boldsymbol{b},\boldsymbol{y})]^s. \tag{4.44}$$

For any $\boldsymbol{y} \in \mathbb{R}^N$, $P_{\boldsymbol{B}}(\boldsymbol{b})[\overline{\mathbb{q}}(\boldsymbol{b},\boldsymbol{y})]^s \geq 0$ and

$$\sum_{\boldsymbol{b}\in\mathbb{B}^m} P_{\boldsymbol{B}}(\boldsymbol{b})[\overline{\mathbb{q}}(\boldsymbol{b},\boldsymbol{y})]^s = [\Upsilon(\boldsymbol{y})]^s \sum_{\boldsymbol{b}\in\mathbb{B}^m} P_{\boldsymbol{B}}(\boldsymbol{b})[\mathbb{q}(\boldsymbol{b},\boldsymbol{y})]^s = 1, \tag{4.45}$$

and thus, $P_{\boldsymbol{B}}(\boldsymbol{b})[\overline{\mathbb{q}}(\boldsymbol{b},\boldsymbol{y})]^s$ can be interpreted as a PMF for the variable $\boldsymbol{B}$. Therefore, (4.42) is the relative entropy between the PMFs $P_{\boldsymbol{B}|\boldsymbol{Y}}(\boldsymbol{B}|\boldsymbol{y})$ and $P_{\boldsymbol{B}}(\boldsymbol{B})[\overline{\mathbb{q}}(\boldsymbol{B},\boldsymbol{y})]^s$. Owing to the nonnegativity of the entropy, $t(\boldsymbol{y}) \geq 0$, and because of (4.41), the relationship $I(\boldsymbol{B};\boldsymbol{Y}) \geq I^{\text{gmi}}_{\mathbb{q},s}(\boldsymbol{B};\boldsymbol{Y})$ holds for any $\mathbb{q}$ and s. The proof is completed by noting that because of (4.36)–(4.37), equality is obtained when $\mathbb{q}(\boldsymbol{b},\boldsymbol{y}) \propto p_{\boldsymbol{Y}|\boldsymbol{B}}(\boldsymbol{y}|\boldsymbol{b})$ and $s = 1$. □

The following theorem gives an expression for $I^{\text{gmi}}_{\mathbb{q},s}(\boldsymbol{B};\boldsymbol{Y})$ in (4.35) when the decoder uses a symbol metric $\mathbb{q}$ given by (4.32) and an arbitrary bit metric $\mathbb{q}_k$.

Theorem 4.11 *If the bits $B_1, \ldots, B_m$ are independent and the decoding is based on metrics satisfying (4.32), the following holds for any $\mathbb{q}_k$*

$$I^{\text{gmi}}_{\mathbb{q},s}(\boldsymbol{B};\boldsymbol{Y}) = \sum_{k=1}^{m} I^{\text{gmi}}_{\mathbb{q}_k,s}(B_k;\boldsymbol{Y}), \tag{4.46}$$

where

$$I^{\text{gmi}}_{\mathbb{q}_k,s}(B_k;\boldsymbol{Y}) = \mathbb{E}_{B_k,\boldsymbol{Y}}\left[\log_2 \frac{[\mathbb{q}_k(B_k,\boldsymbol{Y})]^s}{\sum_{b\in\mathbb{B}} P_{B_k}(b)[\mathbb{q}_k(b,\boldsymbol{Y})]^s}\right]. \tag{4.47}$$

Proof: Using (4.32) and $P_{\boldsymbol{B}}(\boldsymbol{b}) = \prod_{k=1}^{m} P_{B_k}(B_k)$ (see (2.72)) in (4.35), we obtain

$$I_{\mathbb{q},s}^{\text{gmi}}(\boldsymbol{B};\boldsymbol{Y}) = \mathbb{E}_{\boldsymbol{B},\boldsymbol{Y}}\left[\log_2 \frac{\prod_{k=1}^{m}[\mathbb{q}_k(B_k,\boldsymbol{Y})]^s}{\sum_{\boldsymbol{b}\in\mathbb{B}^m}\prod_{k=1}^{m}P_{B_k}(B_k)[\mathbb{q}_k(B_k,\boldsymbol{Y})]^s}\right] \tag{4.48}$$

$$= \mathbb{E}_{\boldsymbol{B},\boldsymbol{Y}}\left[\log_2 \frac{\prod_{k=1}^{m}[\mathbb{q}_k(B_k,\boldsymbol{Y})]^s}{\prod_{k=1}^{m}\sum_{b\in\mathbb{B}}P_{B_k}(b)[\mathbb{q}_k(b,\boldsymbol{Y})]^s}\right] \tag{4.49}$$

$$= \sum_{k=1}^{m}\mathbb{E}_{B_1,\ldots,B_m,\boldsymbol{Y}}\left[\log_2 \frac{[\mathbb{q}_k(B_k,\boldsymbol{Y})]^s}{\sum_{b\in\mathbb{B}}P_{B_k}(b)[\mathbb{q}_k(b,\boldsymbol{Y})]^s}\right] \tag{4.50}$$

$$= \sum_{k=1}^{m}\mathbb{E}_{B_k,\boldsymbol{Y}}\left[\log_2 \frac{[\mathbb{q}_k(B_k,\boldsymbol{Y})]^s}{\sum_{b\in\mathbb{B}}P_{B_k}(b)[\mathbb{q}_k(b,\boldsymbol{Y})]^s}\right], \tag{4.51}$$

where to pass from (4.48) to (4.49) we use the relationship

$$\sum_{a_1\in\mathcal{A}}\sum_{a_2\in\mathcal{A}}\cdots\sum_{a_m\in\mathcal{A}}[a_1\cdot\ldots\cdot a_m] = \left[\sum_{a_1\in\mathcal{A}}a_1\right]\cdot\left[\sum_{a_2\in\mathcal{A}}a_2\right]\cdot\ldots\cdot\left[\sum_{a_m\in\mathcal{A}}a_m\right]. \tag{4.52}$$

□

Corollary 4.12 *If the metrics $\mathbb{q}_k(\cdot)$ enter the decoding via the product* (4.32)*, the maximum GMI given by*

$$\max_{\mathbb{q}_1,\ldots,\mathbb{q}_m,s\geq 0}\sum_{k=1}^{m}I_{\mathbb{q}_k,s}^{\text{gmi}}(B_k;\boldsymbol{Y}) = \sum_{k=1}^{m}I(B_k;\boldsymbol{Y}) \tag{4.53}$$

is achieved by $\mathbb{q}_k(b,\boldsymbol{y}) \propto p_{\boldsymbol{Y}|B_k}(\boldsymbol{y}|b)$ and $s=1$.

Proof: By applying Theorem 4.10 to each of the terms in the *right-hand side* (r.h.s.) of (4.46). □

Corollary 4.12 shows that when the symbol metrics are constrained to follow (4.32), the best we can do is to use the bits metrics (4.33). The resulting achievable rates lead to the following definition of the *BICM generalized mutual information* (BICM-GMI) for fading and nonfading channels as

$$\bar{\mathsf{I}}^{\text{bi}}(\overline{\mathsf{snr}},\mathbf{X},\boldsymbol{p}_{\text{b}},\mathbf{Q}) \triangleq \sum_{k=1}^{m}I(B_k;\boldsymbol{Y}|H), \tag{4.54}$$

$$\mathsf{I}^{\text{bi}}(\mathsf{snr},\mathbf{X},\boldsymbol{p}_{\text{b}},\mathbf{Q}) \triangleq \sum_{k=1}^{m}I(B_k;\boldsymbol{Y}), \tag{4.55}$$

where the dependence on the constellation symbols $\mathbf{X}$, their labeling $\mathbf{Q}$, and the bits' PMF $\boldsymbol{p}_{\text{b}} = [P_{B_1}(0),\ldots,P_{B_m}(0)]^{\text{T}}$ is made explicit in the argument of $\mathsf{I}^{\text{bi}}(\cdot)$. The bitwise MIs necessary to calculate the BICM-GMI in (4.54)–(4.55) are given by

$$I(B_k;\boldsymbol{Y}|H) = \mathbb{E}_{B_k,H,\boldsymbol{Y}}\left[\log_2 \frac{p_{\boldsymbol{Y}|H,B_k}(\boldsymbol{Y}|H,B_k)}{p_{\boldsymbol{Y}|H}(\boldsymbol{Y}|H)}\right], \tag{4.56}$$

$$I(B_k;\boldsymbol{Y}) = \mathbb{E}_{B_k,\boldsymbol{Y}|H=h}\left[\log_2 \frac{p_{\boldsymbol{Y}|H,B_k}(\boldsymbol{Y}|h,B_k)}{p_{\boldsymbol{Y}|H}(\boldsymbol{Y}|h)}\right]. \tag{4.57}$$

At this point, some observations are in order:

- The MI in (4.22) is an achievable rate for any CM transmission, provided that the receiver implements the ML decoding rule. The BICM-GMI in (4.54) and (4.55) is an achievable rate for the (suboptimal) BICM decoder.
- The BICM-GMI was originally derived without the notion of mismatched decoding, but based on an equivalent channel model where the interface between the interleaver and deinterleaver in Fig. 2.7 is replaced by m parallel memoryless *binary-input continuous-output* (BICO) channels. Such a model requires the assumption of a quasi-random interleaver. Hence, the value of Theorem 4.11 is that it holds independently of the interleaver's presence.
- No converse decoding theorem exists, i.e., while the BICM-GMI defines an achievable rate, it has not been proved to be the *largest* achievable rate when using a mismatched (BICM) decoder. This makes difficult a rigorous comparison of BICM transceivers because we cannot guarantee that differences in the BICM-GMI will translate into differences between the maximum achievable rates. However, practical coding schemes most often closely follow and do not exceed the BICM-GMI. This justifies its practical use as a design rule for the BICM. We illustrate this in the following example.

Example 4.13 (BICM with Turbo Encoders) *Consider a BICM transmission with a* 4PAM *or an* 8PAM *constellation labeled by the binary reflected Gray code (BRGC) over the AWGN channel. The encoder is based on a parallel concatenated convolutional encoder (PCCENC) (i.e., a turbo encoder (TENC) like the one shown in Example 2.31) with feedforward and feedback generator polynomials* 11 *and* 15*, respectively. In the numerical simulations, we used* $N_\mathrm{s} = 1548$ *and a pseudorandom interleaver* Π_tc*. The decoder was based on the Log-MAP decoding algorithm with 10 iterations. In order to obtain a family of 11 different code rates* $R_\mathrm{c} \in \{1/3, 4/11, 2/5, 4/9, 1/2, 5/9, 6/10, 2/3, 3/4, 4/5, 5/6\}$*, the parity bits were periodically punctured. To obtain an estimate of the word-error probability (WEP), which we denote by* WEP*, at least* 200 *blocks in error were counted for each value of SNR.*

We consider the effective throughput $T \triangleq mR_\mathrm{c}(1 - \mathsf{WEP})$*, which may be interpreted as an achievable rate for small values of WEP. In other words, when* WEP *is very small, the transmission is, for all practical purposes, error-free, and thus,* $T \approx mR_\mathrm{c}$*. We show the estimated throughput in Fig. 4.9, where we see that* T *is below the BICM-GMI curve, but follows it relatively closely. As a reference, we also include the CM-MI. The horizontal gap between the throughput of the TENC-BICM system and the BICM-GMI is approximately* 1.5 dB*, while the vertical gap is approximately* 0.2 bit/symbol.

While the quantities (4.54) and (4.55) were originally called the *BICM capacity*, we avoid using the term "capacity" to point out that no optimization of the input distribution is carried out. Using (4.56), the BICM-GMI in (4.54) can be expressed as

$$\bar{\mathsf{I}}^{\mathrm{bi}}(\mathsf{snr}, \mathbf{X}, \boldsymbol{p}_\mathrm{b}, \mathbf{Q}) = \sum_{k=1}^{m} \sum_{b\in\mathbb{B}} P_{B_k}(b) \mathbb{E}_{H,\boldsymbol{Y}|B_k=b}\left[\log_2 \frac{p_{\boldsymbol{Y}|H,B_k}(\boldsymbol{Y}|H,b)}{p_{\boldsymbol{Y}|H}(\boldsymbol{Y}|H)}\right] \tag{4.58}$$

$$= \sum_{k=1}^{m} \sum_{b\in\mathbb{B}} P_{B_k}(b) \sum_{i\in\mathcal{I}} P_{\boldsymbol{X}|B_k}(\boldsymbol{x}_i|b) \cdot \mathbb{E}_{H,\boldsymbol{Y}|\boldsymbol{X}=\boldsymbol{x}_i,B_k=b}\left[\log_2 \frac{p_{\boldsymbol{Y}|H,\boldsymbol{X},B_k}(\boldsymbol{Y}|H,\boldsymbol{x}_i,b)}{p_{\boldsymbol{Y}|\boldsymbol{X},H}(\boldsymbol{Y}|\boldsymbol{x}_i,H)}\right], \tag{4.59}$$

where to pass from (4.58) to (4.59) we used the law of total probability applied to expectations. Moreover, by using (2.75) and by expanding the expectation $\mathbb{E}_{H,\boldsymbol{Y}|\boldsymbol{X},B_k}[\cdot]$ over H and then over $\boldsymbol{Y}$, we can express (4.59) as

$$\bar{\mathsf{I}}^{\mathrm{bi}}(\overline{\mathsf{snr}}, \mathbf{X}, \boldsymbol{p}_{\mathrm{b}}, \mathbf{Q}) = \sum_{k=1}^{m} \sum_{b \in \mathbb{B}} \sum_{i \in \mathcal{I}_{k,b}} p_{\mathrm{s},i} \int_{\mathbb{R}} p_H(h) \cdot \mathbb{E}_{\boldsymbol{Y}|\boldsymbol{X}=\boldsymbol{x}_i, B_k=b, H=h} \left[\log_2 \frac{p_{\boldsymbol{Y}|\boldsymbol{X},B_k,H}(\boldsymbol{Y}|\boldsymbol{x}_i, b, h)}{p_{\boldsymbol{Y}|\boldsymbol{X},H}(\boldsymbol{Y}|\boldsymbol{x}_i, h)} \right] \mathrm{d}h \tag{4.60}$$

$$= \sum_{k=1}^{m} \sum_{b \in \mathbb{B}} \sum_{i \in \mathcal{I}_{k,b}} p_{\mathrm{s},i} \int_{\mathbb{R}} p_H(h) \int_{\mathbb{R}^N} p_{\boldsymbol{Y}|\boldsymbol{X},H}(\boldsymbol{y}|\boldsymbol{x}_i, h) \cdot \log_2 \frac{\frac{1}{P_{B_k}(b)} \sum_{j \in \mathcal{I}_{k,b}} p_{\mathrm{s},j} p_{\boldsymbol{Y}|\boldsymbol{X},H}(\boldsymbol{y}|\boldsymbol{x}_j, h)}{\sum_{j' \in \mathcal{I}} p_{\mathrm{s},j'} p_{\boldsymbol{Y}|\boldsymbol{X},H}(\boldsymbol{y}|\boldsymbol{x}_{j'}, h)} \, \mathrm{d}\boldsymbol{y} \, \mathrm{d}h, \tag{4.61}$$

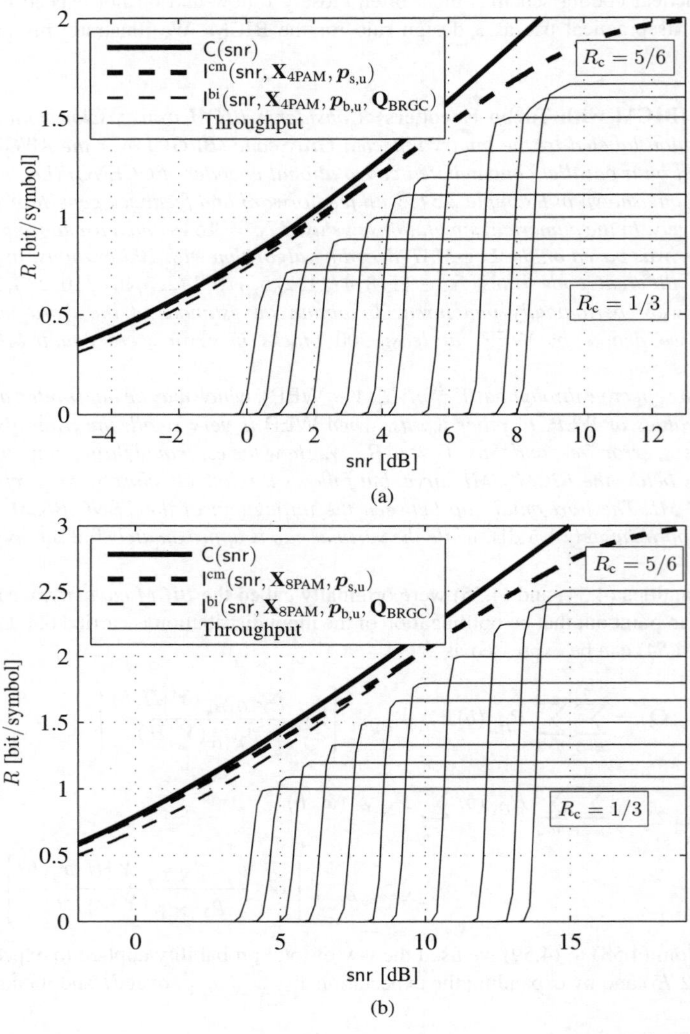

(b)

Figure 4.9 Throughputs T obtained for the AWGN channel and a PCCENC TENC with 11 different code rates R_c for (a) 4PAM and (b) 8PAM labeled by the BRGC. The CM-MI, BICM-GMI, and AWGN capacity $\mathsf{C}(\mathsf{snr})$ in (4.9) are also shown where to pass from (4.60) to (4.61), we used (2.75) and the fact that the value of B_k does not affect the conditional channel transition probability, i.e., $p_{\boldsymbol{Y}|\boldsymbol{X},B_k,H}(\boldsymbol{y}|\boldsymbol{x},b,h) = p_{\boldsymbol{Y}|\boldsymbol{X},H}(\boldsymbol{y}|\boldsymbol{x},h)$. The dependence of the BICM-GMI on the labeling $\mathbf{Q}$ becomes evident because the sets $\mathcal{I}_{k,b}$ appear in (4.61), and as we showed in Section 2.5.2, these sets define the subconstellations generated by the labeling.

For AWGN channels ($h = 1$), the BICM-GMI in (4.61) can be expressed as

$$\mathsf{I}^{\mathrm{bi}}(\mathsf{snr},\mathbf{X},\boldsymbol{p}_{\mathrm{b}},\mathbf{Q}) = \int_{\mathbb{R}^N} \sum_{k=1}^{m}\sum_{b\in\mathbb{B}}\sum_{i\in\mathcal{I}_{k,b}} p_{\mathrm{s},i}\left(\frac{\mathsf{snr}}{\pi}\right)^{\frac{N}{2}} \exp(-\mathsf{snr}\|\boldsymbol{y}-\boldsymbol{x}_i\|^2) \cdot \log_2 \frac{\frac{1}{P_{B_k}(b)}\sum_{j\in\mathcal{I}_{k,b}} p_{\mathrm{s},j}\exp(-\mathsf{snr}\|\boldsymbol{y}-\boldsymbol{x}_j\|^2)}{\sum_{j'\in\mathcal{I}} p_{j'}\exp(-\mathsf{snr}\|\boldsymbol{y}-\boldsymbol{x}_{j'}\|^2)}\,\mathrm{d}\boldsymbol{y}. \tag{4.62}$$

Furthermore, for uniformly distributed bits, $P_{B_k}(0) = 1/2$ for $k = 1,\ldots,m$, and thus, the BICM-GMI in (4.62) becomes

$$\mathsf{I}^{\mathrm{bi}}(\mathsf{snr},\mathbf{X},\boldsymbol{p}_{\mathrm{b,u}},\mathbf{Q}) = m - \int_{\mathbb{R}^N} \frac{1}{M}\left(\frac{\mathsf{snr}}{\pi}\right)^{\frac{N}{2}} \sum_{k=1}^{m}\sum_{b\in\mathbb{B}}\sum_{i\in\mathcal{I}_{k,b}} \exp(-\mathsf{snr}\|\boldsymbol{y}-\boldsymbol{x}_i\|^2) \cdot \log_2 \frac{\sum_{j\in\mathcal{I}_{k,b}}\exp(-\mathsf{snr}\|\boldsymbol{y}-\boldsymbol{x}_j\|^2)}{\sum_{j'\in\mathcal{I}}\exp(-\mathsf{snr}\|\boldsymbol{y}-\boldsymbol{x}_{j'}\|^2)}\,\mathrm{d}\boldsymbol{y}, \tag{4.63}$$

where we use $\boldsymbol{p}_{\mathrm{b,u}}$ to denote uniformly distributed bits, i.e., $\boldsymbol{p}_{\mathrm{b,u}} = [0.5, 0.5, \ldots, 0.5]^{\mathrm{T}}$. In what follows, we show examples of the BICM-GMI for different constellations $\mathbf{X}$ and labelings $\mathbf{Q}$ with uniform input distributions.

Example 4.14 (BICM-GMI for 8PSK and 16PSK constellations) *In Fig. 4.10 (a) we show the CM-MI and BICM-GMI in* (4.63) *for* 8PSK *and* 16PSK. *The BICM-GMI is shown for two binary labelings, the BRGC and the natural binary code (NBC), given by Definitions 2.10 and 2.11, respectively. We can clearly observe the effect of the labeling on the BICM-GMI as well as the suboptimality of the latter with respect to the CM-MI (see* (4.104)*). To better appreciate these two observations, in Fig. 4.10 (b) we show the same results in a plot versus* $\mathsf{E_b}/\mathsf{N_0}$. *This figure also shows the suboptimality of BICM in the low SNR regime, i.e., the function* $f_{\mathsf{I}^{\mathrm{bi}}}(R,\cdot)$ *does not achieve the SL when* $R \to 0$.

Example 4.15 (Binary Labelings for 8PAM**)** *In Fig. 4.11 (a), we show the BICM-GMI in* (4.63) *and the CM-MI in* (4.24) *for* 8PAM *with a uniform distribution and for the four binary labelings in Fig. 2.13. From these curves, we can see that the difference between the CM capacity and the BICM-GMI is small if the binary labeling is properly selected. From the four labelings we analyzed, those offering the highest MI are the NBC for low SNR (*$R \leq 0.43$ bit/symbol*),*[4] *the folded binary code (FBC) for medium SNR (*$0.43 \leq R \leq 1.09$ bit/symbol*), and the BRGC for high SNR (*$R \geq 1.09$ bit/symbol*). On the other hand, the gap between the CM-MI and the BICM-GMI for the binary semi-Gray code (BSGC) is quite large. Fig. 4.11 shows that the BRGC is largely suboptimal, not being the best choice of labeling for at least* 36% *of the achievable rates R (*$R \leq 1.09$ bit/symbol*). In Fig. 4.11 (b), we show the CM capacity and BICM-GMI versus* $\mathsf{E_b}/\mathsf{N_0}$*, for the same cases as in Fig. 4.11 (a). This figure clearly shows how the NBC and the FBC outperform the BRGC in the low SNR region and it also shows a particular property of the BSGC. When* $R \to 0$ *(or equivalently when* $\mathsf{snr} \to 0$*), the* $\mathsf{E_b}/\mathsf{N_0}$ *needed for reliable transmission grows to infinity. Moreover, this figure also shows that the NBC achieves the SL, which is not the case for any of the other labelings.*

[4] In fact, the NBC was recently shown to be the unique optimal labeling for MPAM constellations and asymptotically low SNR.

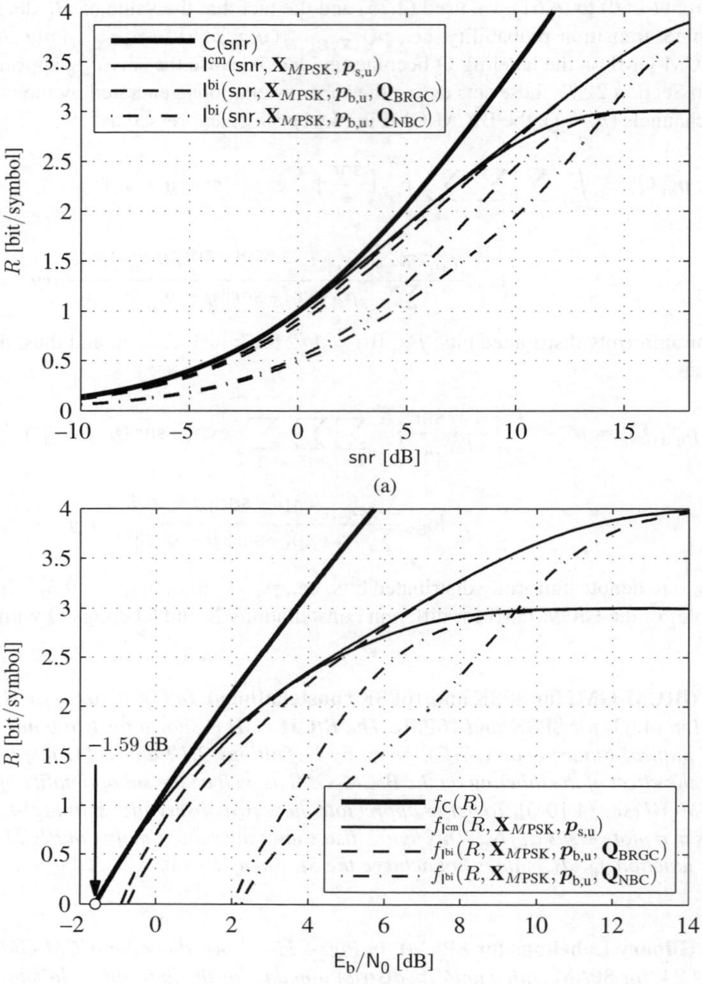

Figure 4.10 BICM-GMI and CM-MI for 8PSK and 16PSK (see Fig. 2.9) over the AWGN channel (a) and the corresponding functions $f_{\mathsf{I}}(R, \cdot)$ in (4.16) (b). The AWGN capacity $\mathsf{C}(\mathsf{snr})$ in (4.9) ($N = 2$) (a) and the function $f_{\mathsf{C}}(R)$ in (4.17) together with the SL in (4.18) (b) are also shown

Example 4.16 (Binary Labelings for PSK) *In Fig. 4.12, we show the BICM-GMI in* (4.63) *and* 4PSK *and* 8PSK *constellations. This figure shows the BICM-GMI for all binary labelings that give a different BICM-GMI. For* 4PSK*, there are only two (the BRGC and the NBC), and for* 8PSK *there are* 49*. The results in Fig. 4.12 show that the optimum binary labeling for* 4PSK *is the BRGC, but for* 8PSK*, this depends on the SNR. For low SNR, the optimum binary labeling for* 8PSK *is the FBC and for high SNR, the optimum binary labeling is the BRGC. Moreover, this figure shows that no binary labeling for* 8PSK *achieves the SL. This result is, in fact, a particular case of a more general result which says that no binary labeling achieves the SL if MPSK constellations with $M > 4$ are used. More importantly, if we allow the system to switch the constellation (from $M = 4$ to $M = 8$) and the binary labeling, the BRGC is the*

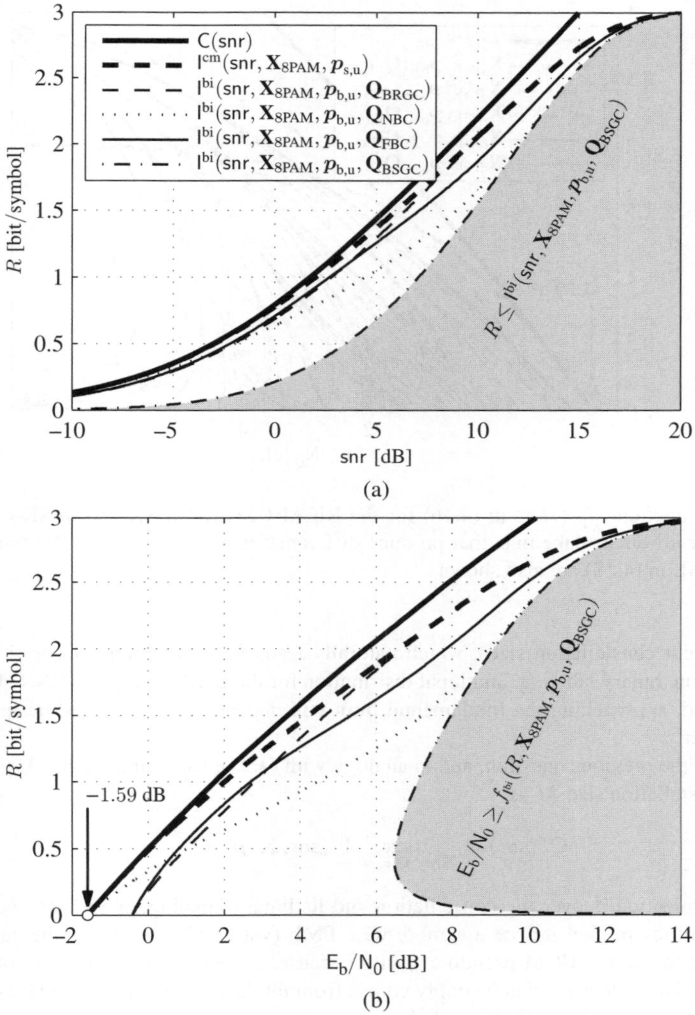

Figure 4.11 BICM-GMI and CM-MI for 8PAM over the AWGN channel (a) and the corresponding functions $f_{\mathsf{I}}(R,\cdot)$ in (4.16) (b). The BICM-GMI is shown for the four binary labelings in Fig. 2.13. The shadowed regions shows the inequalities (4.7) (a) and (4.16) (b) for the BSGC. The AWGN capacity $\mathsf{C}(\mathsf{snr})$ in (4.9) $(N=1)$ (a) and the function $f_{\mathsf{C}}(R)$ in (4.17) together with the SL in (4.18) (b) are also shown

optimum binary labeling for this particular case and any SNR. In addition, we note that the BICM-GMI curves for 8PSK *and high SNR merge into seven different groups.*

4.3.2 BICM Pseudo-Capacity

The results from the previous section point out the fact that binary labeling strongly affects the BICM-GMI. Moreover, we have seen that the BICM-GMI curves for a given binary labeling intersect each

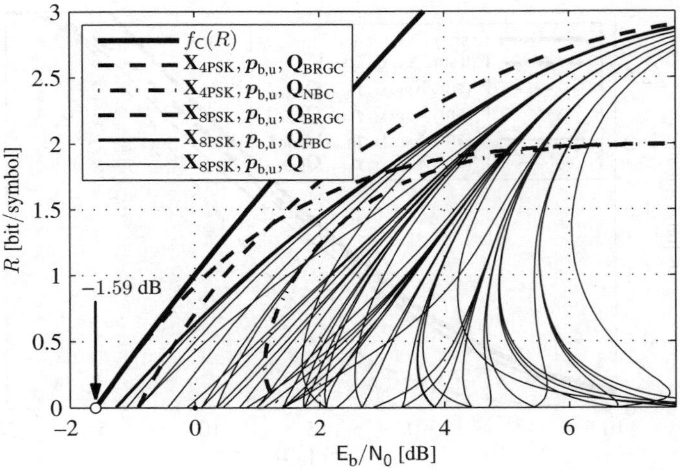

Figure 4.12 Functions $f_{\mathsf{I}^{\mathrm{bi}}}(R, \cdot)$ in (4.16) for the BICM-GMI for 4PSK and 8PSK over the AWGN channel and for all binary labelings that produce different BICM-GMI curves. The function $f_{\mathsf{C}}(R)$ in (4.17) and the SL in (4.18) are also shown

other for different constellation sizes, which naturally leads to a quite general question: What is the best constellation, binary labeling, and input distribution for the BICM at a given SNR? Once this question is answered, approaching the fundamental limit will depend only on a good design of the binary encoder/decoder.

To formalize the previous question, and in analogy with (4.25), we define the *BICM pseudo-capacity* for a given constellation size M as

$$\mathsf{C}^{\mathrm{bi}}(\mathsf{snr}) \triangleq \max_{\mathbf{X}, \boldsymbol{p}_{\mathrm{b}}, \mathbf{Q}: \mathsf{E}_{\mathrm{s}}=1} \mathsf{I}^{\mathrm{bi}}(\mathsf{snr}, \mathbf{X}, \boldsymbol{p}_{\mathrm{b}}, \mathbf{Q}), \tag{4.64}$$

where the optimization is over the constellation and its binary labeling, as well as over the vector of bitwise probabilities $\boldsymbol{p}_{\mathrm{b}}$ that induce a symbol-wise PMF (via (2.72)) satisfying the energy constraint $\mathsf{E}_{\mathrm{s}} = 1$. We use the name BICM pseudo-capacity because, in theory, the quantity in (4.64) might not be the *largest achievable rate*, which simply comes from the fact that the BICM-GMI is not the largest achievable rate for given $\mathbf{X}, \boldsymbol{p}_{\mathrm{b}}, \mathbf{Q}$. Nevertheless, the numerical results we present in this section will, in fact, show that the BICM pseudo-capacity can close the gap to the AWGN capacity, and thus, it is indeed a meaningful quantity to study.

Similarly to the CM capacity, the BICM pseudo-capacity in (4.64) represents an upper bound on the number of bits per symbol that can be reliably communicated using BICM, i.e., when for each SNR, the constellation, its labeling, and the bits' distribution are optimized.

In general, solutions of the problem in (4.64) are difficult to find, even when both the constellation and the labeling are fixed. This is because, unlike the CM-MI, the BICM-GMI is not a concave function of the input distribution. To clarify this, consider the following example.

Example 4.17 (BICM-GMI as a Function of the Bit Probabilities) *In Fig. 4.13, we show the BICM-GMI* $\mathsf{I}^{\mathrm{bi}}(\mathsf{snr}, \mathbf{X}, \boldsymbol{p}_{\mathrm{b}}, \mathbf{Q})$ *for* 8PAM *labeled by the BRGC as a function of the input bit probabilities for* $\mathsf{snr} = 0$ dB. *In order to obtain a distribution* $\boldsymbol{p}_{\mathrm{s}}$ *symmetric with respect to zero, and because of the structure of the BRGC, we set* $P_{B_1}(0) = 1/2$, *and thus, we plot the BICM-GMI as a function of the other two variables, i.e.,* $P_{B_2}(0)$ *and* $P_{B_3}(0)$. *This figure illustrates well that the*

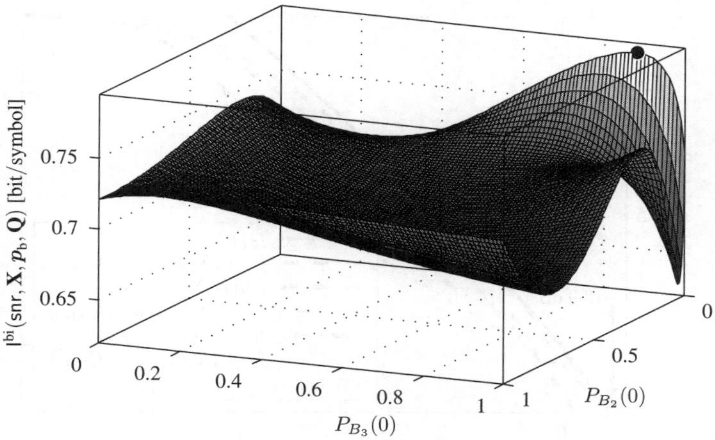

Figure 4.13 BICM-GMI $\mathsf{I}^{\mathrm{bi}}(\mathsf{snr}, \mathbf{X}, \boldsymbol{p}_{\mathrm{b}}, \mathbf{Q})$ for 8PAM labeled by the BRGC as a function of the bit probabilities for $k = 2, 3$ and $\mathsf{snr} = 0$ dB ($P_{B_1}(0) = 1/2$). The filled black circle shows the optimal value

BICM-GMI is a nonconcave and nonconvex function of the bits' PMF. We can find the optimal input distribution for $P_{B_1}(0) = 1/2$ *to be* $[P_{B_2}(0), P_{B_3}(0)] = [0, 0.88]$, *which induces symbol probabilities* $\boldsymbol{p}_{\mathrm{s}} = [0, 0, 0.06, 0.44, 0.44, 0.06, 0, 0]^{\mathrm{T}}$; *this input PMF corresponds to a nonequally probable* 4PAM *constellation.*

Although not explicitly mentioned, the results in Fig. 4.13 are obtained by scaling the constellation for each $\boldsymbol{p}_{\mathrm{b}}$ so that $\mathsf{E}_{\mathrm{s}} = 1$. To formally define the problem in Example 4.17, we define the BICM constellation and labeling-constrained pseudo-capacity as

$$\mathsf{C}^{\mathrm{bi}}(\mathsf{snr}, \mathbf{X}, \mathbf{Q}) \triangleq \max_{\boldsymbol{p}_{\mathrm{b}}, \eta : \mathsf{E}_{\mathrm{s}} = 1} \mathsf{I}^{\mathrm{bi}}(\eta \mathbf{X}, \boldsymbol{p}_{\mathrm{b}}, \mathbf{Q}, \mathsf{snr}), \tag{4.65}$$

where the optimization is over all the $\boldsymbol{p}_{\mathrm{b}}$ and $\eta > 0$ is chosen so that $\mathsf{E}_{\mathrm{s}} = 1$ is satisfied.

For constellations with a small number of points with a fixed binary labeling, we may use an exhaustive search together with an efficient numerical implementation of the BICM-GMI (which we show in Section 4.5) to solve the problem in (4.65). While this approach has obvious complexity limitations, i.e., the complexity grows exponentially with the number of bits per symbol m, it allows us to obtain an insight into the importance of the bit-level probabilistic shaping in BICM.

Example 4.18 (BICM Pseudo-capacity for 8PAM **and Different Labelings)** *In Fig. 4.14, we show the BICM constellation and labeling-constrained pseudo-capacity in* (4.65) *for an* 8PAM *constellation labeled by the BRGC, NBC, and FBC. We performed a grid search (with steps of 0.01) over the three variables defining the input distribution:* $P_{B_1}(0)$, $P_{B_2}(0)$, *and* $P_{B_3}(0)$. *For each SNR value, we selected the input distribution that maximizes the BICM-GMI . The results in this figure show how, by properly selecting the input distribution, the BICM-GMI increases and gets very close to the AWGN capacity for* $R \leq 2$ bit/symbol. *Interestingly, the results in Fig. 4.14 show that if the input distribution is optimized, the NBC is not the optimal binary labeling for low SNR anymore. The figure also shows that the BRGC with an optimized input distribution achieves the SL.*

On the basis of the results in Examples 4.15–4.18, some concluding remarks about the BICM pseudo-capacity can be made. First of all, these results show the suboptimality of the BRGC in terms

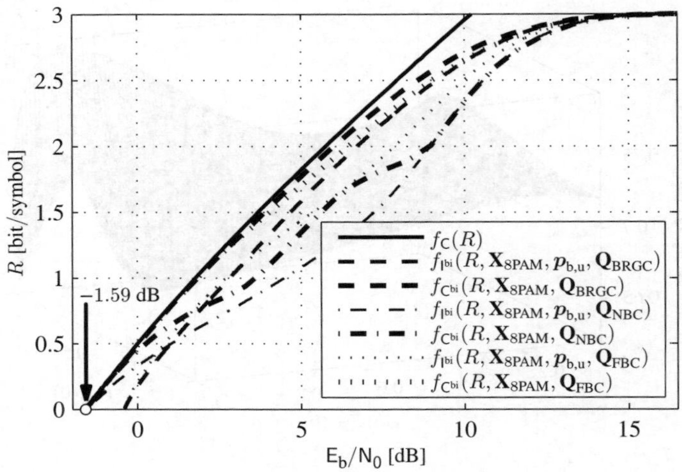

Figure 4.14 Functions $f_{\mathsf{I}^{\mathrm{bi}}}(R, \cdot)$ and $f_{\mathsf{C}^{\mathrm{bi}}}(R, \cdot)$ in (4.16) for 8PAM over the AWGN channel and the BRGC, NBC, and FBC. The function $f_{\mathsf{C}}(R)$ in (4.17) and the SL in (4.18) are also shown

of maximizing the BICM-GMI for a given equally spaced constellation and uniform input distributions. For 8PAM and for low and medium SNR values, the NBC and the FBC give higher BICM capacities than the BRGC. Moreover, for asymptotically low rates, the optimum binary labeling for MPAM constellations is the NBC. The results in these examples also show that a BICM system can be optimum for asymptotically low rates (i.e., it can achieve the SL). This can be obtained by using probabilistic shaping or by properly selecting the binary labeling.

4.3.3 Useful Relationships

In this section, we provide some useful relationships for the analysis of information-theoretic aspects of BICM transmission.

Theorem 4.19 *The BICM-GMI can be expressed as*

$$\mathsf{I}^{\mathrm{bi}}(\overline{\mathsf{snr}}, \mathbf{X}, \boldsymbol{p}_{\mathrm{b}}, \mathbf{Q}) = mI(\boldsymbol{X}; \boldsymbol{Y}|H) - \sum_{k=1}^{m} I(\boldsymbol{X}; \boldsymbol{Y}|H, B_k), \tag{4.66}$$

where $I(\boldsymbol{X}; \boldsymbol{Y}|H)$ and $I(\boldsymbol{X}; \boldsymbol{Y}|H, B_k)$ are given by (4.3) *and* (4.4), *respectively.*

Proof: We express the MI between $\boldsymbol{X}$, B_k and $\boldsymbol{Y}$ conditioned on H as

$$\begin{aligned} I(\boldsymbol{X}, B_k; \boldsymbol{Y}|H) &= I(B_k; \boldsymbol{Y}|H) + I(\boldsymbol{X}; \boldsymbol{Y}|H, B_k) && (4.67)\\ &= I(\boldsymbol{X}; \boldsymbol{Y}|H) + I(B_k; \boldsymbol{Y}|H, \boldsymbol{X}) && (4.68)\\ &= I(\boldsymbol{X}; \boldsymbol{Y}|H), && (4.69) \end{aligned}$$

where both (4.67) and (4.68) follow from the chain rule of MI applied to the *left-hand side* (l.h.s.) of (4.67) and (4.69) from the fact that B_k and $\boldsymbol{Y}$ are independent for given $\boldsymbol{X}$ and H. The proof is completed by combining the r.h.s. of (4.67), the r.h.s. of (4.69), and (4.54). □

To show the usefulness of (4.66), consider (for simplicity) the AWGN channel. In this case, the conditional MI $I(\boldsymbol{X};\boldsymbol{Y}|H,B_k)$ in (4.66) is given by (4.4) and can be expressed as

$$I(\boldsymbol{X};\boldsymbol{Y}|B_k) = \mathbb{E}_{\boldsymbol{X},\boldsymbol{Y},B_k}\left[\log_2 \frac{p_{\boldsymbol{Y}|\boldsymbol{X},B_k}(\boldsymbol{Y}|\boldsymbol{X},B_k)}{p_{\boldsymbol{Y}|B_k}(\boldsymbol{Y}|B_k)}\right] \tag{4.70}$$

$$= \sum_{b\in\mathbb{B}}\sum_{i\in\mathcal{I}} P_{B_k,\boldsymbol{X}}(b,\boldsymbol{x}_i)\mathbb{E}_{\boldsymbol{Y}|B_k=b,\,\boldsymbol{X}=\boldsymbol{x}_i}\left[\log_2 \frac{p_{\boldsymbol{Y}|\boldsymbol{X},B_k}(\boldsymbol{Y}|\boldsymbol{x}_i,b)}{p_{\boldsymbol{Y}|B_k}(\boldsymbol{Y}|b)}\right] \tag{4.71}$$

$$= \sum_{b\in\mathbb{B}} P_{B_k}(b) I(\boldsymbol{X}_{k,b};\boldsymbol{Y}), \tag{4.72}$$

where

$$I(\boldsymbol{X}_{k,b};\boldsymbol{Y}) \triangleq \sum_{i\in\mathcal{I}_{k,b}} P_{\boldsymbol{X}|B_k}(\boldsymbol{x}_i|b)\mathbb{E}_{\boldsymbol{Y}|\boldsymbol{X}=\boldsymbol{x}_i}\left[\log_2 \frac{p_{\boldsymbol{Y}|\boldsymbol{X}}(\boldsymbol{Y}|\boldsymbol{x}_i)}{\sum_{j\in\mathcal{I}_{k,b}} P_{\boldsymbol{X}|B_k}(\boldsymbol{x}_j|b)p_{\boldsymbol{Y}|\boldsymbol{X}}(\boldsymbol{Y}|\boldsymbol{x}_j)}\right], \tag{4.73}$$

and where (4.72) follows from (2.75) and from the fact that $p_{\boldsymbol{Y}|\boldsymbol{X},B_k}(\boldsymbol{y}|\boldsymbol{x},b) = p_{\boldsymbol{Y}|\boldsymbol{X}}(\boldsymbol{y}|\boldsymbol{x}), \forall \boldsymbol{x}\in\mathcal{S}_{k,b}$.

In (4.73), we use $\boldsymbol{X}_{k,b}$ to denote a random variable with support $\mathcal{S}_{k,b}$ and PMF $P_{\boldsymbol{X}|B_k}(\boldsymbol{x}|b)$ given by (2.75), i.e., when the symbols are drawn from the subconstellation $\mathcal{S}_{k,b}$ created by the binary labeling (see, e.g., Fig. 2.16). As the r.h.s. of (4.73) is an MI between $\mathcal{S}_{k,b}$ and $\boldsymbol{Y}$, we use the notation $I(\boldsymbol{X}_{k,b};\boldsymbol{Y})$. Using (4.72) in (4.66) we obtain

$$\mathsf{I}^{\text{bi}}(\mathsf{snr},\mathbf{X},\boldsymbol{p}_{\text{b}},\mathbf{Q}) = \sum_{k=1}^{m}\left(I(\boldsymbol{X};\boldsymbol{Y}) - \sum_{b\in\mathbb{B}} P_{B_k}(b) I(\boldsymbol{X}_{k,b};\boldsymbol{Y})\right). \tag{4.74}$$

The expression in (4.74) shows that the BICM-GMI can be expressed as a difference of CM-MIs. This interpretation is useful because it shows that a good characterization of MIs for arbitrary (sub)constellations leads to a good characterization of the BICM-GMI. This is particularly interesting as plenty of results already exist in the literature regarding CM-MIs, which, because of (4.74), can be "reused" to study the BICM-GMI. This approach has been taken, e.g., to study the low- and high-SNR behavior of BICM, where the asymptotic behavior of CM-MIs for arbitrary constellations is used to study the asymptotic behavior of the BICM-GMI.

Theorem 4.20 *For the GMI in* (4.46) *and arbitrary mismatched symbol metrics* $\mathfrak{q}_k(\boldsymbol{b},\boldsymbol{y})$ *in* (4.32) *the following hold*

$$I^{\text{gmi}}_{\mathfrak{q}_k,s}(B_k;\boldsymbol{Y}) = \mathcal{H}(B_k) - \sum_{b\in\mathbb{B}} P_{B_k}(b)\mathbb{E}_{\tilde{L}_k|B_k=b}\left[\log_2\left(1+\mathrm{e}^{(-1)^b\cdot(s\tilde{L}_k+l^{\text{a}}_k)}\right)\right], \tag{4.75}$$

where $\mathcal{H}(B) \triangleq -\mathbb{E}_B[\log_2(P_B(B))]$ *is the entropy of* B, l^{a}_k *is the a priori L-value given in* (3.32), *and*

$$\tilde{L}_k \triangleq \log\frac{\mathfrak{q}_k(1,\boldsymbol{Y})}{\mathfrak{q}_k(0,\boldsymbol{Y})} \tag{4.76}$$

is a "generalized" L-value.

Proof: Using (4.47) we obtain

$$I^{\text{gmi}}_{\mathfrak{q}_k,s}(B_k;\boldsymbol{Y}) = -\sum_{b\in\mathbb{B}} P_{B_k}(b)\mathbb{E}_{\boldsymbol{Y}|B_k=b}\left[\log_2\frac{\sum_{u\in\mathbb{B}} P_{B_k}(u)[\mathfrak{q}_k(u,\boldsymbol{Y})]^s}{[\mathfrak{q}_k(b,\boldsymbol{Y})]^s}\right] \tag{4.77}$$

$$= -\sum_{b\in\mathbb{B}} P_{B_k}(b)\mathbb{E}_{\boldsymbol{Y}|B_k=b}\left[\log_2\left(P_{B_k}(b) + P_{B_k}(\bar{b})\frac{[\mathbb{q}_k(\bar{b},\boldsymbol{Y})]^s}{[\mathbb{q}_k(b,\boldsymbol{Y})]^s}\right)\right] \tag{4.78}$$

$$= \mathcal{H}(B_k) - \sum_{b\in\mathbb{B}} P_{B_k}(b)\mathbb{E}_{\boldsymbol{Y}|B_k=b}\left[\log_2\left(1+\frac{P_{B_k}(\bar{b})}{P_{B_k}(b)}\mathrm{e}^{(-1)^b s\tilde{L}_k(\boldsymbol{Y})}\right)\right] \tag{4.79}$$

$$= \mathcal{H}(B_k) - \sum_{b\in\mathbb{B}} P_{B_k}(b)\mathbb{E}_{\boldsymbol{Y}|B_k=b}\left[\log_2\left(1+\mathrm{e}^{(-1)^b[s\tilde{L}_k(\boldsymbol{Y})+l_k^{\mathrm{a}}]}\right)\right], \tag{4.80}$$

where we used the notation $\tilde{L}_k = \tilde{L}_k(\boldsymbol{Y})$ to emphasize that $\tilde{L}_k$ is a function of the random variable $\boldsymbol{Y}$; this allows us to replace the expectation over $\boldsymbol{Y}$ in (4.80) with the expectation over $\tilde{L}_k$, which leads directly to (4.75). □

We note that (4.76) generalizes the L-values defined in Section 3.3 allowing us to deal with arbitrary mismatched metrics $\mathbb{q}_k(b,\boldsymbol{y})$. Theorem 4.20 demonstrates also that the distribution of the generalized L-values (4.76) suffices to calculate the BICM-GMI.

An immediate consequence of Theorem 4.20 is that when $\mathbb{q}_k(b,\boldsymbol{y})$ is given by (4.33), $s=1$, $I^{\mathrm{gmi}}_{\mathbb{q}_k,s}(B_k;\boldsymbol{Y}) = I(B_k;\boldsymbol{Y})$, and $\tilde{L}_k = L_k$. Thus,

$$I(B_k;\boldsymbol{Y}) = \mathcal{H}(B_k) - \sum_{b\in\mathbb{B}} P_{B_k}(b)\mathbb{E}_{L_k|B_k=b}\left[\log_2\left(1+\mathrm{e}^{(-1)^b\cdot(L_k+l_k^{\mathrm{a}})}\right)\right]. \tag{4.81}$$

In the case of uniformly distributed bits $P_{B_k}(b)=\frac{1}{2}$ and for any *generalized logarithmic-likelihood ratio* (LLR), (4.75) becomes

$$I^{\mathrm{gmi}}_{\mathbb{q}_k,s}(B_k;\boldsymbol{Y}) = 1 - \frac{1}{2}\sum_{b\in\mathbb{B}}\mathbb{E}_{\tilde{L}_k|B_k=b}\left[\log_2\left(1+\mathrm{e}^{(-1)^b s\tilde{L}_k}\right)\right]. \tag{4.82}$$

Furthermore, if we assume the symmetry condition (3.74) is satisfied[5] (i.e., $p_{\tilde{L}_k|B_k}(l|b) = p_{\tilde{L}_k|B_k}(-l|\bar{b})$, we obtain from (4.82)

$$I^{\mathrm{gmi}}_{\mathbb{q}_k,s}(B_k;\boldsymbol{Y}) = 1 - \mathbb{E}_{\tilde{L}_k|B_k=0}\left[\log_2(1+\mathrm{e}^{s\tilde{L}_k})\right] \tag{4.83}$$

$$= \mathbb{E}_{\tilde{L}_k|B_k=0}\left[-0.5s\tilde{L}_k - \log_2\cosh\left(\tfrac{1}{2}s\tilde{L}_k\right)\right]. \tag{4.84}$$

Equivalently, we can express (4.83) as

$$I^{\mathrm{gmi}}_{\mathbb{q}_k,s}(B_k;\boldsymbol{Y}) = 1 - \mathbb{E}_{\tilde{L}_k|B_k=0}\left[T(s\tilde{L}_k)\right], \tag{4.85}$$

where

$$T(z) \triangleq \log_2(1+\mathrm{e}^z) \tag{4.86}$$

$$= \log_2(\mathrm{e})\max\{z,0\} + \log_2\left(1+\mathrm{e}^{-|z|}\right), \tag{4.87}$$

and (4.87), inspired by (3.94), is a numerically stable[6] version of (4.86).

Knowing the observations

$$\tilde{l}_k[n] = \log\frac{\mathbb{q}_k(1,\boldsymbol{y}[n])}{\mathbb{q}_k(0,\boldsymbol{y}[n])}, \quad n=1,\ldots,N_{\mathrm{s}}, \tag{4.88}$$

---

[5] In a natural way, or enforced via scrambling as shown in Section 3.3.2.

[6] We avoid calculating e^x for $x\to\infty$.

and the transmitted bits $b_k[n]$, (4.85) can be transformed into a simple unbiased estimator of $I^{\text{gmi}}_{\mathfrak{q}_k,s}(B_k;\boldsymbol{Y})$ as

$$I^{\text{gmi}}_{\mathfrak{q}_k,s}(B_k;\boldsymbol{Y}) \approx \hat{I}^{\text{gmi}}_{\mathfrak{q}_k,s}(B_k;\boldsymbol{Y}) = 1 - \frac{1}{N_{\text{s}}}\sum_{n=1}^{N_{\text{s}}} T\left((-1)^{b_k[n]} s\tilde{l}_k[n]\right). \tag{4.89}$$

In the case of matched bit metrics $\mathfrak{q}_k(b,\boldsymbol{y}) \propto p_{\boldsymbol{Y}|B_k}(\boldsymbol{y}|b)$ with symmetric L-values and uniformly distributed bits, we obtain the estimate of the bit-level GMI via Monte Carlo integration

$$I(B_k;\boldsymbol{Y}) \approx \hat{I}(B_k;\boldsymbol{Y}) = 1 - \frac{1}{N_{\text{s}}}\sum_{n=1}^{N_{\text{s}}} T\left((-1)^{b_k[n]} l_k[n]\right). \tag{4.90}$$

In a simple case of transmission over the scalar channel, i.e., when the model relating the channel outcome $\boldsymbol{Y}$ and the input $\boldsymbol{X}$ is well known, we may calculate the GMI via the formulas provided in Section 4.5 with much better accuracy than the one obtained from (4.89) and (4.90). On the other hand, as it happens very often, the L-values are calculated using nonlinear expressions, and are affected by various random variables. Then, the direct integrals of Section 4.5 are too involved to derive and/or tedious to use. Moreover, because the PDFs $p_{L_k|B_k}(l|b)$ or $p_{\tilde{L}_k|B_k}(l|b)$ are also difficult to estimate (analytically or via histograms), the estimates (4.89) and (4.90) become very convenient.

Theorem 4.21 (L-values are MI-lossless) *If the L-values are calculated using the matched bit metrics $\mathfrak{q}_k(b,\boldsymbol{y}) \propto p_{\boldsymbol{Y}|B_k}(\boldsymbol{y}|b)$, then*

$$I(B_k;\boldsymbol{Y}) = I(B_k;L_k). \tag{4.91}$$

Proof: By definition, and following similar steps to those in the proof of Theorem 4.20 we obtain

$$\begin{aligned} I(B_k;L_k) &= -\sum_{b\in\mathbb{B}} P_{B_k}(b)\mathbb{E}_{L_k|B_k=b}\left[\log_2 \frac{\sum_{u\in\mathbb{B}} P_{B_k}(u) p_{L_k|B_k}(L_k|u)}{p_{L_k|B_k}(L_k|b)}\right] \\ &= \mathcal{H}(B_k) - \sum_{b\in\mathbb{B}} P_{B_k}(b)\mathbb{E}_{L_k|B_k=b}\left[\log_2\left(1 + \frac{p_{L_k|B_k}(L_k|\bar{b})}{p_{L_k|B_k}(L_k|b)} \mathrm{e}^{(-1)^b l_k^{\text{a}}}\right)\right]. \end{aligned} \tag{4.92}$$

As $l_k[n]$ are calculated using the matched metrics, they satisfy the consistency condition (3.67), *thus*

$$p_{L_k|B_k}(l|\bar{b}) = p_{L_k|B_k}(l|b)\mathrm{e}^{(-1)^b l}, \quad \forall l \in \mathbb{R}. \tag{4.93}$$

Using (4.93) *in* (4.92) *yields the same expression as the one in* (4.81), *which completes the proof.* □

Theorem 4.22 (Concavity of GMI) *For any fixed input distribution, the functions $I^{\text{gmi}}_{\mathfrak{q}_k,s}(B_k;\boldsymbol{Y})$ and $I^{\text{gmi}}_{\mathfrak{q},s}(\boldsymbol{B};\boldsymbol{Y})$ are concave in s.*

Proof: The first and second derivatives of $I^{\text{gmi}}_{\mathfrak{q}_k,s}(B_k;\boldsymbol{Y})$ in (4.47) with respect to s are calculated from (4.75).

$$\frac{\mathrm{d}}{\mathrm{d}s} I^{\text{gmi}}_{\mathfrak{q}_k,s}(B_k;\boldsymbol{Y}) = -\frac{1}{\log(2)}\sum_{b\in\mathbb{B}} P_{B_k}(b)\mathbb{E}_{\tilde{L}_k|B_k=b}\left[\frac{(-1)^b \tilde{L}_k \mathrm{e}^{(-1)^b(s\tilde{L}_k + l_k^{\text{a}})}}{1+\mathrm{e}^{(-1)^b(s\tilde{L}_k + l_k^{\text{a}})}}\right], \tag{4.94}$$

$$\frac{\mathrm{d}^2}{\mathrm{d}s^2} I^{\text{gmi}}_{\mathfrak{q}_k,s}(B_k;\boldsymbol{Y}) = -\frac{1}{\log(2)}\sum_{b\in\mathbb{B}} P_{B_k}(b)\mathbb{E}_{\tilde{L}_k|B_k=b}\left[\frac{\tilde{L}_k^2 \mathrm{e}^{(-1)^b(s\tilde{L}_k + l_k^{\text{a}})}}{\left(1+\mathrm{e}^{(-1)^b(s\tilde{L}_k + l_k^{\text{a}})}\right)^2}\right]. \tag{4.95}$$

As all the functions under integral in (4.95) are positive, then for any finite s we have

$$\frac{\mathrm{d}^2}{\mathrm{d}s^2} I_k^{\mathrm{gmi}}(s) < 0, \tag{4.96}$$

and thus, $I_{\mathrm{q}_k,s}^{\mathrm{gmi}}(B_k; \boldsymbol{Y})$ is concave, from which the concavity of $I_{\mathrm{q},s}^{\mathrm{gmi}}(\boldsymbol{B}; \boldsymbol{Y}) = \sum_{k=1}^{m} I_{\mathrm{q}_k,s}^{\mathrm{gmi}}(B_k; \boldsymbol{Y})$ follows immediately. □

4.4 BICM, CM, and MLC: A Comparison

4.4.1 Achievable Rates

Having analyzed CM and BICM, it is interesting to compare and relate them to *multilevel coding* (MLC), another popular coded modulation scheme we presented in Section 2.3.2. To this end, we start rewriting the CM-MI using the random variables B_k, i.e.,

$$\mathsf{I}^{\mathrm{cm}}(\mathsf{snr}, \mathbf{X}, \boldsymbol{p}_{\mathrm{s}}) = I(\boldsymbol{B}; \boldsymbol{Y}|H) \tag{4.97}$$

$$= \sum_{k=1}^{m} I(B_k; \boldsymbol{Y}|H, B_1, \ldots, B_{k-1}). \tag{4.98}$$

To obtain (4.97), we used the fact that the mapping $\Phi : \boldsymbol{B} \mapsto \boldsymbol{X}$ is bijective, and to obtain (4.98), we used the chain rule of mutual information.

The quantity $I(B_k; \boldsymbol{Y}|H, B_1, \ldots, B_{k-1})$ in (4.98) represents an MI between the bit B_k and the output, conditioned on the knowledge of the bits $B_1, \ldots, B_{k-1}$. This *bit-level* MI represents the maximum rate that can be used at the kth bit position, given a perfect knowledge of the previous $k-1$ bits.

MLC with *multistage decoding* (MSD) (see Fig. 2.5) is in fact a direct application of (4.98), i.e., m parallel encoders send the encoded oddata using bits' positions $1, \ldots, m$, each of the encoders uses a coding rate $R_k \leq I(B_k; \boldsymbol{Y}|H, B_1, \ldots, B_{k-1})$. At the receiver side, the first bit level is decoded and the decisions are passed to the second decoder, which then passes the decisions to the third decoder, and so on, as shown in Fig. 2.6 (a).

Although different mapping rules produce different conditional MIs $I(B_k; \boldsymbol{Y}|H, B_1, \ldots, B_{k-1})$, the sum (4.98) remains constant. In other words, the sum of achievable rates using MLC encoding/decoding does not depend on the labeling and is equal to CM-MI. On the other hand, when MLC is based on *parallel decoding of the individual levels* (PDL), i.e., when no information is passed between the m decoders (see Fig. 2.6b), BICM and MLC are equivalent in terms of achievable rates, although the main difference still remains: instead of m encoder–decoder pairs, BICM uses only one pair.

Using (4.98) and (4.54), we can express the MI loss caused by BICM as

$$\mathsf{I}^{\mathrm{cm}}(\mathsf{snr}, \mathbf{X}, \boldsymbol{p}_{\mathrm{s}}) - \mathsf{I}^{\mathrm{bi}}(\mathsf{snr}, \mathbf{X}, \boldsymbol{p}_{\mathrm{b}}, \mathbf{Q}) = \sum_{k=1}^{m} I(B_k; \boldsymbol{Y}|H, B_1, \ldots, B_{k-1}) - I(B_k; \boldsymbol{Y}|H), \tag{4.99}$$

which clearly shows that the BICM-GMI will be approximately equal to the CM if the differences $I(B_k; \boldsymbol{Y}|H, B_1, \ldots, B_{k-1}) - I(B_k; \boldsymbol{Y}|H)$ are small for all $k = 1, \ldots, m$.

Example 4.23 (CM, MLC, and BICM for 8PAM**)** *Consider an* 8PAM *constellation labeled by the BRGC and BSGC shown in Fig. 2.13 over the AWGN channel. In Fig. 4.15, we show the conditional MIs $I(B_k; \boldsymbol{Y}|B_1, \ldots, B_{k-1})$ for the BRGC (Fig. 4.15 (a)) and the BSGC (Fig. 4.15 (b)). The AWGN capacity* $\mathsf{C}(\mathsf{snr})$ *in* (4.9)*, the CM-MI, and BICM-GMI are also shown. While the obtained conditional MIs are quite different, from* (4.98)*, we know that their sum yields nothing but the CM-MI. In this figure,*

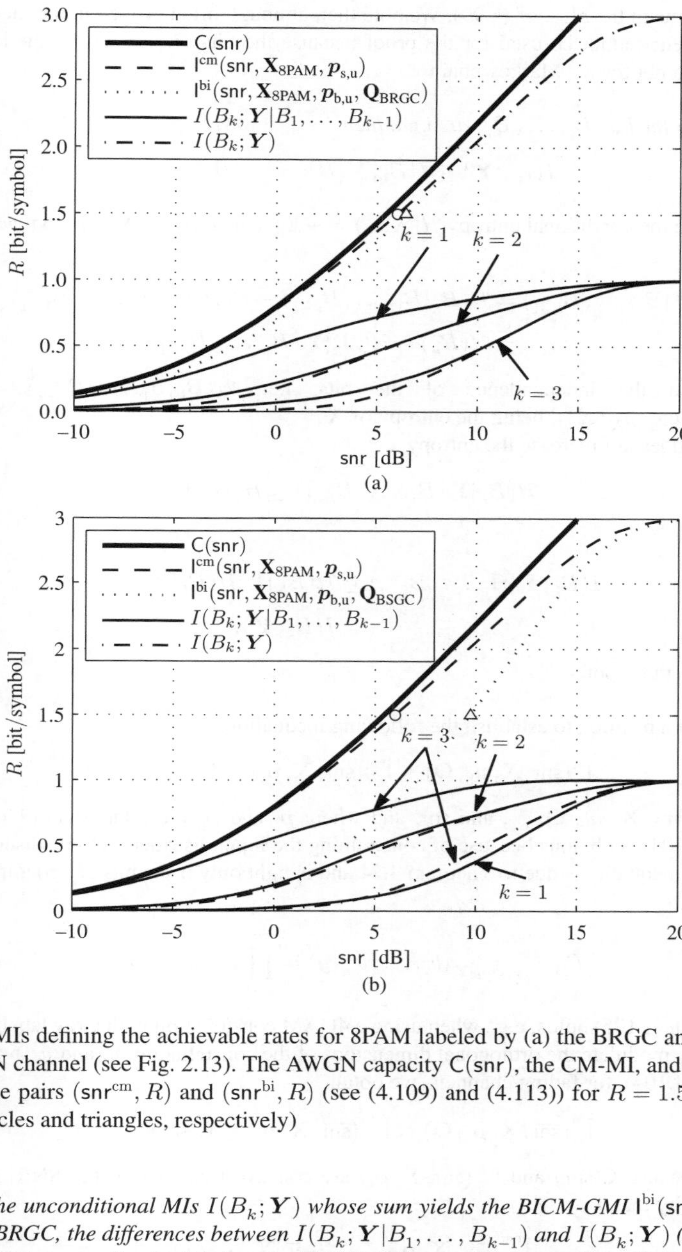

Figure 4.15 MIs defining the achievable rates for 8PAM labeled by (a) the BRGC and (b) the BSGC over the AWGN channel (see Fig. 2.13). The AWGN capacity $\mathsf{C}(\mathsf{snr})$, the CM-MI, and BICM-GMI are also shown. The pairs $(\mathsf{snr}^{\mathrm{cm}}, R)$ and $(\mathsf{snr}^{\mathrm{bi}}, R)$ (see (4.109) and (4.113)) for $R = 1.5$ bit/symbol are also shown (circles and triangles, respectively)

we also show the unconditional MIs $I(B_k; \boldsymbol{Y})$ whose sum yields the BICM-GMI $\mathsf{I}^{\mathrm{bi}}(\mathsf{snr}, \mathbf{X}, \boldsymbol{p}_{\mathrm{b}}, \mathbf{Q})$. In the case of the BRGC, the differences between $I(B_k; \boldsymbol{Y}|B_1, \ldots, B_{k-1})$ and $I(B_k; \boldsymbol{Y})$ (i.e., the terms in the r.h.s. of (4.99)*) are small, which brings the BICM-GMI very close to the CM-MI. On the other hand, using the labeling BSGC, the difference becomes significant for the bit position $k = 3$, and this causes a significant loss in terms of achievable rates when using BICM rather than MLC.*

We have already demonstrated in Theorem 4.10 that the GMI is upper bounded by the MI, which indicates that the BICM-GMI is upper bounded by the CM-MI. This property can be also demonstrated analyzing the terms in the r.h.s. of (4.99). We note that, although often evoked in the literature, the data processing theorem cannot be used for the proof because the bits $B_1, \dots, B_{k-1}$, the bit B_k, and the observation $\boldsymbol{Y}$ do not form a Markov chain.

Theorem 4.24 *If the bits* $B_1, \dots, B_m$ *are i.i.d. then*

$$I(B_k; \boldsymbol{Y}) \leq I(B_k; \boldsymbol{Y}|B_{k-1}, \dots, B_1). \tag{4.100}$$

Proof: We use the conditional entropy $\mathcal{H}(X|Y) \triangleq -\mathbb{E}_{X,Y}[\log_2 p_{X|Y}(X|Y)]$, to rewrite the r.h.s. of (4.100) as

$$I(B_k; \boldsymbol{Y}|B_1, \dots, B_{k-1}) = \mathcal{H}(B_k|B_1, \dots, B_{k-1}) - \mathcal{H}(B_k|\boldsymbol{Y}, B_1, \dots, B_{k-1}) \tag{4.101}$$

$$= \mathcal{H}(B_k) - \mathcal{H}(B_k|\boldsymbol{Y}, B_1, \dots, B_{k-1}), \tag{4.102}$$

where we used the independence of the bits, i.e., $\mathcal{H}(B_k|B_1, \dots, B_{k-1}) = \mathcal{H}(B_k)$, with $\mathcal{H}(X) = -\mathbb{E}_X[\log_2(p_X(X))]$ being the entropy of X.

Conditioning does not increase the entropy, i.e.,

$$\mathcal{H}(B_k|\boldsymbol{Y}, B_1, \dots, B_{k-1}) \leq \mathcal{H}(B_k|\boldsymbol{Y}).$$

Thus,

$$I(B_k; \boldsymbol{Y}|B_1, \dots, B_{k-1}) \geq \mathcal{H}(B_k) - \mathcal{H}(B_k|\boldsymbol{Y})$$

$$= I(B_k; \boldsymbol{Y}), \tag{4.103}$$

which completes the proof. □

We are now in a position to establish the following inequalities

$$\mathsf{I}^{\text{bi}}(\mathsf{snr}, \mathbf{X}, \boldsymbol{p}_{\text{b}}, \mathbf{Q}) \leq \mathsf{I}^{\text{cm}}(\mathsf{snr}, \mathbf{X}, \boldsymbol{p}_{\text{s}}) \leq \mathsf{C}(\mathsf{snr}), \tag{4.104}$$

which hold for any $\mathbf{X}$, $\boldsymbol{p}_{\text{b}}$, $\mathbf{Q}$, $\boldsymbol{p}_{\text{s}}$, and snr, and where $\boldsymbol{p}_{\text{s}}$ and $\boldsymbol{p}_{\text{b}}$ are related via (2.72). The second inequality in (4.104) is obvious because of constraining the input distribution to be discrete and unoptimized. The first inequality is due to Theorem 4.24 and is tight only if the bits B_k conditioned on $\boldsymbol{Y}$ are independent, i.e.,

$$P_{B_1, \dots, B_m|\boldsymbol{Y}}(b_1, \dots, b_m|\boldsymbol{y}) = \prod_{k=1}^{m} P_{B_k|\boldsymbol{Y}}(b_k|\boldsymbol{y}). \tag{4.105}$$

The expression in (4.105) holds, e.g., when using a 4QAM constellation with Gray labeling. In this case the bits b_1 and b_2 modulate the orthogonal dimensions of the constellation: y_1 and y_2, respectively.

Similarly to (4.104), for fading channels, we obtain

$$\bar{\mathsf{I}}^{\text{bi}}(\overline{\mathsf{snr}}, \mathbf{X}, \boldsymbol{p}_{\text{b}}, \mathbf{Q}) \leq \bar{\mathsf{I}}^{\text{cm}}(\overline{\mathsf{snr}}, \mathbf{X}, \boldsymbol{p}_{\text{s}}) \leq \overline{\mathsf{C}}(\overline{\mathsf{snr}}). \tag{4.106}$$

Furthermore, because $\mathsf{C}(\mathsf{snr})$ and $\mathsf{I}^{\text{cm}}(\mathsf{snr}, \mathbf{X}, \boldsymbol{p}_{\text{s}})$ are concave functions of the SNR, it follows from Jensen's inequality that

$$\bar{\mathsf{I}}^{\text{cm}}(\overline{\mathsf{snr}}, \mathbf{X}, \boldsymbol{p}_{\text{s}}) \leq \mathsf{I}^{\text{cm}}(\overline{\mathsf{snr}}, \mathbf{X}, \boldsymbol{p}_{\text{s}}), \tag{4.107}$$

$$\overline{\mathsf{C}}(\overline{\mathsf{snr}}) \leq \mathsf{C}(\overline{\mathsf{snr}}). \tag{4.108}$$

On the other hand, the BICM-GMI is not always a concave function of the SNR,[7] and thus, we cannot guarantee an inequality similar to (4.107).

[7] The concavity of the BICM-GMI depends on the binary labeling.

4.4.2 Robustness of the Encoding Scheme

While the rates achievable using MLC are never smaller than those attainable using BICM, it is important to note that the design of MLC, i.e., the choice of the coding rates R_k for each of the coding levels, requires the knowledge of all the conditional MIs $I(B_k; \boldsymbol{Y}|H, B_1, \ldots, B_{k-1})$. This brings up an issue related to the robustness of the design with respect to lack of complete knowledge of the channel model.

To clarify this issue, we now assume that the MLC encoding/decoding scheme is designed for the AWGN channel for a coding rate R. We also assume that the design is done so as to operate at the SNR decoding threshold $\mathsf{I}^{\text{cm}}(\mathsf{snr}, \mathbf{X}, \boldsymbol{p}_{\text{s}}) = R$, i.e., we design the encoders to ensure successful decoding at the SNR limit defined by the CM-MI, which we denote as

$$\mathsf{snr}^{\text{cm}} \triangleq f_{\mathsf{I}^{\text{cm}}}(R, \mathbf{X}, \mathbf{Q}, \boldsymbol{p}_{\text{b}}), \tag{4.109}$$

where $f_{\mathsf{I}^{\text{cm}}}(R, \mathbf{X}, \boldsymbol{p}_{\text{s}})$ is the inverse of the MI function $\mathsf{I}^{\text{cm}}(\mathsf{snr}, \mathbf{X}, \boldsymbol{p}_{\text{s}})$ defined in Section 4.1.

The decoding threshold of each of the MLC levels is given by

$$\mathsf{snr}_k^{\text{mlc}} \triangleq f_{\mathsf{I}_k^{\text{mlc}}}(R_k, \mathbf{X}, \mathbf{Q}, \boldsymbol{p}_{\text{b}}), \tag{4.110}$$

where $f_{\mathsf{I}_k^{\text{mlc}}}(R_k, \mathbf{X}, \mathbf{Q}, \boldsymbol{p}_{\text{b}})$ is the inverse of the conditional MI $I(B_k; \boldsymbol{Y}|B_1, \ldots, B_{k-1})$.

We can thus define the SNR decoding threshold as the value of SNR, for which the message at all MLC levels can be decoded

$$\mathsf{snr}^{\text{mlc}} \triangleq \max_{k=1,\ldots,m} \{\mathsf{snr}_k^{\text{mlc}}\}. \tag{4.111}$$

In order to operate at the CM-MI limit with $R = \sum_{k=1}^{m} R_k$ we need to satisfy the following:

$$\mathsf{snr}^{\text{cm}} = \mathsf{snr}^{\text{mlc}} = \mathsf{snr}_k^{\text{mlc}}, \tag{4.112}$$

that is, the SNR decoding thresholds for each of the decoders must be the same.

If a BICM scheme is designed for the same coding rate R, the corresponding SNR decoding threshold is given by

$$\mathsf{snr}^{\text{bi}} \triangleq f_{\mathsf{I}^{\text{bi}}}(R, \mathbf{X}, \mathbf{Q}, \boldsymbol{p}_{\text{b}}), \tag{4.113}$$

where $f_{\mathsf{I}^{\text{bi}}}(R, \mathbf{X}, \mathbf{Q}, \boldsymbol{p}_{\text{b}})$ is the inverse function of the BICM-GMI.

To illustrate the previous definitions, we show in Fig. 4.15 the pairs $(\mathsf{snr}^{\text{cm}}, R)$ and $(\mathsf{snr}^{\text{bi}}, R)$ for a rate $R = 1.5$ bit/symbol. The difference $(\mathsf{snr}^{\text{bi}} - \mathsf{snr}^{\text{cm}})$ can be understood as the SNR penalty caused by using a BICM scheme for a given coding rate R. For the BRGC, the penalty for $R = 1.5$ bit/symbol is close to zero, however, for the BSGC it increases up to about 4 dB.

We are now ready to see what happens if MLC and BICM are used when transmitting over a non-AWGN channel. In other words, we are interested in studying how the SNR decoding thresholds change if the design made for the AWGN channel is used in a different channel. This can be the case, e.g., when transmitting over a fading channel which was not foreseen during the design.

If the encoders with the rates R_k with $k = 1, \ldots, m$ (designed so as to satisfy (4.112) for the AWGN channel) are used in a fading channel, the corresponding SNR decoding thresholds are given by

$$\overline{\mathsf{snr}}_k^{\text{mlc}} \triangleq f_{\bar{\mathsf{I}}_k^{\text{mlc}}}(R_k, \mathbf{X}, \mathbf{Q}, \boldsymbol{p}_{\text{b}}), \quad k = 1, \ldots, m, \tag{4.114}$$

where $f_{\bar{\mathsf{I}}_k^{\text{mlc}}}(R_k, \mathbf{X}, \mathbf{Q}, \boldsymbol{p}_{\text{b}})$ is the inverse of the conditional MI $I(B_k; \boldsymbol{Y}|B_1, \ldots, B_{k-1}, H)$.

Again, successful decoding at all decoding levels of MLC, is possible if $\overline{\mathsf{snr}}$ is greater than all the individual SNR decoding thresholds in (4.114), i.e.,

$$\overline{\mathsf{snr}}^{\text{mlc}} \triangleq \max_{k=1,\ldots,m} \{\overline{\mathsf{snr}}_k^{\text{mlc}}\}. \tag{4.115}$$

Since the form of $I(B_k;\boldsymbol{Y}|B_1,\ldots,B_{k-1})$ varies with k, the values of the corresponding inverse functions $f_{\mathsf{I}_k^{\mathrm{mlc}}}(R_k,\mathbf{X},\mathbf{Q},\boldsymbol{p}_{\mathrm{b}})$ will not be the same. Consequently, we cannot guarantee that the condition for the average SNR, similar to (4.112), will be satisfied, and, in general, we will have $\overline{\mathsf{snr}}_1^{\mathrm{mlc}}\neq\overline{\mathsf{snr}}_2^{\mathrm{mlc}}\neq\cdots\neq\overline{\mathsf{snr}}_m^{\mathrm{mlc}}$.

Therefore, the following relationship holds

$$\overline{\mathsf{snr}}^{\mathrm{cm}}\leq\overline{\mathsf{snr}}^{\mathrm{mlc}}, \tag{4.116}$$

and the difference $\overline{\mathsf{snr}}^{\mathrm{mlc}}-\overline{\mathsf{snr}}^{\mathrm{cm}}$ should be interpreted as the SNR penalty because of a "code design mismatch".

We emphasize that the mismatch happens because the coding rates R_k were selected using the functions $I(B_k;\boldsymbol{Y}|B_1,\ldots,B_{k-1})$. If, on the other hand, we rather used $I(B_k;\boldsymbol{Y}|B_1,\ldots,B_{k-1},H)$, then the code designed would be matched to the fading channel. In such a case, the effect of the code design mismatch would appear if MLC were used over the AWGN channel. More generally, the code design mismatch phenomenon will be observed when the model of the channel used for the design is mismatched with respect to the actual channel.

Of course, in BICM transmission, the SNR decoding threshold is similar to (4.113), i.e.,

$$\overline{\mathsf{snr}}^{\mathrm{bi}}\triangleq f_{\mathsf{I}^{\mathrm{bi}}}(R,\mathbf{X},\mathbf{Q},\boldsymbol{p}_{\mathrm{b}}). \tag{4.117}$$

In many cases of practical interest, $\overline{\mathsf{snr}}^{\mathrm{bi}}\geq\mathsf{snr}^{\mathrm{bi}}$ but in general, this relationship depends on the mapping as we discussed after (4.108). Of course, the relationship $\overline{\mathsf{snr}}^{\mathrm{bi}}\geq\overline{\mathsf{snr}}^{\mathrm{cm}}$ always holds, and we interpret the difference $\overline{\mathsf{snr}}^{\mathrm{bi}}-\overline{\mathsf{snr}}^{\mathrm{cm}}$ as the SNR gap because of the suboptimality of BICM. Unlike in the case of MLC, there is no effect of the code design mismatch: BICM remains "matched" to the channel independently of the underlying channel model.

Example 4.25 (Code Mismatch for 8PAM**)** *Let us investigate the effect of the code design mismatch for an* 8PAM *constellation labeled by the BRGC. Figure 4.16 shows the rates achievable by MLC and BICM together with the SNR decoding thresholds for MLC. The pairs* $(\mathsf{snr}_k^{\mathrm{mlc}},R_k)$ *as well as* $(\mathsf{snr}^{\mathrm{mlc}},R)$, *where* $R=R_1+R_2+R_3$ *are shown by circles. The rates* R_k *are chosen for operation over the AWGN channel, so the vertical alignment of the circles means that* $\mathsf{snr}_k^{\mathrm{mlc}}=\mathsf{snr}^{\mathrm{mlc}}=\mathsf{snr}^{\mathrm{cm}}$. *Thus, the code is designed to operate at the CM–MI curve. The pair* $(\mathsf{snr}^{\mathrm{bi}},R)$ *is shown by a triangle and we see that the SNR penalty* $\mathsf{snr}^{\mathrm{bi}}-\mathsf{snr}^{\mathrm{cm}}$ *is relatively small.*

If we now consider the case where the same MLC encoders are used in a Rayleigh fading channel, we see the mismatch: the pairs $(\overline{\mathsf{snr}}_k^{\mathrm{mlc}},R_k)$ *shown as squares are not aligned vertically. This results in the SNR decoding thresholds of MLC (the pair* $(\overline{\mathsf{snr}}^{\mathrm{mlc}},R)$ *is shown by the top square) being larger than the SNR decoding threshold for BICM (the pair* $(\overline{\mathsf{snr}}^{\mathrm{bi}},R)$ *is shown by a diamond).*

By varying the value of R, *we can find pairs* $(\overline{\mathsf{snr}}^{\mathrm{mlc}},R)$, *which we show in Fig. 4.16 using a dashed line. For clarity of presentation, we only consider* $R\in(1.0,2.5)$ bit/symbol. *These results illustrate that while MLC has the potential to operate at the limits of CM-MI, and thus, also to outperform BICM, the latter, having a binary encoder independent of the channel model, is more robust to variations in the conditions of operation.*

4.5 Numerical Calculation of Mutual Information

In Sections 4.2 and 4.3, we have shown various examples of the CM-MI and the BICM-GMI calculated for the AWGN channel. While we cannot avoid the problem of enumerating the constellation points, we may efficiently calculate the multidimensional integrals (over $\mathbb{R}^N$) appearing in the MI expressions using *Gauss–Hermite* (GH) quadratures. The methods we discussed in Section 4.3.3 may also be used but the

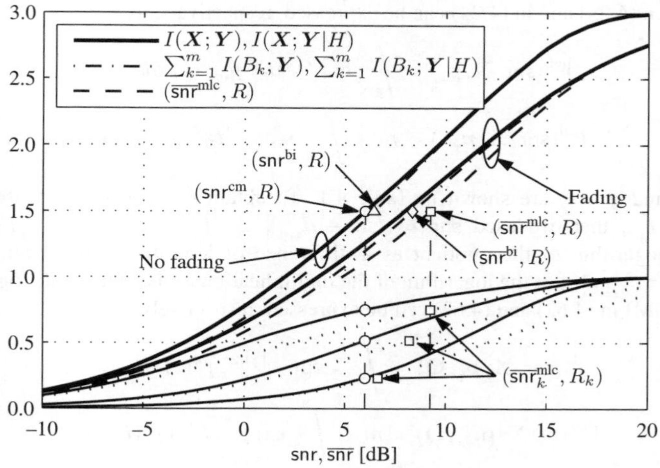

Figure 4.16 Achievable rates for 8PAM using the BRGC. In MLC, the rates R_k are designed for the AWGN channel and then used in a Rayleigh fading channel. For $R = 1.5$ bit/symbol, the circles correspond to the pairs $(\mathsf{snr}^{\mathrm{cm}}, R_k)$ and $(\mathsf{snr}^{\mathrm{cm}}, R)$, the triangle shows $(\mathsf{snr}^{\mathrm{bi}}, R)$, the diamond shows $(\overline{\mathsf{snr}}^{\mathrm{bi}}, R)$ and the squares indicate $(\overline{\mathsf{snr}}^{\mathrm{mlc}}, R_k)$ and $(\overline{\mathsf{snr}}^{\mathrm{mlc}}, R)$. The dashed line corresponds to the set of pairs $(\overline{\mathsf{snr}}^{\mathrm{mlc}}, R)$ for $R \in (1.0, 2.5)$ bit/symbol

numerical quadratures are much more accurate for small values of N and, what is more important, they are much more reliable for being used in a numerical optimization.

For any function $g(t)$ with bounded $(2J)$th derivative, the J-points GH quadrature

$$\int_{-\infty}^{\infty} \exp(-t^2) g(t)\, \mathrm{d}t \approx \sum_{k=1}^{J} \alpha_k g(\xi_k) \tag{4.118}$$

is asymptotically ($J \to \infty$) exact if we use the quadrature nodes ξ_k as the kth root of the Hermite polynomial

$$H_J(x) = J! \sum_{r=0}^{\lfloor J/2 \rfloor} \frac{(-1)^r}{r!(J-2r)!} (2x)^{J-2r}, \tag{4.119}$$

and the weights defined as

$$\alpha_k = \frac{2^{J-1} J! \sqrt{\pi}}{[J H_{J-1}(\xi_k)]^2}. \tag{4.120}$$

The values of α_k and ξ_k can be easily found for different values of J, which determines the trade-off between the computation speed and the accuracy of the quadrature. The expression in (4.118) can be straightforwardly expanded to a quadrature over an N-dimensional space as

$$\int_{\mathbb{R}^N} \exp(-\|\boldsymbol{t}\|^2) g(\boldsymbol{t})\, \mathrm{d}\boldsymbol{t} \approx \sum_{k_1=1}^{J} \cdots \sum_{k_N=1}^{J} g(\boldsymbol{\xi}) \prod_{n=1}^{N} \alpha_{k_n}, \tag{4.121}$$

where $\boldsymbol{t} = [t_1, \ldots, t_N]^{\mathrm{T}} \in \mathbb{R}^N$ and $\boldsymbol{\xi} = [\xi_{k_1}, \ldots, \xi_{k_N}]^{\mathrm{T}} \in \mathbb{R}^N$. It is important to note that the expansion we use in (4.121) is not unique; however, it is a simple way of generalizing (4.118) to N dimensions.

The CM-MI in (4.23) and in (4.24) can be expressed, respectively, as

$$\mathsf{I}^{\text{cm}}(\mathsf{snr}, \mathbf{X}, \boldsymbol{p}_{\text{s}}) = \int_{\mathbb{R}^N} \exp(-\|\boldsymbol{t}\|^2) g^{\text{cm}}(\boldsymbol{t}) \, \text{d}\boldsymbol{t}, \tag{4.122}$$

$$\mathsf{I}^{\text{cm}}(\mathsf{snr}, \mathbf{X}, \boldsymbol{p}_{\text{s,u}}) = m + \int_{\mathbb{R}^N} \exp(-\|\boldsymbol{t}\|^2) g_{\text{u}}^{\text{cm}(\boldsymbol{t})} \, \text{d}\boldsymbol{t}, \tag{4.123}$$

where $g^{\text{cm}}(\boldsymbol{t})$ and $g_{\text{u}}^{\text{cm}}(\boldsymbol{t})$ are shown in Table 4.1. To obtain (4.122), we first used the substitution $\boldsymbol{t} = \sqrt{\mathsf{snr}}(\boldsymbol{y} - \boldsymbol{x}_i)$, then we used $\mathsf{snr}\|\boldsymbol{t}/\sqrt{\mathsf{snr}} + \boldsymbol{d}_{i,j}\|^2 = \|\boldsymbol{t}\|^2 + 2\sqrt{\mathsf{snr}}\langle \boldsymbol{t}, \boldsymbol{d}_{i,j}\rangle + \mathsf{snr}\|\boldsymbol{d}_{i,j}\|^2$, and then split the logarithm of the quotient as a difference of logarithms. To obtain (4.123), we used $p_{\text{s},i} = \frac{1}{M}$ in $g^{\text{cm}}(\boldsymbol{t})$ and split the logarithm of the multiplication as the sum of the logarithms.

The BICM-GMI in (4.62) and (4.63) can be expressed, respectively, as

$$\mathsf{I}^{\text{bi}}(\mathsf{snr}, \mathbf{X}, \mathbf{p}_{\text{b}}, \mathbf{Q}) = \int_{\mathbb{R}^N} \exp(-\|\boldsymbol{t}\|^2) g^{\text{bi}}(\boldsymbol{t}) \, \text{d}\boldsymbol{t}, \tag{4.124}$$

$$\mathsf{I}^{\text{bi}}(\mathsf{snr}, \mathbf{X}, \mathbf{p}_{\text{b,u}}, \mathbf{Q}) = m + \int_{\mathbb{R}^N} \exp(-\|\boldsymbol{t}\|^2) g_{\text{u}}^{\text{bi}}(\boldsymbol{t}) \, \text{d}\boldsymbol{t}, \tag{4.125}$$

where $g^{\text{bi}}(\boldsymbol{t})$ and $g_{\text{u}}^{\text{bi}}(\boldsymbol{t})$ are shown in Table 4.1. To obtain (4.124), we first used (4.66), then the substitution $\boldsymbol{t} = \sqrt{\mathsf{snr}}(\boldsymbol{y} - \boldsymbol{x}_i)$, and then again split the logarithm of the quotient (in the conditional expectation (4.4)) as a difference of logarithms. The expression (4.125) can be obtained by repeating the procedure and using a uniform input distribution.

Using (4.121)–(4.123), we obtain the following ready-to-use expressions for the CM-MI

$$\mathsf{I}^{\text{cm}}(\mathsf{snr}, \mathbf{X}, \boldsymbol{p}_{\text{s}}) \approx \sum_{k_1=1}^{J} \cdots \sum_{k_N=1}^{J} g^{\text{cm}}(\boldsymbol{\xi}) \prod_{n=1}^{N} \alpha_{k_n}, \tag{4.126}$$

$$\mathsf{I}^{\text{cm}}(\mathsf{snr}, \mathbf{X}, \boldsymbol{p}_{\text{s,u}}) \approx m + \sum_{k_1=1}^{J} \cdots \sum_{k_N=1}^{J} g_{\text{u}}^{\text{cm}}(\boldsymbol{\xi}) \prod_{n=1}^{N} \alpha_{k_n}, \tag{4.127}$$

where (4.126) and (4.127) shall be used to evaluate the CM-MI with arbitrary and uniform input distributions, respectively.

Table 4.1 Six functions used in (4.122)–(4.129) to evaluate the CM-MI and the BICM-GMI

Function	Expression
$g^{\text{cm}}(\boldsymbol{t})$	$-\pi^{-N/2}\sum_{i\in\mathcal{I}} p_{\text{s},i} \cdot \log_2\left(\sum_{j\in\mathcal{I}} p_{\text{s},j} \exp\left(-2\sqrt{\mathsf{snr}}\langle \boldsymbol{t}, \boldsymbol{d}_{i,j}\rangle - \mathsf{snr}\|\boldsymbol{d}_{i,j}\|^2\right)\right)$
$g_{\text{u}}^{\text{cm}}(\boldsymbol{t})$	$-\pi^{-N/2}\sum_{i\in\mathcal{I}} M^{-1} \cdot \log_2\left(\sum_{j\in\mathcal{I}} \exp\left(-2\sqrt{\mathsf{snr}}\langle \boldsymbol{t}, \boldsymbol{d}_{i,j}\rangle - \mathsf{snr}\|\boldsymbol{d}_{i,j}\|^2\right)\right)$
$g^{\text{bi}}(\boldsymbol{t})$	$mg^{\text{cm}}(\boldsymbol{t}) + \pi^{-N/2}\sum_{k=1}^{m}\sum_{b\in\mathbb{B}}\sum_{i\in\mathcal{I}_{k,b}} p_{\text{s},i} \cdot \log_2 \sum_{j\in\mathcal{I}_{k,b}} \frac{p_{\text{s},j}}{P_{C_k}(b)} \exp\left(-2\sqrt{\mathsf{snr}}\langle \boldsymbol{t}, \boldsymbol{d}_{i,j}\rangle - \mathsf{snr}\|\boldsymbol{d}_{i,j}\|^2\right)$
$g_{\text{u}}^{\text{bi}}(\boldsymbol{t})$	$mg_{\text{u}}^{\text{cm}}(\boldsymbol{t}) + M^{-1}\pi^{-N/2}\sum_{k=1}^{m}\sum_{b\in\mathbb{B}}\sum_{i\in\mathcal{I}_{k,b}} \cdot \log_2\left(\sum_{j\in\mathcal{I}_{k,b}} \exp\left(-2\sqrt{\mathsf{snr}}\langle \boldsymbol{t}, \boldsymbol{d}_{i,j}\rangle - \mathsf{snr}\|\boldsymbol{d}_{i,j}\|^2\right)\right)$

Similarly, by using (4.121)–(4.125), we obtain the following ready-to-use expressions for the BICM-GMI

$$\mathsf{I}^{\mathrm{bi}}(\mathsf{snr}, \mathbf{X}, \boldsymbol{p}_{\mathrm{b}}, \mathbf{Q}) \approx \sum_{k_1=1}^{J} \dots \sum_{k_N=1}^{J} g^{\mathrm{bi}}(\boldsymbol{\xi}) \prod_{n=1}^{N} \alpha_{k_n}, \tag{4.128}$$

$$\mathsf{I}^{\mathrm{bi}}(\mathsf{snr}, \mathbf{X}, \mathbf{p}_{\mathrm{b,u}}, \mathbf{Q}) \approx m + \sum_{k_1=1}^{J} \dots \sum_{k_N=1}^{J} g_{\mathrm{u}}^{\mathrm{bi}}(\boldsymbol{\xi}) \prod_{n=1}^{N} \alpha_{k_n}, \tag{4.129}$$

which shall be used to numerically evaluate the BICM-GMI with arbitrary and uniform input distributions, respectively.

Example 4.26 (Accuracy of MI Calculations) *Consider the case of* 2PAM *and* 8PAM *transmission over the AWGN channel. We can compare the accuracy of the MI calculation via* (4.126) *as a function of* J. *Assuming the calculation for* $J = 200$ *as exact, we show in Fig. 4.17 the absolute error of* $\mathsf{I}^{\mathrm{cm}}(\mathsf{snr}, \mathbf{X}, \boldsymbol{p}_{\mathrm{s,u}})$ *for different values of* J.

It is interesting to compare these results with those obtained via the unbiased estimator (4.90) *based on Monte Carlo integration. The latter is a random variable with mean equal to* $I(B_k; \boldsymbol{Y})$ *and variance*

$$\mathbb{V}\mathrm{ar}_{\boldsymbol{Y},B_k}[\hat{I}(B_k; \boldsymbol{Y})] = \frac{1}{N_{\mathrm{s}}} \sigma_T^2, \tag{4.130}$$

where

$$\sigma_T^2 = \mathbb{V}\mathrm{ar}_{\boldsymbol{Y},B_k}[T((-1)^{B_k} L_k)], \tag{4.131}$$

where $T(z)$ *is given by* (4.86).

For large N_{s}, *we can model* $\hat{I}(B_k; \boldsymbol{Y})$ *as a Gaussian random variable and then, the estimation error is approximated as*

$$\Pr\left\{|\hat{I}(B_k; \boldsymbol{Y}) - I(B_k; \boldsymbol{Y})| > \epsilon\right\} \approx 2Q\left(\frac{\sqrt{N_{\mathrm{s}}}\epsilon}{\sigma_T}\right) \le \exp\left(-\frac{N_{\mathrm{s}}\epsilon^2}{2\sigma_T^2}\right), \tag{4.132}$$

which means that to maintain the same (small) probability of error, we have to maintain $N_{\mathrm{s}} \propto \epsilon^{-2}$: *to increase the accuracy by one order of magnitude* (10*-fold decrease of* ϵ), *we have to increase the number of samples* N_{s} *by two orders of magnitude. Thus, in the simple cases we analyzed here, the numerical quadratures offer a clear implementation-complexity advantage over the estimator* (4.90). *However, if the number of dimensions* N *and/or the number of independent variables affecting the observation* $\boldsymbol{Y}$ *are increased, the numerical quadratures quickly become computationally demanding and tedious to implement. Then, the estimators* (4.89) *and* (4.90) *become simpler and viable alternative to the numerical quadratures.*

All the results presented in this chapter were obtained with $J = 10$, which we found to provide an adequate accuracy in all numerical examples.

4.6 Bibliographical Notes

All the concepts in this chapter can be traced back to Shannon's works [1, 2]. For a detailed treatment of MIs, channel capacity, and so on, we refer the reader to standard textbook such as [3, 4].

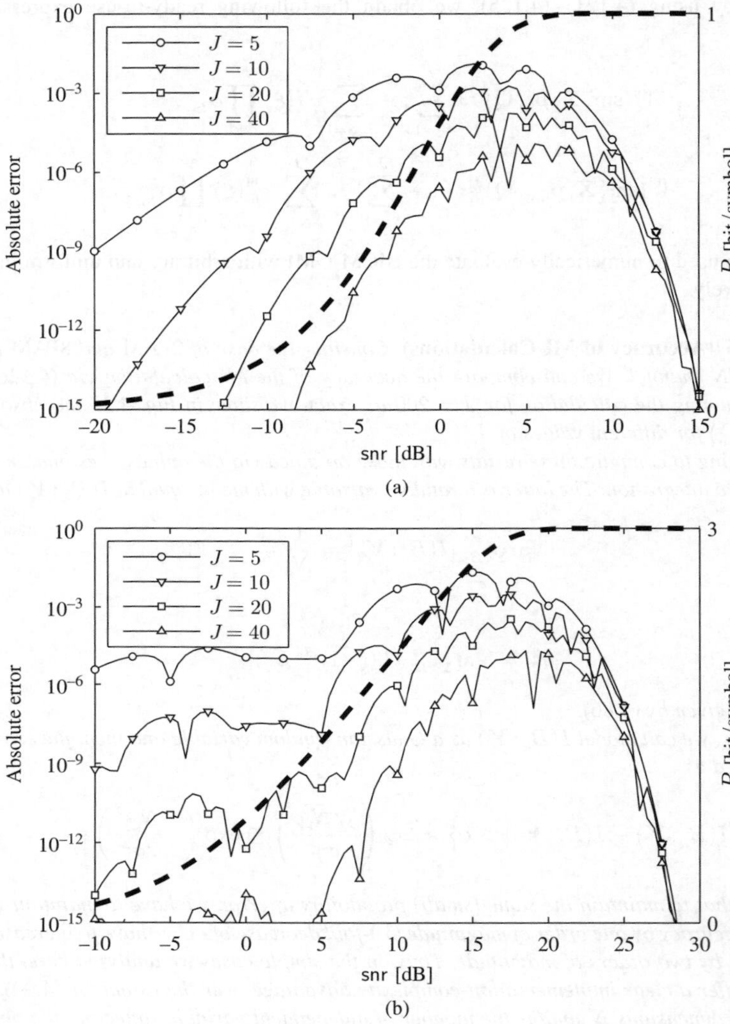

Figure 4.17 Average value of absolute error when evaluating $\mathrm{I}^{\mathrm{cm}}(\mathsf{snr}, \mathbf{X}, \boldsymbol{p}_{\mathrm{s,u}})$ using different number of quadrature points J for (a) 2PAM and (b) 8PAM. The left axis indicates the values of the error (shown with solid lines) while the right axis indicates the values of the MI (thick dashed lines)

The CM-MI has received different names in the literature. It is called *joint capacity* in [5], (constellation) *constrained capacity* in [6, 7], and *coded modulation capacity* in [8–12]. We use the name CM-MI to indicate that no optimization over the input distribution is performed.

The information-theoretical model for BICM was first introduced by Caire *et al.* using m parallel memoryless BICO channels [8, Fig. 3]. Martinez *et al.* later refined this model in [10], where it was shown that the model of parallel channels leads to correct conclusions about achievable rates, but may yield incorrect error exponents/cutoff rates. Error exponents for BICM were also studied in [13].

The BICM-GMI has received different names in the literature. It is called *parallel decoding capacity* in [5], *receiver constrained capacity* in [6], and *BICM capacity* in [8–12]. Again, we refrain from using the word "capacity" to emphasize the lack of optimization over the input distribution. We adopted the name BICM-GMI, which fits well the finding of Martinez *et al.* [10] who recognized the BICM decoder as a mismatched decoder and showed that the BICM-GMI in (4.54) corresponds to an achievable rate of such a decoder. Theorem 4.11 is thus adopted from [10] and was also used in [14]. The results on the optimality of matched processing in Theorem 4.10 and Corollary 4.12 were inspired by [15].

As we indicated, the BICM-GMI is an achievable rate but it has not be proven to be the largest achievable rate. For example, a different achievable rate—the so-called LM rate—has been recently studied in [16, Part I]. Finding the largest achievable rate remains as an open research problem. Despite this cautionary statement, the BICM-GMI predicts well the performance of capacity-approaching codes, as shown in Example 4.13 (see also [17, Section 3], [18, Section IV]). The generator polynomials of the TENC used in Example 4.13 were chosen on the basis of the results in [19] and some of the puncturing patterns used were taken from [19, 20].

Changing the distribution of the constellation symbols—known as *signal/constellation shaping*—has been studied for a number years, see [21–24]. In the context of BICM, geometrical shaping (i.e., design of the constellation under the assumption of uniform distribution of the symbols) was studied in [5, 25, 26], and probabilistic shaping (i.e., varying the probabilities of the symbols and/or bits when using the fixed constellation) was proposed in [27, 28] and developed further in [16, 29–33]. Recently, a numerical algorithm to obtain optimal input distributions in BICM was proposed in [34]. As the BICM-GMI is not the largest achievable rate, probabilistic or geometric shaping in BICM yields what we call in Section 4.3.2 a "pseudo-capacity."

The expression in (4.66) was first introduced in [9, Proposition 1], [7, eq. (65)] and later generalized in [35, Theorem 2] to multidimensional constellations and arbitrary input distributions. An alternative proof for Theorem 4.22 was presented in [16, Appendix B.1.1]. Achievable rates for 4D constellations for BICM in the context of coherent optical communications have been recently studied in [17].

The solution of (4.64) is in general unknown but results are available in the low-SNR regime. The low-SNR regime of the CM-MI has been studied in [36, 37], where conditions for constellations to achieve the SL are given. The fact that the SL is not always achieved for BICM was first noticed in [9] and later shown to be caused by the selection of the binary labeling [11]. The behavior of BICM in the low-SNR regime was studied in [9, 11, 35, 38, 39] as a function of the constellation and the binary labeling, assuming a uniform input distribution. Results for arbitrary input distributions under the indepence assumption of the bits have been recently presented in [40]. It is shown in [40] that probabilistic shaping offers no extra degrees of freedom in addition to what is provided by geometrical shaping for BICM in the low-SNR regime.

In terms of binary labelings, the NBC was shown to be optimal for MPAM in the low-SNR regime [11, 35] and the FBC was analyzed in [41] for uncoded transmission. The 49 labelings that give different BICM-GMI in Fig. 4.12 correspond to the 49 classes in [42, Tables A.1 and A.2] and the 7 different classes of labelings that appear in Fig. 4.12 for high SNR correspond to those in [42, Table A.1]. The optimality of a Gray code was conjectured in [8, Section III-C] in a rather general setup, i.e., without specifying the Gray code, the constellation, or the SNR regime. This conjecture was later disproved in [43] where it is shown that for low and medium SNR, there exist other labelings that give a higher BICM-GMI (see also [12, Chapter 3]). The numerical results presented in [12, Chapter 3], [44] show that in the high-SNR regime, Gray codes are indeed optimal. A formal proof for the optimality of Gray codes in the high-SNR regime for arbitrary 1D constellations was recently given in [45]. A simple approximation for the BICM-GMI that can be used to optimize the binary labeling of the constellation was recently introduced in [18].

GH quadratures can be found in [46, Section 7.3.4] and tables with the coefficient α_k and ξ_k for different values of J in [46, Appendix 7.3(b)]. Ready-to-use expressions for computing the CM-MI and BICM-GMI are given in [44].

References

[1] Shannon, C. E. (1948) A mathematical theory of communications. *Bell Syst. Tech. J.*, **27**, 379–423 and 623–656.
[2] Shannon, C. E. (1949) Communication in the presence of noise. *Proc. IRE*, **37** (1), 10–21.
[3] Cover, T. and Thomas, J. (2006) *Elements of Information Theory*, 2nd edn., John Wiley & Sons, Inc., Hoboken, NJ.
[4] MacKay, D. J. C. (2005) *Information Theory, Inference, and Learning Algorithms*, Cambridge University Press.
[5] Barsoum, M., Jones, C., and Fitz, M. (2007) Constellation design via capacity maximization. IEEE International Symposium on Information Theory (ISIT), June 2007, Nice, France.
[6] Schreckenbach, F. (2007) Iterative decoding of bit-interleaved coded modulation. PhD dissertation, Technische Universität München, Munich, Germany.
[7] Brännström, F. and Rasmussen, L. K. (2009) Classification of unique mappings for 8PSK based on bit-wise distance spectra. *IEEE Trans. Inf. Theory*, **55** (3), 1131–1145.
[8] Caire, G., Taricco, G., and Biglieri, E. (1998) Bit-interleaved coded modulation. *IEEE Trans. Inf. Theory*, **44** (3), 927–946.
[9] Martinez, A., Guillén i Fàbregas, A., and Caire, G. (2008) Bit-interleaved coded modulation in the wideband regime. *IEEE Trans. Inf. Theory*, **54** (12), 5447–5455.
[10] Martinez, A., Guillén i Fàbregas, A., Caire, G., and Willems, F. M. J. (2009) Bit-interleaved coded modulation revisited: a mismatched decoding perspective. *IEEE Trans. Inf. Theory*, **55** (6), 2756–2765.
[11] Stierstorfer, C. and Fischer, R. F. H. (2009) Asymptotically optimal mappings for BICM with M-PAM and M^2-QAM. *IET Electron. Lett.*, **45** (3), 173–174.
[12] Stierstorfer, C. (2009) A bit-level-based approach to coded multicarrier transmission. PhD dissertation, Friedrich-Alexander-Universität Erlangen-Nürnberg, Erlangen, Germany.
[13] Wachsmann, U., Fischer, R. F. H., and Huber, J. B. (1999) Multilevel codes: theoretical concepts and practical design rules. *IEEE Trans. Inf. Theory*, **45** (5), 1361–1391.
[14] Nguyen, T. and Lampe, L. (2011) Bit-interleaved coded modulation with mismatched decoding metrics. *IEEE Trans. Commun.*, **59** (2), 437–447.
[15] Jaldén, J., Fertl, P., and Matz, G. (2010) On the generalized mutual information of BICM systems with approximate demodulation. IEEE Information Theory Workshop (ITW), January 2010, Cairo, Egypt.
[16] Peng, L. (2012) Fundamentals of bit-interleaved coded modulation and reliable source transmission. PhD dissertation, University of Cambridge, Cambridge.
[17] Alvarado, A. and Agrell, E. (2014) Achievable rates for four-dimensional coded modulation with a bit-wise receiver. Optical Fiber Conference (OFC), March 2014, San Francisco, CA.
[18] Alvarado, A., Brännström, F., and Agrell, E. (2014) A simple approximation for the bit-interleaved coded modulation capacity. *IEEE Commun. Lett.*, **18** (3), 495–498.
[19] Açikel, O. and Ryan, W. (1999) Punctured turbo-codes for BPSK/QPSK channels. *IEEE Trans. Commun.*, **47** (9), 1325–1323.
[20] Kousa, M. A. and Mugaibel, A. H. (2002) Puncturing effects on turbo codes. *Proc. IEE*, **149** (3), 132–138.
[21] Calderbank, A. R. and Ozarow, L. H. (1990) Nonequiprobable signaling on the Gaussian channel. *IEEE Trans. Inf. Theory*, **36** (4), 726–740.
[22] Forney, G. D. Jr. and Wei, L.-F. (1989) Multidimensional constellations—Part I: Introduction, figures of merit, and generalized cross constellations. *IEEE J. Sel. Areas Commun.*, **7** (6), 877–892.
[23] Forney, G. D. Jr. (1989) Multidimensional constellations—Part II: voronoi constellations. *IEEE J. Sel. Areas Commun.*, **7** (6), 941–957.
[24] Fischer, R. F. H. (2002) *Precoding and Signal Shaping for Digital Transmission*, John Wiley & Sons, Inc..
[25] Sommer, D. and Fettweis, G. P. (2000) Signal shaping by non-uniform QAM for AWGN channels and applications using turbo coding. International ITG Conference on Source and Channel Coding (SCC), January 2000, Munich, Germany.
[26] Goff, S. Y. L. (2003) Signal constellations for bit-interleaved coded modulation. *IEEE Trans. Inf. Theory*, **49** (1), 307–313.
[27] Le Goff, S. Y., Sharif, B. S., and Jimaa, S. A. (2004) A new bit-interleaved coded modulation scheme using shaping coding. IEEE Global Telecommunications Conference (GLOBECOM), November–December 2004, Dallas, TX.
[28] Raphaeli, D. and Gurevitz, A. (2004) Constellation shaping for pragmatic turbo-coded modulation with high spectral efficiency. *IEEE Trans. Commun.*, **52** (3), 341–345.

[29] Le Goff, S., Sharif, B. S., and Jimaa, S. A. (2005) Bit-interleaved turbo-coded modulation using shaping coding. *IEEE Commun. Lett.*, **9** (3), 246–248.
[30] Le Goff, S., Khoo, B. K., Tsimenidis, C. C., and Sharif, B. S. (2007) Constellation shaping for bandwidth-efficient turbo-coded modulation with iterative receiver. *IEEE Trans. Wirel. Commun.*, **6** (6), 2223–2233.
[31] Guillén i Fàbregas, A. and Martinez, A. (2010) Bit-interleaved coded modulation with shaping. IEEE Information Theory Workshop (ITW), August–September 2010, Dublin, Ireland.
[32] Peng, L., Guillén i Fàbregas, A., and Martinez, A. (2012) Mismatched shaping schemes for bit-interleaved coded modulation. IEEE International Symposium on Information Theory (ISIT), July 2012, Cambridge, MA.
[33] Valenti, M. and Xiang, X. (2012) Constellation shaping for bit-interleaved LDPC coded APSK. *IEEE Trans. Commun.*, **60** (10), 2960–2970.
[34] Böcherer, G., Altenbach, F., Alvarado, A., Corroy, S., and Mathar, R. (2012) An efficient algorithm to calculate BICM capacity. IEEE International Symposium on Information Theory (ISIT), July 2012, Cambridge, MA.
[35] Agrell, E. and Alvarado, A. (2011) Optimal alphabets and binary labelings for BICM at low SNR. *IEEE Trans. Inf. Theory*, **57** (10), 6650–6672.
[36] Verdú, S. (2002) Spectral efficiency in the wideband regime. *IEEE Trans. Inf. Theory*, **48** (6), 1319–1343.
[37] Prelov, V. V. and Verdú, S. (2004) Second-order asymptotics of mutual information. *IEEE Trans. Inf. Theory*, **50** (8), 1567–1580.
[38] Alvarado, A., Agrell, E., Guillén i Fàbregas, A., and Martinez, A. (2010) Corrections to 'Bit-interleaved coded modulation in the wideband regime'. *IEEE Trans. Inf. Theory*, **56** (12), 6513.
[39] Stierstorfer, C. and Fischer, R. F. H. (2008) Mappings for BICM in UWB scenarios. International ITG Conference on Source and Channel Coding (SCC), January 2008, Ulm, Germany.
[40] Agrell, E. and Alvarado, A. (2013) Signal shaping for BICM at low SNR. *IEEE Trans. Inf. Theory*, **59** (4), 2396–2410.
[41] Lassing, J., Ström, E. G., Agrell, E., and Ottosson, T. (2003) Unequal bit-error protection in coherent M-ary PSK. IEEE Vehicular Technology Conference (VTC-Fall), October 2003, Orlando, FL.
[42] Brännström, F. (2004) Convergence analysis and design of multiple concatenated codes. PhD dissertation, Chalmers University of Technology, Göteborg, Sweden.
[43] Stierstorfer, C. and Fischer, R. F. H. (2007) (Gray) Mappings for bit-interleaved coded modulation. IEEE Vehicular Technology Conference (VTC-Spring), April 2007, Dublin, Ireland.
[44] Alvarado, A., Brännström, F., and Agrell, E. (2011) High SNR bounds for the BICM capacity. IEEE Information Theory Workshop (ITW), October 2011, Paraty, Brazil.
[45] Alvarado, A., Brännström, F., Agrell, E., and Koch, T. (2014) High-SNR asymptotics of mutual information for discrete constellations with applications to BICM. *IEEE Trans. Inf. Theory*, **60** (2), 1061–1076.
[46] Hildebrand, F. B. (ed.) (1981) *Handbook of Applicable Mathematics: Numerical Methods*, Vol. **3**, John Wiley & Sons, Inc.

5

Probability Density Functions of L-values

To characterize complex systems such as *bit-interleaved coded modulation* (BICM) receivers, it is convenient to study their building blocks independently. An important building block in a BICM receiver is the demapper Θ, whose role consists in calculating the L-values. In this chapter, we formally describe its behavior from a probabilistic point of view. The models developed here will be used in the chapters that follow, to study the performance of BICM transceivers.

This chapter is organized as follows. Section 5.1 motivates the need for finding the *probability density function* (PDF) of the L-values and shows its challenges. Sections 5.2 and 5.3 explain how to calculate the PDFs for 1D and 2D constellations, respectively. PDFs of the L-values in fading channels are discussed in Section 5.4 and Gaussian approximations are provided in Section 5.5.

5.1 Introduction and Motivation

5.1.1 BICM Channel

The BICM channel is defined as the entity that encompasses the interleaver, the mapper, the channel, the demapper, and the deinterleaver, as shown in Fig. 5.1. As the off-the-shelf binary encoder/decoder used in BICM operates "blindly" with respect to the channel, the BICM channel corresponds to an equivalent *binary-input continuous-output* (BICO) channel that separates the binary encoder and decoder.

At the output of the BICM channel, we obtain a sequence of L-values $\underline{\boldsymbol{\lambda}} = \Pi^{-1}(\underline{\boldsymbol{l}})$. However, because the interleaving is a one-to-one operation, without loss of generality, we can focus on the sequence of L-values at the output of the demapper $\underline{\boldsymbol{l}}$ instead. This observation can lead us to define a different BICM channel as shown in Fig. 5.2. In this new model, the interleaver and the deinterleaver become parts of the binary encoder and the decoder, respectively, showing that setting boundaries of the "BICM channel" interface is somewhat arbitrary. Regardless of whether we use the model in Fig. 5.1 or the one in Fig. 5.2, we recognize the demapper Θ as the key component in the receiver.

The model from Fig. 5.2 indicates that the L-values l_k in (3.50) are functions of the channel outcome $\boldsymbol{y}$, which is a random variable. The L-values are then functions of a random variable, and thus, are also

Bit-Interleaved Coded Modulation: Fundamentals, Analysis, and Design, First Edition.
Leszek Szczecinski and Alex Alvarado.

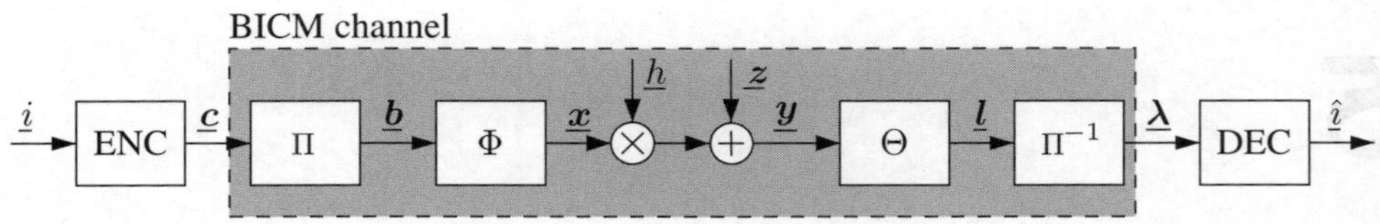

Figure 5.1 The BICM channel provides a single interface, which models the effect of the interleaver, the mapper Φ, the channel, the demapper Θ, and the deinterleaver

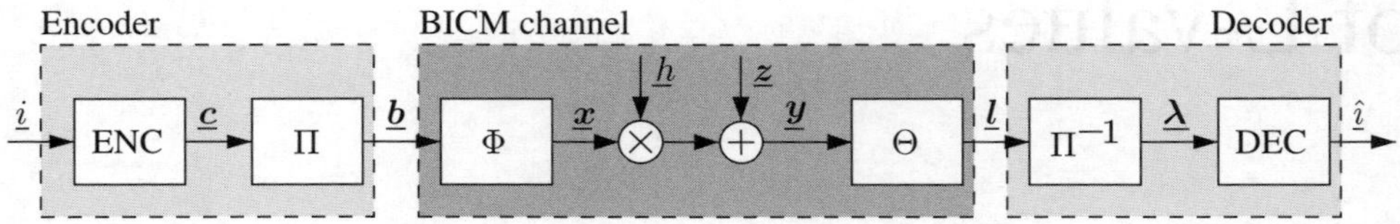

Figure 5.2 The BICM channel here provides the single interface that models the effect of the mapper Φ, the channel, and the demapper Θ. The interleaver and the deinterleaver become parts of the binary encoder and the decoder, respectively

random variables. Therefore, from now on, we use L_k to denote the L-values. To obtain the PDF of L_k conditioned on the bit B_k, we marginalize its PDF over all the symbols in $\mathcal{S}$, i.e.,

$$p_{L_k|B_k}(l|b) = \sum_{\boldsymbol{x}\in\mathcal{S}} p_{L_k|B_k,\boldsymbol{X}}(l|b,\boldsymbol{x})P_{\boldsymbol{X}|B_k}(\boldsymbol{x}|b) \tag{5.1}$$

$$= \frac{2}{M}\sum_{\boldsymbol{x}\in\mathcal{S}_{k,b}} p_{L_k|B_k,\boldsymbol{X}}(l|b,\boldsymbol{x}) \tag{5.2}$$

$$= \frac{2}{M}\sum_{\boldsymbol{x}\in\mathcal{S}_{k,b}} p_{L_k|\boldsymbol{X}}(l|\boldsymbol{x}). \tag{5.3}$$

To pass from (5.1) to (5.2) we used (2.77) and to pass from (5.2) to (5.3) we use the fact that for any $\boldsymbol{x}\in\mathcal{S}_{k,b}$, conditioning on the symbol and the kth bit is equivalent to conditioning on the symbol only.

From the model in Fig. 5.1, we see that the outputs of the BICO channel are the deinterleaved L-values λ, which are next passed to the decoder. These L-values are also random variables, which we denote by Λ. Under some assumptions on the interleaver structure, it can be shown that the L-values passed to the decoder are *independent and identically distributed* (i.i.d.) random variables with PDF[1]

$$p_{\Lambda|C}(\lambda|c) = \frac{2}{mM}\sum_{k=1}^{m}\sum_{\boldsymbol{x}\in\mathcal{S}_{k,c}} p_{L_k|\boldsymbol{X}}(l|\boldsymbol{x}), \tag{5.4}$$

where C is a binary random variable that models the input to the BICM channel in Fig. 5.1.

The rest of this chapter is aimed at characterizing the PDFs $p_{L_k|\boldsymbol{X}}(l|\boldsymbol{x})$, which owing to (5.3) and (5.4), allows us to model the BICM channels in Figs. 5.1 and 5.2. The explicit objective of the modeling is to develop analytical tools for performance evaluation, e.g., in terms of *bit-error probability* (BEP). This is necessary because, even if the performance may be evaluated via Monte Carlo integration, analytical forms simplify the calculations and provide insight into relevant design parameters.

[1] For more details about the analysis leading to (5.4), we refer the reader to Section 6.2.4.

5.1.2 *Case Study: PDF of L-values for 4PAM*

While it is simple to obtain the PDF of the L-values in the case of a 2PAM constellation (see Section 3.3.1), the case of an M-ary constellation is more challenging. In this section, we show an example to illustrate the difficulty of tackling this problem without any simplification.

We consider the *additive white Gaussian noise* (AWGN) channel and the simplest case of a multilevel 1D constellation ($N = 1$), i.e., a constellation with $M = 4$ points, which is well exemplified by a 4PAM constellation defined in Section 2.5. The labeling used is the *binary reflected Gray code* (BRGC),[2] i.e.,

$$\mathbf{X} = \mathbf{X}_{\text{4PAM}} = [-3\Delta, -\Delta, \Delta, 3\Delta], \quad \mathbf{Q} = \mathbf{Q}_{\text{BRGC}} = \begin{bmatrix} 0 & 0 & 1 & 1 \\ 0 & 1 & 1 & 0 \end{bmatrix}. \tag{5.5}$$

The constellation and labeling are shown in Fig. 5.3.

Throughout this chapter, we use $l_k = \theta_k(\boldsymbol{y})$ introduced in Section 3.3.1 to denote the functional relationship between l_k and $\boldsymbol{y}$. We start by calculating the PDF of the L-value L_k for the bit position $k = 1$, for which the relation between l_1 and the observation $\boldsymbol{y}$ in (3.50) is given by[3]

$$\begin{aligned} l_1 &= \theta_1(\boldsymbol{y}) \\ &= \log \frac{\mathrm{e}^{-\mathsf{snr}(\boldsymbol{y}-\Delta)^2} + \mathrm{e}^{-\mathsf{snr}(\boldsymbol{y}-3\Delta)^2}}{\mathrm{e}^{-\mathsf{snr}(\boldsymbol{y}+\Delta)^2} + \mathrm{e}^{-\mathsf{snr}(\boldsymbol{y}+3\Delta)^2}} \\ &= \log \frac{\mathrm{e}^{4\boldsymbol{y}\,\mathsf{snr}\,\Delta}}{\mathrm{e}^{-4\boldsymbol{y}\,\mathsf{snr}\,\Delta}} \frac{\mathrm{e}^{-2\boldsymbol{y}\,\mathsf{snr}\,\Delta + 4\,\mathsf{snr}\,\Delta^2} + \mathrm{e}^{2\boldsymbol{y}\,\mathsf{snr}\,\Delta - 4\,\mathsf{snr}\,\Delta^2}}{\mathrm{e}^{2\boldsymbol{y}\,\mathsf{snr}\,\Delta + 4\,\mathsf{snr}\,\Delta^2} + \mathrm{e}^{-2\boldsymbol{y}\,\mathsf{snr}\,\Delta - 4\,\mathsf{snr}\,\Delta^2}} \frac{\mathrm{e}^{-5\,\mathsf{snr}\,\Delta^2}}{\mathrm{e}^{-5\,\mathsf{snr}\,\Delta^2}} \\ &= 8\gamma \frac{\boldsymbol{y}}{\Delta} + \log \frac{\cosh(2\gamma(\boldsymbol{y}/\Delta - 2))}{\cosh(2\gamma(\boldsymbol{y}/\Delta + 2))}, \end{aligned} \tag{5.6}$$

where

$$\gamma \triangleq \mathsf{snr}\,\Delta^2. \tag{5.7}$$

For the case of a 4PAM constellation, $\gamma = \mathsf{snr}/5$.

The relationship (5.6) is shown in Fig. 5.4 for different values of snr, where the nonlinear behavior of $\theta_1(\boldsymbol{y})$ is evident. This figure also shows that for high *signal-to-noise ratio* (SNR) values, $\theta_1(\boldsymbol{y})$ adopts a piecewise linear form.

As the L-value is a function of the observation $\boldsymbol{Y}$, $L_1 = \theta_1(\boldsymbol{Y})$, its *cumulative distribution function* (CDF) conditioned on the symbol $\boldsymbol{X} = \boldsymbol{x}$ can be calculated by the definition of a CDF, i.e.,

$$F_{L_1|\boldsymbol{X}}(l|\boldsymbol{x}) \triangleq \Pr\{L_1 \le l | \boldsymbol{X} = \boldsymbol{x}\} \tag{5.8}$$

$$= \Pr\{\theta_1(\boldsymbol{Y}) \le l | \boldsymbol{X} = \boldsymbol{x}\} \tag{5.9}$$

$$= \int_{\boldsymbol{y}:\theta_1(\boldsymbol{Y}) \le l} p_{\boldsymbol{Y}|\boldsymbol{X}}(\boldsymbol{y}|\boldsymbol{x})\, \mathrm{d}\boldsymbol{y} \tag{5.10}$$

$$= \Pr\{\boldsymbol{Y} \le \theta_1^{-1}(l) | \boldsymbol{X} = \boldsymbol{x}\} = F_{\boldsymbol{Y}|\boldsymbol{X}}(\theta_1^{-1}(l)|\boldsymbol{x}), \tag{5.11}$$

where we are able to pass from (5.10) to (5.11) because the signal $\boldsymbol{y}$ is 1D ($N = 1$) and $\theta_1(\boldsymbol{y})$ is bijective, and thus, its inverse $\theta_1^{-1}(l)$ exists.

Figure 5.3 4PAM constellation labeled by the BRGC

[2] The use of the BRGC slightly simplifies the derivations; however, any other labeling could be considered.

[3] The vectors used in this section belong to the space of dimension $N = 1$; however, we continue to use the vectorial notation $\boldsymbol{y}$.

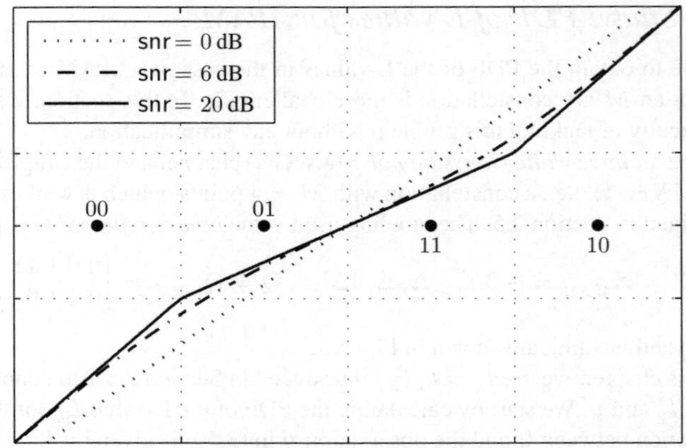

Figure 5.4 Nonlinear relationship $l_1 = \theta_1(\boldsymbol{y})$ given by (5.6) for a 4PAM constellation

The PDF $p_{L_1|\boldsymbol{X}}(l|\boldsymbol{x})$ is obtained via differentiation of (5.11), i.e.,

$$\begin{aligned} p_{L_1|\boldsymbol{X}}(l|\boldsymbol{x}) &= \frac{\mathrm{d}}{\mathrm{d}l} F_{\boldsymbol{Y}|\boldsymbol{X}}(\theta_1^{-1}(l)|\boldsymbol{x}) \\ &= p_{\boldsymbol{Y}|\boldsymbol{X}}(\theta_1^{-1}(l)|\boldsymbol{x}) \frac{\mathrm{d}\theta_1^{-1}(l)}{\mathrm{d}l} \\ &= p_{\boldsymbol{Y}|\boldsymbol{X}}(\theta_1^{-1}(l)|\boldsymbol{x}) \frac{1}{\theta_1'(\theta_1^{-1}(l))}, \end{aligned} \tag{5.12}$$

where

$$\theta_1'(\boldsymbol{y}) = \frac{\mathrm{d}}{\mathrm{d}\boldsymbol{y}}\theta_1(\boldsymbol{y}) = 4\,\mathsf{snr}\,\Delta\left[2 - \frac{\sinh(8\gamma)}{\cosh(4\gamma\boldsymbol{y}/\Delta) + \cosh(8\gamma)}\right] \tag{5.13}$$

is obtained from (5.6). Using (2.31) and (5.13) in (5.12), we obtain the final expression for the PDF

$$p_{L_1|\boldsymbol{X}}(l|\boldsymbol{x}) = \sqrt{\frac{\mathsf{snr}}{\pi}}\exp(-\mathsf{snr}(\theta_1^{-1}(l) - \boldsymbol{x})^2)\frac{1}{\theta_1'(\theta_1^{-1}(l))}. \tag{5.14}$$

The main difficulty in evaluating (5.14) is to obtain the inverse function $\theta_1^{-1}(l)$, which for this case cannot be found analytically. But, for a given l, we can find $\boldsymbol{y}$ by solving $\theta_1(\boldsymbol{y}) = l$. This has to be done numerically. The results obtained are shown in Fig. 5.5.[4]

We can repeat the same analysis for $k = 2$. In this case, we have

$$\begin{aligned} l_2 &= \theta_2(\boldsymbol{y}) \\ &= \log\frac{\mathrm{e}^{-\mathsf{snr}(\boldsymbol{y}-\Delta)^2} + \mathrm{e}^{-\mathsf{snr}(\boldsymbol{y}+\Delta)^2}}{\mathrm{e}^{-\mathsf{snr}(\boldsymbol{y}-3\Delta)^2} + \mathrm{e}^{-\mathsf{snr}(\boldsymbol{y}+3\Delta)^2}} \\ &= \log\frac{\mathrm{e}^{-\mathsf{snr}\,\Delta^2}}{\mathrm{e}^{-9\,\mathsf{snr}\,\Delta^2}}\frac{\mathrm{e}^{2\boldsymbol{y}\,\mathsf{snr}\,\Delta} + \mathrm{e}^{-2\boldsymbol{y}\,\mathsf{snr}\,\Delta}}{\mathrm{e}^{6\boldsymbol{y}\,\mathsf{snr}\,\Delta} + \mathrm{e}^{-6\boldsymbol{y}\,\mathsf{snr}\,\Delta}} \\ &= 8\,\mathsf{snr}\,\Delta^2 - \log\frac{\cosh(6\boldsymbol{y}\,\mathsf{snr}\,\Delta)}{\cosh(2\boldsymbol{y}\,\mathsf{snr}\,\Delta)}, \end{aligned} \tag{5.15}$$

which is shown in Fig. 5.6.

---

[4] The transformation we show in Fig. 5.5 (and also later in Fig. 5.7) are not to scale. The correctly scaled PDFs will be shown later in Fig. 5.12.

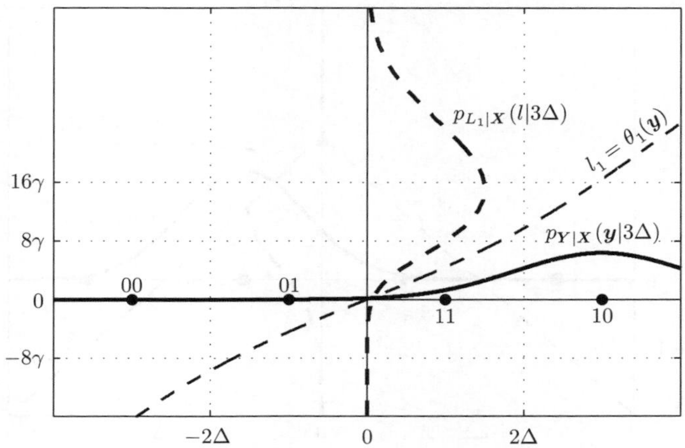

Figure 5.5 PDF (not to scale) of the L-values $p_{L_1|\boldsymbol{X}}(l|3\Delta)$ obtained as a transformation of the variable $\boldsymbol{Y}$ with PDF $p_{\boldsymbol{Y}|\boldsymbol{X}}(\boldsymbol{y}|3\Delta)$ via the function $L_1 = \theta_1(\boldsymbol{Y})$ for a 4PAM constellation; here, snr = 6 dB

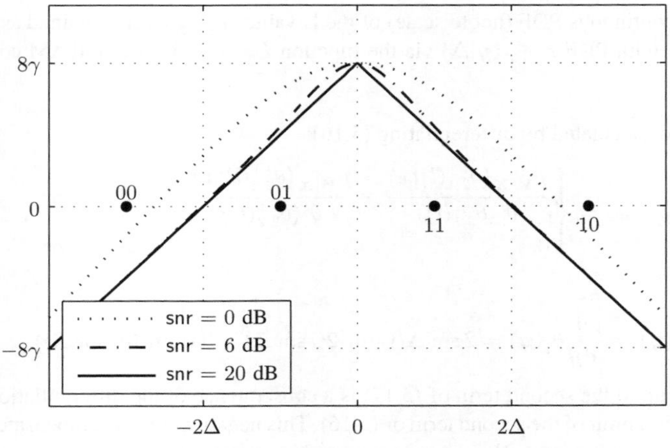

Figure 5.6 Nonlinear relationship $l_2 = \theta_2(\boldsymbol{y})$ given by (5.15) for a 4PAM constellation

The function $\theta_2(\boldsymbol{y})$ shown in Fig. 5.6 has no inverse, which was essential in deriving the PDF for $k = 1$. To deal with this problem, we tessellate the space of $\boldsymbol{y}$ into two disjoint regions $\mathcal{Y}_{2,1}$ and $\mathcal{Y}_{2,2}$ such that $\mathcal{Y}_{2,1} \cup \mathcal{Y}_{2,2} = \mathbb{R}$, where $\mathcal{Y}_{2,1} = \{\boldsymbol{y} : \boldsymbol{y} < 0\}$ and $\mathcal{Y}_{2,2} = \{\boldsymbol{y} : \boldsymbol{y} \geq 0\}$. Over each of the sets, the function $\theta_2(\boldsymbol{y})$ is bijective, and thus, we can define the "pseudoinverse" functions $\theta_{2,1}^{-1}(l)$ and $\theta_{2,2}^{-1}(l)$ that map l to the values $\boldsymbol{y} \in \mathcal{Y}_{2,j}$ with $j = 1, 2$, i.e., $\theta_{2,j}^{-1}(\theta_2(\boldsymbol{y})) = \boldsymbol{y}$ if $\boldsymbol{y} \in \mathcal{Y}_{2,j}$.

As $\theta_2(\boldsymbol{y}) \leq 8\gamma\ \forall \boldsymbol{y} \in \mathbb{R}$ (see Fig. 5.6), we immediately see that the CDF $F_{L_2|\boldsymbol{X}}(l|\boldsymbol{x}) = 1$ for $l \geq 8\gamma$. For $l < 8\gamma$, we obtain

$$\begin{aligned} F_{L_2|\boldsymbol{X}}(l|\boldsymbol{x}) &= \Pr\{\theta_2(\boldsymbol{Y}) \leq l \wedge \boldsymbol{Y} \in \mathcal{Y}_{2,1}|\boldsymbol{X} = \boldsymbol{x}\} + \Pr\{\theta_2(\boldsymbol{Y}) \leq l \wedge \boldsymbol{Y} \in \mathcal{Y}_{2,2}|\boldsymbol{X} = \boldsymbol{x}\} \\ &= \Pr\{\boldsymbol{Y} \leq \theta_{2,1}^{-1}(l)|\boldsymbol{X} = \boldsymbol{x}\} + \Pr\{\boldsymbol{Y} \geq \theta_{2,2}^{-1}(l)|\boldsymbol{X} = \boldsymbol{x}\} \\ &= \int_{-\infty}^{\theta_{2,1}^{-1}(l)} p_{\boldsymbol{Y}|\boldsymbol{X}}(\boldsymbol{y}|\boldsymbol{x})\, \mathrm{d}\boldsymbol{y} + \int_{\theta_{2,2}^{-1}(l)}^{\infty} p_{\boldsymbol{Y}|\boldsymbol{X}}(\boldsymbol{y}|\boldsymbol{x})\, \mathrm{d}\boldsymbol{y}. \end{aligned} \quad (5.16)$$

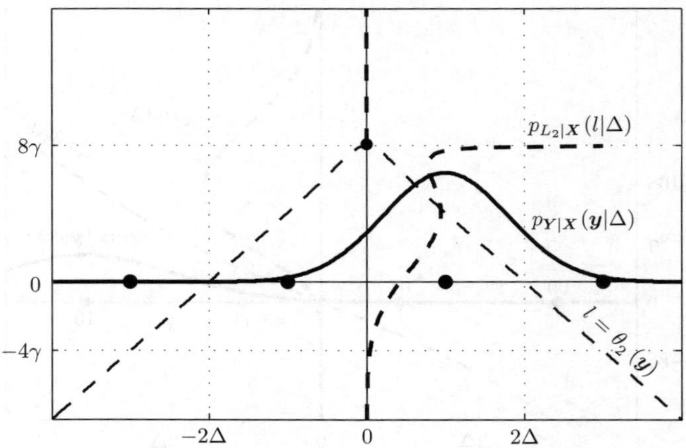

Figure 5.7 Discontinuous PDF (not to scale) of the L-values $p_{L_2|\boldsymbol{X}}(l|\Delta)$ obtained as a transformation of the variable $\boldsymbol{Y}$ with PDF $p_{\boldsymbol{Y}|\boldsymbol{X}}(\boldsymbol{y}|\Delta)$ via the function $L_2 = \theta_2(\boldsymbol{Y})$ for a 4PAM constellation; here, $\mathsf{snr} = 9$ dB

The PDF is now calculated by differentiating (5.16):

$$p_{L_2|\boldsymbol{X}}(l|\boldsymbol{x}) = \begin{cases} \dfrac{p_{\boldsymbol{Y}|\boldsymbol{X}}(\theta_{2,1}^{-1}(l)|\boldsymbol{x})}{\theta_2'(\theta_{2,1}^{-1}(l))} - \dfrac{p_{\boldsymbol{Y}|\boldsymbol{X}}(\theta_{2,2}^{-1}(l)|\boldsymbol{x})}{\theta_2'(\theta_{2,2}^{-1}(l))}, & \text{if } l < 8\gamma \\ 0, & \text{if } l \geq 8\gamma \end{cases}, \tag{5.17}$$

where

$$\theta_2'(\boldsymbol{y}) = \frac{\mathrm{d}}{\mathrm{d}\boldsymbol{y}}\theta_2(\boldsymbol{y}) = 2\,\mathsf{snr}\,\Delta(\tanh(2\boldsymbol{y}\,\mathsf{snr}\,\Delta) - 3\tanh(6\boldsymbol{y}\,\mathsf{snr}\,\Delta)). \tag{5.18}$$

The negative sign in the second term of (5.17) is a consequence of the differentiation with respect to the lower integration limit of the second term of (5.16). This negative sign is compensated by the negative sign of $\theta_2'(\boldsymbol{y})$ with $\boldsymbol{y} = \theta_{2,2}^{-1}(l) \in \mathcal{Y}_{2,2}$, so only nonnegative functions are added.

The transformation of the PDF $p_{\boldsymbol{Y}|\boldsymbol{X}}(\boldsymbol{y}|\boldsymbol{x})$ into the PDF $p_{L_2|\boldsymbol{X}}(l|\boldsymbol{x})$ is shown in Fig. 5.7, where we can observe a "peak" appearing around $l = 8\gamma$. This is perfectly normal and happens because $\theta_2'(0) = 0$, see (5.18). The PDF $p_{L_2|\boldsymbol{X}}(l|\boldsymbol{x})$ is thus undefined for $l = 8\gamma$.

Example 5.1 (Verifying Symmetry and Consistency Conditions) *Using the PDFs in* (5.12) *and* (5.17) *in* (5.3), *we can calculate the PDFs* $p_{L_k|B_k}(l|b)$:

$$p_{L_1|B_1}(l|1) = \frac{1}{2}[p_{L_1|\boldsymbol{X}}(l|\Delta) + p_{L_1|\boldsymbol{X}}(l|3\Delta)], \tag{5.19}$$

$$p_{L_1|B_1}(l|0) = \frac{1}{2}[p_{L_1|\boldsymbol{X}}(l|-\Delta) + p_{L_1|\boldsymbol{X}}(l|-3\Delta)], \tag{5.20}$$

$$p_{L_2|B_2}(l|1) = \frac{1}{2}[p_{L_2|\boldsymbol{X}}(l|-\Delta) + p_{L_2|\boldsymbol{X}}(l|\Delta)], \tag{5.21}$$

$$p_{L_2|B_2}(l|0) = \frac{1}{2}[p_{L_2|\boldsymbol{X}}(l|-3\Delta) + p_{L_2|\boldsymbol{X}}(l|3\Delta)]. \tag{5.22}$$

For $k = 1$, we can verify the consistency condition (3.67) by using (5.19), (5.20), *and* (5.12), *i.e.,*

$$\log \frac{p_{L_k|B_k}(l|1)}{p_{L_k|B_k}(l|0)} = \frac{p_{\mathbf{Y}|\mathbf{X}}(\theta_1^{-1}(l)|\Delta) + p_{\mathbf{Y}|\mathbf{X}}(\theta_1^{-1}(l)|3\Delta)}{p_{\mathbf{Y}|\mathbf{X}}(\theta_1^{-1}(l)|-\Delta) + p_{\mathbf{Y}|\mathbf{X}}(\theta_1^{-1}(l)|-3\Delta)} \tag{5.23}$$

$$= \frac{\sum_{\boldsymbol{x}\in\mathcal{S}_{1,1}} p_{\mathbf{Y}|\mathbf{X}}(\theta_1^{-1}(l)|\boldsymbol{x})}{\sum_{\boldsymbol{x}\in\mathcal{S}_{1,0}} p_{\mathbf{Y}|\mathbf{X}}(\theta_1^{-1}(l)|\boldsymbol{x})} \tag{5.24}$$

$$= \theta_k(\theta_k^{-1}(l)) = l. \tag{5.25}$$

For $k = 2$, a similar result can be obtained (in this case, by using (5.21), (5.22), *and* (5.17)). *These results are, in fact, a particularization of Theorem 3.10.*

We now study the symmetry condition in (3.74). For $k = 1$, we have $\theta_1(\boldsymbol{y}) = -\theta_1(-\boldsymbol{y})$, its inverse is also odd, i.e., $\theta_1^{-1}(l) = -\theta_1^{-1}(-l)$, and thus, using (5.14), *we obtain $p_{L_1|B_1}(l|1) = p_{L_1|B_1}(-l|0)$. Therefore, the PDF for $k = 1$ satisfies the symmetry condition. On the other hand, for $k = 2$, the symmetry condition $p_{L_2|B_2}(l|1) = p_{L_2|B_2}(-l|0)$ is not satisfied because $p_{L_2|B_2}(l|b)$ has a semi-infinite support, i.e., $p_{L_2|B_2}(l|b) = 0$ for $l > 8\gamma$. An intuitive explanation of this behavior is the following: the constellation $\mathcal{S}_{1,b}$ is the reflection of $\mathcal{S}_{1,\bar{b}}$ (i.e., $\forall \boldsymbol{x} \in \mathcal{S}_{1,b}\ \exists \boldsymbol{x}' \in \mathcal{S}_{1,\bar{b}}$, $\boldsymbol{x}' = -\boldsymbol{x}$), while this property does not hold of $k = 2$.*

5.1.3 Local Linearization

While it is definitely possible to calculate the PDF of the L-values (as shown in Section 5.1.2), a significant numerical effort may be required, as we do not known the analytical forms of the inverse or pseudoinverse of $\theta_k(\boldsymbol{y})$. Moreover, such numerical results provide little insight into the properties of BICM systems.

The problem of finding the PDF of the L-values is greatly simplified when we consider the max-log approximation from (3.99), which reduces the function $\theta_k(\boldsymbol{y})$ to piecewise linear functions. This linearization is typically exploited at the receiver to reduce the complexity of the L-values calculation. Here, we take advantage of the max-log approximation to obtain a piecewise linear model for the L-values. This model will be used in the following sections to develop analytical expressions/approximations for the PDF of the L-values.

The linearization caused by the max-log approximation can be formalized by rewriting the L-values in (3.99) as

$$\theta_k(\boldsymbol{y}) = \mathsf{snr}\left(\min_{\boldsymbol{x}_i\in\mathcal{S}_{k,0}} \|\boldsymbol{y}-\boldsymbol{x}_i\|^2 - \min_{\boldsymbol{x}_j\in\mathcal{S}_{k,1}} \|\boldsymbol{y}-\boldsymbol{x}_j\|^2\right)$$

$$= \mathsf{snr} \sum_{j\in\mathcal{I}_{k,1}} \sum_{i\in\mathcal{I}_{k,0}} (\|\boldsymbol{y}-\boldsymbol{x}_i\|^2 - \|\boldsymbol{y}-\boldsymbol{x}_j\|^2)\mathbb{I}_{[\boldsymbol{y}\in\mathcal{Y}_{k;j,i}]}, \tag{5.26}$$

where

$$\mathcal{Y}_{k;j,i} \triangleq \left\{\boldsymbol{y} : \boldsymbol{x}_i = \underset{\boldsymbol{x}\in\mathcal{S}_{k,0}}{\operatorname{argmin}}\|\boldsymbol{y}-\boldsymbol{x}\|^2 \wedge \boldsymbol{x}_j = \underset{\boldsymbol{x}\in\mathcal{S}_{k,1}}{\operatorname{argmin}}\|\boldsymbol{y}-\boldsymbol{x}\|^2\right\} \tag{5.27}$$

for $k = 1, \ldots, m$, $i \in \mathcal{I}_{k,0}$ and $j \in \mathcal{I}_{k,1}$.

The tessellation region $\mathcal{Y}_{k;j,i}$ in (5.27) contains all the observations $\boldsymbol{y}$, for which $\boldsymbol{x}_j$ and $\boldsymbol{x}_i$ are the closest (in the sense of *Euclidean distance* (ED)) symbols to the constellations $\mathcal{S}_{k,1}$ and $\mathcal{S}_{k,0}$, respectively. This tessellation principle is valid for any number of dimensions N. We note that although the sum over i and j in (5.26) covers all possible combinations of the indices (i.e., there are $M^2/4$ sets $\mathcal{Y}_{k;j,i}$), some of the sets $\mathcal{Y}_{k;j,i}$ are empty.

Following the steps we have already taken in (3.51)–(3.53), we express (5.26) as

$$\theta_k(\boldsymbol{y}) = 2\,\mathsf{snr} \sum_{j\in\mathcal{I}_{k,1}} \sum_{i\in\mathcal{I}_{k,0}} \boldsymbol{d}_{j,i}^{\mathrm{T}}(\boldsymbol{y}-\overline{\boldsymbol{x}}_{j,i})\mathbb{I}_{[\boldsymbol{y}\in\mathcal{Y}_{k;j,i}]} \tag{5.28}$$

$$= 2\,\mathsf{snr} \sum_{j\in\mathcal{I}_{k,1}} \sum_{i\in\mathcal{I}_{k,0}} \xi_{j,i}(\boldsymbol{y})\mathbb{I}_{[\boldsymbol{y}\in\mathcal{Y}_{k;j,i}]}, \tag{5.29}$$

where $\overline{\boldsymbol{x}}_{j,i}$ and $\xi_{j,i}(\boldsymbol{y})$ are given by (3.55) and (3.54), respectively.

To clarify the definitions above, consider the following example based on the *natural binary code* (NBC) in Definition 2.11.

Example 5.2 (Linearization for 4PAM Labeled by the NBC) *Consider a 4PAM constellation labeled by the NBC. In Fig. 5.8, we show the piecewise linear functions $\theta_1(\boldsymbol{y})$ and $\theta_2(\boldsymbol{y})$ (where the pieces are labeled as $\xi_{j,i}$) as well as the tessellation regions $\mathcal{Y}_{k;j,i}$. In this case, there are four sets $\mathcal{Y}_{k;j,i}$ for each k, however, two of them are empty, i.e., $\mathcal{Y}_{1;4,1} = \varnothing$ and $\mathcal{Y}_{2;4,1} = \varnothing$.*

The CDF of the L-value L_k conditioned on the symbol $\boldsymbol{X}$ can be written using (5.29) as

$$F_{L_k|\boldsymbol{X}}(l|\boldsymbol{x}) = \Pr\{L_k \le l|\boldsymbol{X}=\boldsymbol{x}\} \tag{5.30}$$

$$= \sum_{j\in\mathcal{I}_{k,1}} \sum_{i\in\mathcal{I}_{k,0}} F_{k;j,i}(l|\boldsymbol{x}), \tag{5.31}$$

where

$$F_{k;j,i}(l|\boldsymbol{x}) \triangleq \Pr\{2\,\mathsf{snr}\,\xi_{j,i}(\boldsymbol{Y}) \le l \wedge \boldsymbol{Y}\in\mathcal{Y}_{k;j,i}|\boldsymbol{X}=\boldsymbol{x}\}. \tag{5.32}$$

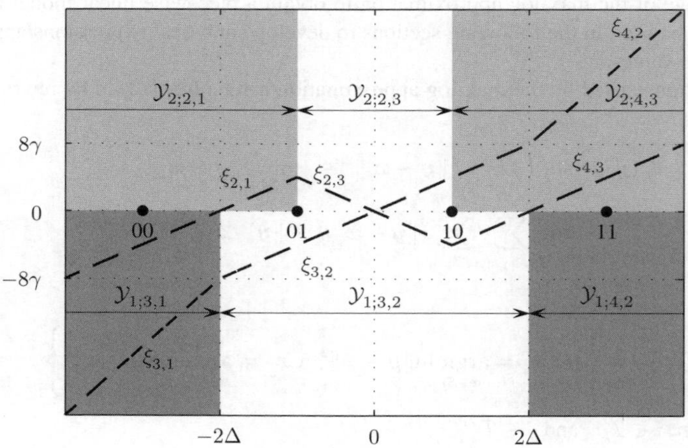

Figure 5.8 Tessellation regions $\mathcal{Y}_{1;j,i}$ (bottom shaded regions) and $\mathcal{Y}_{2;j,i}$ (top shaded regions) for the NBC in Definition 2.11 and a 4PAM constellation. The piecewise linear functions $\theta_k(\boldsymbol{y})$ in (5.29) whose pieces are labeled as $\xi_{j,i}$ are also shown

Differentiation of (5.31) with respect to l produces the conditional PDF

$$p_{L_k|\mathbf{X}}(l|\boldsymbol{x}) = \sum_{j\in\mathcal{I}_{k,1}} \sum_{i\in\mathcal{I}_{k,0}} p_{k;j,i}(l|\boldsymbol{x}), \tag{5.33}$$

where

$$p_{k;j,i}(l|\boldsymbol{x}) \triangleq \frac{\mathrm{d}}{\mathrm{d}l} F_{k;j,i}(l|\boldsymbol{x}). \tag{5.34}$$

We also define the set

$$\mathcal{L}_{k;j,i} \triangleq \{l \in \mathbb{R} : l = \theta_k(\boldsymbol{y}), \boldsymbol{y} \in \mathcal{Y}_{k;j,i}\}, \tag{5.35}$$

which is the "image" of $\mathcal{Y}_{k;j,i}$ after transformation via $\theta_k(\boldsymbol{y}) = 2\,\mathsf{snr}\,\xi_{j,i}(\boldsymbol{y})$. Of course, $\mathcal{L}_{k;j,i}$ is an interval because it is a linear transformation of the convex set $\mathcal{Y}_{k;j,i}$, i.e.,

$$\mathcal{L}_{k;j,i} = \{l \in \mathbb{R} : \mathsf{snr}\check{l}_{k;j,i;1} < l \leq \mathsf{snr}\check{l}_{k;j,i;2}\}, \tag{5.36}$$

where

$$\check{l}_{k;j,i;1} \triangleq \min_{\boldsymbol{y}\in\mathcal{Y}_{k;j,i}} 2\xi_{j,i}(\boldsymbol{y}) \tag{5.37}$$

and

$$\check{l}_{k;j,i;2} \triangleq \max_{\boldsymbol{y}\in\mathcal{Y}_{k;j,i}} 2\xi_{j,i}(\boldsymbol{y}) \tag{5.38}$$

are the (normalized by snr) limits of the interval.

5.2 PDF of L-values for 1D Constellations

In this section, we use the linearization procedure presented in Section 5.1.3 and show how to calculate the PDF of the max-log L-values for arbitrary 1D constellations. In this case, the tessellation regions $\mathcal{Y}_{k;j,i}$ in (5.27) are intervals (see Fig. 5.8 for an example), i.e.,

$$\mathcal{Y}_{k;j,i} = \{\boldsymbol{y} \in \mathbb{R} : \check{\boldsymbol{y}}_{k;j,i;1} \leq \boldsymbol{y} < \check{\boldsymbol{y}}_{k;j,i;2}\}, \tag{5.39}$$

so the limits of the intervals $\mathcal{L}_{k;j,i}$ (5.37) and (5.38) are given by

$$\check{l}_{k;j,i;1} = 2\,\min\{\xi_{j,i}(\check{\boldsymbol{y}}_{k;j,i;1}), \xi_{j,i}(\check{\boldsymbol{y}}_{k;j,i;2})\}, \tag{5.40}$$

$$\check{l}_{k;j,i;2} = 2\,\max\{\xi_{j,i}(\check{\boldsymbol{y}}_{k;j,i;1}), \xi_{j,i}(\check{\boldsymbol{y}}_{k;j,i;2})\}. \tag{5.41}$$

The following theorem gives a closed-form expression for the conditional PDF of the L-values $p_{L_k|\mathbf{X}}(l|\boldsymbol{x})$ for arbitrary 1D constellations.

Theorem 5.3 (PDF of L-values for 1D Constellations) *For any 1D constellation and labeling, the conditional PDF of the max-log L-values $p_{L_k|\mathbf{X}}(l|\boldsymbol{x})$ can be expressed as* (5.33) *where*

$$p_{k;j,i}(l|\boldsymbol{x}) = \Psi(l, \mu_{j,i}(\boldsymbol{x})\,\mathsf{snr}, \sigma_{j,i}^2\,\mathsf{snr}) \cdot \mathbb{I}_{[l\in\mathcal{L}_{k;j,i}]}, \tag{5.42}$$

$\mathcal{L}_{k;j,i}$ is given by (5.36) *with $\check{l}_{k;j,i;1}$ and $\check{l}_{k;j,i;2}$ given in* (5.40) *and* (5.41), *and*

$$\mu_{j,i}(\boldsymbol{x}) \triangleq 2\xi_{j,i}(\boldsymbol{x}), \tag{5.43}$$

$$\sigma_{j,i}^2 \triangleq 2\|\boldsymbol{d}_{j,i}\|^2. \tag{5.44}$$

Proof: For all $\boldsymbol{y} \in \mathcal{Y}_{k;j,i}$, the resulting L-values $l_k = \theta_k(\boldsymbol{y})$ belong to the interval $\mathcal{L}_{k;j,i}$ in (5.35) and assuming $\boldsymbol{d}_{j,i} > 0$, the calculation of the CDF in (5.32) gives

$$F_{k;j,i}(l|\boldsymbol{x}) = \Pr\{2\,\mathsf{snr}\,\xi_{j,i}(\boldsymbol{x}+\boldsymbol{Z}) \leq l \wedge \check{\boldsymbol{y}}_{k;j,i;1} \leq \boldsymbol{x}+\boldsymbol{Z} < \check{\boldsymbol{y}}_{k;j,i;2}\} \tag{5.45}$$

$$= \Pr\left\{\boldsymbol{Z} \leq \frac{l - 2\,\mathsf{snr}\,\xi_{j,i}(\boldsymbol{x})}{2\,\mathsf{snr}\|\boldsymbol{d}_{j,i}\|} \wedge \check{\boldsymbol{y}}_{k;j,i;1} - \boldsymbol{x} \leq \boldsymbol{Z} < \check{\boldsymbol{y}}_{k;j,i;2} - \boldsymbol{x}\right\} \tag{5.46}$$

$$= \begin{cases} 0, & \text{if } l < \check{l}_{k;j,i;1} \\ \Pr\left\{\check{\boldsymbol{y}}_{k;j,i;1} - \boldsymbol{x} \leq \boldsymbol{Z} < \frac{l-2\,\mathsf{snr}\,\xi_{j,i}(\boldsymbol{x})}{2\,\mathsf{snr}\|\boldsymbol{d}_{j,i}\|}\right\}, & \text{if } \check{l}_{k;j,i;1} \leq l < \check{l}_{k;j,i;2} \\ K, & \text{if } l \geq \check{l}_{k;j,i;2} \end{cases} \tag{5.47}$$

$$= \begin{cases} 0, & \text{if } l < \check{l}_{k;j,i;1} \\ Q\left(\sqrt{2\,\mathsf{snr}}\,(\check{\boldsymbol{y}}_{k;j,i;1} - \boldsymbol{x})\right) - Q\left(\frac{l-2\,\mathsf{snr}\,\xi_{j,i}(\boldsymbol{x})}{\sqrt{2\,\mathsf{snr}}\|\boldsymbol{d}_{j,i}\|}\right), & \text{if } \check{l}_{k;j,i;1} \leq l < \check{l}_{k;j,i;2}, \\ K, & \text{if } l \geq \check{l}_{k;j,i;2} \end{cases} \tag{5.48}$$

where to pass from (5.46) to (5.47) we consider the three relevant cases for l, K is a real constant, and where to pass from (5.47) to (5.48) we use (2.8). Differentiation of (5.48) gives the final form of the PDF, i.e.,

$$p_{k;j,i}(l|\boldsymbol{x}) = \frac{\mathrm{d}}{\mathrm{d}l}F_{k;j,i}(l|\boldsymbol{x}) \tag{5.49}$$

$$= -\frac{\mathrm{d}}{\mathrm{d}l}Q\left(\frac{l - 2\,\mathsf{snr}\,\xi_{j,i}(\boldsymbol{x})}{\sqrt{2\,\mathsf{snr}}\|\boldsymbol{d}_{j,i}\|}\right) \cdot \mathbb{I}_{[l \in \mathcal{L}_{k;j,i}]} \tag{5.50}$$

$$= \frac{1}{\sqrt{2\pi}}\exp\left(-\frac{(l - 2\,\mathsf{snr}\,\xi_{j,i}(\boldsymbol{x}))^2}{2 \cdot 2\,\mathsf{snr}\|\boldsymbol{d}_{j,i}\|^2}\right)\frac{1}{\sqrt{2\,\mathsf{snr}}\|\boldsymbol{d}_{j,i}\|}, \tag{5.51}$$

which completes the proof for $\boldsymbol{d}_{j,i} > 0$. The case $\boldsymbol{d}_{j,i} < 0$ can be solved using steps similar to those above. □

In view of (5.33), each linear function $\xi_{j,i}(\boldsymbol{y})$ and the corresponding interval $\mathcal{Y}_{k;j,i}$ will "contribute" to the PDF with a piece of a Gaussian function (truncated over the interval $\mathcal{L}_{k;j,i}$, as shown in (5.42)) and whose mean and variance are given by (5.43) and (5.44), respectively. This transformation is illustrated in Fig. 5.9 and is rather intuitive: the Gaussian random variable $\boldsymbol{Y} \sim \mathcal{N}(\mu, \sigma^2)$, after being transformed via a piecewise linear function $a\boldsymbol{Y} + b$, has its mean transformed to $a\mu + b$ and its variance to $a^2\sigma^2$.

We observe that while the variance of the Gaussian piece in (5.42) depends solely on the distance between the symbols $\boldsymbol{x}_j \in \mathcal{S}_{k,1}$ and $\boldsymbol{x}_i \in \mathcal{S}_{k,0}$ defining the tessellation regions $\mathcal{Y}_{k;j,i}$ (see (5.44)), the mean in (5.43) depends also on the symbol $\boldsymbol{x}$. We also note that the limits of the interval $\mathcal{L}_{k;j,i}$ in (5.35) scale linearly with SNR and so do the mean and the variance in (5.43) and (5.44).

Example 5.4 (PDF of L-values for 2PAM**)** *Let us revisit the* 2PAM *constellation we have already analyzed in Section 3.3.1, i.e.,* $\boldsymbol{x}_2 = -\boldsymbol{x}_1 = 1$*. In this case,* $\mathcal{Y}_{1;2,1} = \mathbb{R}$, $\mathcal{L}_{1;j,i} = \mathbb{R}$, $\mu_{2,1}(\boldsymbol{x}) = \boldsymbol{x}\|\boldsymbol{d}_{2,1}\|^2$, *and* $\sigma^2_{j,i} = 2\|\boldsymbol{d}_{2,1}\|^2$*. Using these values in* (5.42) *gives*

$$p_{1;2,1}(l|\boldsymbol{x}) = \Psi(l, \boldsymbol{x}\,\mathsf{snr}\|\boldsymbol{d}_{2,1}\|^2, 2\,\mathsf{snr}\|\boldsymbol{d}_{2,1}\|^2). \tag{5.52}$$

As there is only Gaussian piece contributing to $p_{L_1|\mathbf{X}}(l|\boldsymbol{x})$ *in* (5.33), (5.52) *has the same form as the one we obtained in* (3.63).

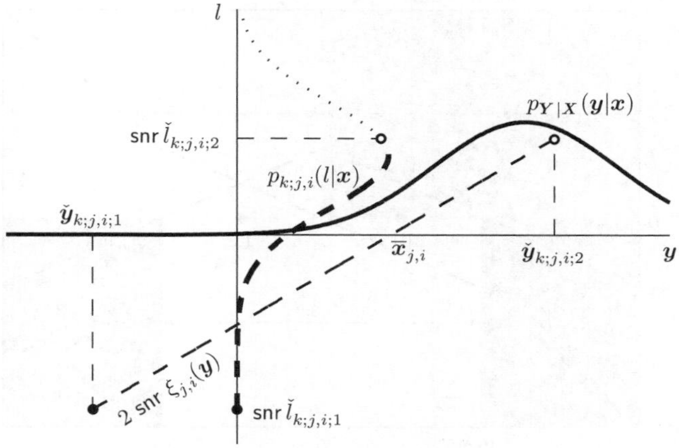

Figure 5.9 A piecewise linear transformation of a Gaussian PDF $p_{\boldsymbol{Y}|\boldsymbol{X}}(\boldsymbol{y}|\boldsymbol{x})$ yields a truncated Gaussian function $p_{k;j,i}(l|\boldsymbol{x})$. The dotted lines show the extension of the function $p_{k;j,i}(l|\boldsymbol{x})$ before its truncation and $\overline{\boldsymbol{x}}_{j,i}$ marks the solution of $2\,\mathsf{snr}\,\xi_{j,i}(\boldsymbol{y}) = 0$

Example 5.5 (4PAM **Labeled by the BRGC**) *Consider the* 4PAM constellation *labeled by the BRGC we have already studied in Section 5.1.2. The functions* $\theta_k(\boldsymbol{y})$ *in this case (using the max-log approximation) are*

$$\theta_1(\boldsymbol{y}) = \mathsf{snr}\cdot\min\{(\boldsymbol{y}+\Delta)^2,(\boldsymbol{y}+3\Delta)^2\} - \mathsf{snr}\cdot\min\{(\boldsymbol{y}-\Delta)^2,(\boldsymbol{y}-3\Delta)^2\}, \tag{5.53}$$

$$\theta_2(\boldsymbol{y}) = \mathsf{snr}\cdot\min\{(\boldsymbol{y}-3\Delta)^2,(\boldsymbol{y}+3\Delta)^2\} - \mathsf{snr}\cdot\min\{(\boldsymbol{y}-\Delta)^2,(\boldsymbol{y}+\Delta)^2\}, \tag{5.54}$$

which we show in Fig. 5.10. These functions, without max-log approximation (i.e., the exact L-values), were shown in Figs. 5.4 and 5.6 for different snr. *Clearly, for* $\mathsf{snr} = 20$ dB, *the exact L-values are practically identical to the results obtained using* (5.53) *and* (5.54)*, which is consistent with the well-known fact that the accuracy of the max-log approximation increases as the SNR increases.*

From Fig. 5.10, we easily deduce the forms of the tessellation regions $\mathcal{Y}_{k;j,i}$ *as well as their images* $\mathcal{L}_{k;j,i}$*, i.e., for* $k = 1$

$$\mathcal{Y}_{1;3,1} = \{\boldsymbol{y} : \boldsymbol{y} < -2\Delta\}, \qquad \mathcal{L}_{1;3,1} = \{l : l < -8\gamma\}, \tag{5.55}$$

$$\mathcal{Y}_{1;3,2} = \{\boldsymbol{y} : -2\Delta \le \boldsymbol{y} < 2\Delta\}, \qquad \mathcal{L}_{1;3,2} = \{l : -8\gamma \le l < 8\gamma\}, \tag{5.56}$$

$$\mathcal{Y}_{1;4,2} = \{\boldsymbol{y} : 2\Delta \le \boldsymbol{y}\}, \qquad \mathcal{L}_{1;4,2} = \{l : 8\gamma \le l\}, \tag{5.57}$$

and for $k = 2$

$$\mathcal{Y}_{2;2,1} = \{\boldsymbol{y} : \boldsymbol{y} \le 0\}, \qquad \mathcal{L}_{2;2,1} = \{l : l \le 8\gamma\}, \tag{5.58}$$

$$\mathcal{Y}_{2;3,4} = \{\boldsymbol{y} : \boldsymbol{y} > 0\}, \qquad \mathcal{L}_{2;3,4} = \{l : l < 8\gamma\}, \tag{5.59}$$

where $\gamma = \mathsf{snr}\,\Delta^2$ *is given by* (5.7). *The interval limits in* (5.40) *and* (5.41) *are thus*

$$\check{l}_{1;3,1;1} = -\infty,\ \check{l}_{1;3,1;2} = -8\gamma = \check{l}_{1;3,2;1},\ \check{l}_{1;3,2;2} = 8\gamma = \check{l}_{1;4,2;1},\ \check{l}_{1;4,2;2} = \infty, \tag{5.60}$$

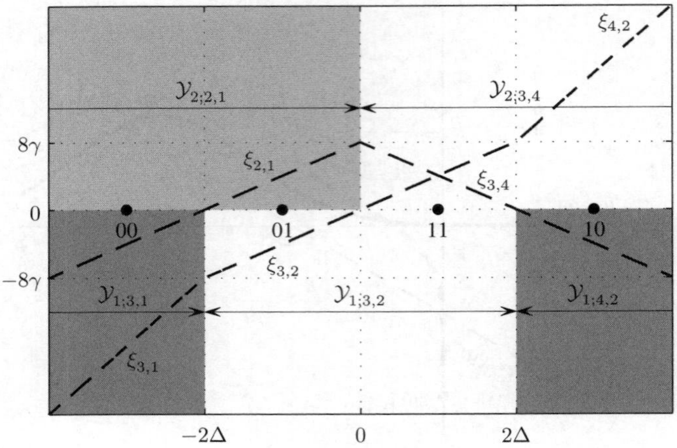

Figure 5.10 Tessellation regions $\mathcal{Y}_{1;j,i}$ (bottom shaded regions) and $\mathcal{Y}_{2;j,i}$ (top shaded regions) for a 4PAM constellation labeled by BRGC. The piecewise linear functions $\theta_k(\boldsymbol{y})$ in (5.29) whose pieces are labeled as $\xi_{j,i}$ are also shown

and

$$\check{l}_{2;2,1;1} = -\infty = \check{l}_{2;3,4;1}, \qquad \check{l}_{2;2,1;2} = 8\gamma = \check{l}_{2;3,4;2}. \tag{5.61}$$

Because $\boldsymbol{x}_j = (2j-5)\Delta$, $\boldsymbol{d}_{j,i} = (j-i)2\Delta$, *and* $\overline{\boldsymbol{x}}_{j,i} = ((j+i)-5)\Delta$, *it follows that*

$$\xi_{j,i}(\boldsymbol{x}_r) = 2\Delta^2(j-i)(2r-j-i). \tag{5.62}$$

Then, using (5.43) *and* (5.44), *we have the following:*

$$\mu_{j,i}(\boldsymbol{x}_r) = 4\Delta^2(j-i)[2r-(j+i)], \tag{5.63}$$

$$\sigma^2_{j,i} = 8\Delta^2(j-i)^2. \tag{5.64}$$

Using (5.33), *the PDF of the L-values can be then represented as*

$$p_{L_1|\boldsymbol{X}}(l|\boldsymbol{x}_r) = \begin{cases} \Psi(l, 8\gamma(2r-4), 32\gamma), & \text{if } l < -8\gamma \\ \Psi(l, 4\gamma(2r-5), 8\gamma), & \text{if } -8\gamma \le l < 8\gamma \\ \Psi(l, 8\gamma(2r-6), 32\gamma), & \text{if } 8\gamma \le l \end{cases}, \tag{5.65}$$

$$p_{L_2|\boldsymbol{X}}(l|\boldsymbol{x}_r) = \begin{cases} \Psi(l, 4\gamma(2r-3), 8\gamma) + \Psi(l, -4\gamma(2r-7), 8\gamma), & \text{if } l \le 8\gamma \\ 0, & \text{if } l > 8\gamma \end{cases}. \tag{5.66}$$

The pieces $p_{1;3,1}(l|\boldsymbol{x})$, $p_{1;3,2}(l|\boldsymbol{x})$, *and* $p_{1;4,2}(l|\boldsymbol{x})$ *appear defined over disjoint intervals in* (5.65), *while* $p_{2;2,1}(l|\boldsymbol{x})$ *and* $p_{2;3,4}(l|\boldsymbol{x})$ *are added in the PDF* $p_{L_2|\boldsymbol{X}}(l|\boldsymbol{x})$ *because the tessellation regions* $\mathcal{Y}_{2;2,1}$ *and* $\mathcal{Y}_{2;3,4}$ *are mapped through the functions* $\theta_1(\cdot)$ *and* $\theta_2(\cdot)$ *into the same image, i.e., into* $\mathcal{L}_{2;2,1} = \mathcal{L}_{2;3,4}$.

The PDFs (5.65) *and* (5.66) *are shown as a result of nonlinear transformation of the Gaussian PDF in Fig. 5.11. Note that the PDF obtained via the max-log approximation for* $k=1$ *is discontinuous because of the change of the sign of the derivatives at the borders of* $\mathcal{Y}_{k;j,i}$, *which is not the case in Fig. 5.5. Moreover, the use of the max-log approximation gives a function* $\theta_2(\boldsymbol{y})$ *with a nonzero first*

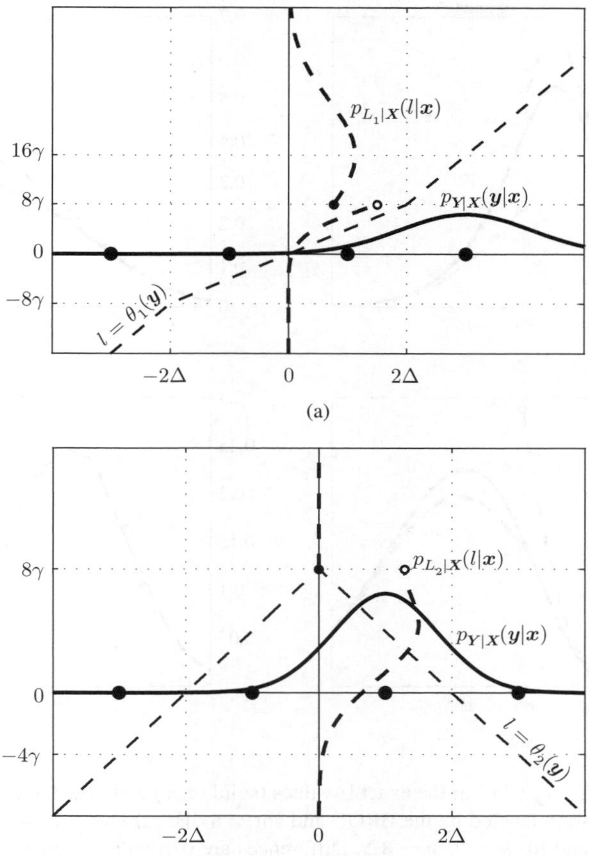

Figure 5.11 PDF (not to scale) of the L-values $p_{L_k|\boldsymbol{X}}(l|\boldsymbol{x})$ obtained as a transformation of the variable $\boldsymbol{Y}$ with PDF $p_{\boldsymbol{Y}|\boldsymbol{X}}(\boldsymbol{y}|\boldsymbol{x})$ via the function $L_k = \theta_k(\boldsymbol{Y})$ based on the max-log approximation for a 4PAM constellation labeled by the BRGC and $\mathsf{snr} = 9$ dB: (a) $k = 1, \boldsymbol{x} = 3\Delta$ and (b) $k = 1, \boldsymbol{x} = \Delta$. The discontinuity of the PDF shown by small white/black circles is due to the piecewise linear form of $\theta_k(\boldsymbol{y})$.

derivative with respect to $\boldsymbol{y}$*, and thus, the "peak" we observed in Fig. 5.7 for the PDF of the exact L-values* $p_{L_2|\boldsymbol{X}}(l|\boldsymbol{x})$ *disappears.*

The PDF conditioned on the bit B_k *can be found using* (5.65) *and* (5.66) *in* (5.3)*. We can verify that* $p_{L_1|B_1}(l|b)$ *satisfies the symmetry condition in* (3.74)*, while* $p_{L_2|B_2}(l|b)$ *does not. More importantly, the consistency condition in* (3.67) *is not satisfied in either case. This comes from the fact that the L-values in this case are obtained through the max-log approximation, thus do not comply with the assumptions made in Theorem 3.10.*

So far, all the PDFs we have shown were presented to facilitate the understanding of the concept of the linear transformations involved, and thus, are not to scale. In Fig. 5.12, we show the true PDFs $p_{L_1|\boldsymbol{X}}(l|\boldsymbol{x})$ and $p_{L_2|\boldsymbol{X}}(l|\boldsymbol{x})$ with $\boldsymbol{x} = \Delta$ and $\boldsymbol{x} = 3\Delta$ for the exact and max-log L-values. This set of PDFs completely characterize the PDFs $p_{L_k|B_k}(l|b)$, as $p_{L_1|\boldsymbol{X}}(l| - \boldsymbol{x}) = p_{L_1|\boldsymbol{X}}(-l|\boldsymbol{x})$ and

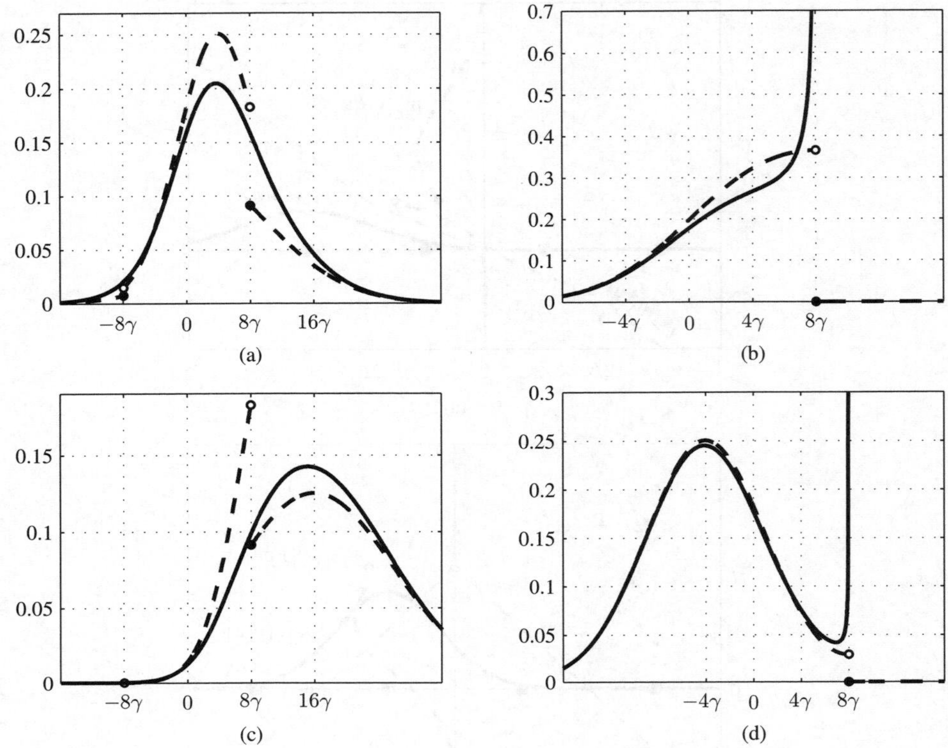

Figure 5.12 PDFs $p_{L_k|\boldsymbol{X}}(l|\boldsymbol{x})$ of the exact L-values (solid lines) and max-log L-values (dashed lines) for a 4PAM constellation labeled by the BRGC and $\mathsf{snr} = 5$ dB: (a) $k = 1$, $\boldsymbol{x} = \Delta$, (b) $k = 2$, $\boldsymbol{x} = \Delta$, (c) $k = 1$, $\boldsymbol{x} = 3\Delta$, and (d) $k = 2$, $\boldsymbol{x} = 3\Delta$. Differences are particularly notable in the vicinity of the borders of the intervals $\mathcal{L}_{k;j,i}$, i.e., around $\pm 8\gamma$. The discontinuities of the PDF are shown as small black/white circles. In the case of max-log L-values, they are a consequence of the piecewise linear form of $\theta_1(\boldsymbol{y})$, while in the case of exact L-value and $k = 2$, the discontinuity is because of the null derivative $\theta_2'(0) = 0$

$p_{L_2|\boldsymbol{X}}(l|-\boldsymbol{x}) = p_{L_2|\boldsymbol{X}}(l|\boldsymbol{x})$. The importance of these PDFs is that they fully characterize the L-values of the practically relevant case of a 16QAM constellation labeled by the BRGC shown in Fig. 2.14(b).

5.3 PDF of L-values for 2D Constellations

The analysis shown in Section 5.2 is slightly more involved in the case of 2D constellations. In what follows, we will first describe how to find the tessellation regions $\mathcal{Y}_{k;j,i}$, and then, we will find the PDF of the L-values for arbitrary 2D constellations.

5.3.1 *Space Tessellation*

Finding the tessellation regions $\mathcal{Y}_{k;j,i}$ for 1D constellations is relatively simple and can be done by inspection, as we did, e.g., in Example 5.5. A more structured approach is required for 2D constellations, because as we will see, even with very regular constellations such as MQAM, the regions $\mathcal{Y}_{k;j,i}$ may significantly change in a function of the binary labeling.

We start by rewriting (5.27) as

$$\mathcal{Y}_{k;j,i} = \{\boldsymbol{y} : \|\boldsymbol{y}-\boldsymbol{x}_i\|^2 \leq \|\boldsymbol{y}-\boldsymbol{x}_{i'}\|^2 \wedge \|\boldsymbol{y}-\boldsymbol{x}_j\|^2 \leq \|\boldsymbol{y}-\boldsymbol{x}_{j'}\|^2, \quad \forall j' \in \mathcal{I}_{k,1}, j' \neq j, \quad i' \in \mathcal{I}_{k,0}, i' \neq i\} \tag{5.67}$$

$$= \{\boldsymbol{y} : 2\langle \boldsymbol{y}, \boldsymbol{d}_{i',i}\rangle + \|\boldsymbol{x}_i\|^2 - \|\boldsymbol{x}_{i'}\|^2 \leq 0 \wedge 2\langle \boldsymbol{y}, \boldsymbol{d}_{j',j}\rangle + \|\boldsymbol{x}_j\|^2 - \|\boldsymbol{x}_{j'}\|^2 \leq 0, \quad \forall j' \in \mathcal{I}_{k,1}, j' \neq j, \quad i' \in \mathcal{I}_{k,0}, i' \neq i\} \tag{5.68}$$

$$= \{\boldsymbol{y} : \xi_{j',j}(\boldsymbol{y}) \leq 0 \wedge \xi_{i',i}(\boldsymbol{y}) \leq 0, \forall j' \in \mathcal{I}_{k,1}, j' \neq j, i' \in \mathcal{I}_{k,0}, i' \neq i\}, \tag{5.69}$$

where $j \in \mathcal{I}_{k,1}$, $i \in \mathcal{I}_{k,0}$, and $\xi_{j,i}(\boldsymbol{y})$ is given by (3.54).

These tessellation regions $\mathcal{Y}_{k;j,i}$ are mutually disjoint

$$\mathcal{Y}_{k;j,i} \cap \mathcal{Y}_{k;j',i'} = \varnothing, \quad \text{if } i' \neq i \text{ or } j \neq j', \tag{5.70}$$

and span the whole space

$$\bigcup_{j,i} \mathcal{Y}_{k;j,i} = \mathbb{R}^N. \tag{5.71}$$

Moreover, because $\mathcal{Y}_{k;j,i}$ in (5.69) is defined by linear inequalities, the tessellation region may be (i) a (convex) polygon, (ii) a region that extends to infinity (i.e., an "infinite" polygon), or (iii) an empty set (if the inequalities in (5.69) are contradictory).

It should be rather clear that the regions $\mathcal{Y}_{k;j,i}$ cannot be deduced from the so-called Voronoi regions of the constellation (see (3.100)). The Voronoi regions depend solely on the constellation $\mathcal{S}$, while $\mathcal{Y}_{k;j,i}$ depends also the on the labeling (via $\mathcal{I}_{k,0}$ and $\mathcal{I}_{k,1}$), or equivalently, on the subconstellations $\mathcal{X}_{k,0}$ and $\mathcal{X}_{k,1}$. However, the tessellation regions $\mathcal{Y}_{k;j,i}$ can be obtained as the intersection of the Voronoi regions of the subconstellations $\mathcal{X}_{k,0}$ and $\mathcal{X}_{k,1}$, as explained in the following example.

Example 5.6 (Tessellation Regions for 16QAM Labeled by the BRGC) *In the case of a 16QAM constellation labeled by the BRGC shown in Fig. 5.13, the tessellation regions $\mathcal{Y}_{k;j,i}$ can be found by inspection. In fact, it is enough to find and intersect the Voronoi regions for the symbols $\boldsymbol{x}_j$ and $\boldsymbol{x}_i$ taken from the subconstellations $\mathcal{S}_{k,0}$ and $\mathcal{S}_{k,1}$ (shown by white and black circles in Fig. 5.13), respectively. Fig. 5.13 shows that indeed three types of regions appear: polygons, "infinite" polygons, and empty sets.*

The tessellation regions in this case can be also found as the Cartesian product of the tessellation regions over the first dimension of $\boldsymbol{y}$ (which we have already found in Fig. 5.10) and the Voronoi regions of the symbols from the 4PAM constellation defined over the second dimension of $\boldsymbol{y}$. This particular property is a consequence of the binary labeling, which is done independently per dimension: the bits b_1, b_2 are mapped to the first dimension of the symbol, while b_3, b_4 are mapped to its second dimension.

Example 5.7 (Tessellation Regions for 8PSK Labeled by the BRGC) *The tessellation regions for an 8PSK constellation can also be obtained by inspection. Because $\|\boldsymbol{x}_i\| = 1, \forall i \in \mathcal{I}$, independently of the labeling, the tessellation regions are always wedges.*[5] *These tessellation regions are shown in Fig. 5.14 for the BRGC. This figure also shows that the subconstellations $\mathcal{S}_{2,b}$ correspond to a rotated version of $\mathcal{S}_{1,b}$, and thus, the tessellation regions are also rotated versions of each other. Consequently, the PDF of the L-values for $k = 1$ and $k = 2$ are the same.*

[5] This is a reminiscence of MPAM constellations, where the tessellation regions are always intervals.

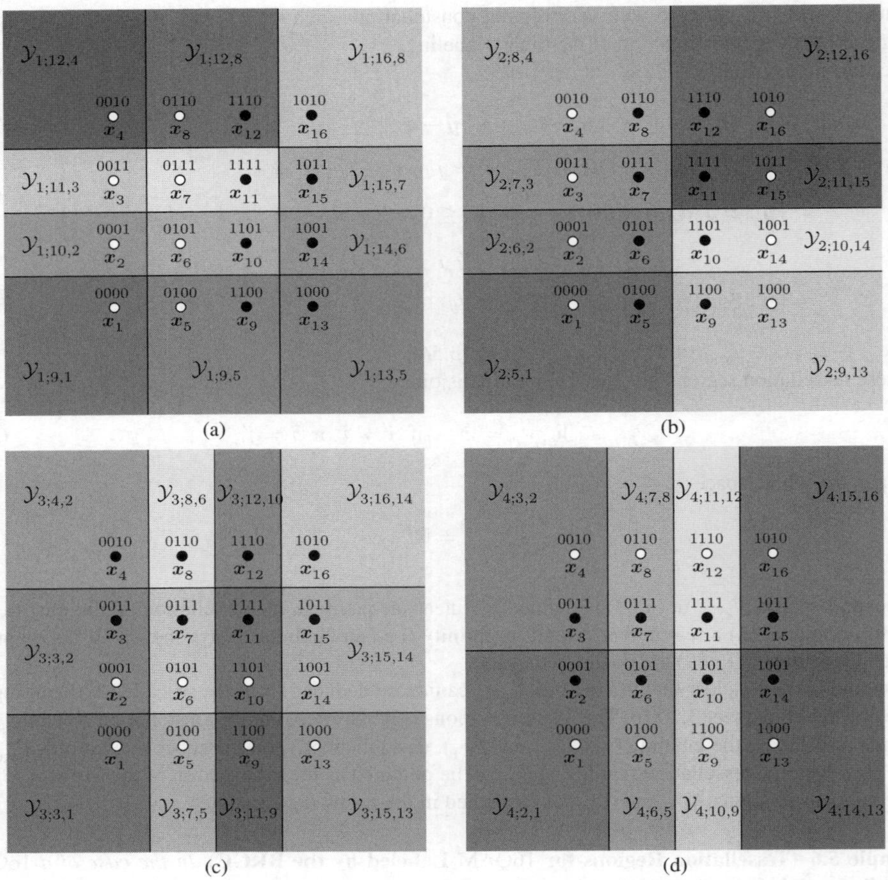

Figure 5.13 Tessellation regions $\mathcal{Y}_{k;j,i}$ in (5.69) for a 16QAM constellation labeled by the BRGC (see Fig. 2.14): (a) $k = 1$, (b) $k = 2$, (c) $k = 3$, and (d) $k = 4$. For clarity, only selected regions $\mathcal{Y}_{k;j,i}$ are identified with labels. The subconstellations $\mathcal{S}_{k,0}$ and $\mathcal{S}_{k,1}$ are indicated by white and black circles, respectively

Finding the tessellation regions in Examples 5.6 and 5.7 is relatively simple, however, this is not always the case. For example, when a 16PAM constellation labeled by the M16 (shown in Fig. 2.15) is considered, the labeling cannot be factorized as the ordered direct product of two labelings, so the tessellation regions are difficult to find by inspection. As we will see in Example 5.8, the regions in this case are very irregular, and thus, to find them, we require an algorithmic support.

In order to find $\mathcal{Y}_{k;j,i}$, we need to remove redundant constraints from (5.69). The $M - 2$ linear inequalities $\xi_{j',j}(\boldsymbol{y}) \leq 0$ and $\xi_{i',i}(\boldsymbol{y}) \leq 0$ in (5.69) can be reformulated in a matrix form[6] as

$$\mathcal{Y}_{k;j,i} = \{\boldsymbol{y}\colon \mathbf{D}\boldsymbol{y} \leq \boldsymbol{e}\}, \tag{5.72}$$

where

$$\mathbf{D} \triangleq [\mathbf{D}_1, \mathbf{D}_0]^{\mathrm{T}} \tag{5.73}$$

[6] The inequality in (5.72) (and all other inequalities involving vectors) are applied element-wise.

is an $(M-2)\times 2$ matrix,[7] and where $\mathbf{D}_1$ and $\mathbf{D}_0$ are $2\times(M/2-1)$ matrices whose columns are taken from the respective sets

$$\mathcal{D}_1 = \{\boldsymbol{d}_{j',j} : j' \in \mathcal{I}_{k,1}, j' \neq j\}, \tag{5.74}$$
$$\mathcal{D}_0 = \{\boldsymbol{d}_{i',i} : i' \in \mathcal{I}_{k,0}, i' \neq i\}. \tag{5.75}$$

Similarly, the $(M-2)\times 1$ vector $\boldsymbol{e}$ in (5.72) is

$$\boldsymbol{e} \triangleq [\boldsymbol{e}_1^{\mathrm{T}}, \boldsymbol{e}_0^{\mathrm{T}}]^{\mathrm{T}}, \tag{5.76}$$

where $\boldsymbol{e}_1$ and $\boldsymbol{e}_0$ are $(M/2-1)$ column vectors whose elements are taken from the sets

$$\mathcal{E}_1 = \{\boldsymbol{d}_{j',j}^{\mathrm{T}} \cdot \overline{\boldsymbol{x}}_{j',j} : j' \in \mathcal{I}_{k,1}, j' \neq j\}, \tag{5.77}$$
$$\mathcal{E}_0 = \{\boldsymbol{d}_{i',i}^{\mathrm{T}} \cdot \overline{\boldsymbol{x}}_{i',i} : i' \in \mathcal{I}_{k,0}, i' \neq i\}, \tag{5.78}$$

respectively. To make (5.72) correspond to (5.69), we also assume that the matrix $\mathbf{D}$ and the vector $\boldsymbol{e}$ are ordered in the same way. In other words, if the column of $\mathbf{D}_1$ is defined by a particular pair of indices (j, j'), the same pair defines the corresponding element of $\boldsymbol{e}_1$. The same applies to $\mathbf{D}_0$ and $\boldsymbol{e}_0$ whose entries are defined by the pairs (i, i').

In general, (5.72) has redundant inequalities, i.e., the number of linear forms defining the region $\mathcal{Y}_{k;j,i}$ is smaller than $M-2$. Removing the redundant inequalities in (5.72) is a "classical" problem in computational geometry, and to solve it, we first require the region $\mathcal{Y}_{k;j,i}$ to be closed, i.e., infinite polygons (which indeed exists, as shown in Fig. 5.13) are artificially closed by adding constraints forming a square with arbitrarily large dimensions $A\times A$

$$\begin{bmatrix} 1 & 0 \\ 0 & 1 \\ -1 & 0 \\ 0 & -1 \end{bmatrix} \boldsymbol{y} = \mathbf{D}_{\mathrm{art}}\boldsymbol{y} \leq A\cdot\mathbf{1}, \tag{5.79}$$

where $\mathbf{1}$ is a 4×1 vector of ones. We can keep track of these artificial constraints and, if they appear in the final solution, we should eliminate them, which would produce an infinite polygon. However, for practical purposes, retaining the artificial inequalities has negligible effect when A is sufficiently large.

We then focus on the augmented problem

$$\overline{\mathbf{D}}\boldsymbol{y} \leq \overline{\boldsymbol{e}}, \tag{5.80}$$

where $\overline{\mathbf{D}} = [\mathbf{D}^{\mathrm{T}}, \mathbf{D}_{\mathrm{art}}^{\mathrm{T}}]^{\mathrm{T}}$ is an $(M+2)\times 2$ matrix and $\overline{\boldsymbol{e}} = [\boldsymbol{e}^{\mathrm{T}}, A\cdot\mathbf{1}^{\mathrm{T}}]^{\mathrm{T}}$ is an $M+2$ column vector. To solve (5.80), we translate the coordinates $\boldsymbol{y}$ via $\boldsymbol{y} = \boldsymbol{y}' + \boldsymbol{y}_{\mathrm{tr}}$ to obtain

$$\overline{\mathbf{D}}(\boldsymbol{y}' + \boldsymbol{y}_{\mathrm{tr}}) \leq \overline{\boldsymbol{e}} \tag{5.81}$$
$$\overline{\mathbf{D}}\boldsymbol{y}' \leq \overline{\boldsymbol{e}} - \overline{\mathbf{D}}\boldsymbol{y}_{\mathrm{tr}} \tag{5.82}$$

so that after the translation, $\boldsymbol{y}' = \mathbf{0}$ satisfies all the inequalities in (5.82). The problem then boils down to finding a solution of

$$\overline{\mathbf{D}}\boldsymbol{y}_{\mathrm{tr}} \leq \overline{\boldsymbol{e}}. \tag{5.83}$$

[7] Or more generally, $(M-2)\times N$ for N-dimensional constellations.

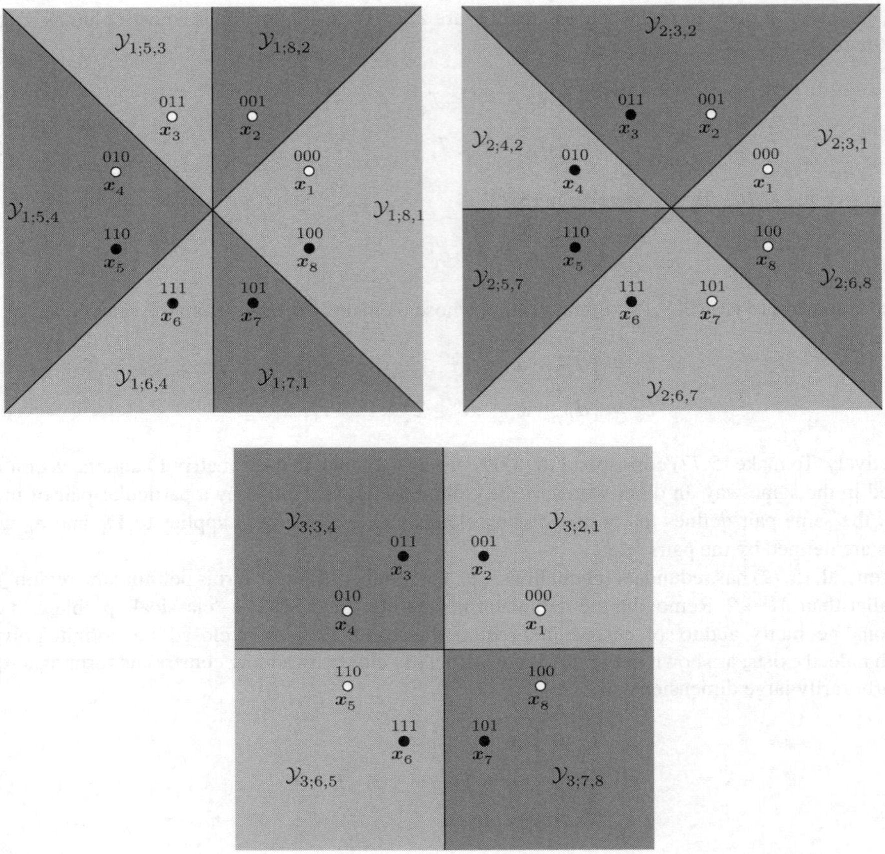

Figure 5.14 Tessellation regions $\mathcal{Y}_{k;j,i}$, for an 8PSK constellation labeled by the BRGC. The subconstellations $\mathcal{S}_{k,0}$ and $\mathcal{S}_{k,1}$ are indicated by white and black circles, respectively

Various methods may be used to find a feasible solution of the set of inequalities in (5.83). Here, we have solved the following quadratic optimization problem:

$$\boldsymbol{y}_{\text{tr}} = \underset{\boldsymbol{y}\in\mathbb{R}^2}{\text{argmin}}\{\|\boldsymbol{y}\|^2 \mid \overline{\mathbf{D}}\boldsymbol{y} < \overline{\boldsymbol{e}}\}. \tag{5.84}$$

Note that the objective function $\|\boldsymbol{y}\|^2$ is defined arbitrarily as we are interested in finding *any* translation point $\boldsymbol{y}_{\text{tr}}$ satisfying the constraints. If the solution of (5.84) does not exist, it means that the inequalities (5.72) are contradictory and the tessellation region $\mathcal{Y}_{k;j,i}$ is empty.

After $\boldsymbol{y}_{\text{tr}}$ has been found, (5.82) can be expressed as

$$\overline{\mathbf{D}}\boldsymbol{y}' \leq \overline{\boldsymbol{e}}', \tag{5.85}$$

where $\overline{\boldsymbol{e}}' = \overline{\boldsymbol{e}} - \overline{\mathbf{D}}\boldsymbol{y}_{\text{tr}}$. The next step is to normalize each inequality in (5.85) to obtain

$$\overline{\mathbf{D}}'\boldsymbol{y}' \leq \mathbf{1}, \tag{5.86}$$

where the normalization is such that the nth row of $\overline{\mathbf{D}}'$ is equal to the nth row of $\overline{\mathbf{D}}$ divided by the (positive by definition) nth entry of $\overline{\boldsymbol{e}}'$.

The original problem in (5.72) expressed as in (5.86) is now in a standard form and can be solved using the duality between the line and points description. This duality states that removing redundant inequalities from (5.86) is equivalent to finding the minimum convex region enclosing the $M-2$ rows of $\overline{\mathbf{D}}'$, where each row is treated as a point in 2D space. We thus have to find a convex hull of these $M-2$ points, which can be done efficiently using numerical routines available, e.g., in MATLAB® or Mathematica®.

As we will see in the next section, to compute the PDF of the L-values, we need not only the linear pieces defining the tessellation regions $\mathcal{Y}_{k;j,i}$ but also the vertices of the polygons. Finding these vertices is again relatively simple and corresponds to a *vertex enumeration problem* treated in computational geometry.

Example 5.8 (Tessellation Regions for 16QAM with the M16 Labeling) *By using the procedure described above for the* 16QAM *constellation with the M16 labeling shown in Fig. 2.15, we obtain*

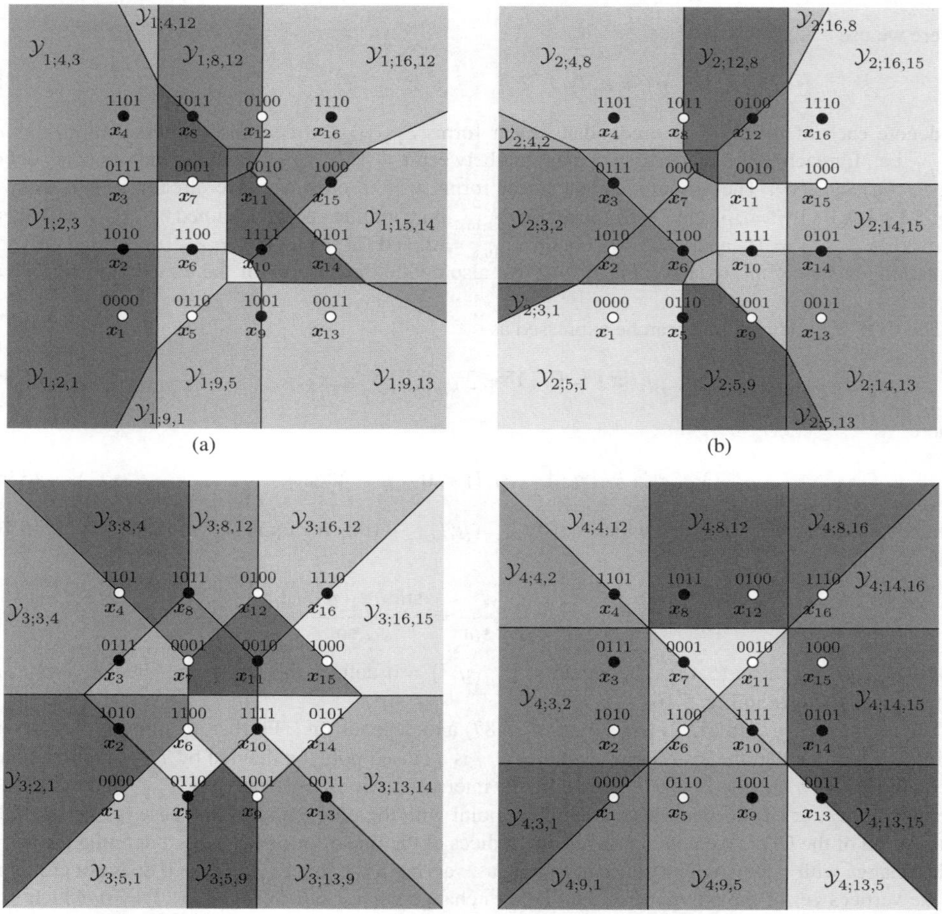

Figure 5.15 Tessellation regions $\mathcal{Y}_{k;j,i}$ in (5.69) for a 16QAM constellation labeled by the M16: (see Fig. 2.15): (a) $k=1$, (b) $k=2$, (c) $k=3$, and (d) $k=4$. For clarity, only selected regions $\mathcal{Y}_{k;j,i}$ are identified with labels. The subconstellations $\mathcal{S}_{k,0}$ and $\mathcal{S}_{k,1}$ are indicated by white and black circles, respectively

the tessellation regions shown in Fig. 5.15. Unlike the results in Fig. 5.14 and Fig. 5.13, in this case, the tessellation regions cannot be expressed as the Cartesian product of regions obtained along each dimension. Moreover, the tessellation regions in this case are rather difficult to find by inspection, which justifies the algorithmic approach presented above. Another particularity of the tessellation regions is that some of them do not contain any constellation symbol, as is the case, e.g., of $\mathcal{Y}_{1;4,12}$ *or* $\mathcal{Y}_{2;16,8}$.

5.3.2 *Finding the PDF*

In the previous section, we have shown how to find the tessellation regions for arbitrary 2D constellations and labelings. The next step toward finding the PDF of the L-values is to find the CDFs $F_{k;j,i}(l|\boldsymbol{x})$ as required by (5.31). We start by rewriting (5.69) as

$$\mathcal{Y}_{k;j,i} = \{\boldsymbol{y} : \mathsf{L}_t(\boldsymbol{y}) \leq 0, \quad t = 1, \dots, T_{k;j,i}\}, \tag{5.87}$$

where we use

$$\mathsf{L}_t(\boldsymbol{y}) \triangleq \boldsymbol{d}_t^{\mathrm{T}}(\boldsymbol{y} - \overline{\boldsymbol{x}}_t), \quad t = 1, \dots, T_{k;j,i}, \tag{5.88}$$

to denote each of the $T_{k;j,i}$ nonredundant linear forms $\xi_{j,i}(\boldsymbol{y}) \leq 0$ defining the tessellation regions $\mathcal{Y}_{k;j,i}$, i.e., for each $\mathcal{Y}_{k;j,i}$ there exists a mapping between $t = 1, \dots, T_{k;j,i}$ and the indices (i, j) defining $\xi_{j,i}(\boldsymbol{y}) = \mathsf{L}_t(\boldsymbol{y})$. These nonredundant linear forms are the ones we have already shown, e.g., in Fig. 5.13: if $k = 1$, $j = 15$, and $i = 7$, the region $\mathcal{Y}_{1;15,7}$ is an infinite polygon defined by $T_{1;15,7} = 3$ linear forms; if $k = 1$, $j = 16$, and $i = 8$, we obtain $T_{1;16,8} = 2$; and for the region $\mathcal{Y}_{1;11,7}$ (the closed polygon containing $\boldsymbol{x}_7$ and $\boldsymbol{x}_{11}$), we have $T_{1;11,7} = 4$. We also define the vertices of the tessellation regions as $\boldsymbol{w}_t, t = 1, \dots, T_{k;j,i}$.

The CDF $F_{k;j,i}(l|\boldsymbol{x})$ (5.32) can be expressed as

$$F_{k;j,i}(l|\boldsymbol{x}) = \Pr\{\boldsymbol{Y} \in \mathcal{Y}_{k;j,i}(l)|\boldsymbol{X} = \boldsymbol{x}\}, \tag{5.89}$$

where

$$\mathcal{Y}_{k;j,i}(l) \triangleq \{\boldsymbol{y} : \mathsf{L}_{j,i}(\boldsymbol{y}; l) < 0 \wedge \boldsymbol{y} \in \mathcal{Y}_{k;j,i}\} \tag{5.90}$$

contains observations $\boldsymbol{y} \in \mathcal{Y}_{k;j,i}$ which satisfy $\mathsf{L}_{j,i}(\boldsymbol{y}; l) \leq 0$ (this region is shown shaded in Fig. 5.16) and

$$\mathsf{L}_{j,i}(\boldsymbol{y}; l) \triangleq \xi_{j,i}(\boldsymbol{y}) - \frac{l}{2\,\mathsf{snr}} = \frac{\mathsf{snr}\,\mu_{j,i}(\boldsymbol{y}) - l}{2\,\mathsf{snr}}, \tag{5.91}$$

where $\mu_{j,i}(\boldsymbol{y})$ given by (5.43). The equation $\mathsf{L}_{j,i}(\boldsymbol{y}; l) = 0$ defines the line perpendicular to $\boldsymbol{d}_{j,i}$, as schematically shown in Fig. 5.16.

The region $\mathcal{Y}_{k;j,i}(l)$ in (5.90) is a subset of (5.87) and depends on l. Further, to simplify the presentation, we assume that the tessellation region $\mathcal{Y}_{k;j,i}$ is a closed polygon defined by $T_{k;j,i}$ nonredundant inequalities. Then the line $\mathsf{L}_{j,i}(\boldsymbol{y}; l) = 0$ either (i) intersects[8] two sides of polygon $\mathcal{Y}_{k;j,i}$ and these sides are then *active*, or (ii) does not have a common point with the polygon. The first case is critical to the calculation of the CDF. We note that while the indices of the linear forms $\mathsf{L}_t(\boldsymbol{y}) = 0$ defining the active sides change with l, they remain constant as long as, varying l, the line $\mathsf{L}_{j,i}(\boldsymbol{y}; l) = 0$ does not cross one of the vertices $\boldsymbol{w}_t$ of the polygon; i.e., the indices change when l satisfies $\mathsf{L}_{j,i}(\boldsymbol{w}_t; l) = 0$, which may occur for $R_{k;j,i} \leq T_{k;l,i}$ distinct values of l defined by

$$\check{l}'_r = 2\,\mathsf{snr}\,\xi_{j,i}(\boldsymbol{w}_r), \quad r = 1, \dots, R_{k;j,i}. \tag{5.92}$$

[8] The intersection with the extreme vertices of the region $\mathcal{Y}_{k;j,i}$ occurs with probability zero, thus may be neglected.

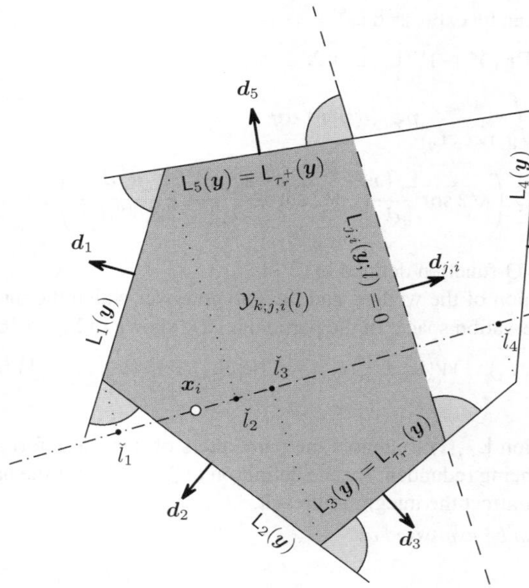

Figure 5.16 The tessellation region $\mathcal{Y}_{k;j,i}$ is a polygon delimited by the lines $\mathsf{L}_t(\boldsymbol{y}) = 0$, where $\mathsf{L}_t(\boldsymbol{y})$ is given by (5.88). The values of $\check{l}_r$ defining the intervals $\mathcal{L}_r$ in (5.93) for $r = 1, \ldots, 5$ together with the line $\mathsf{L}_{j,i}(\boldsymbol{y}; l) = 0$ and the two active linear pieces are also shown. The shaded area corresponds to $\mathcal{Y}_{k;j,i}(l)$

Thus, denoting by $\check{l}_r \leq \check{l}_{r+1}$ the sorted values of $\check{l}'_r$ in (5.92), we find the $R_{k;j,i} - 1$ intervals

$$\mathcal{L}_r \triangleq \{l \in \mathbb{R} : \check{l}_r \leq l < \check{l}_{r+1}\}, \quad r = 1, \ldots, R_{k;j,i} - 1, \tag{5.93}$$

such that

$$\bigcup_{r=1}^{R_{k;j,i}} \mathcal{L}_r = \mathcal{L}_{k;j,i}. \tag{5.94}$$

For a given value of r, we will use the indices τ_r^+ and τ_r^- to denote the two sides of the polygon which remain active for all $l \in \mathcal{L}_r$.

To clarify the above discussion, consider the example shown in Fig. 5.16. In this case, we have $R_{k;j,i} = 5$ and $T_{k;j,i} = 5$, $\tau_r^+ \in \{1, 5, 5, 5\}$, and $\tau_r^- \in \{2, 2, 3, 4\}$. For the particular value of l in Fig. 5.16, $l \in \mathcal{L}_3$ so $\tau_3^+ = 5$ and $\tau_3^- = 3$. Furthermore, in this figure we schematically show the values of $\check{l}_r$ as a projection of the vertices on the line connecting $\boldsymbol{x}_i$ and $\boldsymbol{x}_j$ which becomes the "axis" for the arguments l of the PDF (see also Fig. 3.2). The origin of this axis corresponds to $\overline{\boldsymbol{x}}_{j,i}$. Note that the origin does not necessarily belong to $\mathcal{Y}_{k;j,i}$.

The calculation of (5.89) requires now a 2D-equivalent of the integration shown in (5.48). This is practically the same problem we need to solve when evaluating the bit- and symbol-error rates, so the approach we have adopted here is also explained in detail in Section 6.1.2. More particularly, we decompose the observation space $\mathbb{R}^2$ into elementary *wedges* (see also (6.34)) defined as

$$\mathcal{W}(\mathsf{L}_p, \overline{\mathsf{L}}_q) \triangleq \{\boldsymbol{y} : \mathsf{L}_p(\boldsymbol{y}) \leq 0 \wedge \mathsf{L}_q(\boldsymbol{y}) \geq 0\}. \tag{5.95}$$

Integration over such wedge can be expressed as[9]

$$\mathsf{T}(\mathsf{L}_p, \bar{\mathsf{L}}_q|\boldsymbol{x}) \triangleq \Pr\{\boldsymbol{Y} \in \mathcal{W}(\mathsf{L}_p, \bar{\mathsf{L}}_q)|\boldsymbol{X} = \boldsymbol{x}\} \tag{5.96}$$

$$= \int_{\boldsymbol{y}\in\mathcal{W}(\mathsf{L}_p,\bar{\mathsf{L}}_q)} p_{\boldsymbol{Y}|\boldsymbol{X}}(\boldsymbol{y}|\boldsymbol{x})\, \mathrm{d}\boldsymbol{y} \tag{5.97}$$

$$= \tilde{Q}\left(\sqrt{2\,\mathsf{snr}}\frac{\mathsf{L}_p(\boldsymbol{x})}{\|\boldsymbol{d}_p\|}, \sqrt{2\,\mathsf{snr}}\frac{-\mathsf{L}_q(\boldsymbol{x})}{\|\boldsymbol{d}_q\|}, \frac{-\boldsymbol{d}_p^{\mathrm{T}}\boldsymbol{d}_q}{\|\boldsymbol{d}_p\|\|\boldsymbol{d}_q\|}\right), \tag{5.98}$$

where $\tilde{Q}(a, b, \rho)$ is the bivariate Q-function defined in (2.11).

As shown in Fig. 5.16, the union of the wedges and the polygon over which the integration must be carried out yields the entire observation space. In the particular case shown in Fig. 5.16 we can write

$$\mathbb{R}^2 = \mathcal{Y}_{k;j,i}(l) \cup \mathcal{W}(\mathsf{L}_{j,i}(l), \bar{\mathsf{L}}_3) \cup \mathcal{W}(\mathsf{L}_5, \bar{\mathsf{L}}_{j,i}(l)) \cup \mathcal{W}(\mathsf{L}_1, \bar{\mathsf{L}}_5) \cup \mathcal{W}(\mathsf{L}_2, \bar{\mathsf{L}}_1) \cup \mathcal{W}(\mathsf{L}_3, \bar{\mathsf{L}}_2), \tag{5.99}$$

where we used shorthand notation $\mathsf{L}_{j,i}(l)$ to denote the parameters of the linear form (5.91). We note that the linear form $\mathsf{L}_4(\boldsymbol{y}) = 0$, being redundant for the definition of $\mathcal{Y}_{k;j,i}(l)$ (for the particular value of l used in this example), does not affect the integration result.

Lemma 5.9 *The CDF* (5.89) *can be expressed as*

$$F_{k;j,i}(l|\boldsymbol{x}) = \begin{cases} 0, & \text{if } l < \check{l}_1 \\ 1 - \sum_{r=1}^{R_{k;j,i}} \Big(\mathsf{T}(\mathsf{L}_{j,i}(l), \bar{\mathsf{L}}_{\tau_r^-}|\boldsymbol{x}) & \\ \qquad + \mathsf{T}(\mathsf{L}_{\tau_r^+}, \bar{\mathsf{L}}_{j,i}(l)|\boldsymbol{x}) + A_r\Big)\mathbb{I}_{[l\in\mathcal{L}_r]}, & \text{if } l \in \mathcal{L}_{k;j,i} \\ B, & \text{if } \check{l}_{R_{k;j,i}} < l \end{cases} \tag{5.100}$$

where A_r and B are constants independent of l.

Proof: For $l < \check{l}_1$, we know that $\mathcal{Y}_{k;j,i}(l) = \varnothing$, and thus, $F_{k;j,i}(l|\boldsymbol{x}) = 0$, which gives the first line of (5.100).

For given r and $l \in \mathcal{L}_r$ (i.e., the second line of (5.100)), there are always two sides of the polygon intersected by the line $\mathsf{L}_{j,i}(\boldsymbol{y}; l) = 0$. We thus generalize (5.99) to

$$\mathbb{R}^2 = \mathcal{Y}_{k;j,i}(l) \cup \mathcal{W}(\mathsf{L}_{j,i}(l), \bar{\mathsf{L}}_{\tau_r^-}) \cup \mathcal{W}(\mathsf{L}_{\tau_r^+}, \bar{\mathsf{L}}_{j,i}(l)) \cup \mathcal{A}_r, \tag{5.101}$$

where $\mathcal{A}_r \subset \mathbb{R}^2$ is a set independent of l (for all $l \in \mathcal{L}_r$). We integrate now the PDF $p_{\boldsymbol{Y}|\boldsymbol{X}}(\boldsymbol{y}|\boldsymbol{x})$ over the sets defined in both sides of (5.101), which yields

$$F_{k;j,i}(l|\boldsymbol{x}) = \Pr\{\boldsymbol{Y} \in \mathcal{Y}_{k;j,i}(l)|\boldsymbol{X} = \boldsymbol{x}\} \tag{5.102}$$

$$= 1 - \Pr\{\boldsymbol{Y} \in \mathcal{W}(\mathsf{L}_{j,i}(l), \bar{\mathsf{L}}_{\tau_r^-})|\boldsymbol{X} = \boldsymbol{x}\} - \Pr\{\boldsymbol{Y} \in \mathcal{W}(\mathsf{L}_{\tau_r^+}, \bar{\mathsf{L}}_{j,i}(l))|\boldsymbol{X} = \boldsymbol{x}\}$$
$$- \Pr\{\boldsymbol{Y} \in \mathcal{A}_r|\boldsymbol{X} = \boldsymbol{x}\} \tag{5.103}$$

$$= 1 - \mathsf{T}(\mathsf{L}_{j,i}(l), \bar{\mathsf{L}}_{\tau_r^-}|\boldsymbol{x}) - \mathsf{T}(\mathsf{L}_{\tau_r^+}, \bar{\mathsf{L}}_{j,i}(l)|\boldsymbol{x}) - A_r, \tag{5.104}$$

where (5.103) follows from (5.101), and (5.104) from (5.96) and $A_r = \Pr\{\boldsymbol{Y} \in \mathcal{A}_r|\boldsymbol{X} = \boldsymbol{x}\}$. The second line of (5.100) is obtained by combining (5.104) with the indicator function $\mathbb{I}_{[l\in\mathcal{L}_r]}$.

Finally, for $l > \check{l}_{R_{k;j,i}}$ we have $\mathcal{Y}_{k;j,i}(l) = \mathcal{Y}_{k;j,i}$, and thus,

$$F_{k;j,i}(l|\boldsymbol{x}) = \Pr\{\boldsymbol{Y} \in \mathcal{Y}_{k;j,i}|\boldsymbol{X} = \boldsymbol{x}\} = B, \tag{5.105}$$

which completes the proof. □

We are now ready to find the conditional PDF of the L-values $p_{L_k|\boldsymbol{X}}(l|\boldsymbol{x})$.

---

[9] Note that in the notation $\mathcal{W}(\mathsf{L}_p, \bar{\mathsf{L}}_q)$ and $\mathsf{T}(\mathsf{L}_p, \bar{\mathsf{L}}_q|\boldsymbol{x})$, we do not make explicit the dependency of L_p and $\bar{\mathsf{L}}_q$ on $\boldsymbol{y}$.

Theorem 5.10 (PDF of L-values for 2D Constellations) *For any 2D constellation and labeling, the conditional PDF of the max-log L-values $p_{L_k|\boldsymbol{X}}(l|\boldsymbol{x})$ can be expressed as (5.33), with*

$$p_{k;j,i}(l|\boldsymbol{x}) = \Psi(l, \mathsf{snr}\,\mu_{j,i}(\boldsymbol{x}), \mathsf{snr}\,\sigma_{j,i}^2) \sum_{r=1}^{R_{k;j,i}} (1 - C_{\tau_r^-}(l|\boldsymbol{x}) - C_{\tau_r^+}(l|\boldsymbol{x}))\mathbb{I}_{[l\in\mathcal{L}_r]}, \tag{5.106}$$

where

$$C_t(l|\boldsymbol{x}) \triangleq Q\left(\sqrt{\frac{2\,\mathsf{snr}}{1-\rho_t^2}}\left[\frac{\mu_{j,i}(\boldsymbol{x}) - l/\mathsf{snr}}{\sqrt{2\sigma_{j,i}^2}}\rho_t - \frac{\mathsf{L}_t(\boldsymbol{x})}{\|\boldsymbol{d}_t\|}\right]\right), \tag{5.107}$$

$$\rho_t \triangleq \frac{\boldsymbol{d}_t^{\mathrm{T}}\boldsymbol{d}_{j,i}}{\|\boldsymbol{d}_t\|\|\boldsymbol{d}_{j,i}\|}, \tag{5.108}$$

and $\boldsymbol{d}_t$ are the normal vectors of the linear form $\mathsf{L}_t(\boldsymbol{y})$ in (5.88), $\mu_{j,i}(\boldsymbol{x})$ and $\sigma_{j,i}^2$ are given by (5.43) and (5.44), respectively, the intervals $\mathcal{L}_r, r = 1, \ldots, R_{k;j,i}$ are defined in (5.93), and $\tau_r^-, \tau_r^+, r = 1, \ldots, R_{k;j,i}$ are the indices of the sides of the polygon $\mathcal{Y}_{k;j,i}$ intersected by $\mathsf{L}_{i,j}(\boldsymbol{y}; l)$ when $l \in \mathcal{L}_r$.

Proof: As $\mathsf{T}(\cdot, \cdot|\boldsymbol{x})$ is expressed in (5.98) using the bivariate Q-function, taking the derivative of (5.100) with respect to l, requires differentiation of (2.11) with respect to one of its arguments

$$\frac{\mathrm{d}}{\mathrm{d}a}\tilde{Q}(a, b, \rho) = \frac{-1}{2\pi\sqrt{1-\rho^2}}\int_b^{\infty} \exp\left(-\frac{v^2 - 2va\rho + a^2}{2(1-\rho^2)}\right) \mathrm{d}v \tag{5.109}$$

$$= -\frac{1}{\sqrt{2\pi}}\exp\left(-\frac{a^2}{2}\right) Q\left(\frac{b - a\rho}{\sqrt{1-\rho^2}}\right) \tag{5.110}$$

$$\frac{\mathrm{d}}{\mathrm{d}b}\tilde{Q}(a, b, \rho) = -\frac{1}{\sqrt{2\pi}}\exp\left(-\frac{b^2}{2}\right) Q\left(\frac{a - b\rho}{\sqrt{1-\rho^2}}\right). \tag{5.111}$$

Applying the chain rule for differentiation to (5.100), and using the relationship $\frac{\mathrm{d}}{\mathrm{d}l}\mathsf{L}_{j,i}(\boldsymbol{y}; l) = -\frac{1}{2\,\mathsf{snr}}$ (see (5.91)), after simple algebraic manipulation, we obtain (5.106). □

Analyzing Theorem 5.10, we note that the indicator function in (5.106) takes into account the variation of the indices τ_r^+ and τ_r^- because of the change of l moving throughout the intervals $\mathcal{L}_r$. As we see, the form of the PDF does not depend uniquely on $\boldsymbol{x}$, $\boldsymbol{x}_j$, and $\boldsymbol{x}_i$, but also on other symbols from the constellation which define the polygon's sides. We also note that the Gaussian piece $\Psi(l, \mathsf{snr}\,\mu_{j,i}(\boldsymbol{x}), \mathsf{snr}\,\sigma_{j,i}^2)$ we have already obtained in the case of 1D constellations, is still present, but it is now multiplied by the "correction" terms (5.107).

5.3.3 Case Study: 8PSK Labeled by the BRGC

In this section, we consider an 8PSK constellation, which is probably the simplest nontrivial example of a 2D constellation. Furthermore, because of its practical relevance, we consider the BRGC. The constellation and labeling are shown in Fig. 5.14.

The tessellation regions are simple to find as we have already shown it in Example 5.7: they are always either (i) the Voronoi regions of the constellation symbols (e.g., $\mathcal{Y}_{1;5,3}$, $\mathcal{Y}_{2;4,2}$ in Fig. 5.14) or (ii) a union of two Voronoi regions (e.g., $\mathcal{Y}_{1;5,4}$, or any $\mathcal{Y}_{3;j,i} \neq \varnothing$, see again Fig. 5.14). This property spares us the

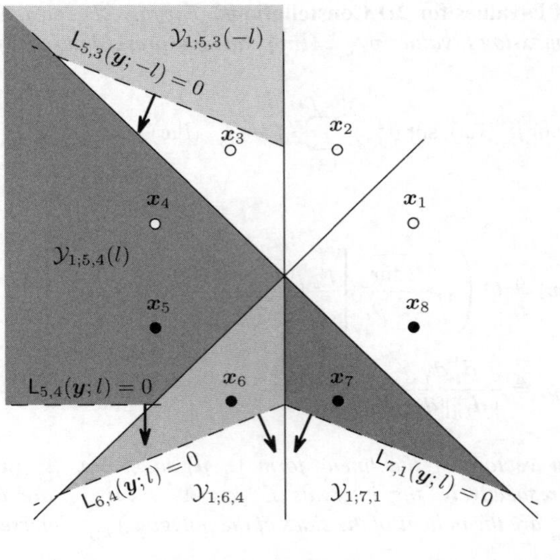

Figure 5.17 Some tessellation regions for 8PSK labeled by the BRGC and $k=1$. To obtain $F_{1;5,4}(l|\boldsymbol{x})$, $F_{1;5,3}(-l|\boldsymbol{x})$, $F_{1;6,4}(l|\boldsymbol{x})$, or $F_{1;7,1}(l|\boldsymbol{x})$ for $l>0$, integration should be carried over the shaded regions

need to apply the algorithmic approach of Section 5.3.1, which means we do not need to set the artificial constraints (5.79). Second, because of symmetry of the BRGC, the regions also have a notable symmetry, which we will exploit to simplify the analysis.

As for MPSK constellations $\boldsymbol{d}_{j,i}^{\mathrm{T}}\overline{\boldsymbol{x}}_{j,i}=\frac{1}{2}(\|\boldsymbol{x}_j\|^2-\|\boldsymbol{x}_i\|^2)=0\ \forall i,j$, we obtain

$$\xi_{j,i}(\boldsymbol{y})=\boldsymbol{d}_{j,i}^{\mathrm{T}}\boldsymbol{y}, \tag{5.112}$$

and thus,

$$\mathsf{L}_{j,i}(\boldsymbol{y};l)=\boldsymbol{d}_{j,i}^{\mathrm{T}}\boldsymbol{y}-\frac{l}{2\,\mathsf{snr}}. \tag{5.113}$$

The expression in (5.112) shows that all the lines delimiting the regions pass through the origin and so does the line $\mathsf{L}_{j,i}(\boldsymbol{y};0)=0$.

Let us focus on the case $k=1$ which also gives us the results for $k=2$ (as explained in Example 5.7). This case is shown in Fig. 5.17. The piece of the PDF $p_{k;j,i}(l|\boldsymbol{x})$ depends on the form of the tessellation region $\mathcal{Y}_{k;j,i}$ as well as on the position of the symbol $\boldsymbol{x}$ with respect to this region. An inspection of the constellation in Fig. 5.17 lets us conclude the following:

- Because of the symmetry of the constellation and the tessellation regions, we may write

$$p_{L_1|\boldsymbol{X}}(l|\boldsymbol{x}_5)=p_{L_1|\boldsymbol{X}}(l|\boldsymbol{x}_8)=p_{L_1|\boldsymbol{X}}(-l|\boldsymbol{x}_4)=p_{L_1|\boldsymbol{X}}(-l|\boldsymbol{x}_1), \tag{5.114}$$

$$p_{L_1|\boldsymbol{X}}(l|\boldsymbol{x}_6)=p_{L_1|\boldsymbol{X}}(l|\boldsymbol{x}_7)=p_{L_1|\boldsymbol{X}}(-l|\boldsymbol{x}_3)=p_{L_1|\boldsymbol{X}}(-l|\boldsymbol{x}_2), \tag{5.115}$$

  thus, it is sufficient to find $p_{L_1|\boldsymbol{X}}(l|\boldsymbol{x})$ for $l\geq 0$, conditioned on $\boldsymbol{x}\in\{\boldsymbol{x}_5,\boldsymbol{x}_6,\boldsymbol{x}_4,\boldsymbol{x}_3\}$.
- For $l\geq 0$ the lines $\mathsf{L}_{j,i}(\boldsymbol{y};l)=0$ always intersect with the same sides of the tessellation regions, therefore we do not need to take into account intervals changing with r as in (5.106). This is obvious from Fig. 5.17: the line $\mathsf{L}_{6,4}(\boldsymbol{y};l)=0$ crosses the same two sides of the region $\mathcal{Y}_{1;6,4}$, the line $\mathsf{L}_{5,4}(\boldsymbol{y};l)=0$ crosses only one side of $\mathcal{Y}_{1,5,4}$, and so on. Considering only $l\geq 0$ thus simplifies the analysis.

- Inspecting the regions $\mathcal{Y}_{1;6,4}(l)$ and $\mathcal{Y}_{1;7,1}(l)$ (shown shaded), we conclude that

$$\Pr\{\boldsymbol{Y} \in \mathcal{Y}_{1;7,1}(l)|\boldsymbol{x}_6\} = \Pr\{\boldsymbol{Y} \in \mathcal{Y}_{1;6,4}(l)|\boldsymbol{x}_7\},$$

i.e., $F_{1;7,1}(l|\boldsymbol{x}_6) = F_{1;6,4}(l|\boldsymbol{x}_7)$.
Moreover, for $l \geq 0$ we observe the following relationship:

$$\Pr\{\boldsymbol{Y} \in \mathcal{Y}_{1;5,3}(-l)|\boldsymbol{x}_6\} = \Pr\{\boldsymbol{Y} \in \mathcal{Y}_{1;6,4}|\boldsymbol{x}_3\} - \Pr\{\boldsymbol{Y} \in \mathcal{Y}_{1;6,4}(l)|\boldsymbol{x}_3\},$$

i.e., $F_{1;5,3}(-l|\boldsymbol{x}_6) = \Pr\{\boldsymbol{Y} \in \mathcal{Y}_{1;6,4}|\boldsymbol{x}_3\} - F_{1;6,4}(l|\boldsymbol{x}_3)$.

After differentiation and generalization to other regions we obtain the following relationships:

$$p_{1;5,4}(l|\boldsymbol{x}_5) = p_{1;5,4}(-l|\boldsymbol{x}_4), \qquad p_{1;8,1}(l|\boldsymbol{x}_5) = p_{1;5,4}(-l|\boldsymbol{x}_1), \tag{5.116}$$

$$p_{1;7,1}(l|\boldsymbol{x}_5) = p_{1;6,4}(l|\boldsymbol{x}_8), \qquad p_{1;8,1}(l|\boldsymbol{x}_5) = p_{1;5,4}(l|\boldsymbol{x}_8), \tag{5.117}$$

$$p_{1;5,3}(l|\boldsymbol{x}_5) = p_{1;6,4}(-l|\boldsymbol{x}_4), \qquad p_{1;8,2}(l|\boldsymbol{x}_5) = p_{1;6,4}(-l|\boldsymbol{x}_1). \tag{5.118}$$

The advantage of (5.116)–(5.118) is that we can now express the PDF $p_{L_1|\boldsymbol{X}}(l|\boldsymbol{x})$ using (5.33) as

$$p_{L_1|\boldsymbol{X}}(l|\boldsymbol{x}_5) = \begin{cases} p_{1;5,4}(l|\boldsymbol{x}_5) + p_{1;8,1}(l|\boldsymbol{x}_5) + p_{1;6,4}(l|\boldsymbol{x}_5) + p_{1;7,1}(l|\boldsymbol{x}_5), & \text{if } l \geq 0 \\ p_{1;5,4}(l|\boldsymbol{x}_5) + p_{1;8,1}(l|\boldsymbol{x}_5) + p_{1;5,3}(l|\boldsymbol{x}_5) + p_{1;8,2}(l|\boldsymbol{x}_5), & \text{if } l < 0 \end{cases} \tag{5.119}$$

$$= \begin{cases} p_{1;5,4}(l|\boldsymbol{x}_5) + p_{1;5,4}(l|\boldsymbol{x}_8) + p_{1;6,4}(l|\boldsymbol{x}_5) + p_{1;6,4}(l|\boldsymbol{x}_8), & \text{if } l \geq 0 \\ p_{1;5,4}(-l|\boldsymbol{x}_4) + p_{1;5,4}(-l|\boldsymbol{x}_1) + p_{1;6,4}(-l|\boldsymbol{x}_4) + p_{1;6,4}(-l|\boldsymbol{x}_1), & \text{if } l < 0 \end{cases}. \tag{5.120}$$

The same can be done for the PDF pieces conditioned on $\boldsymbol{x} = \boldsymbol{x}_6$, namely,

$$p_{1;5,4}(l|\boldsymbol{x}_6) = p_{1;5,4}(-l|\boldsymbol{x}_3), \qquad p_{1;8,1}(l|\boldsymbol{x}_6) = p_{1;5,4}(-l|\boldsymbol{x}_2), \tag{5.121}$$

$$p_{1;7,1}(l|\boldsymbol{x}_6) = p_{1;6,4}(l|\boldsymbol{x}_7), \qquad p_{1;8,1}(l|\boldsymbol{x}_6) = p_{1;5,4}(l|\boldsymbol{x}_7), \tag{5.122}$$

$$p_{1;5,3}(l|\boldsymbol{x}_6) = p_{1;6,4}(-l|\boldsymbol{x}_3), \qquad p_{1;8,2}(l|\boldsymbol{x}_6) = p_{1;6,4}(-l|\boldsymbol{x}_2), \tag{5.123}$$

which yields

$$p_{L_1|\boldsymbol{X}}(l|\boldsymbol{x}_6) = \begin{cases} p_{1;5,4}(l|\boldsymbol{x}_6) + p_{1;8,1}(l|\boldsymbol{x}_6) + p_{1;6,4}(l|\boldsymbol{x}_6) + p_{1;7,1}(l|\boldsymbol{x}_6), & \text{if } l \geq 0 \\ p_{1;5,4}(l|\boldsymbol{x}_6) + p_{1;8,1}(l|\boldsymbol{x}_6) + p_{1;5,3}(l|\boldsymbol{x}_6) + p_{1;8,2}(l|\boldsymbol{x}_6), & \text{if } l < 0 \end{cases} \tag{5.124}$$

$$= \begin{cases} p_{1;5,4}(l|\boldsymbol{x}_6) + p_{1;5,4}(l|\boldsymbol{x}_7) + p_{1;6,4}(l|\boldsymbol{x}_6) + p_{1;6,4}(l|\boldsymbol{x}_7), & \text{if } l \geq 0 \\ p_{1;5,4}(-l|\boldsymbol{x}_3) + p_{1;5,4}(-l|\boldsymbol{x}_2) + p_{1;6,4}(-l|\boldsymbol{x}_3) + p_{1;6,4}(-l|\boldsymbol{x}_2), & \text{if } l < 0 \end{cases}. \tag{5.125}$$

Thus, instead of varying the regions $\mathcal{Y}_{k;j,i}$ for a constant $\boldsymbol{x}$ as suggested by (5.33), in (5.120) and (5.125), we limit our considerations to two regions $\mathcal{Y}_{k;j,i}$ and change the symbols $\boldsymbol{x}$. We emphasize that this is possible in the particular case of an *M*PSK constellation labeled by the BRGC because of the symmetries of the constellation and the binary labeling. This is very convenient because we only need to find $p_{1;5,4}(l|\boldsymbol{x})$ and $p_{1;6,4}(l|\boldsymbol{x})$ for $l \geq 0$.

The same can be done for $k = 3$. In this case, only one symbol needs to be considered, namely, $\boldsymbol{x}_6$. It can be shown that in this case

$$p_{L_3|\boldsymbol{X}}(l|\boldsymbol{x}_6) = \begin{cases} \sum_{\boldsymbol{x} \in \mathcal{S}_{3,1}} p_{3;6,5}(l|\boldsymbol{x}), & \text{if } l \geq 0 \\ \sum_{\boldsymbol{x} \in \mathcal{S}_{3,0}} p_{3;6,5}(-l|\boldsymbol{x}), & \text{if } l < 0 \end{cases}. \tag{5.126}$$

In order to reuse the results obtained for $k = 1$, we can further exploit the equivalence (up to a rotation) of the tessellation regions $\mathcal{Y}_{1;5,4}$ and $\mathcal{Y}_{3;6,5}$ (see Fig. 5.13), i.e., $p_{3;6,5}(l|\boldsymbol{x}_i) = p_{1;5,4}(l|\boldsymbol{x}_{i-1})$ for $i > 1$ and $p_{3;6,5}(l|\boldsymbol{x}_1) = p_{1;5,4}(l|\boldsymbol{x}_8)$.

Up to now, we have shown the PDF $p_{L_k|\boldsymbol{X}}(l|\boldsymbol{x})$ can be expressed as a function of $p_{1;5,4}(l|\boldsymbol{x})$ and $p_{1;6,4}(l|\boldsymbol{x})$. The next step is to use Theorem 5.10. Before doing this, however, we study some properties of MPSK constellations.

More specifically, using $\boldsymbol{x}_k = [\Re[\mathrm{e}^{\mathrm{j}\alpha(2k-1)}], \Im[\mathrm{e}^{\mathrm{j}\alpha(2k-1)}]]^{\mathrm{T}}$, where $\alpha = \pi/M$ (see (2.52)), we find

$$\begin{aligned}\xi_{j,i}(\boldsymbol{x}_r) &= (\boldsymbol{x}_j - \boldsymbol{x}_i)^{\mathrm{T}}\boldsymbol{x}_r && (5.127)\\ &= \Re[(\mathrm{e}^{-\mathrm{j}\alpha(2j-1)} - \mathrm{e}^{-\mathrm{j}\alpha(2i-1)})\mathrm{e}^{\mathrm{j}\alpha(2r-1)}] && (5.128)\\ &= \Re[\mathrm{e}^{-\mathrm{j}\alpha(j+i)}(\mathrm{e}^{-\mathrm{j}\alpha((j-i)} - \mathrm{e}^{\mathrm{j}\alpha(j-i)})\mathrm{e}^{\mathrm{j}\alpha 2r}] && (5.129)\\ &= -2\Re[\mathrm{e}^{-\mathrm{j}\alpha(j+i-2r)}\mathrm{j}\sin(\alpha(j-i))] && (5.130)\\ &= 2\Im[\mathrm{e}^{-\mathrm{j}\alpha(j+i-2r)}]\sin(\alpha(j-i)) && (5.131)\\ &= -2S_{M,i+j-2r}S_{M,j-i}, && (5.132)\end{aligned}$$

where

$$S_{M,t} \triangleq \sin\frac{t\pi}{M}. \quad (5.133)$$

Similarly, we can find

$$\begin{aligned}\boldsymbol{d}_{r,l}^{\mathrm{T}}\boldsymbol{d}_{j,i} &= 4S_{M,r-l}S_{M,j-i}S_{M,M/2+r+l-j-i}, && (5.134)\\ \|\boldsymbol{d}_{j,i}\| &= 2|S_{M,j-i}|. && (5.135)\end{aligned}$$

We can now analyze the contribution of the region $\mathcal{Y}_{1;5,4}$. From (5.106), we obtain

$$p_{1;5,4}(l|\boldsymbol{x}) = \Psi(l, \mathsf{snr}\,\mu_{5,4}(\boldsymbol{x}), \mathsf{snr}\,\sigma_{5,4}^2)(1 - C_1(l|\boldsymbol{x}))\mathbb{I}_{[l\geq 0]}, \quad (5.136)$$

where $\mu_{5,4}(\boldsymbol{x}_i) = -4S_{8,2i-1}S_{8,1}$, and $\sigma_{5,4}^2 = 8S_{8,1}^2$. Here, $\tau_r = 1$, $\mathsf{L}_1(\boldsymbol{y}) = \xi_{6,5}(\boldsymbol{y})$ and $\rho_1 = \frac{1}{\sqrt{2}}$, so (5.107) becomes

$$C_1(l|\boldsymbol{x}_i) = Q\left(\sqrt{\mathsf{snr}}\left[-\sqrt{2}S_{8,2i-1} + 2S_{8,2i-3} - \frac{l}{\mathsf{snr}\sqrt{4-2\sqrt{2}}}\right]\right), \quad (5.137)$$

where we used $S_{8,1} = \frac{1}{2}\sqrt{2-\sqrt{2}}$.

Note that there is only one correction term $C_1(l|\boldsymbol{x})$ in (5.136) because the region $\mathcal{Y}_{1;5,4}(l)$ is a semi-infinite polygon and the line $\mathsf{L}_{5,4}(\boldsymbol{y}; l) = 0$ intersects only one of its sides. Equivalently, we might set an artificial constraint to close the polygon, e.g., $\mathsf{L}_2(\boldsymbol{y}) = \boldsymbol{d}_2^{\mathrm{T}}\boldsymbol{y} - A$, where $\boldsymbol{d}_2 = [-1, 0]^{\mathrm{T}}$ and $A \to \infty$. Then $\mathsf{L}_2(\boldsymbol{x}_i) \to -\infty$ and $C_2(l|\boldsymbol{x}_i) \to 0$, cf. (5.107).

Similarly, for $\mathcal{Y}_{1;6,4}$ we obtain

$$p_{1;6,4}(l|\boldsymbol{x}) = \Psi(l, \mathsf{snr}\,\mu_{6,4}(\boldsymbol{x}), \mathsf{snr}\,\sigma_{6,4}^2)(1 - C_1(l|\boldsymbol{x}) - C_2(l|\boldsymbol{x})), \quad (5.138)$$

where $\mu_{6,4}(\boldsymbol{x}_i) = -2\sqrt{2}S_{4,i-1}$, $\sigma_{6,4}^2 = 4$, and $\mathsf{L}_1(\boldsymbol{y}) = \xi_{7,6}(\boldsymbol{y})$ and $\mathsf{L}_2(\boldsymbol{y}) = \xi_{5,6}(\boldsymbol{y})$, so

$$C_1(l|\boldsymbol{x}_i) = Q\left(\frac{\sqrt{2\,\mathsf{snr}}}{\sqrt{1-\rho_1^2}}\left(\left(-S_{4,i-1} - \frac{l}{\sqrt{8\,\mathsf{snr}}}\right)\rho_1 + S_{8,2i-5}\right)\right), \quad (5.139)$$

$$C_2(l|\boldsymbol{x}_i) = Q\left(\frac{\sqrt{2\,\mathsf{snr}}}{\sqrt{1-\rho_2^2}}\left(\left(-S_{4,i-1} - \frac{l}{\sqrt{8\,\mathsf{snr}}}\right)\rho_2 - S_{8,2i-3}\right)\right), \quad (5.140)$$

with $\rho_1 = S_{8,1}$, and $\rho_2 = -1/(2\sqrt{2}S_{8,1})$.

The form of $p_{L_1|\boldsymbol{X}}(l|\boldsymbol{x})$ is shown in Fig. 5.18 for $\boldsymbol{x} = \boldsymbol{x}_5$ and $\boldsymbol{x} = \boldsymbol{x}_6$, where we identify the pieces $p_{1;j,i}(l|\boldsymbol{x})$. Similarly, we show the corresponding results for $p_{L_3|\boldsymbol{X}}(l|\boldsymbol{x})$ in Fig. 5.19. We can observe

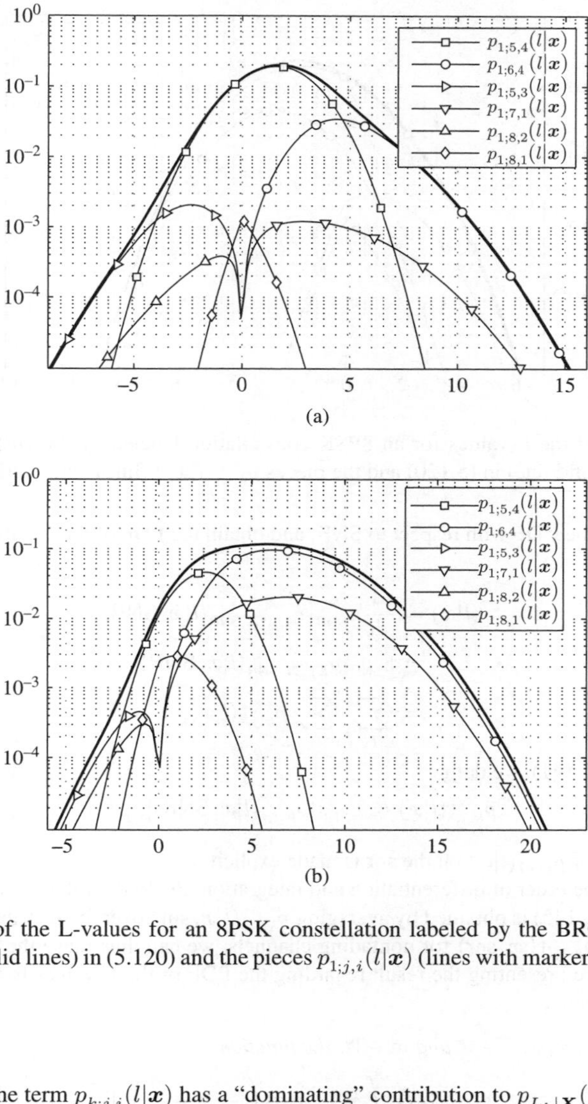

Figure 5.18 PDF of the L-values for an 8PSK constellation labeled by the BRGC and $\mathsf{snr} = 5$ dB: $p_{L_1|\boldsymbol{X}}(l|\boldsymbol{x})$ (thick solid lines) in (5.120) and the pieces $p_{1;j,i}(l|\boldsymbol{x})$ (lines with markers) for (a) $\boldsymbol{x} = \boldsymbol{x}_5$ and (b) $\boldsymbol{x} = \boldsymbol{x}_6$

that for $\boldsymbol{x} \in \mathcal{Y}_{k;j,i}$, the term $p_{k;j,i}(l|\boldsymbol{x})$ has a "dominating" contribution to $p_{L_k|\boldsymbol{X}}(l|\boldsymbol{x})$. In particular, in Fig. 5.18 (a), we see that for $\boldsymbol{x} = \boldsymbol{x}_5 \in \mathcal{Y}_{1;5,4}$, $p_{1;5,4}(l|\boldsymbol{x})$ takes on the largest values, while in Fig. 5.18 (b), we see that for $\boldsymbol{x} = \boldsymbol{x}_6 \in \mathcal{Y}_{1;6,4}$, the piece $p_{1;6,4}(l|\boldsymbol{x})$ dominates the others. Finally, in Fig. 5.19 we see that for $\boldsymbol{x} = \boldsymbol{x}_6 \in \mathcal{Y}_{3;6,5}$, $p_{3;6,5}(l|\boldsymbol{x})$ is again the dominating piece.

5.4 Fading Channels

Up to now, we considered the case of transmission over fixed-gain channels. To address the problem of finding the distribution of the L-values in fading channels, we define the CDF in (5.31) explicitly conditioned on the SNR, i.e.,

$$F_{L_k|\boldsymbol{X},\mathsf{SNR}}(l|\boldsymbol{x},\mathsf{snr}) = \Pr\{L_k \leq l|\boldsymbol{X} = \boldsymbol{x}, \mathsf{SNR} = \mathsf{snr}\}. \tag{5.141}$$

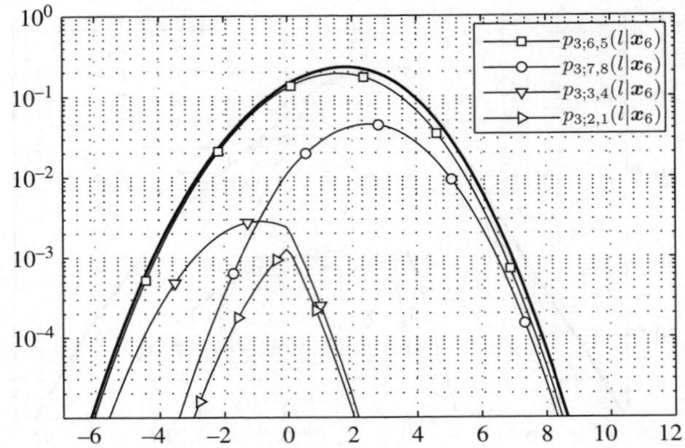

Figure 5.19 PDF of the L-values for an 8PSK constellation labeled by the BRGC and $\mathsf{snr} = 5$ dB: $p_{L_3|\boldsymbol{X}}(l|\boldsymbol{x}_6)$ (thick solid line) in (5.120) and the pieces $p_{3;j,i}(l|\boldsymbol{x}_6)$ (lines with markers)

Further, we average the CDF with respect to SNR, and obtain the PDF of the L-values by differentiating with respect to l, i.e.,

$$\overline{p}_{L_k|\boldsymbol{X}}(l|\boldsymbol{x}) \triangleq \frac{\mathrm{d}}{\mathrm{d}l}\mathbb{E}_{\mathsf{SNR}}[F_{L_k|\boldsymbol{X},\mathsf{SNR}}(l|\boldsymbol{x},\mathsf{SNR})] \tag{5.142}$$

$$= \mathbb{E}_{\mathsf{SNR}}[p_{L_k|\boldsymbol{X},\mathsf{SNR}}(l|\boldsymbol{x},\mathsf{SNR})] \tag{5.143}$$

$$= \sum_{j\in\mathcal{I}_{k,1}}\sum_{i\in\mathcal{I}_{k,0}} \overline{p}_{k;j,i}(l|\boldsymbol{x}), \tag{5.144}$$

where, similarly to (5.141), we define

$$\overline{p}_{k;j,i}(l|\boldsymbol{x}) \triangleq \mathbb{E}_{\mathsf{SNR}}[p_{k;j,i}(l|\boldsymbol{x},\mathsf{SNR})], \tag{5.145}$$

i.e., the dependence of $p_{k;j,i}(l|\boldsymbol{x})$ on the snr is made explicit.

Having changed the order of differentiation and integration, the PDF of the L-values for fading channels in (5.144) and (5.145) is obtained by averaging $p_{k;j,i}(l|\boldsymbol{x},\mathsf{snr})$ over the distribution of the SNR. As we already derived $p_{k;j,i}(l|\boldsymbol{x},\mathsf{snr})$ for nonfading channels, we can thus reuse the results from the previous sections. Before presenting the result regarding the PDF of the L-values for fading channels, we introduce a useful lemma.

Lemma 5.11 *For any $x, \alpha, \beta \in \mathbb{R}$ and $\mathsf{m} \in \mathbb{N}$, the function*

$$G_{\mathsf{m}}(x,\alpha,\beta) \triangleq \int x^{\mathsf{m}-\frac{3}{2}} \exp\left(-\alpha x - \frac{\beta}{x}\right)\mathrm{d}x \tag{5.146}$$

can be expressed using following recursion:

$$G_{\mathsf{m}}(x,\alpha,\beta) = \begin{cases} \frac{1}{2}\sqrt{\frac{\pi}{\alpha}}\sum_{t=0}^{1}(-1)^{t+1}\mathrm{e}^{(-1)^t 2\sqrt{\alpha\beta}}\cdot\mathrm{erfc}\left(\sqrt{\frac{\beta}{x}}+(-1)^t\sqrt{\alpha x}\right), & \text{if } \mathsf{m}=1 \\ -\frac{\sqrt{x}}{\alpha}\mathrm{e}^{-\alpha x-\frac{\beta}{x}} + \frac{1}{4\alpha}\sqrt{\frac{\pi}{\alpha}}\sum_{t=0}^{1}(-1)^t\mathrm{e}^{(-1)^t 2\sqrt{\alpha\beta}} \\ \quad\cdot\,\mathrm{erfc}\left(\sqrt{\alpha x}+(-1)^t\sqrt{\frac{\beta}{x}}\right)(2\sqrt{\alpha\beta}-(-1)^t), & \text{if } \mathsf{m}=2 \\ -\frac{x^{\mathsf{m}-\frac{3}{2}}}{\alpha}\mathrm{e}^{-\alpha x-\frac{\beta}{x}} + \frac{\mathsf{m}-\frac{3}{2}}{\alpha}G_{\mathsf{m}-1}(x,\alpha,\beta) + \frac{\beta}{\alpha}G_{\mathsf{m}-2}(x,\alpha,\beta), & \text{if } \mathsf{m}>2 \end{cases} \tag{5.147}$$

Proof: For $\mathsf{m} = 1$ and $\mathsf{m} = 2$, the integral in (5.146) can be obtained using a symbolic integration software. For $\mathsf{m} > 2$, we use integration by parts to obtain

$$\int x^{\mathsf{m}-\frac{3}{2}} \mathrm{e}^{-\alpha x-\frac{\beta}{x}}\, \mathrm{d}x = \frac{x^{\mathsf{m}+1-\frac{3}{2}}}{\mathsf{m}+1-\frac{3}{2}} \mathrm{e}^{-\alpha x-\frac{\beta}{x}} - \int \frac{x^{\mathsf{m}+1-\frac{3}{2}}}{\mathsf{m}+1-\frac{3}{2}} \mathrm{e}^{-\alpha x-\frac{\beta}{x}} \left(-\alpha + \frac{\beta}{x^2}\right) \mathrm{d}x, \tag{5.148}$$

which results in

$$\left(\mathsf{m}-\tfrac{1}{2}\right) G_{\mathsf{m}}(x,\alpha,\beta) = x^{\mathsf{m}-\frac{1}{2}} \mathrm{e}^{-\alpha x-\frac{\beta}{x}} + \alpha G_{\mathsf{m}+1}(x,\alpha,\beta) - \beta G_{\mathsf{m}-1}(x,\alpha,\beta), \tag{5.149}$$

$$\left(\mathsf{m}-\tfrac{3}{2}\right) G_{\mathsf{m}-1}(x,\alpha,\beta) = x^{\mathsf{m}-\frac{3}{2}} \mathrm{e}^{-\alpha x-\frac{\beta}{x}} + \alpha G_{\mathsf{m}}(x,\alpha,\beta) - \beta G_{\mathsf{m}-2}(x,\alpha,\beta), \tag{5.150}$$

where, to pass from (5.149) to (5.150), we use the substitution $\mathsf{m} - 1 \to \mathsf{m}$. The recursive expression for $\mathsf{m} > 2$ in (5.147) is obtained straightforwardly from (5.150). □

Theorem 5.12 (PDF of L-values for 1D Constellations in Fading Channels) *For any 1D constellation and labeling, the conditional PDF of the max-log L-values in fading channels $\overline{p}_{L_k|\boldsymbol{X}}(l|\boldsymbol{x})$ can be expressed as* (5.144)*, where*

$$\overline{p}_{k;j,i}(l|\boldsymbol{x}) = \frac{1}{\sqrt{2\pi\sigma_{j,i}^2}\Gamma(\mathsf{m})} \left(\frac{\mathsf{m}}{\overline{\mathsf{snr}}}\right)^{\mathsf{m}} \exp\left(\frac{\mu_{j,i}(\boldsymbol{x})l}{\sigma_{j,i}^2}\right) W(l), \tag{5.151}$$

where $\mu_{j,i}(\boldsymbol{x})$ and $\sigma_{j,i}^2$ by (5.43) *and* (5.44)*, respectively,*

$$W(l) \triangleq \begin{cases} V(l/\check{l}_1) - V(l/\check{l}_2), & \text{if } 0 \le \check{l}_1 \le \check{l}_2 \text{ and } l \ge 0 \\ V\left(|l/\check{l}_2|\right) - V(|l/\check{l}_1|), & \text{if } \check{l}_1 \le \check{l}_2 \le 0 \text{ and } l < 0 \\ V(\infty) - V(l/\check{l}_2), & \text{if } \check{l}_1 < 0 < \check{l}_2 \text{ and } l \ge 0\ , \\ V(\infty) - V(l/\check{l}_1), & \text{if } \check{l}_1 < 0 < \check{l}_2 \text{ and } l < 0 \\ 0, & \text{otherwise} \end{cases} \tag{5.152}$$

$$V(t) \triangleq G_{\mathsf{m}}\left(t, \frac{\mu_{j,i}^2(\boldsymbol{x})}{2\sigma_{j,i}^2} + \frac{\mathsf{m}}{\overline{\mathsf{snr}}}, \frac{l^2}{2\sigma_{j,i}^2}\right), \tag{5.153}$$

where $G_{\mathsf{m}}(x,\alpha,\beta)$ is given by (5.147)*, and where $\check{l}_1 = \check{l}_{k;j,i;1}$ and $\check{l}_2 = \check{l}_{k;j,i;2}$ define the limits of the support $\mathcal{L}_{k;j,i}$ in* (5.36) *of the Gaussian piece $p_{k;j,i}(l|\boldsymbol{x}, \mathsf{snr})$.*

Proof: For 1D constellations, the PDF pieces are given by (5.42), i.e.,

$$p_{k;j,i}(l|\boldsymbol{x}) = \Psi(l, \mu_{j,i}(\boldsymbol{x})\,\mathsf{snr}, \sigma_{j,i}^2\,\mathsf{snr})\mathbb{I}_{[l\in(\mathsf{snr}\,\check{l}_1, \mathsf{snr}\,\check{l}_2)]}. \tag{5.154}$$

We start with the case $0 \le \check{l}_1 < \check{l}_2$. We know that $\overline{p}_{k;j,i}(l|\boldsymbol{x}) = 0$ if $l < 0$, so we focus on $l \ge 0$. The average (5.154) over the Nakagami-m distribution (2.35) is

$$\overline{p}_{k;j,i}(l|\boldsymbol{x}) = \mathbb{E}_{\mathsf{SNR}}[\Psi(l, \mu_{j,i}(\boldsymbol{x})\,\mathsf{SNR}, \sigma_{j,i}^2\,\mathsf{SNR})\mathbb{I}_{[\mathsf{SNR}\,\check{l}_1 \le l < \mathsf{SNR}\,\check{l}_2]}] \tag{5.155}$$

$$= \int_{l/\check{l}_2}^{l/\check{l}_1} \Psi(l, \mu_{j,i}(\boldsymbol{x})\,\mathsf{snr}, \sigma_{j,i}^2\,\mathsf{snr}) p_{\mathsf{SNR}}(\mathsf{snr}, \mathsf{m}, \overline{\mathsf{snr}})\, \mathrm{d}\mathsf{snr} \tag{5.156}$$

$$= \frac{1}{\sqrt{2\pi}\sigma_{j,i}\Gamma(\mathsf{m})} \left(\frac{\mathsf{m}}{\overline{\mathsf{snr}}}\right)^{\mathsf{m}} \exp\left(\frac{\mu_{j,i}(\boldsymbol{x})l}{\sigma_{j,i}^2}\right)$$

$$\cdot \int_{l/\check{l}_2}^{l/\check{l}_1} \mathsf{snr}^{\mathsf{m}-\frac{3}{2}} \exp\left(-\frac{l^2}{2\sigma_{j,i}^2}\frac{1}{\mathsf{snr}} - \left(\frac{\mu_{j,i}^2(\boldsymbol{x})}{2\sigma_{j,i}^2} + \frac{\mathsf{m}}{\overline{\mathsf{snr}}}\right)\mathsf{snr}\right) \mathrm{d}\mathsf{snr}, \tag{5.157}$$

which is the same as (5.151) considering the first case in the definition of the function $W(l)$ in (5.152). The case $\check{l}_1 \leq \check{l}_2 \leq 0$ is obtained straightforwardly following similar steps to those above with the difference that, in this case, we need to study the case $l < 0$.

The case $\check{l}_1 \leq 0 < \check{l}_2$ can be reduced to the previous ones by decomposing the support of $p_{k;j,i}(l|\boldsymbol{x})$ into two subintervals $\mathcal{L}_{k;j,i} = (\check{l}_1, 0) \cup [0, \check{l}_2)$. In the interval $(\check{l}_1, 0)$, we only need to consider $l < 0$ and in the interval $[0, \check{l}_2)$, only $l \geq 0$; thus, we obtain the two remaining cases in the definition of $W(l)$ in (5.152). □

Example 5.13 (2PAM **in Fading Channels**) *The PDF we derived on analyzing* 2PAM *transmission is given by* (3.66), *i.e., for* $\boldsymbol{x} \in \mathcal{S}_{k,b}$,

$$p_{L_1|\mathbf{X}}(l|\boldsymbol{x}) = \Psi(l, (-1)^{\bar{b}} 4\,\mathsf{snr}, 8\,\mathsf{snr}). \tag{5.158}$$

The PDF in (5.158) *has only one nontruncated Gaussian piece with* $\mu_{2,1}(\boldsymbol{x}) = (-1)^{\bar{b}}4$, $\sigma_{2,1}^2 = 8$. *As* $\check{l}_1 = -\check{l}_2 = \infty$, *we analyze the third and fourth cases of* (5.152), *which boil down to calculating* $V(\infty) - V(0) = [G_{\mathsf{m}}(t, \alpha, \beta)]_{t=0}^{t=\infty}$ *and, for,* $\mathsf{m} = 1$, *we obtain it as follows using* (5.147):

$$[G_1(t, \alpha, \beta)]_{t=0}^{t=\infty} = \sqrt{\frac{\pi}{\alpha}} \exp(-2\sqrt{\alpha\beta}). \tag{5.159}$$

We then apply (5.159) *in* (5.151) *to obtain*

$$\overline{p}_{L_1|\mathbf{X}}(l|\boldsymbol{x}) = \frac{1}{4\sqrt{\mathsf{snr}(1+\mathsf{snr})}} \exp\left(-\frac{|l|}{2}\left[\sqrt{\frac{\mathsf{snr}+1}{\mathsf{snr}}} + (-1)^b \cdot \mathrm{sign}(l)\right]\right). \tag{5.160}$$

For $\mathsf{m} = 2$, *we obtain from* (5.147)

$$\begin{aligned}[G_2(t, \alpha, \beta)]_{t=0}^{t=\infty} &= \frac{1}{2\alpha}\sqrt{\frac{\pi}{\alpha}} \exp(-2\sqrt{\alpha\beta})(1 + 2\sqrt{\alpha\beta}) \\ &= \frac{1 + 2\sqrt{\alpha\beta}}{2\alpha}[G_1(t, \alpha, \beta)]_{t=0}^{t=\infty},\end{aligned} \tag{5.161}$$

which is given by (5.159). *Using this in* (5.151) *yields*

$$\begin{aligned}\overline{p}_{L_1|\mathbf{X}}(l|\boldsymbol{x}) = \frac{1}{2\sqrt{\mathsf{snr}(\mathsf{snr}+2)^3}} &\exp\left(-\frac{|l|}{2}\left[\sqrt{\frac{\mathsf{snr}+2}{\mathsf{snr}}} + (-1)^b \cdot \mathrm{sign}(l)\right]\right) \\ &\cdot \left(1 + \frac{|l|}{2}\sqrt{\frac{\mathsf{snr}+2}{\mathsf{snr}}}\right).\end{aligned} \tag{5.162}$$

For $\mathsf{m} > 2$, *we apply the recursive formula* (5.150) *to obtain*

$$[G_{\mathsf{m}}(t, \alpha, \beta)]_{t=0}^{t=\infty} = \frac{1}{\alpha}\left(\left(\mathsf{m} - \frac{3}{2}\right)[G_{\mathsf{m}-1}(t, \alpha, \beta)]_{t=0}^{t=\infty} + \beta[G_{\mathsf{m}-2}(t, \alpha, \beta)]_{t=0}^{t=\infty}\right) \tag{5.163}$$

and use it in (5.151) *to get*

$$\overline{p}_{L_1|\mathbf{X}}(l|\boldsymbol{x}) = \frac{1}{4\Gamma(\mathsf{m})\sqrt{\pi}}\left(\frac{\mathsf{m}}{\mathsf{snr}}\right)^{\mathsf{m}} \mathrm{e}^{(-1)^{\bar{b}}l/2}[G_{\mathsf{m}}(t, \alpha, \beta)]_{t=0}^{t=\infty}. \tag{5.164}$$

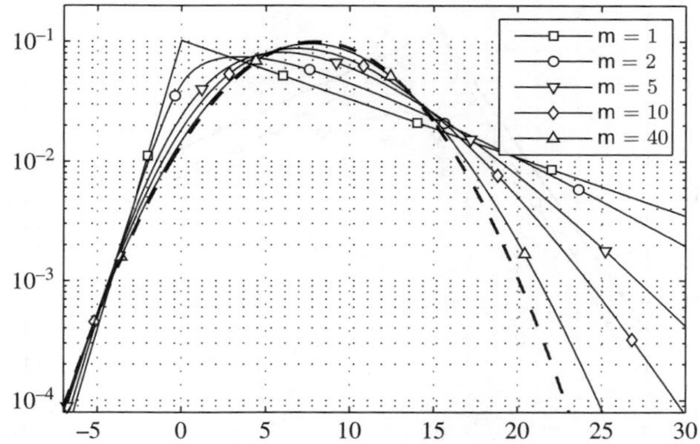

Figure 5.20 PDF of the L-values $\overline{p}_{L_1|\mathbf{X}}(l|\boldsymbol{x}_2)$ for 2PAM in Nakagami-m fading channel for $\overline{\mathsf{snr}} = 3$ dB given by (5.165) and (5.166). The thick dashed line indicates the PDF of the L-values for the AWGN channel with $\mathsf{snr} = \overline{\mathsf{snr}}$

Further, using (5.163), (5.161), *and* (5.159) *we obtain*

$$\overline{p}_{L_1|\mathbf{X}}(l|\boldsymbol{x}) = \frac{\mathsf{m}^{\mathsf{m}}}{4\Gamma(\mathsf{m})\overline{\mathsf{snr}}^{\mathsf{m}-1}\sqrt{\overline{\mathsf{snr}}(\overline{\mathsf{snr}}+\mathsf{m})}} \cdot \tau_{\mathsf{m}}(l) \cdot \exp\left(-\frac{|l|}{2}\left[\sqrt{\frac{\overline{\mathsf{snr}}+\mathsf{m}}{\overline{\mathsf{snr}}}} + (-1)^b \cdot \text{sign}(l)\right]\right), \tag{5.165}$$

where

$$\tau_t(l) = \begin{cases} 1, & \text{if } t = 1 \\ \dfrac{\overline{\mathsf{snr}}}{2(\overline{\mathsf{snr}}+\mathsf{m})}\left[1 + \dfrac{|l|}{2}\sqrt{\dfrac{\overline{\mathsf{snr}}+\mathsf{m}}{\overline{\mathsf{snr}}}}\right], & \text{if } t = 2 \\ \dfrac{\overline{\mathsf{snr}}}{\overline{\mathsf{snr}}+\mathsf{m}}\left(\left(t-\dfrac{3}{2}\right)\tau_{t-1}(l) + \dfrac{l^2}{16}\tau_{t-2}(l)\right), & \text{if } t > 2 \end{cases}. \tag{5.166}$$

The results obtained are shown in Fig. 5.20. We can observe that a double-sided exponential function obtained for $\mathsf{m} = 1$ *(i.e., Rayleigh fading) tends to the Gaussian distribution (shown by dashed lines) when* m *grows.*

Example 5.14 (4PAM **Labeled by the BRGC in Fading Channels)** *We take advantage of the expressions for the PDF in the nonfading channel we have already derived in Example 5.5. For* $k = 1$*, the indices yielding nonempty tessellation regions are* $(j, i) \in \{(3,1), (3,2), (4,2)\}$*, and for* $k = 2$*,* $(j, i) \in \{(2,1), (3,4)\}$*. We then apply directly Theorem 5.12. For* $\mathsf{m} = 1$*, we obtain the results we show in Fig. 5.21, where we observe that for* $k = 1$*,* $\overline{p}_{1;3,2}(l|\boldsymbol{x}_3)$ *is the "dominating" piece of the PDF* $\overline{p}_{L_1|\mathbf{X}}(l|\boldsymbol{x}_3)$*, while it is not obvious how to identify such a piece in* $\overline{p}_{L_1|\mathbf{X}}(l|\boldsymbol{x}_4)$*.*

5.5 Gaussian Approximations

Although, as explained in the previous sections, we can find the PDF through the explicit derivation of the CDF, such an approach has two main disadvantages: (i) the expressions require analysis of the tessellation space and (ii) the resulting forms are piecewise functions. The former is critical when we have to deal with arbitrary constellations, particularly when the dimensionality of the signal space increases beyond $N = 2$. The latter complicates the analysis of the operations on the PDFs. For example, recall

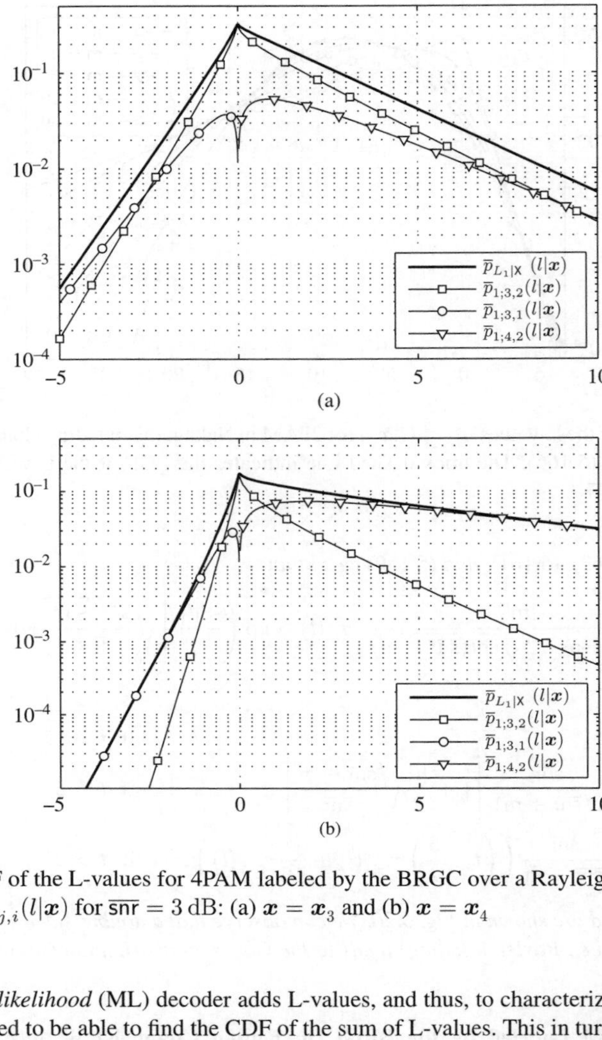

Figure 5.21 PDF of the L-values for 4PAM labeled by the BRGC over a Rayleigh fading channel and the PDF pieces $\overline{p}_{k;j,i}(l|\boldsymbol{x})$ for $\overline{\mathsf{snr}} = 3$ dB: (a) $\boldsymbol{x} = \boldsymbol{x}_3$ and (b) $\boldsymbol{x} = \boldsymbol{x}_4$

that the *maximum likelihood* (ML) decoder adds L-values, and thus, to characterize the performance of the decoder, we need to be able to find the CDF of the sum of L-values. This in turn requires finding the convolution of the PDFs and the integration of its tail,[10] which becomes difficult for PDFs defined via piecewise functions.

The main objective of this section is to provide a model for the PDF of the L-values, such that (i) its parameters are easy to find, (ii) its form simplifies operations on the PDF of the L-values (such as convolution), and (iii) the integration over its tail accurately estimates the actual value of the probability of error.

Assuming that the L-values are Gaussian is convenient because it will, indeed, simplify the operations of convolution (necessary to find the PDF of the sum of the L-values) and integration (necessary to find the error probability). The Gaussian model is also interesting because the expression we derived for nonfading channels already contains Gaussian pieces whose parameters (mean and variance) depend in closed form on the constellation and labeling. The main idea is thus to take only the Gaussian piece from the piecewise model and to ignore windowing and weighting (correction) factors. The resulting

[10] More information on this subject may be found in Section 6.2.

simplicity is indeed appealing and, in such a case, the corresponding approximate PDF conditioned on a transmitted symbol will be

$$\tilde{p}_{L_k|\boldsymbol{X}}(l|\boldsymbol{x}) \triangleq \Psi(l, \mathsf{snr}\tilde{\mu}_k(\boldsymbol{x}), \mathsf{snr}\tilde{\sigma}_k^2(\boldsymbol{x})), \tag{5.167}$$

where the mean and variance are the parameters related to a Gaussian piece $p_{k;\hat{\jmath},\hat{\imath}}(l|\boldsymbol{x})$, indexed by $\hat{\jmath}$ and $\hat{\imath}$, i.e.,

$$\tilde{\mu}_k(\boldsymbol{x}) = \mu_{\hat{\jmath},\hat{\imath}}(\boldsymbol{x}), \tag{5.168}$$

$$\tilde{\sigma}_k^2(\boldsymbol{x}) = \sigma_{\hat{\jmath},\hat{\imath}}^2, \tag{5.169}$$

where $\mu_{j,i}(\boldsymbol{x})$ and $\sigma_{j,i}^2$ are given by (5.43) and (5.44), respectively. The main problem consists then in finding the indices $\hat{\jmath}$ and $\hat{\imath}$ such that (5.167) represents "well" the PDF $p_{L_k|\boldsymbol{X}}(l|\boldsymbol{x})$.

We continue to keep in mind that the particular choice of $\hat{\jmath}$ and $\hat{\imath}$ should be assessed using the criterion related to the objective we fixed, namely, an the accurate evaluation of the integration over the tails of the PDF. As the details on the performance evaluation of BICM receivers are mostly given in Chapter 6, here, we will only highlight the most important features of the two proposed heuristics with respect to their impact on the accuracy of evaluation of the BEP for uncoded transmission. While passing through Section 6.1 may be helpful, the following contents can be understood without such additional reading.

5.5.1 Consistent Model

For notational convenience, we will denote the transmitted symbol by $\boldsymbol{x}_r$ and assume that it belongs to the subconstellation $\mathcal{S}_{k,b}$, i.e., b is the value of the kth bit in the label $\boldsymbol{q}_r$, ($q_{r,k} = b$, see Section 2.5.2). The next definition introduces the first Gaussian model for the PDF of the L-values, namely, the so-called *consistent model* (CoM).

Definition 5.15 (Consistent Model) *The "consistent" model of the PDF of the L-values is defined via the following assignment of the indices:*

$$\hat{\jmath} = r, \quad \hat{\imath} = \underset{i\in\mathcal{I}_{k,0}}{\operatorname{argmin}}\|\boldsymbol{x}_i - \boldsymbol{x}_r\|^2, \quad \text{if } \boldsymbol{x}_r \in \mathcal{S}_{k,1}, \tag{5.170}$$

$$\hat{\imath} = r, \quad \hat{\jmath} = \underset{j\in\mathcal{I}_{k,1}}{\operatorname{argmin}}\|\boldsymbol{x}_j - \boldsymbol{x}_r\|^2, \quad \text{if } \boldsymbol{x}_r \in \mathcal{S}_{k,0}. \tag{5.171}$$

Definition 5.15 states that the tessellation region $\mathcal{Y}_{k,\hat{\jmath},\hat{\imath}}$ that provides the most "representative" Gaussian piece (i.e., the one whose parameters will be used for the approximation in (5.167)) is obtained by considering the transmitted symbol $\boldsymbol{x}_r \in \mathcal{S}_{k,b}$ and the closest symbol from $\mathcal{S}_{k,\bar{b}}$. From (5.43), we know that

$$\mu_{\hat{\jmath},\hat{\imath}}(\boldsymbol{x}_r) = 2\xi_{\hat{\jmath},\hat{\imath}}(\boldsymbol{x}_r) \tag{5.172}$$

$$= 2\boldsymbol{d}_{\hat{\jmath},\hat{\imath}}^{\mathrm{T}}\left(\boldsymbol{x}_r - \frac{1}{2}(\boldsymbol{x}_{\hat{\imath}} + \boldsymbol{x}_{\hat{\jmath}})\right) \tag{5.173}$$

$$= \begin{cases} +\|\boldsymbol{d}_{\hat{\jmath},\hat{\imath}}\|^2, & \text{if } \boldsymbol{x}_r \in \mathcal{S}_{k,1} \\ -\|\boldsymbol{d}_{\hat{\jmath},\hat{\imath}}\|^2, & \text{if } \boldsymbol{x}_r \in \mathcal{S}_{k,0} \end{cases} \tag{5.174}$$

and

$$\sigma_{\hat{\jmath},\hat{\imath}}^2 = 2\|\boldsymbol{d}_{\hat{\jmath},\hat{\imath}}\|^2, \quad \forall \boldsymbol{x}_r \in \mathcal{S}. \tag{5.175}$$

Using (5.174) and (5.175) in (5.168) and (5.169), we conclude that for any $\boldsymbol{x}_r \in \mathcal{S}_{k,b}$,

$$\tilde{\mu}_k(\boldsymbol{x}_r) = (-1)^{\overline{b}}\overline{\delta}_{r,k}^2, \tag{5.176}$$

$$\tilde{\sigma}_k^2(\boldsymbol{x}_r) = 2\overline{\delta}_{r,k}^2, \tag{5.177}$$

where

$$\overline{\delta}_{k,r} = \min_{i \in \mathcal{I}_{k,\overline{b}}} \|\boldsymbol{d}_{r,i}\| \tag{5.178}$$

is the distance between $\boldsymbol{x}_r \in \mathcal{S}_{k,b}$ and the closest symbol having a different bit at position k. Therefore, (5.167) becomes

$$p_{L_k|\boldsymbol{X}}^{\text{CoM}}(l|\boldsymbol{x}_r) = \Psi(l, (-1)^{\overline{b}}\,\mathsf{snr}\,\overline{\delta}_{k,r}^2, 2\,\mathsf{snr}\,\overline{\delta}_{k,r}^2), \tag{5.179}$$

which has the same form we would obtain for a 2PAM constellation if we used $\boldsymbol{x}_1 = \boldsymbol{x}_{\hat{j}}$ and $\boldsymbol{x}_0 = \boldsymbol{x}_{\hat{i}}$ in (5.52).

The PDF in (5.179) also satisfies the consistency condition in Definition 3.8. This particular property explains the name we gave to the model that also corresponds to decomposing the BICM into "virtual" 2PAM transmissions, each characterized by the distance between the corresponding virtual 2PAM symbols.

Example 5.16 (CoM in 4PAM **Labeled by the BRGC)** *In the case of a* 4PAM *constellation labeled by the BRGC, we analyzed in Example 5.5 (see Fig. 5.3), we easily conclude that*

$$\overline{\delta}_{1,1} = \overline{\delta}_{1,4} = 4\Delta, \tag{5.180}$$

$$\overline{\delta}_{1,2} = \overline{\delta}_{1,3} = 2\Delta, \tag{5.181}$$

$$\overline{\delta}_{2,r} = 2\Delta, \quad r = 1, \ldots, 4. \tag{5.182}$$

Using (5.180)–(5.182) *in* (5.179), *we obtain*

$$p_{L_1|\boldsymbol{X}}^{\text{CoM}}(l|\boldsymbol{x}) = \begin{cases} \Psi(l, -16\gamma, 32\gamma), & \text{if } \boldsymbol{x} = -3\Delta \\ \Psi(l, -4\gamma, 8\gamma), & \text{if } \boldsymbol{x} = -\Delta \\ \Psi(l, 4\gamma, 8\gamma), & \text{if } \boldsymbol{x} = \Delta \\ \Psi(l, 16\gamma, 32\gamma), & \text{if } \boldsymbol{x} = 3\Delta \end{cases}, \tag{5.183}$$

$$p_{L_2|\boldsymbol{X}}^{\text{CoM}}(l|\boldsymbol{x}) = \begin{cases} \Psi(l, -4\gamma, 8\gamma), & \text{if } \boldsymbol{x} \in \{3\Delta, -3\Delta\} \\ \Psi(l, 4\gamma, 8\gamma), & \text{if } \boldsymbol{x} \in \{\Delta, -\Delta\} \end{cases}, \tag{5.184}$$

where, as defined in (5.7), $\gamma = \Delta^2\,\mathsf{snr} = \mathsf{snr}/5$. *The resulting PDFs correspond to those that would be obtained using equivalent* 2PAM *constellations composed of* $\boldsymbol{x}_{\hat{j}}$ *and* $\boldsymbol{x}_{\hat{i}}$ *(i.e., ignoring the other* $M-2$ *symbols in the constellation).*

Example 5.17 (CoM in 8PSK **Labeled by the BRGC)** *Using Fig. 5.14 and the relationship* $\|\boldsymbol{d}_{j,i}\|^2 = 4S_{M,j-i}^2$, *see* (5.135), *we easily obtain from* (5.179)

$$p_{L_1|\boldsymbol{X}}^{\text{CoM}}(l|\boldsymbol{x}) = \begin{cases} \Psi(l, 4\,\mathsf{snr}\,S_{8,1}^2, 8\,\mathsf{snr}\,S_{8,1}^2), & \text{if } \boldsymbol{x} \in \{\boldsymbol{x}_5, \boldsymbol{x}_8\} \\ \Psi(l, 4\,\mathsf{snr}\,S_{8,2}^2, 8\,\mathsf{snr}\,S_{8,2}^2), & \text{if } \boldsymbol{x} \in \{\boldsymbol{x}_6, \boldsymbol{x}_7\} \\ \Psi(l, -4\,\mathsf{snr}\,S_{8,1}^2, 8\,\mathsf{snr}\,S_{8,1}^2), & \text{if } \boldsymbol{x} \in \{\boldsymbol{x}_4, \boldsymbol{x}_1\} \\ \Psi(l, -4\,\mathsf{snr}\,S_{8,2}^2, 8\,\mathsf{snr}\,S_{8,2}^2), & \text{if } \boldsymbol{x} \in \{\boldsymbol{x}_2, \boldsymbol{x}_3\} \end{cases}, \tag{5.185}$$

$$p_{L_3|\boldsymbol{X}}^{\text{CoM}}(l|\boldsymbol{x}) = \begin{cases} \Psi(l, 4\,\mathsf{snr}\,S_{8,1}^2, 8\,\mathsf{snr}\,S_{8,1}^2), & \text{if } \boldsymbol{x} \in \mathcal{S}_{3,1}, \\ \Psi(l, -4\,\mathsf{snr}\,S_{8,1}^2, 8\,\mathsf{snr}\,S_{8,1}^2), & \text{if } \boldsymbol{x} \in \mathcal{S}_{3,0} \end{cases}. \tag{5.186}$$

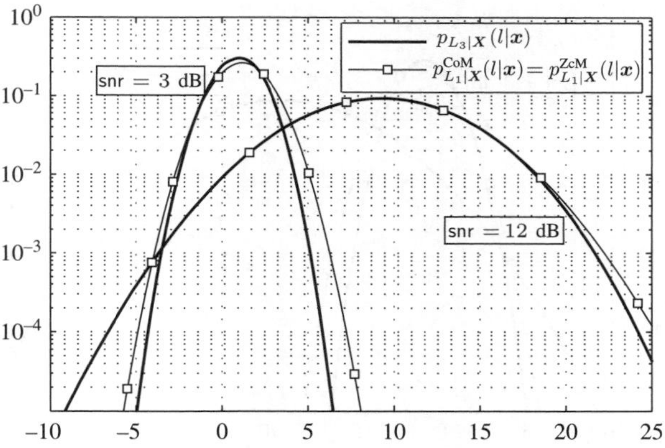

(a)

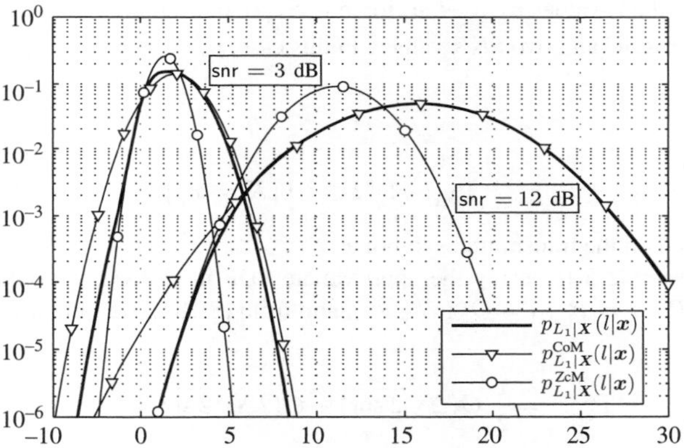

(b)

Figure 5.22 PDF of the L-values $p_{L_1|\boldsymbol{X}}(l|\boldsymbol{x})$ and its approximations based on the CoM $p^{\text{CoM}}_{L_1|\boldsymbol{X}}(l|\boldsymbol{x})$ and ZcM $p^{\text{ZcM}}_{L_1|\boldsymbol{X}}(l|\boldsymbol{x})$ for an 8PSK constellation labeled by the BRGC and two SNR values: (a) $k = 1, \boldsymbol{x} \in \{\boldsymbol{x}_5, \boldsymbol{x}_8\}$ and (b) $k = 1, \boldsymbol{x} \in \{\boldsymbol{x}_6, \boldsymbol{x}_7\}$

In Figs. 5.22 and 5.23, we show the PDF $p^{\text{CoM}}_{L_k|\boldsymbol{X}}(l|\boldsymbol{x})$ together with the corresponding exact form whose expressions we derived in Section 5.3.3. These results show that the form of PDF $p^{\text{CoM}}_{L_3|\boldsymbol{X}}(l|\boldsymbol{x})$ accurately approximates the true PDF around its mean value, which is particularly noticeable at high SNR.

Let us now verify how the obtained models affect the calculation of the BEP, which is an important parameter characterizing digital communications.[11] Here, the BEP for uncoded transmission in (3.110) can be obtained directly from the definition of the L-value.

$$P_{\text{b},k} = \frac{1}{2}(\Pr\{L_k > 0|B_k = 0\} + \Pr\{L_k < 0|B_k = 1\}) \tag{5.187}$$

$$= \frac{1}{M}\sum_{b\in\mathbb{B}}\sum_{r\in\mathcal{I}_{k,b}} P_{\text{b},k,r}, \tag{5.188}$$

[11] For more details, see Section 6.1.1.

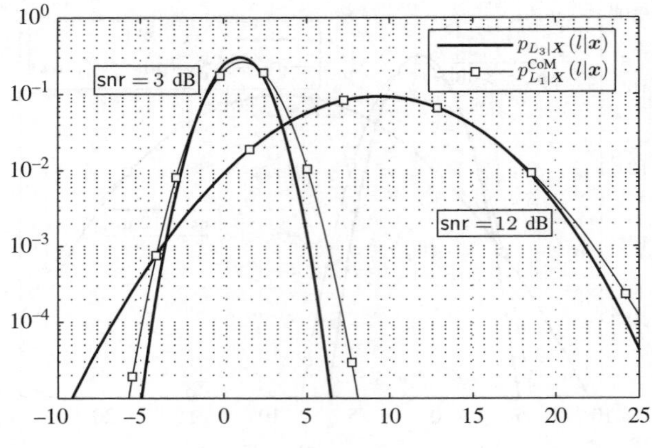

Figure 5.23 PDF of the L-values $p_{L_3|\boldsymbol{X}}(l|\boldsymbol{x})$ for $\boldsymbol{x} \in \mathcal{S}_{3,1}$ and its approximations for an 8PSK constellation labeled by the BRGC and two SNR values. In this case, $p^{\text{CoM}}_{L_3|\boldsymbol{X}}(l|\boldsymbol{x}) = p^{\text{ZcM}}_{L_3|\boldsymbol{X}}(l|\boldsymbol{x})$

where

$$P_{\text{b},k,r} \triangleq \Pr\{(-1)^{\bar{b}} L_k < 0 | \boldsymbol{X} = \boldsymbol{x}_r\} \tag{5.189}$$

is the BEP conditioned on the transmitted symbol $\boldsymbol{x}_r$.

For the case in Example 5.16, using the approximate PDF in (5.183), we obtain $P_{\text{b},1,1} = P_{\text{b},1,4} = Q(16\gamma/\sqrt{32\gamma})$ and $P_{\text{b},1,2} = P_{\text{b},1,3} = Q(4\gamma/\sqrt{8\gamma})$, and from (5.184), we obtain $P_{\text{b},2,r} = Q(4\gamma/\sqrt{8\gamma})$. Thus,

$$P^{\text{CoM}}_{\text{b},1} = \frac{1}{2}(Q(\Delta\sqrt{2\,\text{snr}}) + Q(2\Delta\sqrt{2\,\text{snr}})), \tag{5.190}$$

$$P^{\text{CoM}}_{\text{b},2} = Q(\Delta\sqrt{2\,\text{snr}}). \tag{5.191}$$

We note that the expressions in (5.190) and (5.191) are similar to the exact error probability expressions given in (6.58) and (6.60), respectively. Although the dominant Q-function $Q(\Delta\sqrt{2\,\text{snr}})$ correctly appear in the expressions derived using the CoM, differences appears when other terms are compared. We also note that for sufficiently high SNR, $P^{\text{CoM}}_{\text{b},1} < P^{\text{CoM}}_{\text{b},2}$. Thus, depending on the position k, the transmitted bits are more (or less) prone to errors. This so-called *unequal error protection* (UEP) will be analyzed in Chapter 8.

For the case in Example 5.17, we obtain

$$P^{\text{CoM}}_{\text{b},1} = \frac{1}{2}(Q(S_{8,1}\sqrt{2\,\text{snr}}) + Q(S_{8,2}\sqrt{2\,\text{snr}})), \tag{5.192}$$

$$P^{\text{CoM}}_{\text{b},3} = Q(S_{8,1}\sqrt{2\,\text{snr}}). \tag{5.193}$$

The first (dominating) terms in (5.192) and (5.193) are again the same as the dominating terms in the exact expressions (6.65) and (6.66), and the differences appear only in the remaining terms.

Finally, it interesting to analyze the CoM at high SNR. When $\text{snr} \to \infty$, $p_{\boldsymbol{Y}|\boldsymbol{X}}(\boldsymbol{y}|\boldsymbol{x}_r)$ is increasingly narrow and centered around $\boldsymbol{y} = \boldsymbol{x}_r$. After the transformation $\theta_k(\boldsymbol{y})$, the L-values are most likely to lie in the vicinity of $\theta_k(\boldsymbol{x}_r)$. In other words, for high SNR, $p_{L_k|\boldsymbol{X}}(l|\boldsymbol{x}_r)$ is centered around $\text{snr}\tilde{\mu}_k(\boldsymbol{x}_r)$ and the components $p_{k;j,i}(l|\boldsymbol{x}_r)$ for $j \neq \hat{\jmath}$ or $i \neq \hat{\imath}$ vanish. This is in fact what is observed in Figs. 5.22 and 5.23 for $\text{snr} = 12$ dB, i.e., the CoM approximates well the true PDF around its mean value. This observation has often been used as a justification to use the CoM to approximate the PDF of the L-values.

5.5.2 Zero-Crossing Model

For any $\boldsymbol{x}_r \in \mathcal{S}_{k,1}$, the conditional BEP in (5.189) can be expressed as

$$P_{\mathrm{b},k,r} = \int_{-\infty}^{0} p_{L_k|\mathbf{X}}(l|\boldsymbol{x}_r)\, \mathrm{d}l, \tag{5.194}$$

which shows that to have a good approximation of the conditional BEP, an accurate model of the tail of the PDF is necessary. When observing the results obtained with the CoM (see e.g., Fig. 5.22 (b)), we note that the (left) tail of the PDF is not well approximated. Consequently, the second terms in the BEP expressions obtained in Examples 5.16 and 5.17 do not match the terms that appear in the exact evaluation.

In this section, we introduce the *zero-crossing model* (ZcM) which aims at approximating the PDF well around $l = 0$, and thus, removing some of the discrepancies (in terms of BEP) observed when using the CoM. Before giving a formal definition, we introduce this model using an example. Consider an 8PSK constellation labeled by the BRGC, $k = 1$, $\boldsymbol{x} \in \{\boldsymbol{x}_6, \boldsymbol{x}_7\}$, and $\mathsf{snr} = 3$ dB. The exact PDF and the CoM approximation are shown in Fig. 5.24, where the shaded regions represent the integral in (5.194). This figure clearly shows that the CoM approximation fails to predict well the conditional BEP in (5.194). On the other hand, this figure also shows the ZcM defined below, which uses the Gaussian piece around $l = 0$. This results in a PDF that approximates much better the tail of the exact PDF, and thus, the conditional BEP in (5.194).

Definition 5.18 (Zero-crossing Model) *The "zero-crossing" model of the PDF of the L-values is defined as follows:*

$$\hat{\imath} = \underset{i\in\mathcal{I}_{k,0}}{\mathrm{argmin}}\|\boldsymbol{x}_r - \boldsymbol{x}_i\|^2,\ \hat{\jmath} = \underset{j\in\mathcal{I}_{k,1}}{\mathrm{argmin}}\|\boldsymbol{x}_r - \overline{\boldsymbol{x}}_{j,\hat{\imath}}\|^2,\ \text{if}\ \boldsymbol{x}_r \in \mathcal{S}_{k,1}, \tag{5.195}$$

$$\hat{\jmath} = \underset{j\in\mathcal{I}_{k,1}}{\mathrm{argmin}}\|\boldsymbol{x}_r - \boldsymbol{x}_j\|^2,\ \hat{\imath} = \underset{i\in\mathcal{I}_{k,0}}{\mathrm{argmin}}\|\boldsymbol{x}_r - \overline{\boldsymbol{x}}_{\hat{\jmath},i}\|^2,\ \text{if}\ \boldsymbol{x}_r \in \mathcal{S}_{k,0}. \tag{5.196}$$

We can now compare the ZcM in Definition 5.18 with the CoM in Definition 5.15. First of all, we see that both ZcM and CoM require finding the symbol from $\mathcal{S}_{k,\bar{b}}$ which is closest to $\boldsymbol{x}_r$. This symbol

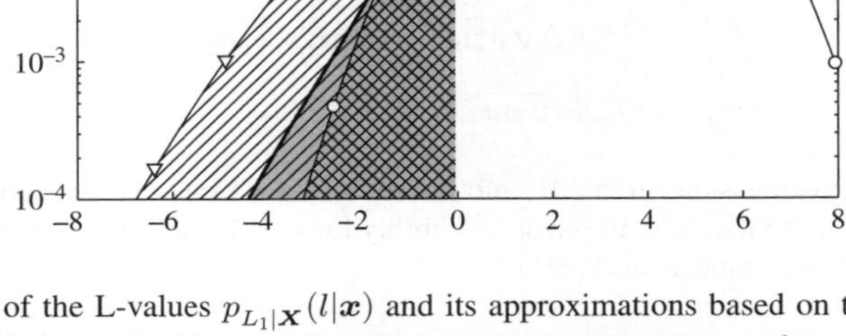

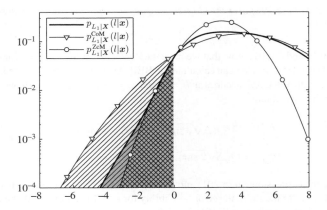

Figure 5.24 PDF of the L-values $p_{L_1|\mathbf{X}}(l|\boldsymbol{x})$ and its approximations based on the CoM $p^{\mathrm{CoM}}_{L_1|\mathbf{X}}(l|\boldsymbol{x})$ and ZcM $p^{\mathrm{ZcM}}_{L_1|\mathbf{X}}(l|\boldsymbol{x})$ for an 8PSK constellation labeled by the BRGC, $\boldsymbol{x} \in \{\boldsymbol{x}_6, \boldsymbol{x}_7\}$ and $\mathsf{snr} = 3$ dB. The shaded regions represent the conditional BEP in (5.194)

(and the corresponding index) are thus common in both models. The difference is how to determine the complementary index $\hat{\jmath}$ (when $b = 1$) or $\hat{\imath}$ (when $b = 0$). In Definition 5.15, this index is simply taken as equal to r, while in Definition 5.18, we perform a search over $\mathcal{S}_{k,b}$ to find $\overline{\boldsymbol{x}}_{\hat{\jmath},\hat{\imath}}$ closest to $\boldsymbol{x}_r$. In some cases, both definitions yield exactly the same results, namely, when the symbols $\boldsymbol{x}_j$ and $\boldsymbol{x}_i$ belong to the tessellation region $\mathcal{Y}_{k;j,i}$ they define, i.e., $\boldsymbol{x}_j, \boldsymbol{x}_i \in \mathcal{Y}_{k;j,i}$. In these cases, the PDF conditioned on $\boldsymbol{x}_r = \boldsymbol{x}_j$ or $\boldsymbol{x}_r = \boldsymbol{x}_i$ obtained from ZcM and CoM will be same.

Example 5.19 (ZcM in 4PAM **Labeled with the BRGC)** *Let us consider again 4PAM constellation labeled by the BRGC we analyzed in Example 5.5. From Fig. 5.10, we see that for* $k = 1$ *and any* $\boldsymbol{x}_r$, *we get* $\hat{\jmath} = 3$ *and* $\hat{\imath} = 2$ *(i.e., there is only one tessellation region* $\mathcal{Y}_{1;3,2}$ *where* $\theta_1(\boldsymbol{y}) = 0$*). For* $k = 2$, *when* $\boldsymbol{x}_r \in \{-3\Delta, -\Delta\}$, *we obtain* $\hat{\jmath} = 2$ *and* $\hat{\imath} = 1$, *and when* $\boldsymbol{x}_r \in \{\Delta, 3\Delta\}$, *we obtain* $\hat{\jmath} = 3$ *and* $\hat{\imath} = 4$. *In summary, the expressions for the PDFs are given by*

$$p^{\text{ZcM}}_{L_1|\mathbf{X}}(l|\boldsymbol{x}) = \begin{cases} \Psi(l, -12\gamma, 8\gamma), & \text{if } \boldsymbol{x} = -3\Delta \\ \Psi(l, -4\gamma, 8\gamma), & \text{if } \boldsymbol{x} = -\Delta \\ \Psi(l, 4\gamma, 8\gamma), & \text{if } \boldsymbol{x} = \Delta \\ \Psi(l, 12\gamma, 8\gamma), & \text{if } \boldsymbol{x} = 3\Delta \end{cases}, \tag{5.197}$$

$$p^{\text{ZcM}}_{L_2|\mathbf{X}}(l|\boldsymbol{x}) = \begin{cases} \Psi(l, -4\gamma, 8\gamma), & \text{if } \boldsymbol{x} \in \{-3\Delta, 3\Delta\} \\ \Psi(l, 4\gamma, 8\gamma), & \text{if } \boldsymbol{x} \in \{-\Delta, \Delta\} \end{cases}. \tag{5.198}$$

We first note that the PDF conditioned on a transmitted symbol does not satisfy the consistency condition. By comparing (5.197) *and* (5.198) *with* (5.183) *and* (5.184), *we also note that* $p^{\text{ZcM}}_{L_1|\mathbf{X}}(l|\boldsymbol{x}) = p^{\text{CoM}}_{L_1|\mathbf{X}}(l|\boldsymbol{x})$ *except for* $k = 1$ *and* $\boldsymbol{x} = \pm 3\Delta$. *It is also instructive to compare the expressions in* (5.197) *and* (5.198) *with the exact PDF expressions in* (5.65) *and* (5.66). *For* $k = 1$, *the ZcM in* (5.197) *is based on the mean and variance of the zero-crossing Gaussian piece defined for* $\lambda \in (-8\gamma, 8\gamma)$ *in* (5.65). *For* $k = 2$, *however, there are two zero-crossing Gaussian pieces:*

$$p_{L_2|\mathbf{X}}(l|\pm 3\Delta) = [\Psi(l, -4\gamma, 8\gamma) + \Psi(l, 20\gamma, 8\gamma)]\mathbb{I}_{[l \le 8\gamma]}, \tag{5.199}$$

$$p_{L_2|\mathbf{X}}(l|\pm \Delta) = [\Psi(l, 4\gamma, 8\gamma) + \Psi(l, 12\gamma, 8\gamma)]\mathbb{I}_{[l \le 8\gamma]}. \tag{5.200}$$

In this case, the ZcM in (5.198) *uses the Gaussian piece corresponding to the closest zero-crossing to the transmitted symbol.*

Similarly to (5.190) and (5.191), now that we have the ZcM PDF of the L-values for 4PAM labeled by the BRGC in (5.197) and (5.198), we can calculate the BEP. As we have already stated in Example 5.19, the only difference with the CoM occurs for $k = 1$ and $\boldsymbol{x} = \pm 3\Delta$, where we have $P_{\text{b},1,1} = P_{\text{b},1,4} = Q(12\gamma/\sqrt{8\gamma})$ and then we obtain

$$P^{\text{ZcM}}_{\text{b},1} = \frac{1}{2}(Q(\Delta\sqrt{2\,\mathsf{snr}}) + Q(3\Delta\sqrt{2\,\mathsf{snr}})), \tag{5.201}$$

$$P^{\text{ZcM}}_{\text{b},2} = Q(\Delta\sqrt{2\,\mathsf{snr}}). \tag{5.202}$$

Again we compare the expressions in (5.201) and (5.202) with the exact expression in (6.58) and (6.60). Unlike in the CoM case, for the ZcM, the error probability for $k = 1$ coincides with the exact calculation. For $k = 2$, the result is the same as in (5.191).

Example 5.20 (ZcM in 8PSK **Labeled by the BRGC)** *Using Fig. 5.14 and the symmetries of the constellation, we can easily determine the "zero-crossing" tessellation regions for each* *of the transmitted symbols, as well as the corresponding parameters of the mean and variance. We obtain*

$$p_{L_1|\mathbf{X}}^{\text{ZcM}}(l|\boldsymbol{x}) = \begin{cases} \Psi(l, \mathsf{snr}\,\mu_{5,4}(\boldsymbol{x}_5), \mathsf{snr}\,\sigma_{5,4}^2), & \text{if } \boldsymbol{x} \in \{\boldsymbol{x}_5, \boldsymbol{x}_8\} \\ \Psi(l, \mathsf{snr}\,\mu_{5,4}(\boldsymbol{x}_6), \mathsf{snr}\,\sigma_{5,4}^2), & \text{if } \boldsymbol{x} \in \{\boldsymbol{x}_6, \boldsymbol{x}_7\} \\ \Psi(l, \mathsf{snr}\,\mu_{5,4}(\boldsymbol{x}_4), \mathsf{snr}\,\sigma_{5,4}^2), & \text{if } \boldsymbol{x} \in \{\boldsymbol{x}_4, \boldsymbol{x}_1\} \\ \Psi(l, \mathsf{snr}\,\mu_{5,4}(\boldsymbol{x}_3), \mathsf{snr}\,\sigma_{5,4}^2), & \text{if } \boldsymbol{x} \in \{\boldsymbol{x}_3, \boldsymbol{x}_2\} \end{cases}, \tag{5.203}$$

$$p_{L_3|\mathbf{X}}^{\text{ZcM}}(l|\boldsymbol{x}) = \begin{cases} \Psi(l, \mathsf{snr}\,\mu_{5,4}(\boldsymbol{x}_5), \mathsf{snr}\,\sigma_{5,4}^2), & \text{if } \boldsymbol{x} \in \mathcal{S}_{3,1} \\ \Psi(l, \mathsf{snr}\,\mu_{5,4}(\boldsymbol{x}_4), \mathsf{snr}\,\sigma_{5,4}^2), & \text{if } \boldsymbol{x} \in \mathcal{S}_{3,0} \end{cases}. \tag{5.204}$$

Then, using the parameters defined after (5.136) *we obtain*

$$p_{L_1|\mathbf{X}}^{\text{ZcM}}(l|\boldsymbol{x}) = \begin{cases} \Psi(l, 4\,\mathsf{snr}\,S_{8,1}^2, 8\,\mathsf{snr}\,S_{8,1}^2), & \text{if } \boldsymbol{x} \in \{\boldsymbol{x}_5, \boldsymbol{x}_8\} \\ \Psi(l, 4\,\mathsf{snr}\,S_{8,3}\cdot S_{8,1}, 8\,\mathsf{snr}\,S_{8,1}^2), & \text{if } \boldsymbol{x} \in \{\boldsymbol{x}_6, \boldsymbol{x}_7\} \\ \Psi(l, -4\,\mathsf{snr}\,S_{8,1}^2, 8\,\mathsf{snr}\,S_{8,1}^2), & \text{if } \boldsymbol{x} \in \{\boldsymbol{x}_4, \boldsymbol{x}_1\} \\ \Psi(l, -4\,\mathsf{snr}\,S_{8,3}\cdot S_{8,1}, 8\,\mathsf{snr}\,S_{8,1}^2), & \text{if } \boldsymbol{x} \in \{\boldsymbol{x}_3, \boldsymbol{x}_2\} \end{cases}, \tag{5.205}$$

$$p_{L_3|\mathbf{X}}^{\text{ZcM}}(l|\boldsymbol{x}) = \begin{cases} \Psi(l, 4\,\mathsf{snr}\,S_{8,1}^2, 8\,\mathsf{snr}\,S_{8,1}^2), & \text{if } \boldsymbol{x} \in \mathcal{S}_{3,1} \\ \Psi(l, -4\,\mathsf{snr}\,S_{8,1}^2, 8\,\mathsf{snr}\,S_{8,1}^2), & \text{if } \boldsymbol{x} \in \mathcal{S}_{3,0} \end{cases}. \tag{5.206}$$

The forms of the PDF are shown in Figs. 5.22 and 5.23, where we observe that the ZcM yields very accurate approximations of the tail of the true PDF, which translates into more accurate BEP estimates. We can also observe how the accuracy of the approximate PDF form increases with SNR. This is to be expected because with increasing snr*, the effect of multiple Gaussian pieces becomes negligible and a single piece is sufficient to accurately calculate the BEP. On the other hand, the approximation obtained using the CoM becomes less accurate when* snr *grows.*

As in Example 5.19, the ZcM and the CoM give different BEP expressions. In particular, if we compare (5.185) *and* (5.186) *with* (5.203) *and* (5.204)*, we note that the differences come from* $k = 1$ *and the cases* $\boldsymbol{x} \in \{\boldsymbol{x}_6, \boldsymbol{x}_7, \boldsymbol{x}_3, \boldsymbol{x}_2\}$*, which are symbols that increase the "protection" of the bits. More importantly, in this case, the BEP expression for* $k = 1$ *is*

$$P_{\text{b},1}^{\text{ZcM}} = \frac{1}{2}(S_{8,1}Q(\sqrt{2\,\mathsf{snr}}) + Q(S_{8,3}\sqrt{2\,\mathsf{snr}})), \tag{5.207}$$

which corresponds to the exact expression derived in (6.65).

5.5.3 *Fading Channels*

The extension of the developed formulas to the case of the fading channel may be done by averaging the approximate Gaussian PDF in (5.167) over the distribution of the fading

$$\overline{p}_{L_k|\mathbf{X}}(l|\boldsymbol{x}_r) = \mathbb{E}_{\mathsf{SNR}}[\Psi(l, \mathsf{SNR}\,\tilde{\mu}_k(\boldsymbol{x}_r), \mathsf{SNR}\,\tilde{\sigma}_k^2(\boldsymbol{x}_r))]. \tag{5.208}$$

In the particular case of Rayleigh fading, using the same approach as those leading to (5.160), we obtain

$$\overline{p}_{L_k|\mathbf{X}}(l|\boldsymbol{x}) = \frac{1}{\sqrt{\overline{\mathsf{snr}}(\tilde{\mu}_{\hat{\jmath},\hat{\imath}}^2(\boldsymbol{x})\,\overline{\mathsf{snr}} + 2\sigma_{\hat{\jmath},\hat{\imath}}^2)}} \cdot \exp\left(\frac{(-1)^{\overline{b}}\mu_{\hat{\jmath},\hat{\imath}}(\boldsymbol{x})\cdot l}{\sigma_{\hat{\jmath},\hat{\imath}}^2} - \frac{|l|}{\sigma_{\hat{\jmath},\hat{\imath}}^2}\sqrt{\mu_{\hat{\jmath},\hat{\imath}}^2(\boldsymbol{x}) + \frac{2\sigma_{\hat{\jmath},\hat{\imath}}^2}{\overline{\mathsf{snr}}}}\right), \tag{5.209}$$

where $\hat{\jmath}$ and $\hat{\imath}$ depend also on the Gaussian simplification strategy (CoM vs. ZcM).

Example 5.21 (Rayleigh Fading for 4PAM Labeled by the BRGC) *Consider the case of a 4PAM constellation labeled by the BRGC and the results in Examples 5.16 and 5.19. Here we extend those results to a Rayleigh fading channel and compare them with the exact solutions we obtained in Example 5.14.*

The results obtained using (5.209) are shown in Fig. 5.25 for $k = 1$ and $\boldsymbol{x} = \boldsymbol{x}_3$ and $\boldsymbol{x} = \boldsymbol{x}_4$. In both cases $\boldsymbol{x} \in \mathcal{S}_{1,1}$, and thus, an accurate approximation of the PDF for $l \leq 0$ is critical for approximating the BEP. When $\boldsymbol{x} = \boldsymbol{x}_3$, both models are equivalent. However, when $\boldsymbol{x} = \boldsymbol{x}_4$, the ZcM PDF $\overline{p}^{\text{ZcM}}_{L_1|\boldsymbol{X}}(l|\boldsymbol{x})$ provides a better approximation of the tail of the PDF than the CoM PDF $\overline{p}^{\text{CoM}}_{L_1|\boldsymbol{X}}(l|\boldsymbol{x})$.

5.5.4 *QAM and PSK Constellations*

Two of the most practically relevant constellations are the *M*PAM and *M*PSK constellations labeled by the BRGC. In these cases, it is possible to determine the forms of the approximate PDF without

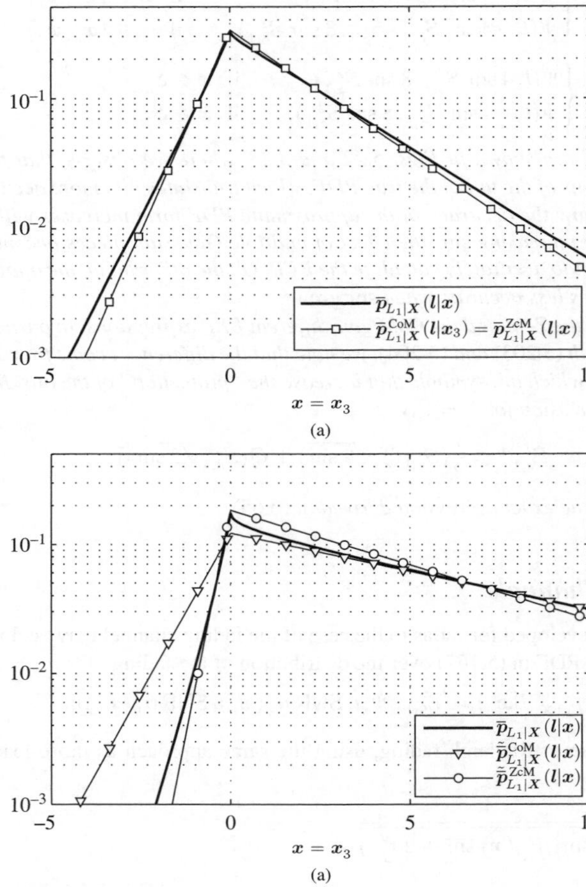

Figure 5.25 PDF of the L-values for a 4PAM constellation labeled by the BRGC in Rayleigh fading channel and $\overline{\mathsf{snr}} = 3$ dB. The exact expression $\overline{p}_{L_1|\boldsymbol{X}}(l|\boldsymbol{x})$ and the approximations via the ZcM $\overline{p}^{\text{ZcM}}_{L_1|\boldsymbol{X}}(l|\boldsymbol{x})$ and the CoM $\overline{p}^{\text{CoM}}_{L_1|\boldsymbol{X}}(l|\boldsymbol{x})$ for $\boldsymbol{x} \in \{\boldsymbol{x}_3, \boldsymbol{x}_4\}$ are shown for (a) $\boldsymbol{x} = \boldsymbol{x}_3$ and (b) $\boldsymbol{x} = \boldsymbol{x}_4$

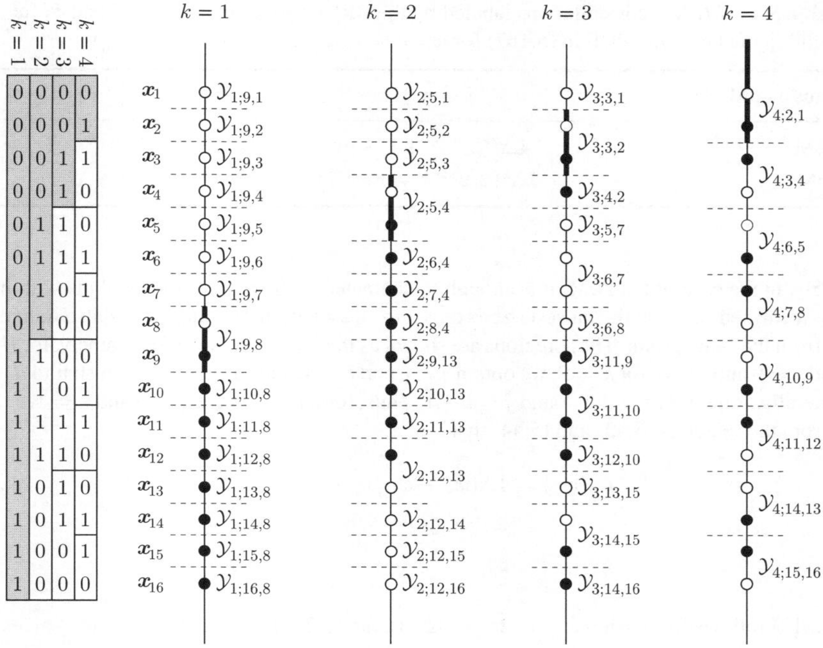

Figure 5.26 Tessellation regions for a 16PAM constellation labeled by the BRGC. The shaded rectangles gather the symbols that are relevant for the PDF computation. The subconstellations $\mathcal{S}_{k,0}$ and $\mathcal{S}_{k,1}$ are indicated by white and black circles, respectively, and the tessellation regions that define the zero-crossing for the relevant group of bits are shown by a thick line

algorithmic steps. This is thanks to the structure of the binary labeling as well as to the regular geometry of the constellation. In other words, in these cases we are able to determine the parameters of the approximate PDF (i.e., the indices $\hat{\imath}$ and $\hat{\jmath}$ of the representative tessellation regions $\mathcal{Y}_{k;\hat{\jmath},\hat{\imath}}$) as a function of the index r of the transmitted symbol $\boldsymbol{x}_r$.

We start by considering the BRGC for 16PAM shown in Fig. 5.26. We note that for a given k, the constellation may be split into 2^{k-1} groups of $\frac{M}{2^{k-1}} = 2^{m-k+1}$ consecutive symbols. These groups are identified by rectangles on the *left-hand side* (l.h.s.) of Fig. 5.26. Furthermore, because of the symmetry of the labeling, it is enough to consider the PDF conditioned on $B_k = 0$, which implies that only the symbols $\boldsymbol{x}_1, \ldots, \boldsymbol{x}_{2^{m-k}}$ are relevant for the analysis. For any symbol $\boldsymbol{x}_r \in \mathcal{S}_{k,0}$ within each group, the closest symbol from $\mathcal{S}_{k,1}$ lies also within the group. Because of the symmetries of the constellation, it is enough to analyze the first group, shown as shaded rectangles in Fig. 5.26.

To obtain the CoM, we note that for any $\boldsymbol{x}_r$ (because we analyze $r = 1, \ldots, 2^{m-k}$, it follows that $\boldsymbol{x}_r \in \mathcal{S}_{k,0}$), the closest symbol from $\mathcal{S}_{k,1}$ is $\boldsymbol{x}_{\hat{\jmath}} = \boldsymbol{x}_{2^{m-k}+1}$ (e.g., we have $\boldsymbol{x}_{\hat{\jmath}} = \boldsymbol{x}_9$ for $k = 1$, $\boldsymbol{x}_{\hat{\jmath}} = \boldsymbol{x}_5$ for $k = 2$, etc.). The distance between the symbols is given $|\boldsymbol{x}_{\hat{\jmath}} - \boldsymbol{x}_r| = 2\Delta|\hat{\jmath} - r|$, so we obtain

$$\overline{\delta}_{k,r} = 2\Delta|2^{m-k} + 1 - r|, \quad r = 1, \ldots, 2^{m-k}. \tag{5.210}$$

By using (5.210) in (5.179), we obtain the CoM PDF for an MPAM constellation labeled by the BRGC, namely, for any $\boldsymbol{x}_r$, $r = 1, \ldots, 2^{m-k}$ ($\boldsymbol{x}_r \in \mathcal{S}_{k,0}$),

$$p^{\mathrm{CoM}}_{L_k|\mathbf{X}}(l|\boldsymbol{x}_r) = \Psi(l, -4\Delta^2 \,\mathsf{snr}(2^{m-k} + 1 - r)^2, 8\Delta^2 \,\mathsf{snr}(2^{m-k} + 1 - r)^2). \tag{5.211}$$

Table 5.1 *M*PAM constellations labeled by the BRGC: parameters $\tilde{\mu}_k(\boldsymbol{x}_r)$ and $\tilde{\sigma}_k^2(\boldsymbol{x}_r)$ defining the Gaussian PDF in (5.167) for $\boldsymbol{x}_r, r = 1, \ldots, 2^{m-k}$(i.e., $\boldsymbol{x}_r \in \mathcal{S}_{k,0}$)

Gaussian Model	$\tilde{\mu}_k(\boldsymbol{x}_r)$	$\tilde{\sigma}_k^2(\boldsymbol{x}_r)$
CoM	$-4\Delta^2(2^{m-k}+1-r)^2$	$8\Delta^2(2^{m-k}+1-r)^2$
ZcM	$-4\Delta^2(2(2^{m-k}-r)+1)$	$8\Delta^2$

Similarly, in the case of the ZcM, it is enough to characterize the PDF of the symbols within the first group we identified because the closest "zero-crossing" tessellation region $\mathcal{Y}_{k;\hat{\jmath},\hat{\imath}}$ is also defined by the symbols from the same group. These regions are shown as thick lines in Fig. 5.26. Namely, for $k = 1$, we easily find $\hat{\jmath} = 9$ and $\hat{\imath} = 8$; for $k = 2$, we obtain $\hat{\jmath} = 5$ and $\hat{\imath} = 4$; and so on; then, we also find $\boldsymbol{d}_{\hat{\jmath},\hat{\imath}} = 2\Delta$. More generally, for any k, $\hat{\imath} = 2^{m-k}$ and $\hat{\jmath} = \hat{\imath} + 1$. Thus, for any $\boldsymbol{x}_r$ in the group and $r = 1, \ldots, 2^{m-k}$, we find from the relations (5.43) and (5.44) that

$$\begin{aligned}\tilde{\mu}_k(\boldsymbol{x}_r) &= 4\Delta(\boldsymbol{x}_r - \overline{\boldsymbol{x}}_{\hat{\jmath},\hat{\imath}})\\ &= -4\Delta^2(2(2^{m-k}-r)+1), \qquad (5.212)\\ \tilde{\sigma}_k^2(\boldsymbol{x}_r) &= 8\Delta^2, \qquad (5.213)\end{aligned}$$

where (5.212) follows from using $\boldsymbol{x}_r = (2r-5)\Delta$. Using (5.212) and (5.213) in (5.167) gives

$$p_{L_k|\boldsymbol{X}}^{\text{ZcM}}(l|\boldsymbol{x}_r) = \Psi(l, -4\Delta^2\,\mathsf{snr}(2(2^{m-k}-r)+1), 8\Delta^2\,\mathsf{snr}). \qquad (5.214)$$

The results in (5.211) and (5.214) are summarized in Table 5.1.

The case of *M*PSK shown in Fig. 5.27 may be treated in a similar way. The main difference is that the L-value L_1 and L_2 will have the same distribution because the groups of the symbols for $k = 1$ and $k = 2$ have the same form due to the circular symmetry of the constellation, i.e., the first and the last symbols ($\boldsymbol{x}_1$ and $\boldsymbol{x}_M$ in Fig. 5.27) become adjacent. It is thus enough to calculate the parameters of the simplified form of the PDF for $k = 2, \ldots, m$ and consider only symbol $\boldsymbol{x}_r \in \mathcal{S}_{k,0}$, i.e., $\boldsymbol{x}_r, r = 1, \ldots, 2^{m-k}$. Again, for the CoM, we have $\hat{\jmath} = 2^{m-k}+1$ and $\hat{\imath} = r$, so from (5.135), we obtain $\overline{\delta}_{k,r} = \|\boldsymbol{x}_{\hat{\jmath}} - \boldsymbol{x}_{\hat{\imath}}\| = 2S_{M,\hat{\jmath}-\hat{\imath}}$. For the ZcM, we use $\hat{\jmath} = 2^{m-k}+1$ and $\hat{\imath} = 2^{m-k}$ and thus $\tilde{\sigma}_k^2 = 8S_{M,1}$ and from (5.132), we get

$$\tilde{\mu}_k(\boldsymbol{x}_r) = 2\xi_{\hat{\jmath},\hat{\imath}}(\boldsymbol{x}_r) = -4S_{M,2(2^{m-k}-r)+1}S_{M,1}. \qquad (5.215)$$

These results are summarized in Table 5.2.

Table 5.2 *M*PSK constellation labeled by the BRGC: parameters $\tilde{\mu}_k(\boldsymbol{x}_r)$ and $\tilde{\sigma}_k^2(\boldsymbol{x}_r)$ defining the Gaussian PDF in (5.167) for $k = 2, \ldots, m$, $\boldsymbol{x}_r, r = 1, \ldots, 2^{m-k}$ (i.e., $\boldsymbol{x}_r \in \mathcal{S}_{k,0}$) and $S_{M,t} = \sin\frac{t\pi}{M}$

Gaussian Model	$\tilde{\mu}_k(\boldsymbol{x}_r)$	$\tilde{\sigma}_k^2(\boldsymbol{x}_r)$
CoM	$-4S_{M,2^{m-k}+1-r}^2$	$8S_{M,2^{m-k}+1-r}^2$
ZcM	$-4S_{M,2(2^{m-k}-r)+1}S_{M,1}$	$8S_{M,1}^2$

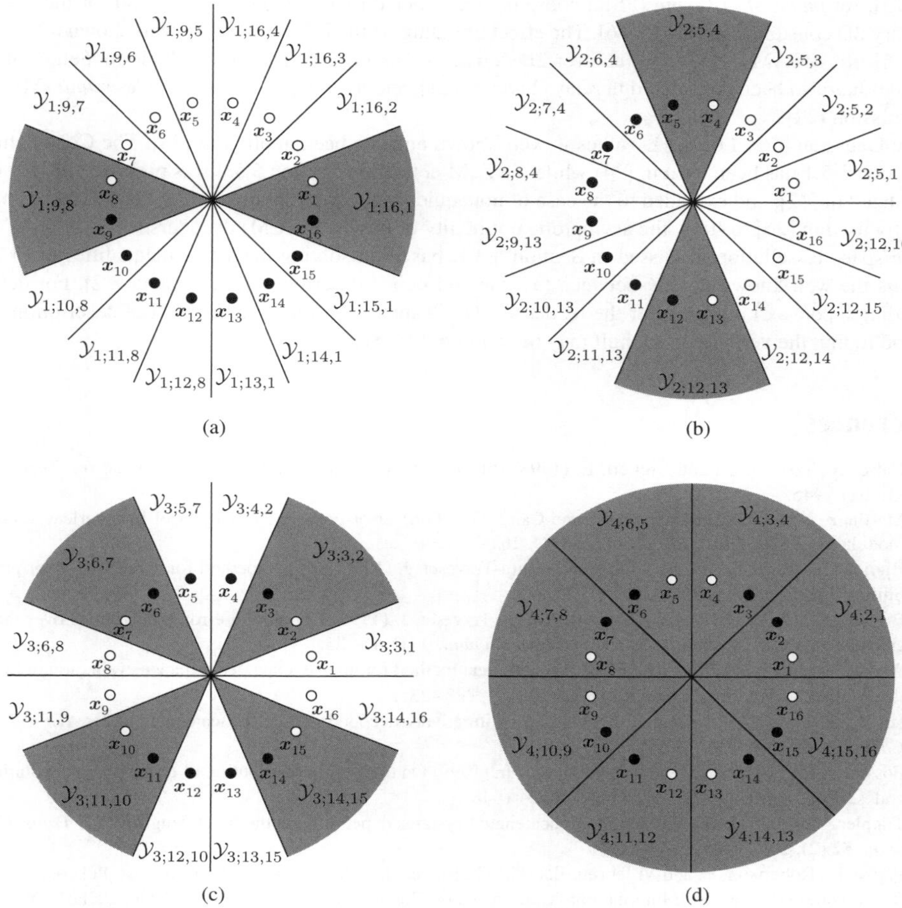

Figure 5.27 Tessellation regions $\mathcal{Y}_{k;j,i}$, for a 16PSK constellation labeled by the BRGC. The subconstellations $\mathcal{S}_{k,0}$ and $\mathcal{S}_{k,1}$ are indicated by white and black circles, respectively, and the regions $\mathcal{Y}_{k;j,i}$ in which the zero-crossing occurs are shaded: (a) $k = 1$, (b) $k = 2$, (c) $k = 3$, and (d) $k = 4$

5.6 Bibliographical Notes

The idea of the BICM channel can be tracked to [1], where the model from Fig. 5.2 corresponds to the model in [1, Fig. 3]. The model in Fig. 5.1 corresponds to the one introduced and formalized later in [2, Fig. 1]. The statistics of the L-values for the performance evaluation have been used in various works, e.g., in [1–4]. The formalism of using the PDF to evaluate the performance of the decoder in BICM transmissions was introduced in [5]. A probabilistic description of the L-values was also considered for the analysis of the decoding in *turbo codes* (TCs), *low-density parity-check* (LDPC) codes, or turbo-like processing in [6–8], respectively.

The probabilistic model for the L-values and 2PAM is well known [6, 8, 9]. The explicit modeling of the PDF for BICM based on *quadrature amplitude modulation* (QAM) constellations appeared in

[10–13], for *phase shift keying* (PSK) constellations labeled by the BRGC in [14], and for the case of arbitrary 2D constellations in [15, 16]. The effect of fading on the PDF of L-values was shown for 2PAM in [2, 6], for MPAM in [17–19], and for 2D constellations in [20]. The probabilistic modeling of the L-values has also been considered in relay channels [21], and in *multiple-input multiple-output* (MIMO) transmission [22].

The Gaussian model of the L-values is well known and has been applied in [23]. The CoM defined in Section 5.5.1 has been used in [24] while the ZcM defined in Section 5.5.2 was proposed in [10, 11], formalized in [12], and extended to the case of nonequidistant constellations in [25]. The ZcM has been recently used in [26] to study the asymptotic optimality of bitwise (BICM) decoders.

The space tessellation discussed in Section 5.3.1, based on solving the set of inequalities in (5.72), exploits the well-known duality between the line and points description [27, Chapter 8.2]. For details of finding a convex hull, we refer the reader to [27, Chapter 1.1], [28, Sec. 2.10]. The description of a method to find the vertices of the hull may be found in [28, Sec. 2.12].

References

[1] Caire, G., Taricco, G., and Biglieri, E. (1998) Bit-interleaved coded modulation. *IEEE Trans. Inf. Theory*, **44** (3), 927–946.

[2] Martinez, A., Guillén i Fàbregas, A., and Caire, G. (2006) Error probability analysis of bit-interleaved coded modulation. *IEEE Trans. Inf. Theory*, **52** (1), 262–271.

[3] Biglieri, E., Caire, G., Taricco, G., and Ventura-Traveset, J. (1996) Simple method for evaluating error probabilities. *Electron. Lett.*, **32** (2), 191–192.

[4] Biglieri, E., Caire, G., Taricco, G., and Ventura-Traveset, J. (1998) Computing error probabilities over fading channels: a unified approach. *Eur. Trans. Telecommun.*, **9** (1), 15–25.

[5] Abedi, A. and Khandani, A. K. (2004) An analytical method for approximate performance evaluation of binary linear block codes. *IEEE Trans. Commun.*, **52** (2), 228–235.

[6] ten Brink, S. (2001) Convergence behaviour of iteratively decoded parallel concatenated codes. *IEEE Trans. Commun.*, **49** (10), 1727–1737.

[7] Fossorier, M. P. C., Lin, S., and Costello, D.J. Jr. (1999) On the weight distribution of terminated convolutional codes. *IEEE Trans. Inf. Theory*, **45** (5), 1646–1648.

[8] Tüchler, M. (2004) Design of serially concatenated systems depending on the block length. *IEEE Trans. Commun.*, **52** (2), 209–218.

[9] Papke, L., Robertson, P., and Villebrun, E. (1996) Improved decoding with the SOVA in a parallel concatenated (turbo-code) scheme. IEEE International Conference on Communications (ICC), June 1996, Dallas, TX.

[10] Benjillali, M., Szczecinski, L., and Aissa, S. (2006) Probability density functions of logarithmic likelihood ratios in rectangular QAM. 23rd Biennial Symposium on Communications, May 2006, Kingston, ON, Canada

[11] Benjillali, M., Szczecinski, L., Aissa, S., and Gonzalez, C. (2008) Evaluation of bit error rate for packet combining with constellation rearrangement. *Wiley J Wirel Commun. Mob. Comput.*, **8**, 831–844.

[12] Alvarado, A., Szczecinski, L., Feick, R., and Ahumada, L. (2009) Distribution of L-values in Gray-mapped M^2-QAM: closed-form approximations and applications. *IEEE Trans. Commun.*, **57** (7), 2071–2079.

[13] Alvarado, A., Szczecinski, L., and Feick, R. (2007) On the distribution of extrinsic L-values in gray-mapped 16-QAM. AMC International Wireless Communications and Mobile Computing Conference (IWCMC), August 2007, Honolulu, HI.

[14] Szczecinski, L. and Benjillali, M. (2006) Probability density functions of logarithmic likelihood ratios in phase shift keying BICM. IEEE Global Telecommunications Conference (GLOBECOM), November 2006, San Francisco, CA.

[15] Szczecinski, L., Bettancourt, R., and Feick, R. (2006) Probability density function of reliability metrics in BICM with arbitrary modulation: closed-form through algorithmic approach. IEEE Global Telecommunications Conference (GLOBECOM), November 2006, San Francisco, CA.

[16] Szczecinski, L., Bettancourt, R., and Feick, R. (2008) Probability density function of reliability metrics in BICM with arbitrary modulation: closed-form through algorithmic approach. *IEEE Trans. Commun.*, **56** (5), 736–742.

[17] Szczecinski, L., Alvarado, A., and Feick, R. (2007) Probability density functions of reliability metrics for 16-QAM-based BICM transmission in Rayleigh channel. IEEE International Conference on Communications (ICC), June 2007, Glasgow, UK.

[18] Szczecinski, L., Alvarado, A., and Feick, R. (2009) Distribution of max-log metrics for QAM-based BICM in fading channels. *IEEE Trans. Commun.*, **57** (9), 2558–2563.
[19] Szczecinski, L., Xu, H., Gao, X., and Bettancourt, R. (2007) Efficient evaluation of BER for arbitrary modulation and signaling in fading channels. *IEEE Trans. Commun.*, **55** (11), 2061–2064.
[20] Kenarsari-Anhari, A. and Lampe, L. (2010) An analytical approach for performance evaluation of BICM over Nakagami-m fading channels. *IEEE Trans. Commun.*, **58** (4), 1090–1101.
[21] Benjillali, M. and Szczecinski, L. (2010) Detect-and-forward in two-hop relay channels: a metrics-based analysis. *IEEE Trans. Commun.*, **58** (6), 1729–1732.
[22] Čirkić, M., Persson, D., Larsson, J., and Larsson, E. G. (2012) Approximating the LLR distribution for a class of soft-output mimo detectors. *IEEE Trans. Signal Process.*, **60** (12), 6421–6434.
[23] Guillén i Fàbregas, A., Martinez, A., and Caire, G. (2004) Error probability of bit-interleaved coded modulation using the Gaussian approximation. Conference on Information Sciences and Systems (CISS), March 2004, Princeton University, NJ.
[24] Hermosilla, C. and Szczecinski, L. (2005) Performance evaluation of linear turbo-receivers using analytical extrinsic information transfer functions. *EURASIP J. Appl. Signal Process.*, **2005** (6), 892–905.
[25] Hossain, Md.J., Alvarado, A., and Szczecinski, L. (2011) Towards fully optimized BICM transceivers. *IEEE Trans. Commun.*, **59** (11), 3027–3039.
[26] Ivanov, M., Alvarado, A., Brännström, F., and Agrell, E. (2014) On the asymptotic performance of bit-wise decoders for coded modulation. *IEEE Trans. Inf. Theory*, **60** (5), 2796–2804.
[27] de Berg, M., van Kreveld, M., Overmars, O., and Schwartzkopf, O. (2000) *Computational Geometry, Algorithms and Applications*, Springer.

6

Performance Evaluation

This chapter describes different methods that can be used to evaluate the performance of *bit-interleaved coded modulation* (BICM) receivers. We focus on *bit-error probability* (BEP), *symbol-error probability* (SEP), and *word-error probability* (WEP) as performance metrics, as these are often considered relevant in practical systems. We pay special attention to techniques based on the knowledge of the *probability density function* (PDF) of the L-values we developed in Chapter 5, and in particular to the Gaussian models introduced in Section 5.5.

This chapter is organized as follows. In Section 6.1 we review methods to evaluate the performance for uncoded transmission and in Section 6.2 we study the performance of decoders (i.e., coded transmission). Section 6.3 is devoted to analyzing the *pairwise-error probability* (PEP), which appears as the key quantity when evaluating coded transmission. Section 6.4 studies the performance of BICM based on the Gaussian approximations of the PDF of the L-values.

6.1 Uncoded Transmission

By uncoded transmission we mean that the channel coding is absent (or is ignored), and thus, the decision on the transmitted symbols is made without taking into account the *sequence* of symbols (i.e., the constraints on the codewords). We are instead interested in deciding which symbol was transmitted at a given time instant n using the channel output $\boldsymbol{y}[n]$. These are "hard decisions" (briefly discussed in Section 3.4) which might eventually be fed to the channel decoder.

The criterion for making a hard decision depends on how it will be used. Probably the simplest and most intuitive approach is to minimize the SEP. Assuming an *additive white Gaussian noise* (AWGN) channel and equally likely symbols, we then obtain the expression we already showed in (3.100), i.e.,

$$\begin{aligned}\hat{\boldsymbol{x}} &= \underset{\boldsymbol{x}\in\mathcal{S}}{\operatorname{argmax}}\{\log p_{\boldsymbol{Y}|\boldsymbol{X}}(\boldsymbol{y}|\boldsymbol{x})\}\\ &= \underset{\boldsymbol{x}\in\mathcal{S}}{\operatorname{argmin}}\{\|\boldsymbol{y}-\boldsymbol{x}\|^2\}.\end{aligned} \tag{6.1}$$

Similarly, assuming equally likely bits, to minimize the BEP, the decision on the transmitted bits is

$$\hat{b}_k^{\mathrm{ml}} = \underset{b\in\mathbb{B}}{\operatorname{argmax}}\{\log P_{\boldsymbol{Y}|B_k}(\boldsymbol{y}|b)\}, \tag{6.2}$$

Bit-Interleaved Coded Modulation: Fundamentals, Analysis, and Design, First Edition.
Leszek Szczecinski and Alex Alvarado.

which can also be expressed as

$$\hat{b}_k^{\text{ml}} = \underset{b\in\mathbb{B}}{\text{argmax}} \left\{ \log \sum_{\boldsymbol{x}\in\mathcal{S}_{k,b}} p_{\boldsymbol{Y}|\boldsymbol{X}}(\boldsymbol{y}|\boldsymbol{x}) \right\}$$

$$= \underset{b\in\mathbb{B}}{\text{argmax}} \left\{ \log \sum_{\boldsymbol{x}\in\mathcal{S}_{k,b}} \exp(-\mathsf{snr}\|\boldsymbol{y}-\boldsymbol{x}\|^2) \right\}, \tag{6.3}$$

or in terms of L-values, as

$$\hat{b}_k^{\text{ml}} = \begin{cases} 1, & \text{if } \log P_{\boldsymbol{Y}|B_k}(\boldsymbol{y}|1) \geq \log P_{\boldsymbol{y}|B_k}(\boldsymbol{y}|0) \\ 0, & \text{otherwise} \end{cases}$$

$$= \begin{cases} 1, & \text{if } l_k \geq 0 \\ 0, & \text{if } l_k < 0 \end{cases}. \tag{6.4}$$

From (6.4) we conclude that a hard decision on the transmitted bits is equivalent to making a decision on the sign of the L-values in (3.50).[1]

The implementation of (6.3) has a complexity dominated by the calculation of the sum of exponential functions so simplifications may be sought. In particular, we may apply the max-log approximation, which used in (6.3) gives

$$\hat{b}_k = \underset{b\in\mathbb{B}}{\text{argmin}} \left\{ \min_{\boldsymbol{x}\in\mathcal{S}_{k,b}} \|\boldsymbol{y}-\boldsymbol{x}\|^2 \right\}. \tag{6.5}$$

The hard decision on the bits shown in (6.5) is also equivalent to determining the sign of the L-values calculated via the max-log approximation (i.e., when l_k in (6.4) are max-log L-values). This was already shown in (3.106).

As we showed in Theorem 3.17, the decision in (6.5) is equivalent to the decision in (6.1), and thus, we can consider the vector of hard decisions on the bits

$$\hat{\boldsymbol{b}} = \underset{\boldsymbol{b}\in\mathbb{B}^m}{\text{argmin}}\{\|\boldsymbol{y}-\Phi(\boldsymbol{b})\|^2\}, \tag{6.6}$$

where $\hat{\boldsymbol{b}} \triangleq [\hat{b}_1, \ldots, \hat{b}_m]^T$. Owing to this equivalence we focus on the symbol-wise hard decisions in (6.1), or equivalently, on the bitwise hard decisions in (6.6) based on max-log L-values.

Since the hard decisions are equivalent to a one-bit (sign only) quantization of the L-values, we may continue to view them as a part of the BICM channel, which now becomes a binary-input binary-output channel. Methods for analyzing the performance of such a BICM channel in terms of SEP or BEP for uncoded transmission have been studied in the communication theory community over decades. The main reason for focusing on these quantities is that SEP/BEP for uncoded transmission are relatively easy to obtain compared to similar quantities in coded transmission. The underlying assumption is that systems with the same SEP/BEP will perform similarly when coding/decoding is added. As we will see in Chapter 8, this is, in general not true, as the design of the interleaver may change the performance of the coded transmission, while uncoded transmission is unaffected by interleaving.

6.1.1 Decision Regions

The SEP is defined as

$$P_{\text{s}} \triangleq \Pr\{\hat{\boldsymbol{X}} \neq \boldsymbol{X}\}, \tag{6.7}$$

[1] An expression similar to (6.4) (for nonequiprobable bits) was discussed in Example 3.6.

where

$$P_{\mathrm{s}} = \frac{1}{M}\sum_{j\in\mathcal{I}} \Pr\{\boldsymbol{Y} \notin \mathcal{Y}_j | \boldsymbol{X} = \boldsymbol{x}_j\} \tag{6.8}$$

$$= \frac{1}{M}\sum_{j\in\mathcal{I}} (1 - \Pr\{\boldsymbol{Y} \in \mathcal{Y}_j | \boldsymbol{X} = \boldsymbol{x}_j\}), \tag{6.9}$$

where the decision region of the symbol $\boldsymbol{x}_j$ is defined using (6.1) as

$$\mathcal{Y}_j \triangleq \{\boldsymbol{y} : \|\boldsymbol{y} - \boldsymbol{x}_j\|^2 \le \|\boldsymbol{y} - \boldsymbol{x}_i\|^2, i \in \mathcal{I}, i \ne j\} \tag{6.10}$$

$$= \{\boldsymbol{y} : \xi_{i,j}(\boldsymbol{y}) \le 0, i \in \mathcal{I}, i \ne j\}, \tag{6.11}$$

and where $\xi_{i,j}(\boldsymbol{y})$ is given by (3.54).

The set $\mathcal{Y}_j$ is the so-called *Voronoi region* of the symbol $\boldsymbol{x}_j$. In many practically relevant cases, the form of the Voronoi regions can be found by inspection. For example, when the constellation points are located on a rectangular grid (i.e., *for M-ary quadrature amplitude modulation* (MQAM) constellations) or on a circle (i.e., for *M-ary phase shift keying* (MPSK) constellations), finding the decision regions $\mathcal{Y}_j$ consists in determining the closest constellation points to the symbol $\boldsymbol{x}_j$ along each dimension of the grid. In the case of MQAM there are at most four such symbols, and in the case of MPSK constellations there are only two of them. In the more general case of irregular constellations, the Voronoi regions may be found using numerical routines.

Example 6.1 (Voronoi Regions for 16QAM **and** 8PSK**)** *The Voronoi regions for* 16QAM *and* 8PSK *are shown in Fig. 6.1, where the Voronoi region for the symbol $\boldsymbol{x}_7$ is highlighted.*

Let $P_{i\to j}$ be the probability of detecting $\boldsymbol{x}_j$ when $\boldsymbol{x}_i$ was transmitted, i.e.,

$$P_{i\to j} \triangleq \Pr\{\boldsymbol{Y} \in \mathcal{Y}_j | \boldsymbol{X} = \boldsymbol{x}_i\} \tag{6.12}$$

$$= \int_{\boldsymbol{y}\in\mathcal{Y}_j} p_{\boldsymbol{Y}|\boldsymbol{X}}(\boldsymbol{y}|\boldsymbol{x}_i)\, \mathrm{d}\boldsymbol{y} \tag{6.13}$$

$$= \left(\frac{\mathsf{snr}}{\pi}\right)^{\frac{N}{2}} \int_{\boldsymbol{y}\in\mathcal{Y}_j} \exp(-\mathsf{snr}\|\boldsymbol{y} - \boldsymbol{x}_i\|^2)\, \mathrm{d}\boldsymbol{y}. \tag{6.14}$$

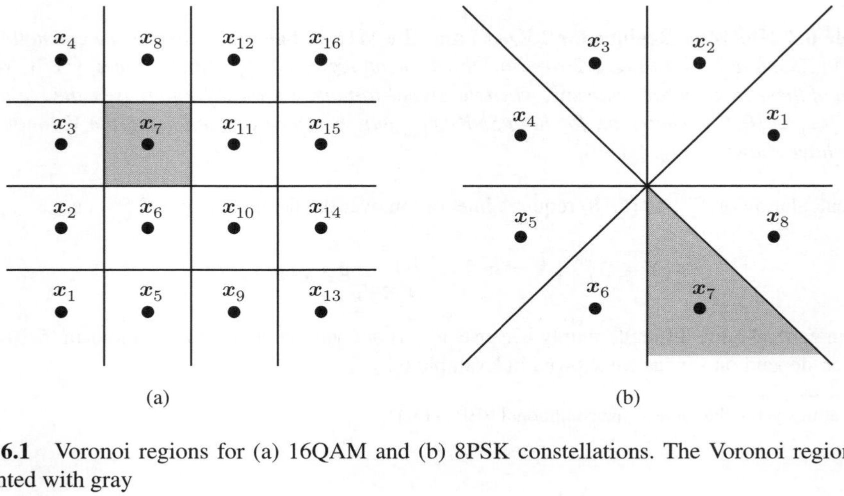

Figure 6.1 Voronoi regions for (a) 16QAM and (b) 8PSK constellations. The Voronoi region $\mathcal{Y}_7$ is highlighted with gray

From now on, we refer to $P_{i\to j}$ as the *transition probability* (TP).

Using (6.12), the SEP (6.8) can then be expressed as

$$P_{\mathrm{s}} = \frac{1}{M}\sum_{i\in\mathcal{I}}\sum_{j\in\mathcal{I},j\neq i} P_{i\to j} \tag{6.15}$$

$$= 1 - \frac{1}{M}\sum_{j\in\mathcal{I}} P_{j\to j}, \tag{6.16}$$

where to pass from (6.15) to (6.16) we used (6.9).

The BEP for the kth bit was already defined in (3.110) for max-log metrics.[2] For exact L-values, we have the following analogous definition

$$P_{\mathrm{b},k}^{\mathrm{ml}} \triangleq \Pr\{\hat{B}_k^{\mathrm{ml}} \neq B_k\} \tag{6.17}$$

$$= \frac{1}{2}\sum_{b\in\mathbb{B}} \Pr\left\{\boldsymbol{Y}\in\mathcal{Y}_{k,\bar{b}}^{\mathrm{ml}} \,\middle|\, B_k = b\right\} \tag{6.18}$$

$$= \frac{1}{M}\sum_{b\in\mathbb{B}}\sum_{i\in\mathcal{I}_{k,b}} \Pr\left\{\boldsymbol{Y}\in\mathcal{Y}_{k,\bar{b}}^{\mathrm{ml}} \,\middle|\, \boldsymbol{X} = \boldsymbol{x}_i\right\}, \tag{6.19}$$

where $\hat{B}_k^{\mathrm{ml}}$ are given by (6.3) and (6.4) and the "decision regions" for the kth bit are thus

$$\mathcal{Y}_{k,b}^{\mathrm{ml}} \triangleq \left\{\boldsymbol{y} : \log\sum_{\boldsymbol{x}\in\mathcal{S}_{k,b}} \exp(-\mathsf{snr}\|\boldsymbol{y}-\boldsymbol{x}\|^2) \geq \log\sum_{\boldsymbol{x}\in\mathcal{S}_{k,\bar{b}}} \exp(-\mathsf{snr}\|\boldsymbol{y}-\boldsymbol{x}\|^2)\right\}. \tag{6.20}$$

Similarly, in the case of max-log L-values, for the kth bit in (3.110) with $\hat{B}_k$ given by (6.5), the BEP is expressed as

$$P_{\mathrm{b},k} = \frac{1}{2}\sum_{b\in\mathbb{B}} \Pr\{\boldsymbol{Y}\in\mathcal{Y}_{k,\bar{b}} \,|\, B_k = b\}, \tag{6.21}$$

where the decision regions in this case are

$$\mathcal{Y}_{k,b} \triangleq \bigcup_{j\in\mathcal{I}_{k,b}} \mathcal{Y}_j, \tag{6.22}$$

which follows from the fact that $\hat{B}_k$ is the kth bit of the label of the detected symbol via (6.1) (see (6.6)).

Example 6.2 (Decision Regions for 16QAM **and the M16 Labeling)** *Consider the constellation we studied in Example 2.17. In Fig. 6.2 we show the decision regions $\mathcal{Y}_{k,b}^{\mathrm{ml}}$ in (6.20) for $k = 1, 3$. Note that the form of these regions vary with snr: when the signal-to-noise ratio (SNR) increases, the regions $\mathcal{Y}_{k,b}^{\mathrm{ml}}$ tend to $\mathcal{Y}_{k,b}$ in (6.22). Therefore, for high SNR, $\mathcal{Y}_{k,b}^{\mathrm{ml}}$ may be approximated using the Voronoi regions $\mathcal{Y}_{k,b}$ we have shown in Fig. 6.1 (a).*

The calculation of $P_{\mathrm{b},k}^{\mathrm{ml}}$ in (6.18) requires integration over the decision region $\mathcal{Y}_{k,\bar{b}}^{\mathrm{ml}}$, i.e.,

$$\Pr\{\boldsymbol{Y}\in\mathcal{Y}_{k,\bar{b}}^{\mathrm{ml}} \,|\, \boldsymbol{X} = \boldsymbol{x}_i\} = \int_{\boldsymbol{Y}\in\mathcal{Y}_{k,\bar{b}}} p_{\boldsymbol{Y}|\boldsymbol{X}}(\boldsymbol{y}|\boldsymbol{x}_i)\,\mathrm{d}\boldsymbol{y}. \tag{6.23}$$

This is in general quite difficult, mainly because $\mathcal{Y}_{k,b}^{\mathrm{ml}}$ is defined via nonlinear equations in (6.20) which, moreover, depend on snr (as we showed in Example 6.2).

[2] Note that this is not the same as the conditional BEP in (3.41).

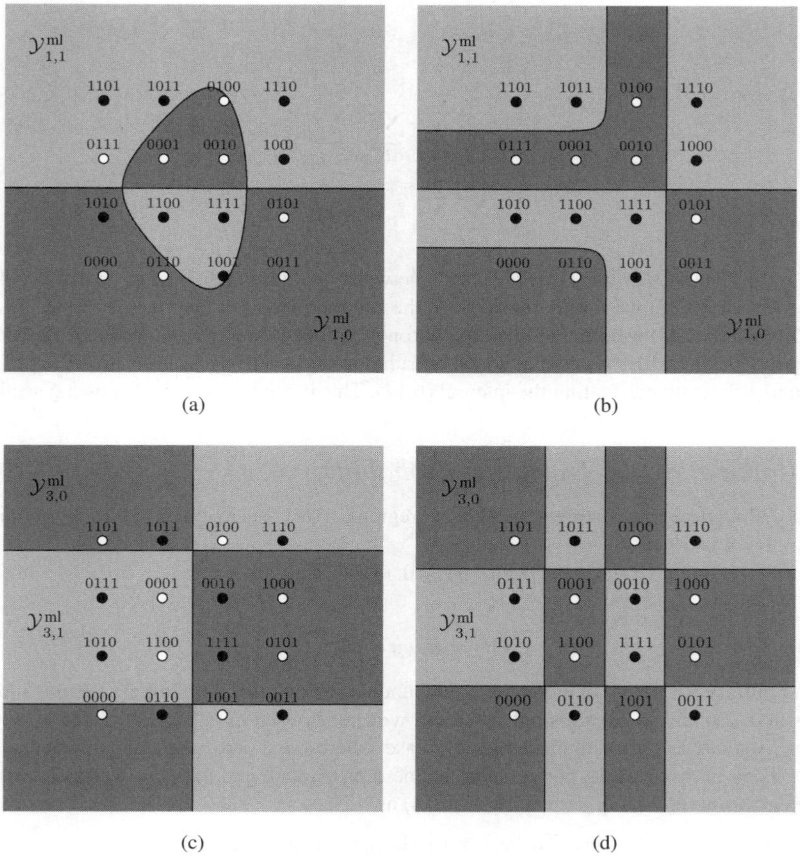

Figure 6.2 Decision regions $\mathcal{Y}^{\text{ml}}_{k,1}$ (light gray) and $\mathcal{Y}^{\text{ml}}_{k,0}$ (dark gray) in (6.20) for 16QAM with the M16 labeling shown in Fig. 2.5: (a) $k = 1$, $\mathsf{snr} = 0$ dB; (b) $k = 1$, $\mathsf{snr} = 10$ dB; (c) $k = 3$, $\mathsf{snr} = 0$ dB; and (d) $k = 3$, $\mathsf{snr} = 10$ dB. The subconstellations $\mathcal{S}_{k,0}$ and $\mathcal{S}_{k,1}$ are indicated, respectively, with white and black circles

On the other hand, with the max-log approximation, the decision region $\mathcal{Y}_{k,b}$ is the union of the Voronoi regions $\mathcal{Y}_j$ in (6.10), and thus, (i) it is independent of the SNR and (ii) it is defined via linear equations only. The BEP in this case is calculated as

$$P_{\text{b},k} = \frac{1}{M} \sum_{b \in \mathbb{B}} \sum_{i \in \mathcal{I}_{k,b}} \Pr\left\{ \boldsymbol{Y} \in \mathcal{Y}_{k,\bar{b}} | \boldsymbol{X} = \boldsymbol{x}_i \right\} \tag{6.24}$$

$$= \frac{1}{M} \sum_{b \in \mathbb{B}} \sum_{i \in \mathcal{I}_{k,b}} \sum_{j \in \mathcal{I}_{k,\bar{b}}} P_{i \to j}, \tag{6.25}$$

where (6.24) follows from the law of total probability, (6.25) from (6.10) and (6.21), and where the $P_{i \to j}$ is given by (6.12). The BEP averaged over the bits' positions is then defined as

$$P_{\mathrm{b}} \triangleq \frac{1}{m}\sum_{k=1}^{m} P_{\mathrm{b},k} \tag{6.26}$$

$$= \frac{1}{mM}\sum_{k=1}^{m}\sum_{b\in\mathbb{B}}\sum_{i\in\mathcal{I}_{k,b}}\sum_{j\in\mathcal{I}_{k,\overline{b}}} P_{i\to j} \tag{6.27}$$

$$= \frac{1}{mM}\sum_{i\in\mathcal{I}}\sum_{j\in\mathcal{I}} d_{\mathrm{H}}(\boldsymbol{q}_i,\boldsymbol{q}_j)P_{i\to j}, \tag{6.28}$$

where $d_{\mathrm{H}}(\boldsymbol{q}_i,\boldsymbol{q}_j)$ is the *Hamming distance* (HD) between the binary labels of $\boldsymbol{x}_i$ and $\boldsymbol{x}_j$, and where to pass from (6.27) to (6.28) we simply reorganized the summations.

From (6.16) and (6.28) we see that once the Voronoi regions $\mathcal{Y}_j$ in (6.10) are found, the challenge of calculating the SEP and BEP is reduced to the calculation of the TPs $P_{i\to j}$ in (6.12). The problem then boils down to efficiently calculating the integral (6.14). This is the focus of the following sections.

6.1.2 Calculating the Transition Probability

Before proceeding further, we express the Voronoi regions in (6.11) using the $T_j \le M-1$ nonredundant inequalities defining $\mathcal{Y}_j$, i.e.,

$$\mathcal{Y}_j = \{\boldsymbol{y} : \mathsf{L}_t(\boldsymbol{y}) \le 0, t = 1,\ldots,T_j\}, \tag{6.29}$$

where the linear forms in this case are

$$\mathsf{L}_t(\boldsymbol{y}) = \boldsymbol{d}_t^{\mathrm{T}}(\boldsymbol{y} - \overline{\boldsymbol{x}}_t). \tag{6.30}$$

We use an arbitrary variable t to index the nonredundant inequalities defining the region, and thus, the definitions in (6.29) and (6.30) are similar to those we already used in (5.87) and (5.88).[3]

We start by analyzing the simplest possible case where only one nonredundant inequality $\mathsf{L}_1(\boldsymbol{y})$ exists (i.e., $T_j = 1$). In this case, the problem is one-dimensional, and thus, the Voronoi region $\mathcal{Y}_j$ is a half-space. In this case, we express the TP in (6.14) as

$$P_{i\to j} = \mathsf{T}(\mathsf{L}_1|\boldsymbol{x}_i) \triangleq \Pr\{\mathsf{L}_1(\boldsymbol{y}) \le 0|\boldsymbol{X} = \boldsymbol{x}_i\}, \tag{6.31}$$

where with slight abuse of notation we use L_t in $\mathsf{T}(\mathsf{L}_t|\boldsymbol{x}_i)$ (without the argument $\boldsymbol{y}$) to denote the parameters $\boldsymbol{d}_t$ and $\overline{\boldsymbol{x}}_t$ defining the linear form $\mathsf{L}_t(\boldsymbol{y})$ in (6.30). The TP in (6.31) can be expressed as

$$\begin{aligned}\mathsf{T}(\mathsf{L}_1|\boldsymbol{x}_i) &= \Pr\{\mathsf{L}_1(\boldsymbol{x}_i + \boldsymbol{Z}) \le 0\}\\ &= \Pr\left\{-\frac{\boldsymbol{d}_1^{\mathrm{T}}\sqrt{2\,\mathsf{snr}}}{\|\boldsymbol{d}_1\|}\boldsymbol{Z} \ge \frac{\mathsf{L}_1(\boldsymbol{x}_i)}{\|\boldsymbol{d}_1\|}\sqrt{2\,\mathsf{snr}}\right\}\\ &= Q\left(\frac{\mathsf{L}_1(\boldsymbol{x}_i)}{\|\boldsymbol{d}_1\|}\sqrt{2\,\mathsf{snr}}\right),\end{aligned} \tag{6.32}$$

where (6.32) follows from the fact that $-\boldsymbol{d}_1^{\mathrm{T}}\sqrt{2\,\mathsf{snr}}\,\boldsymbol{Z}/\|\boldsymbol{d}_1\|$ is a zero-mean, unit variance Gaussian random variable.

Example 6.3 (BEP for 2PAM) *For a* 2-ary pulse amplitude modulation *(*2PAM*) constellation,* $\boldsymbol{x}_1 = -\Delta = -1$ *and* $\mathsf{L}_1(\boldsymbol{y}) = -2\boldsymbol{y}\Delta$ *(i.e.,* $\boldsymbol{d}_1 = -2$*), so* (6.32) *gives the well-known expression*

$$P_{1\to 2} = \mathsf{T}(\mathsf{L}_1|\boldsymbol{x}_1) = Q\left(\sqrt{2\,\mathsf{snr}}\right). \tag{6.33}$$

In addition, from the symmetry of the constellation, $P_{1\to 2} = P_{2\to 1}$*, and thus,* $P_{\mathrm{b}} = Q\left(\sqrt{2\,\mathsf{snr}}\right)$.

[3] One difference between the notation in (5.87) and the one in (6.29) is that the number of nonredundant inequalities T_j is not indexed by k and i. This comes from the fact that the Voronoi region $\mathcal{Y}_j$ depends only on j, unlike the *tessellation* regions $\mathcal{Y}_{k;j,i}$, which also depend on k and i.

The result shown in Example 6.3 can also be obtained by simple inspection: because the symbol $\boldsymbol{x}_1$ is at distance Δ from the limit of the decision region $\mathcal{Y}_2$, and the noise $\boldsymbol{Z}$ is Gaussian with variance $\mathsf{N}_0/2$, the calculation of $\Pr\{\boldsymbol{Z} > \Delta\}$ yields (6.33). Nevertheless, we use the formalism of the decision region defined by the line $\mathsf{L}_t(\boldsymbol{y}) = 0$ for illustrative purposes.

The next step is to consider the region defined by two nonredundant inequalities $\mathsf{L}_1(\boldsymbol{y}) \leq 0$ and $\mathsf{L}_2(\boldsymbol{y}) \leq 0$ (i.e., $T_j = 2$). In this case, the Voronoi region $\mathcal{Y}_j$ is a *wedge*, which we define in analogy to (5.95) as

$$\mathcal{W}(\mathsf{L}_1, \mathsf{L}_2) \triangleq \{\boldsymbol{y} : \mathsf{L}_1(\boldsymbol{y}) \leq 0 \wedge \mathsf{L}_2(\boldsymbol{y}) \leq 0\}. \tag{6.34}$$

For convenience, we also define a "complementary" wedge

$$\mathcal{W}(\overline{\mathsf{L}}_1, \mathsf{L}_2) \triangleq \{\boldsymbol{y} : \mathsf{L}_1(\boldsymbol{y}) \geq 0 \wedge \mathsf{L}_2(\boldsymbol{y}) \leq 0\}. \tag{6.35}$$

The wedges $\mathcal{W}(\mathsf{L}_1, \mathsf{L}_2)$ and $\mathcal{W}(\overline{\mathsf{L}}_1, \mathsf{L}_2)$ are schematically shown in Fig. 6.3. These simple forms of $\mathcal{Y}_j$ are important because they appear, e.g., in the case of MPSK constellations (see Fig. 6.1). As we will see later, the analysis of a wedge also leads to expressions that can be used to study MPAM, MQAM, as well as arbitrary constellations.

For the case of the Voronoi region being a wedge, we denote the TP in (6.14) as

$$P_{i \to j} = \mathsf{T}(\mathsf{L}_1, \mathsf{L}_2 | \boldsymbol{x}_i), \tag{6.36}$$

where in analogy to (5.96)

$$\mathsf{T}(\mathsf{L}_1, \mathsf{L}_2 | \boldsymbol{x}_i) \triangleq \Pr\{\boldsymbol{Y} \in \mathcal{W}(\mathsf{L}_1, \mathsf{L}_2) | \boldsymbol{X} = \boldsymbol{x}_i\} \tag{6.37}$$

$$= \int_{\boldsymbol{y} \in \mathcal{W}(\mathsf{L}_1, \mathsf{L}_2)} p_{\boldsymbol{Y}|\boldsymbol{X}}(\boldsymbol{y}|\boldsymbol{x}_i)\, \mathrm{d}\boldsymbol{y}. \tag{6.38}$$

The TP in (6.37) can be expressed as

$$\mathsf{T}(\mathsf{L}_1, \mathsf{L}_2 | \boldsymbol{x}_i) = \Pr\{\mathsf{L}_1(\boldsymbol{x}_i + \boldsymbol{Z}) \leq 0 \wedge \mathsf{L}_2(\boldsymbol{x}_i + \boldsymbol{Z}) \leq 0\} \tag{6.39}$$

$$= \Pr\{-\boldsymbol{d}_1^{\mathrm{T}} \boldsymbol{Z} \geq \mathsf{L}_1(\boldsymbol{x}_i) \wedge -\boldsymbol{d}_2^{\mathrm{T}} \boldsymbol{Z} \geq \mathsf{L}_2(\boldsymbol{x}_i)\} \tag{6.40}$$

$$= \Pr\left\{\sqrt{2\,\mathsf{snr}}\frac{\mathsf{L}_1(\boldsymbol{x}_i)}{\|\boldsymbol{d}_1\|} \leq Z' \wedge \sqrt{2\,\mathsf{snr}}\frac{\mathsf{L}_2(\boldsymbol{x}_i)}{\|\boldsymbol{d}_2\|} \leq Z''\right\}, \tag{6.41}$$

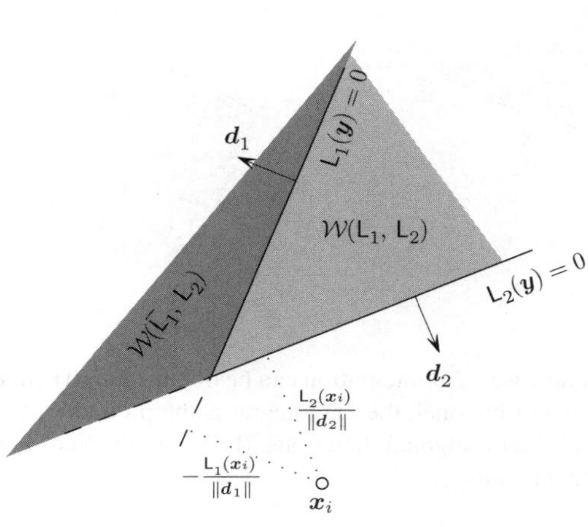

Figure 6.3 The wedges $\mathcal{W}(\mathsf{L}_1, \mathsf{L}_2)$ and $\mathcal{W}(\overline{\mathsf{L}}_1, \mathsf{L}_2)$ are delimited by the lines $\mathsf{L}_1(\boldsymbol{y}) = 0$ and $\mathsf{L}_2(\boldsymbol{y}) = 0$. The distances from $\boldsymbol{x}_i$ to the lines (the length of the dotted lines) are $\frac{\mathsf{L}_1(\boldsymbol{x}_i)}{\|\boldsymbol{d}_1\|}$ and $\frac{\mathsf{L}_2(\boldsymbol{x}_i)}{\|\boldsymbol{d}_2\|}$

where

$$Z' = -\boldsymbol{d}_1^{\mathrm{T}} \boldsymbol{Z} \frac{\sqrt{2\,\mathsf{snr}}}{\|\boldsymbol{d}_1\|}, \tag{6.42}$$

$$Z'' = -\boldsymbol{d}_2^{\mathrm{T}} \boldsymbol{Z} \frac{\sqrt{2\,\mathsf{snr}}}{\|\boldsymbol{d}_2\|}. \tag{6.43}$$

Owing to the normalization, Z' and Z'' can be shown to be zero-mean, unit-variance Gaussian random variables with correlation

$$\rho = \mathbb{E}_{\boldsymbol{Z}}[Z'Z''] = \frac{\boldsymbol{d}_1^{\mathrm{T}}\boldsymbol{d}_2}{\|\boldsymbol{d}_1\|\|\boldsymbol{d}_2\|}. \tag{6.44}$$

Using (6.41)–(6.44), (6.37) can be expressed as

$$\begin{aligned}\mathsf{T}(\mathsf{L}_1,\mathsf{L}_2|\boldsymbol{x}_i) &= \Pr\left\{\boldsymbol{Y} \in \mathcal{W}(\mathsf{L}_1,\mathsf{L}_2)|\boldsymbol{X} = \boldsymbol{x}_i\right\}\\ &= \tilde{Q}\left(\sqrt{2\,\mathsf{snr}}\frac{\mathsf{L}_1(\boldsymbol{x}_i)}{\|\boldsymbol{d}_1\|}, \sqrt{2\,\mathsf{snr}}\frac{\mathsf{L}_2(\boldsymbol{x}_i)}{\|\boldsymbol{d}_2\|}, \frac{\boldsymbol{d}_1^{\mathrm{T}}\boldsymbol{d}_2}{\|\boldsymbol{d}_1\|\|\boldsymbol{d}_2\|}\right),\end{aligned} \tag{6.45}$$

where $\tilde{Q}(a,b,\rho)$ is given by (2.11).[4]

In the same way we defined a complementary wedge in (6.35), we define a "complementary" TP as

$$\mathsf{T}(\overline{\mathsf{L}}_1,\mathsf{L}_2|\boldsymbol{x}_i) \triangleq \Pr\left\{\boldsymbol{Y} \in \mathcal{W}(\overline{\mathsf{L}}_1,\mathsf{L}_2)|\boldsymbol{X} = \boldsymbol{x}_i\right\}. \tag{6.46}$$

Following similar steps to those in (6.39)–(6.45), we find that the complementary TP in (6.46) is given by

$$\mathsf{T}(\overline{\mathsf{L}}_1,\mathsf{L}_2|\boldsymbol{x}_i) = \tilde{Q}\left(-\sqrt{2\,\mathsf{snr}}\frac{\mathsf{L}_1(\boldsymbol{x}_i)}{\|\boldsymbol{d}_1\|}, \sqrt{2\,\mathsf{snr}}\frac{\mathsf{L}_2(\boldsymbol{x}_i)}{\|\boldsymbol{d}_2\|}, -\frac{\boldsymbol{d}_1^{\mathrm{T}}\boldsymbol{d}_2}{\|\boldsymbol{d}_1\|\|\boldsymbol{d}_2\|}\right). \tag{6.47}$$

The general expression (6.45) for the TP when the Voronoi region is a wedge particularizes to two interesting cases, which are shown schematically in Fig. 6.4.

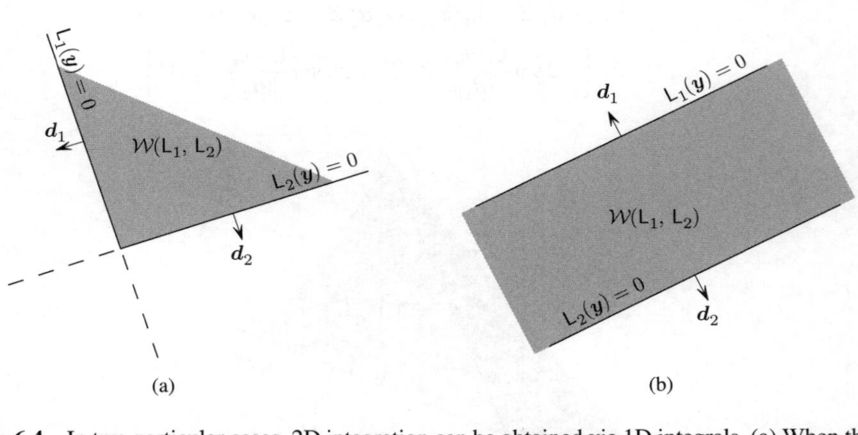

Figure 6.4 In two particular cases, 2D integration can be obtained via 1D integrals. (a) When the lines $\mathsf{L}_1(\boldsymbol{y}) = 0$ and $\mathsf{L}_2(\boldsymbol{y}) = 0$ are orthogonal, the 2D integral is the product of two 1D integrals calculated independently along each of the orthogonal directions. (b) When the lines are parallel, the 2D integral becomes a difference of 1D integrals

[4] In analogy to the Q-function, we consider the bivariate Q-function to be a "closed form," even if it requires numerical integration.

- When $\rho = 0$, the lines $\mathsf{L}_1(\boldsymbol{y})$ and $\mathsf{L}_2(\boldsymbol{y})$ are orthogonal, and thus, the variables Z'', Z' in (6.41) are independent. We then obtain $\tilde{Q}(a, b, 0) = Q(a)Q(b)$, and thus,

$$P_{i\to j} = Q\left(\sqrt{2\,\mathsf{snr}}\frac{\mathsf{L}_1(\boldsymbol{x}_i)}{\|\boldsymbol{d}_1\|}\right) Q\left(\sqrt{2\,\mathsf{snr}}\frac{\mathsf{L}_2(\boldsymbol{x}_i)}{\|\boldsymbol{d}_2\|}\right). \tag{6.48}$$

- When $\rho = -1$, the lines $\mathsf{L}_1(\boldsymbol{y})$ and $\mathsf{L}_2(\boldsymbol{y})$ are parallel, and thus, the wedge $\mathcal{W}(\mathsf{L}_1, \mathsf{L}_2)$ "degenerates" into a stripe. In this case, the variables in (6.41) satisfy $Z'' = -Z'$, and thus, using $\rho = -1$ in (2.12), we obtain

$$P_{i\to j} = Q\left(\sqrt{2\,\mathsf{snr}}\frac{\mathsf{L}_1(\boldsymbol{x}_i)}{\|\boldsymbol{d}_1\|}\right) - Q\left(-\sqrt{2\,\mathsf{snr}}\frac{\mathsf{L}_2(\boldsymbol{x}_i)}{\|\boldsymbol{d}_2\|}\right) \tag{6.49}$$

$$= Q\left(\sqrt{2\,\mathsf{snr}}\frac{\mathsf{L}_2(\boldsymbol{x}_i)}{\|\boldsymbol{d}_2\|}\right) - Q\left(-\sqrt{2\,\mathsf{snr}}\frac{\mathsf{L}_1(\boldsymbol{x}_i)}{\|\boldsymbol{d}_1\|}\right), \tag{6.50}$$

which is valid for $\frac{\mathsf{L}_1(\boldsymbol{x}_i)}{\|\boldsymbol{d}_1\|} + \frac{\mathsf{L}_2(\boldsymbol{x}_i)}{\|\boldsymbol{d}_2\|} \leq 0$ (see (2.12)), i.e., when the linear inequalities are not contradictory. The 2D integral is thus reduced to a 1D integration of a zero-mean Gaussian PDF along a line orthogonal to the stripe's limits.

With the expressions we have developed in (6.32) and (6.49), we are in a position to compute the SEP and BEP for any 1D constellation. In the following example we show how to do this for 4PAM and the *binary reflected Gray code* (BRGC).

Example 6.4 (BEP for 4PAM **Labeled by the BRGC)** *For a* 4PAM *constellation shown in Fig. 5.3,* $\mathcal{Y}_1 = \{\boldsymbol{y} : \boldsymbol{y} \leq -2\Delta\}$, $\mathcal{Y}_2 = \{\boldsymbol{y} : -2\Delta < \boldsymbol{y} \leq 0\}$, $\mathcal{Y}_3 = \{\boldsymbol{y} : 0 < \boldsymbol{y} \leq 2\Delta\}$, *and* $\mathcal{Y}_4 = \{\boldsymbol{y} : 2\Delta < \boldsymbol{y}\}$. *In principle, there are* $M(M-1) = 12$ *TPs to compute. However, because the constellation is 1D, we can classify them into two groups: open intervals (for which we can use* (6.32)*) and closed intervals (for which we can use* (6.49)*). For the closed intervals we have*

$$P_{1\to 2} = P_{2\to 3} = P_{3\to 2} = P_{4\to 3}, = Q(\Delta\sqrt{2\,\mathsf{snr}}) - Q(3\Delta\sqrt{2\,\mathsf{snr}}), \tag{6.51}$$

$$P_{1\to 3} = P_{4\to 2} = Q(3\Delta\sqrt{2\,\mathsf{snr}}) - Q(5\Delta\sqrt{2\,\mathsf{snr}}), \tag{6.52}$$

and for the open intervals

$$P_{1\to 4} = P_{4\to 1} = Q(5\Delta\sqrt{2\,\mathsf{snr}}), \tag{6.53}$$

$$P_{2\to 4} = P_{3\to 1} = Q(3\Delta\sqrt{2\,\mathsf{snr}}), \tag{6.54}$$

$$P_{2\to 1} = P_{3\to 4} = Q(\Delta\sqrt{2\,\mathsf{snr}}). \tag{6.55}$$

The SEP (6.15) *is then given by*

$$P_{\mathrm{s}} = \frac{3}{2}Q(\Delta\sqrt{2\,\mathsf{snr}}). \tag{6.56}$$

Next, exploiting the symmetries of the BRGC (see Fig. 5.3), we calculate BEP per position k *using* (6.25)*:*

$$P_{\mathrm{b},1} = \frac{1}{2}(P_{1\to 3} + P_{1\to 4} + P_{2\to 3} + P_{2\to 4}) \tag{6.57}$$

$$= \frac{1}{2}(Q(3\Delta\sqrt{2\,\mathsf{snr}}) + Q(\Delta\sqrt{2\,\mathsf{snr}})), \tag{6.58}$$

$$P_{\mathrm{b},2} = \frac{1}{2}(P_{1\to 2} + P_{1\to 3} + P_{2\to 4} + P_{2\to 1}) \tag{6.59}$$

$$= Q(\Delta\sqrt{2\,\mathsf{snr}}) + \frac{1}{2}(Q(3\Delta\sqrt{2\,\mathsf{snr}}) - Q(5\Delta\sqrt{2\,\mathsf{snr}})), \tag{6.60}$$

and then, the BEP in (6.26) *as*

$$P_{\mathrm{b}} = \frac{1}{2}\left(P_{\mathrm{b},1} + P_{\mathrm{b},2}\right)$$
$$= \frac{3}{4}Q(\Delta\sqrt{2\,\mathsf{snr}}) + \frac{1}{2}Q(3\Delta\sqrt{2\,\mathsf{snr}}) - \frac{1}{4}Q(5\Delta\sqrt{2\,\mathsf{snr}}). \tag{6.61}$$

As we showed in Example 6.4, the TP need not always be calculated for all pairs of the transmitted symbols. Instead, the regularity and the symmetry of the constellation can be exploited to simplify the calculations. This can also be done for 2D constellation as we show in the following example.

Example 6.5 (BEP for M**PSK)** *As shown in Fig. 6.1 and reproduced in Fig. 6.5, the decision region* $\mathcal{Y}_j = \{\boldsymbol{y} : \xi_{j+1,j}(\boldsymbol{y}) \leq 0 \wedge \xi_{j-1,j}(\boldsymbol{y}) \leq 0\}$ *is a wedge. Moreover, because of the circular symmetry of the constellation, for any symbol pair* $(\boldsymbol{x}_i, \boldsymbol{x}_j)$, $P_{i\to j}$ *depends solely on the absolute value of the difference between their indices via*

$$P_{i\to j} = \begin{cases} P_{|j-i|}, & \text{if } |j-i| \leq M/2 \\ P_{M-|j-i|}, & \text{if } |j-i| > M/2 \end{cases}, \tag{6.62}$$

where the M*PSK-specific TP can be written as*

$$P_l = \tilde{Q}\left(\sqrt{2\,\mathsf{snr}}\frac{\xi_{l+2,l+1}(\boldsymbol{x}_1)}{\|\boldsymbol{d}_{l+2,l+1}\|}, \sqrt{2\,\mathsf{snr}}\frac{\xi_{l,l+1}(\boldsymbol{x}_1)}{\|\boldsymbol{d}_{l,l+1}\|}, \frac{\boldsymbol{d}_{l+2,l+1}^{\mathrm{T}}\boldsymbol{d}_{l,l+1}}{\|\boldsymbol{d}_{l+2,l+1}\|\|\boldsymbol{d}_{l,l+1}\|}\right), \quad l = 1, \ldots, M/2. \tag{6.63}$$

From the trigonometric relationships (5.132) *and* (5.133) *we obtain*

$$P_l = \tilde{Q}\left(-\sqrt{2\,\mathsf{snr}}\,S_{M,2l+1}, \sqrt{2\,\mathsf{snr}}\,S_{M,2l-1}, -\cos\frac{2\pi}{M}\right). \tag{6.64}$$

The expressions we derived in the previous example are entirely general; yet, particular cases of MPSK constellations lead to simpler results that do not need bivariate Q-functions or, in fact, exploit the simplicity of the particular cases we analyzed in (6.48) and (6.49). We show this in the following example.

Example 6.6 (BEP for 8PSK **Labeled by the BRGC)** *In Fig. 6.6 we show the decision regions* $\mathcal{Y}_{k,b}$ *for an* 8PSK *constellation labeled by the BRGC for* $k = 1$ *and* $k = 3$. *Owing to symmetries of the constellation and labeling, the form of* $\mathcal{Y}_{2,b}$ *is the same as* $\mathcal{Y}_{1,b}$.

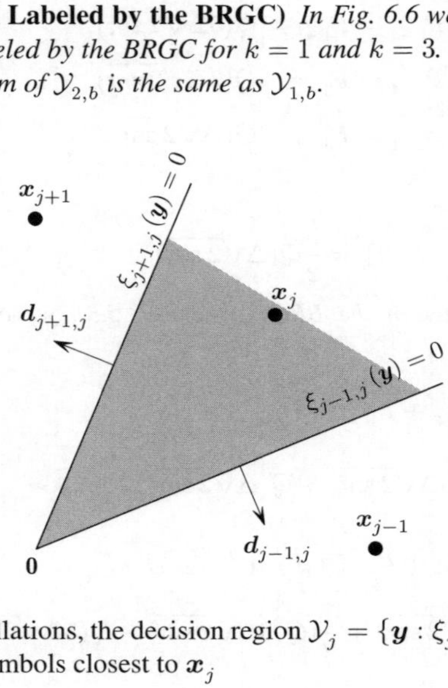

Figure 6.5 In MPSK constellations, the decision region $\mathcal{Y}_j = \{\boldsymbol{y} : \xi_{j+1,j}(\boldsymbol{y}) \leq 0 \wedge \xi_{j-1,j}(\boldsymbol{y}) \leq 0\}$ is a wedge defined by the two symbols closest to $\boldsymbol{x}_j$

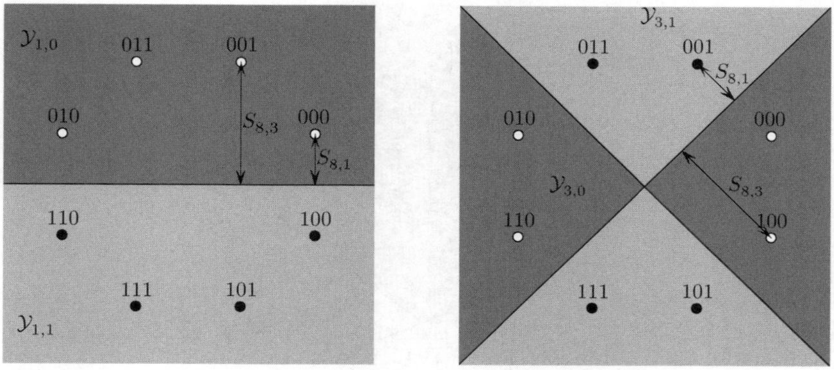

(a) (b)

Figure 6.6 Decision regions $\mathcal{Y}_{k,b}$ for an 8PSK constellation labeled by the BRGC: (a) $k = 1$ and (b) $k = 3$

For $k = 1$ the decision regions $\mathcal{Y}_{1,b}$ are half-planes so only 1D integration is needed. In particular, for $i = 1, 2, 3, 4$, adding the terms in (6.27) *we obtain*

$$\sum_{j \in \mathcal{S}_{1,1}} P_{i \to j} = \Pr\{\boldsymbol{Y} \in \mathcal{Y}_{1,1} | \boldsymbol{X} = \boldsymbol{x}_i\}.$$

Similarly, for $i = 5, 6, 7, 8$,

$$\sum_{j \in \mathcal{S}_{1,0}} P_{i \to j} = \Pr\{\boldsymbol{Y} \in \mathcal{Y}_{1,1} | \boldsymbol{X} = \boldsymbol{x}_i\}.$$

In all these cases, we have only one linear form $\mathsf{L}_1(\boldsymbol{y}) = 0$, shown in Fig. 6.6 (a), and thus, we can use (6.32) *to obtain*

$$P_{\mathrm{b},1} = \frac{1}{2}\left(Q(S_{8,1}\sqrt{2\,\mathsf{snr}}) + Q(S_{8,3}\sqrt{2\,\mathsf{snr}})\right), \tag{6.65}$$

where $S_{8,1}$ and $S_{8,3}$ are distances from the symbol to the frontier of the decision region, as shown in Fig. 6.6 (a). Because of the symmetries of the constellation and labeling, we have $P_{\mathrm{b},1} = P_{\mathrm{b},2}$.

For $k = 3$, the lines defining the bit decision regions $\mathcal{Y}_{3,b}$ are orthogonal so we use (6.48) *to obtain*

$$P_{\mathrm{b},3} = Q(S_{8,1}\sqrt{2\,\mathsf{snr}}) + Q(S_{8,3}\sqrt{2\,\mathsf{snr}}) - 2Q(S_{8,1}\sqrt{2\,\mathsf{snr}})Q(S_{8,3}\sqrt{2\,\mathsf{snr}}). \tag{6.66}$$

The final expression for the BEP for an 8PSK *labeled by the BRGC is then obtained as*

$$P_{\mathrm{b}} = \frac{2}{3}\left(Q(S_{8,1}\sqrt{2\,\mathsf{snr}} + Q(S_{8,3}\sqrt{2\,\mathsf{snr}}) - Q(S_{8,1}\sqrt{2\,\mathsf{snr}})Q(S_{8,3}\sqrt{2\,\mathsf{snr}})\right). \tag{6.67}$$

In Example 6.5 we were able to express $P_{i \to j}$ for MPSK constellation using bivariate Q-functions. Our objective now is to do the same in the case of an arbitrary 2D constellation. First, we assume that $\mathcal{Y}_j$ is a closed polygon defined by T_j linear forms $\mathsf{L}_t(\boldsymbol{y}) = 0$ corresponding to the polygon's sides. We assume that the linear pieces are enumerated counter-clockwise, as shown in Fig. 6.7. In such a case, using the complementary wedge we defined in (6.35), the whole space can be expressed as a union of disjoint sets, i.e.,

$$\mathbb{R}^N = \mathcal{Y}_j \cup \mathcal{W}(\overline{\mathsf{L}}_1, \mathsf{L}_{T_j}) \cup \bigcup_{t=1}^{T_j - 1} \mathcal{W}(\overline{\mathsf{L}}_{t+1}, \mathsf{L}_t). \tag{6.68}$$

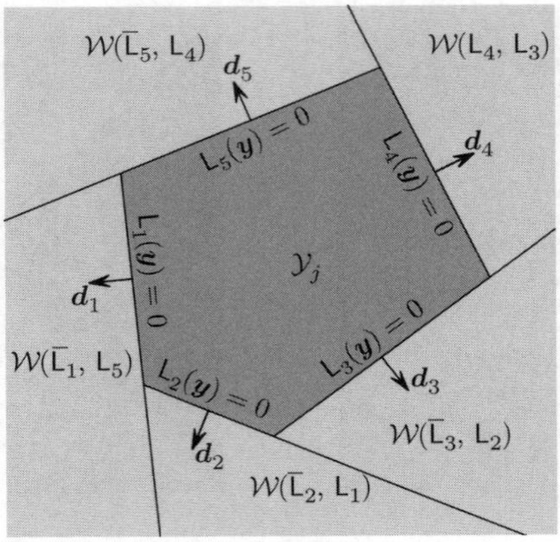

Figure 6.7 If the decision region $\mathcal{Y}_j$ is a polygon delimited by the lines $\mathsf{L}_t(\boldsymbol{y})$, the observation space can be represented as a union of $\mathcal{Y}_j$ and complementary wedges

The TP in (6.12) can then be expressed as

$$\begin{aligned} P_{i\to j} &= \Pr\left\{\boldsymbol{Y} \in \mathcal{Y}_j | \boldsymbol{X} = \boldsymbol{x}_i\right\} \\ &= 1 - \Pr\left\{\boldsymbol{Y} \notin \mathcal{Y}_j | \boldsymbol{X} = \boldsymbol{x}_i\right\} \\ &= 1 - \mathsf{T}(\overline{\mathsf{L}}_1, \mathsf{L}_{T_j} | \boldsymbol{x}_i) - \sum_{t=1}^{T_j-1} \mathsf{T}(\overline{\mathsf{L}}_{t+1}, \mathsf{L}_t | \boldsymbol{x}_i), \end{aligned} \tag{6.69}$$

where $\mathsf{T}(\overline{\mathsf{L}}_t, \mathsf{L}_{t'} | \boldsymbol{x}_i)$ is given by (6.46).

Similarly, if $\mathcal{Y}_j$ is an "infinite" polygon as shown in Fig. 6.8, we can write

$$\mathcal{W}(\mathsf{L}_1, \mathsf{L}_2) = \mathcal{Y}_j \cup \bigcup_{t=2}^{T_j-1} \mathcal{W}(\overline{\mathsf{L}}_{t+1}, \mathsf{L}_t), \tag{6.70}$$

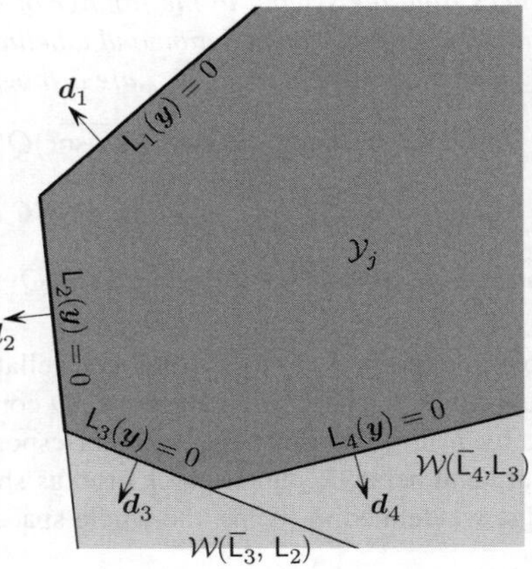

Figure 6.8 If the decision region is an infinite polygon, the TP can be expressed using a sum of integrals over wedges and complementary wedges

and then,

$$P_{i\to j} = \mathsf{T}(\mathsf{L}_1, \mathsf{L}_2|\boldsymbol{x}_i) - \sum_{t=2}^{T_j-1} \mathsf{T}(\overline{\mathsf{L}}_{t+1}, \mathsf{L}_t|\boldsymbol{x}_i). \tag{6.71}$$

We conclude then that the TPs for any 2D constellation can be calculated using only functions $\mathsf{T}(\cdot,\cdot|\boldsymbol{x}_i)$, which as we showed before, are expressed in terms of bivariate Q-function. These bivariate Q-functions now play the same role that the Q-function has when calculating $P_{i\to j}$ for 1D constellations.

6.1.3 *Bounding Techniques*

The approach we used to evaluate the TP in 2D constellations can be extended to the general N-dimensional case. However, the geometric considerations get more tedious as the dimension of the integration space increases. Therefore, simple approximations are often used, which may also be sufficient for the purpose of the analysis and/or design. A popular approach relies on finding an upper bound on the TP $P_{i\to j}$, which we do in the following.

For a given j, and for any $l \in \mathcal{I}$ with $l \neq j$, we have

$$\mathcal{Y}_j \subset \{\boldsymbol{y} : \xi_{l,j}(\boldsymbol{y}) \leq 0\}, \tag{6.72}$$

which follows from the fact that, by using one inequality, we relax the constraint imposed by T_j inequalities defining $\mathcal{Y}_j$. The TP is therefore upper bounded, for any i, as

$$P_{i\to j} \leq \Pr\left\{\xi_{l,j}(\boldsymbol{Y}) \leq 0 | \boldsymbol{X} = \boldsymbol{x}_i\right\}. \tag{6.73}$$

Setting $l = i$ we obtain the bound

$$P_{i\to j} \leq \mathrm{UB}^{\mathrm{PEP}}_{i\to j}, \tag{6.74}$$

where

$$\mathrm{UB}^{\mathrm{PEP}}_{i\to j} \triangleq \Pr\left\{\xi_{i,j}(\boldsymbol{Y}) \leq 0 | \boldsymbol{X} = \boldsymbol{x}_i\right\}. \tag{6.75}$$

It is easy to see that

$$\mathrm{UB}^{\mathrm{PEP}}_{i\to j} = \Pr\left\{p_{\boldsymbol{Y}|\boldsymbol{X}}(\boldsymbol{Y}|\boldsymbol{x}_j) \geq p_{\boldsymbol{Y}|\boldsymbol{X}}(\boldsymbol{Y}|\boldsymbol{x}_i) | \boldsymbol{X} = \boldsymbol{x}_i\right\} \tag{6.76}$$

is the PEP, i.e., the probability that after transmitting $\boldsymbol{X} = \boldsymbol{x}_i$, the likelihood $p_{\boldsymbol{Y}|\boldsymbol{X}}(\boldsymbol{y}|\boldsymbol{x}_j)$ is larger than the likelihood $p_{\boldsymbol{Y}|\boldsymbol{X}}(\boldsymbol{y}|\boldsymbol{x}_i)$ (or equivalently, that the received signal $\boldsymbol{Y}$ is closer to $\boldsymbol{x}_j$ than to $\boldsymbol{x}_i$). We can calculate (6.75) as

$$\mathrm{UB}^{\mathrm{PEP}}_{i\to j} = \Pr\left\{\boldsymbol{d}_{i,j}^{\mathrm{T}}(\boldsymbol{Z} + \boldsymbol{x}_i - \overline{\boldsymbol{x}}_{i,j}) \leq 0\right\} \tag{6.77}$$

$$= \Pr\left\{\boldsymbol{d}_{i,j}^{\mathrm{T}}\left(\boldsymbol{Z} + \frac{1}{2}\boldsymbol{d}_{i,j}\right) \leq 0\right\} \tag{6.78}$$

$$= Q\left(\frac{\sqrt{2\,\mathsf{snr}}\,\|\boldsymbol{d}_{i,j}\|}{2}\right). \tag{6.79}$$

To tighten the bound in (6.74), we note that (6.73) is true for any l, so a better bound is obtained via

$$P_{i\to j} \leq \mathrm{UB}^{\mathrm{1D}}_{i\to j}, \tag{6.80}$$

where

$$\mathrm{UB}^{\mathrm{1D}}_{i\to j} \triangleq \min_{l\in\mathcal{I},l\neq j} \left\{\Pr\left\{\xi_{l,j}(\boldsymbol{Y})\leq 0|\boldsymbol{X}=\boldsymbol{x}_i\right\}\right\} \tag{6.81}$$

$$= Q\left(\sqrt{2\,\mathsf{snr}}\,\frac{\xi_{\hat{l},j}(\boldsymbol{x}_i)}{\|\boldsymbol{d}_{\hat{l},j}\|}\right), \tag{6.82}$$

with

$$\hat{l} = \operatorname*{argmax}_{l\in\mathcal{I},l\neq j}\left\{\frac{\xi_{l,j}(\boldsymbol{x}_i)}{\|\boldsymbol{d}_{l,j}\|}\right\}. \tag{6.83}$$

In other words, we find the index l of the linear form $\xi_{l,j}(\boldsymbol{y})$ defining the Voronoi region $\mathcal{Y}_j$ such that $\xi_{l,j}(\boldsymbol{x}_i) > 0$, so as to maximize the distance between the half-space $\{\boldsymbol{y} : \xi_{l,j}(\boldsymbol{y}) \leq 0\}$ and $\boldsymbol{x}_i$, and next, we calculate the probability that, conditioned on $\boldsymbol{X} = \boldsymbol{x}_i$, the observation $\boldsymbol{Y}$ falls into the half-space we found, i.e., that the 1D projection of the zero-mean Gaussian random variable $\boldsymbol{Z}$ falls in the interval $[\xi_{l,j}(\boldsymbol{x}_i)/\|\boldsymbol{d}_{l,j}\|, \infty)$.

To tighten the bound even further, instead of finding the half-space that contains $\mathcal{Y}_j$, we can consider finding the wedge containing $\mathcal{Y}_j$, i.e., we generalize the bound in (6.80) as

$$P_{i\to j} \leq \mathrm{UB}^{\mathrm{2D}}_{i\to j}, \tag{6.84}$$

where

$$\mathrm{UB}^{\mathrm{2D}}_{i\to j} \triangleq \min_{l,l'\in\mathcal{I},\ l,l'\neq j} \left\{\Pr\left\{\boldsymbol{Y}\in\mathcal{W}(\xi_{l,j},\xi_{l',j})|\boldsymbol{X}=\boldsymbol{x}_i\right\}\right\} \tag{6.85}$$

$$= \tilde{Q}\left(\sqrt{2\,\mathsf{snr}}\frac{\xi_{\hat{l},j}(\boldsymbol{x}_i)}{\|\boldsymbol{d}_{\hat{l},j}\|}, \sqrt{2\,\mathsf{snr}}\frac{\xi_{\hat{l}',j}(\boldsymbol{x}_i)}{\|\boldsymbol{d}_{\hat{l}',j}\|}, \frac{\boldsymbol{d}^{\mathrm{T}}_{\hat{l},j}\boldsymbol{d}_{\hat{l}',j}}{\|\boldsymbol{d}_{\hat{l},j}\|\|\boldsymbol{d}_{\hat{l}',j}\|}\right), \tag{6.86}$$

where $\hat{l}$ and $\hat{l}'$ are the indices minimizing (6.85).[5]

It is easy to see that the three bounds presented above satisfy

$$P_{i\to j} \leq \mathrm{UB}^{\mathrm{2D}}_{i\to j} \leq \mathrm{UB}^{\mathrm{1D}}_{i\to j} \leq \mathrm{UB}^{\mathrm{PEP}}_{i\to j}. \tag{6.87}$$

Among the three bounds, the PEP-based bound in (6.74) is the simplest one. Not surprisingly, however, with decreasing implementation complexity, the accuracy of the approximation also decreases. On the other hand, the two bounds in (6.81) and (6.85) are tighter, but *algorithmic*, i.e., they require optimizations to find the relevant linear forms. The advantage of (6.81) and (6.85) is that they can be used for any N-dimensional constellation, but at the same time, they are based on 1D or 2D integrals (and not on N-dimensional ones).

Example 6.7 (BEP Bounds for 4PAM**)** *We continue here with the case analyzed in Example 6.4. For the bound in* (6.84)*, we note that the Voronoi regions of the symbol $\boldsymbol{x}_j$ are defined by at most two inequalities, and thus, $\hat{l}$ and $\hat{l}'$ in* (6.86) *are simply the borders of the Voronoi region of the symbol $\boldsymbol{x}_j$. We therefore conclude that in this case, the bound in* (6.84) *holds with equality, i.e., $P_{i\to j} = \mathrm{UB}^{\mathrm{2D}}_{i\to j}$. For the bound in* (6.80)*, we note that $\hat{l}$ in* (6.83) *corresponds to the border of the Voronoi region of $\boldsymbol{x}_j$ farthest apart from $\boldsymbol{x}_i$. This distance is given by $(2|i-j|-1)\Delta$, and thus,*

$$\mathrm{UB}^{\mathrm{1D}}_{i\to j} = Q\left((2|i-j|-1)\Delta\sqrt{2\,\mathsf{snr}}\right). \tag{6.88}$$

[5] Note that we used the wedge notation $\mathcal{W}(\xi_{l,j},\xi_{l',j}) = \{\boldsymbol{y} : \xi_{l,j}(\boldsymbol{y}) \leq 0 \wedge \xi_{l',j}(\boldsymbol{y}) \leq 0\}$.

Finally, for the bound in (6.74), *we have that* $\|\boldsymbol{d}_{i,j}\|/2 = |j-i|\Delta$, *and thus,* (6.79) *gives*

$$\mathrm{UB}_{i\to j}^{\mathrm{PEP}} = Q\left(|j-i|\Delta\sqrt{2\,\mathsf{snr}}\right). \tag{6.89}$$

To obtain bounds on the average BEP, we replace $P_{i\to j}$ *in* (6.57) *and* (6.59) *by* (6.88) *and* (6.89) *to obtain*

$$P_{\mathrm{b}} \leq \frac{3}{4}Q(\Delta\sqrt{2\,\mathsf{snr}}) + Q(3\Delta\sqrt{2\,\mathsf{snr}}) + \frac{1}{4}Q(5\Delta\sqrt{2\,\mathsf{snr}}), \tag{6.90}$$

and

$$P_{\mathrm{b}} \leq \frac{3}{4}Q(\Delta\sqrt{2\,\mathsf{snr}}) + Q(2\Delta\sqrt{2\,\mathsf{snr}}) + \frac{1}{4}Q(3\Delta\sqrt{2\,\mathsf{snr}}), \tag{6.91}$$

for the 1D and PEP-based bounds, respectively.

It is immediately seen that although the bounds (6.90) *and* (6.91) *are different, they approach the exact BEP* P_{b} *in* (6.61) *for large values of* snr. *This is because the BEP at high SNR is dominated by the Q-functions with the smallest argument, which is the same in all the considered cases.*

Example 6.8 (Bounds for M**PSK)** *In the case of an* MPSK *constellation, the TP* $P_{i\to j}$ *depends only on the distance between the symbols. This is reflected in the notation where we replaced* $i \longrightarrow j$ *by the value corresponding to the difference* $l = |j-i|$ *(see* (6.62)*). The distance between symbols is given by* $2S_{M,l}$ *(see* (5.135)*), so the PEP-based bound is obtained as*

$$\mathrm{UB}_{l}^{\mathrm{PEP}} = Q(\sqrt{2\,\mathsf{snr}}\,S_{M,l}). \tag{6.92}$$

Because the decision regions of an MPSK *constellation are defined by two linear constraints, the 2D bound* (6.85) *will produce the same result as the exact evaluation* (6.64), *i.e.,* $\mathrm{UB}_{l}^{\mathrm{2D}} = P_{l}$.

The 1D bound (6.81) *is found as*

$$\mathrm{UB}_{l}^{\mathrm{1D}} = Q(S_{M,\hat{t}+l-1}), \tag{6.93}$$

where

$$\hat{t} = \underset{1\leq t\leq M}{\mathrm{argmin}}\left\{S_{M,t-i+l}\ \mathrm{sign}(S_{M,t+i-l})\right\}, \tag{6.94}$$

which can be shown to be

$$\mathrm{UB}_{l}^{\mathrm{1D}} = Q\left(\sqrt{2\,\mathsf{snr}}\frac{\xi_{\hat{t},l+i}(\boldsymbol{x}_i)}{\|\boldsymbol{d}_{\hat{t},l+i}\|}\right) \tag{6.95}$$

$$= \begin{cases} Q(\sqrt{2\,\mathsf{snr}}\,S_{M,2l-1}), & \text{if } 1\leq l\leq M/4 \\ Q(\sqrt{2\,\mathsf{snr}}), & \text{if } M/4 < l \leq M/2 \end{cases}. \tag{6.96}$$

Comparing (6.96) *with* (6.92) *we see that* $\mathrm{UB}_{l}^{\mathrm{1D}} = \mathrm{UB}_{l}^{\mathrm{PEP}}$ *for* $l=1$ *and* $l = M/2$ *but, for* $1 < l < M/2$ $\mathrm{UB}_{l}^{\mathrm{1D}} < \mathrm{UB}_{l}^{\mathrm{PEP}}$. *Fig. 6.9 compares the bounds and the exact evaluation, showing that* $\mathrm{UB}_{1}^{\mathrm{1D}} = \mathrm{UB}_{1}^{\mathrm{PEP}}$ *tends to the exact value* P_1 *for high SNR. This expression is then useful to evaluate the probability of wrongly detecting the closest symbol. Moreover,* $\mathrm{UB}_{2}^{\mathrm{1D}}$ *also provides a tight approximation of* P_2 *whereas* $\mathrm{UB}_{2}^{\mathrm{PEP}}$ *fails to do so. More generally, the approximation by* $\mathrm{UB}_{l}^{\mathrm{1D}}$ *will tend to* P_l *for any* $1 < l \leq M/4$.

6.1.4 Fading Channels

The previous analysis is valid for the case of nonfading channel, i.e., for transmission with fixed SNR. Although we did not make it explicit in the notation, the BEP depends on the SNR, i.e., $P_{\mathrm{b}} = P_{\mathrm{b}}(\mathsf{snr})$.

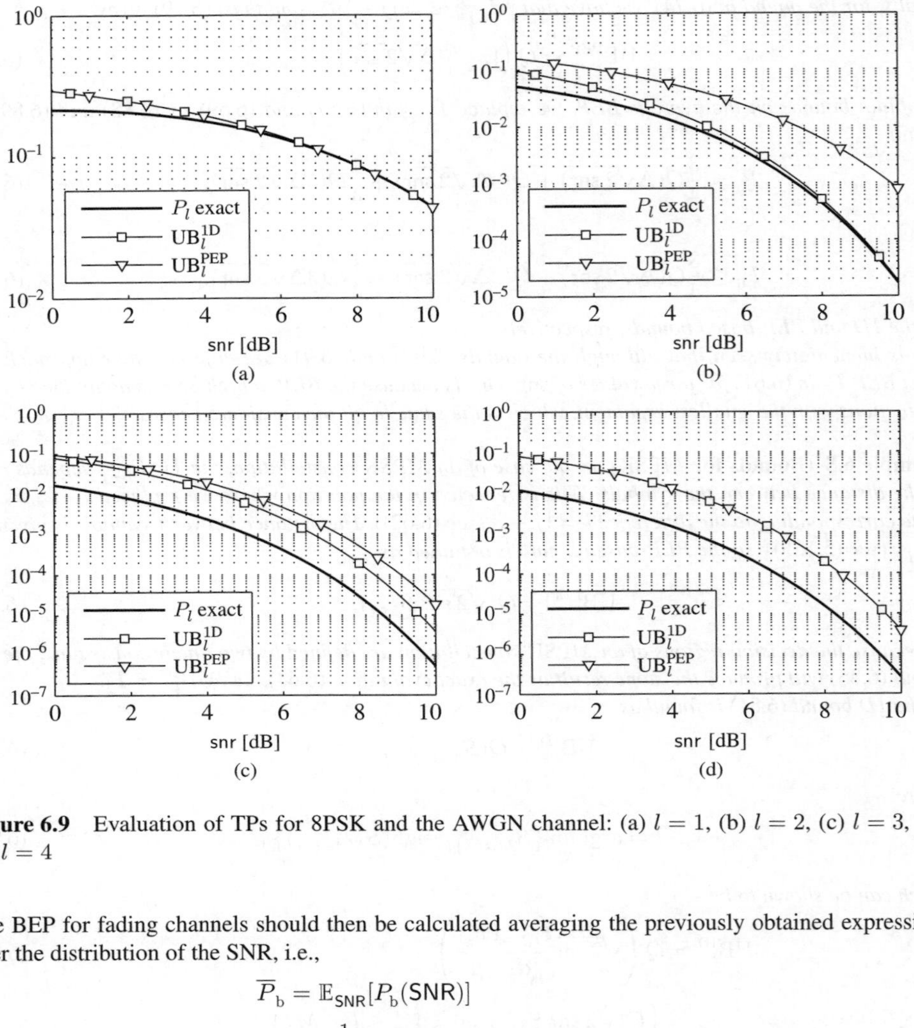

Figure 6.9 Evaluation of TPs for 8PSK and the AWGN channel: (a) $l = 1$, (b) $l = 2$, (c) $l = 3$, and (d) $l = 4$

The BEP for fading channels should then be calculated averaging the previously obtained expressions over the distribution of the SNR, i.e.,

$$\begin{aligned}\overline{P}_{\mathrm{b}} &= \mathbb{E}_{\mathsf{SNR}}[P_{\mathrm{b}}(\mathsf{SNR})]\\ &= \frac{1}{mM}\sum_{i\in\mathcal{I}}\sum_{j\in\mathcal{I}} d_{\mathrm{H}}(\boldsymbol{q}_i, \boldsymbol{q}_j)\overline{P}_{i\to j},\end{aligned} \tag{6.97}$$

where

$$\overline{P}_{i\to j} = \mathbb{E}_{\mathsf{SNR}}[P_{i\to j}(\mathsf{SNR})]. \tag{6.98}$$

In the previous sections we have shown that the TP $P_{i\to j}(\mathsf{snr})$ can always be expressed via Q-functions $Q(a\sqrt{\mathsf{snr}})$ or via bivariate Q-functions $\tilde{Q}(a\sqrt{\mathsf{snr}}, b\sqrt{\mathsf{snr}}, \rho)$. But because $Q(a\sqrt{\mathsf{snr}}) = \tilde{Q}(a\sqrt{\mathsf{snr}}, a\sqrt{\mathsf{snr}}, 1)$, to obtain (6.98) it is enough to find $\mathbb{E}_{\mathsf{SNR}}[Q(a\sqrt{\mathsf{SNR}}, b\sqrt{\mathsf{SNR}}, \rho)]$. In order to calculate the latter we will exploit the following alternative form of the bivariate Q-function (2.11) valid for any $a, b \geq 0$

$$\begin{aligned}\tilde{Q}(a, b, \rho) &= \frac{1}{2\pi}\int_0^{\vartheta(a,b,\rho)} \exp\left(-\frac{a^2}{2\sin^2(\theta)}\right)\mathrm{d}\theta\\ &\quad + \frac{1}{2\pi}\int_0^{\vartheta(b,a,\rho)} \exp\left(-\frac{b^2}{2\sin^2(\theta)}\right)\mathrm{d}\theta,\end{aligned} \tag{6.99}$$

where

$$\vartheta(a, b, \rho) \triangleq \begin{cases} \tan^{-1}\left(\sqrt{\frac{1+\rho}{1-\rho}}\right), & \text{if } a = 0, b = 0 \\ \tan_{\pi}^{-1}\left(\frac{a\sqrt{1-\rho^2}}{b - \rho a}\right), & \text{otherwise} \end{cases} \tag{6.100}$$

and

$$\tan_{\pi}^{-1}(x) \triangleq \begin{cases} \tan^{-1}(x), & \text{if } x \geq 0 \\ \tan^{-1}(x) + \pi, & \text{if } x < 0 \end{cases}. \tag{6.101}$$

The expression in (6.99) can be used for any $a, b \geq 0$; however, some of the expressions we developed for the TP have negative arguments. For these cases, we use the identity

$$\tilde{Q}(a, b, \rho) = \begin{cases} Q(a) - \tilde{Q}(a, -b, -\rho), & \text{if } a \geq 0, b < 0 \\ Q(b) - \tilde{Q}(-a, b, -\rho), & \text{if } b \geq 0, a < 0 \\ Q(a) - Q(-b) + \tilde{Q}(-a, -b, \rho), & \text{if } a < 0, b < 0 \end{cases}. \tag{6.102}$$

The results in (6.102) show that to evaluate (6.98), it is enough to consider the case $a, b \geq 0$. The following theorem gives a general expression for the expectation $\mathbb{E}_{\mathsf{SNR}}[\tilde{Q}(a\sqrt{\mathsf{SNR}}, b\sqrt{\mathsf{SNR}}, \rho)]$ for Nakagami-m fading channels.

Theorem 6.9 *For Nakagami-*m *fading with integer* m *and any* $a, b \geq 0$,

$$\mathbb{E}_{\mathsf{SNR}}[\tilde{Q}(a\sqrt{\mathsf{SNR}}, b\sqrt{\mathsf{SNR}}, \rho)] = A_{\mathsf{m}}\left(\frac{a^2\,\overline{\mathsf{snr}}}{2\mathsf{m}}, \vartheta(a, b, \rho)\right) + A_{\mathsf{m}}\left(\frac{b^2\,\overline{\mathsf{snr}}}{2\mathsf{m}}, \vartheta(b, a, \rho)\right), \tag{6.103}$$

$$\mathbb{E}_{\mathsf{SNR}}[Q(a\sqrt{\mathsf{SNR}})] = 2A_{\mathsf{m}}\left(\frac{a^2\,\overline{\mathsf{snr}}}{2\mathsf{m}}, \frac{\pi}{2}\right), \tag{6.104}$$

where $\vartheta(b, a, \rho)$ *is given by* (6.100) *and*

$$A_{\mathsf{m}}(c, \phi) \triangleq \frac{1}{2\pi}\int_0^{\phi}\left(\frac{\sin^2\theta}{\sin^2\theta + c}\right)^{\mathsf{m}} \mathrm{d}\theta. \tag{6.105}$$

Proof: Using (6.99) yields

$$\begin{aligned} \mathbb{E}_{\mathsf{SNR}}[\tilde{Q}(a\sqrt{\mathsf{SNR}}, b\sqrt{\mathsf{SNR}}, \rho)] &= \frac{1}{2\pi}\mathbb{E}_{\mathsf{SNR}}\left[\int_0^{\vartheta(a,b,\rho)} \exp\left(-\frac{a^2\mathsf{SNR}}{2\sin^2(\theta)}\right)\mathrm{d}\theta\right] \\ &\quad + \frac{1}{2\pi}\mathbb{E}_{\mathsf{SNR}}\left[\int_0^{\vartheta(b,a,\rho)} \exp\left(-\frac{b^2\mathsf{SNR}}{2\sin^2(\theta)}\right)\mathrm{d}\theta\right] \end{aligned} \tag{6.106}$$

$$\begin{aligned} &= \frac{1}{2\pi}\int_0^{\vartheta(a,b,\rho)} \mathsf{P}_{\mathsf{SNR}}\left(-\frac{a^2}{2\sin^2(\theta)}\right)\mathrm{d}\theta \\ &\quad + \frac{1}{2\pi}\int_0^{\vartheta(b,a,\rho)} \mathsf{P}_{\mathsf{SNR}}\left(-\frac{b^2}{2\sin^2(\theta)}\right)\mathrm{d}\theta, \end{aligned} \tag{6.107}$$

where to pass from (6.106) to (6.107) we changed the integration order and used the *moment-generating function* (MGF) of a random variable defined in (2.6). Further, using the form of the MGF of SNR in Nakagami-m fading in (6.107) we obtain (6.105). To obtain (6.104) we use the well-known alternative form of the Q-function[6]

$$\tilde{Q}(a, a, 1) = Q(a) = \frac{1}{\pi}\int_0^{\pi/2} \exp\left(-\frac{a^2}{2\sin^2(\theta)}\right)\mathrm{d}\theta. \tag{6.108}$$

□

[6] Which can be derived from (6.99) setting $a = b$ and $\rho \to 1$.

The main challenge now consists in calculating $A_{\mathsf{m}}(c,\phi)$. In what follows we provide a closed-form expression without giving a proof. Such a proof can be found in the references we cite in Section 6.5. The closed-form expression for $A_{\mathsf{m}}(c,\phi)$ and any $\mathsf{m} \in \mathbb{N}_+$ is

$$A_{\mathsf{m}}(c,\phi) \triangleq \frac{\phi}{2\pi} + \sum_{n=1}^{\mathsf{m}} \frac{(-1)^n}{2\pi} \sum_{r=0}^{n-1} \frac{c^{n-r-\frac{1}{2}}}{(1+c)^{n-\frac{1}{2}}} \binom{\mathsf{m}}{n} \binom{n-1}{r} T_{\mathrm{A}}\left(\sqrt{\frac{1+c}{c}}\tan\phi, 1+r\right), \tag{6.109}$$

where

$$T_{\mathrm{A}}(q,n) \triangleq \frac{(2n-1)!!}{(n-1)!} \left(\frac{\tan^{-1}(q)}{2^{n-1}} + \sum_{k=1}^{n-1} \frac{q(n-k-1)!}{(1+q^2)^{n-1} 2^k (2n-2k-1)!!} \right), \tag{6.110}$$

$(2n-1)!! \triangleq 1 \cdot 3 \cdot \ldots \cdot (2n-5) \cdot (2n-1)$ for $n \geq 1$ and $(-1)!! \triangleq 1$.

Example 6.10 (BEP for 2PAM in Rayleigh Fading) *From Example 6.3 we know that for 2PAM, the BEP is given by*

$$P_{\mathrm{b}}(\mathsf{snr}) = Q(\sqrt{2\,\mathsf{snr}}). \tag{6.111}$$

The somewhat lengthy formula of (6.109) *simplifies in Rayleigh fading (i.e.,* $\mathsf{m} = 1$*), as follows:*

$$A_1(c,\phi) = \frac{1}{2\pi}\left(\phi - \sqrt{\frac{c}{c+1}} \tan_{\pi}^{-1}\left(\sqrt{\frac{c+1}{c}}\tan(\phi)\right)\right). \tag{6.112}$$

Then

$$\mathbb{E}_{\mathsf{SNR}}[Q(a\sqrt{\mathsf{SNR}})] = 2A_1\left(\frac{a^2\,\overline{\mathsf{snr}}}{2\mathsf{m}}, \frac{\pi}{2}\right) = \frac{1}{2}\left(1 - \sqrt{\frac{a^2\,\overline{\mathsf{snr}}}{a^2\,\overline{\mathsf{snr}}+2}}\right) \tag{6.113}$$

and thus, we obtain the well-known expression

$$\overline{P}_{\mathrm{b}} = \mathbb{E}_{\mathsf{SNR}}\left[Q(\sqrt{2\,\mathsf{SNR}})\right] = \frac{1}{2}\left(1 - \sqrt{\frac{\overline{\mathsf{snr}}}{\overline{\mathsf{snr}}+1}}\right), \tag{6.114}$$

which for $\overline{\mathsf{snr}} \to \infty$ *can be approximated as*[7]

$$\overline{P}_{\mathrm{b}} \approx \frac{1}{4}\frac{1}{\overline{\mathsf{snr}}}. \tag{6.115}$$

Example 6.11 (TP for MPSK in Rayleigh Fading) *To evaluate the TP for MPSK constellation in Rayleigh fading channel, we rewrite* (6.64) *as*

$$P_l = \tilde{Q}\left(-\sqrt{2\,\mathsf{snr}}\,S_{M,2l+1}, \sqrt{2\,\mathsf{snr}}\,S_{M,2l-1}, -\cos\frac{2\pi}{M}\right) \tag{6.116}$$

$$= \begin{cases} Q\left(\sqrt{2\,\mathsf{snr}}\,S_{M,2l-1}\right) - \tilde{Q}\left(\sqrt{2\,\mathsf{snr}}\,S_{M,2l+1}, \sqrt{2\,\mathsf{snr}}\,S_{M,2l-1}, \cos\frac{2\pi}{M}\right), & \text{if } l = 1,\ldots,\frac{M}{2}-1 \\ \tilde{Q}\left(\sqrt{2\,\mathsf{snr}}\,S_{M,1}, \sqrt{2\,\mathsf{snr}}\,S_{M,1}, -\cos\frac{2\pi}{M}\right), & \text{if } l = \frac{M}{2} \\ 1 - 2Q\left(\sqrt{2\,\mathsf{snr}}\,S_{M,1}\right) + \tilde{Q}\left(\sqrt{2\,\mathsf{snr}}\,S_{M,1}, \sqrt{2\,\mathsf{snr}}\,S_{M,1}, -\cos\frac{2\pi}{M}\right), & \text{if } l = 0 \end{cases}, \tag{6.117}$$

so the bivariate Q-function is used only for positive values of the first two arguments.

[7] Treating $\overline{P}_{\mathrm{b}}$ as a function of $x = 1/\overline{\mathsf{snr}}$ and computing a Taylor series expansion around $x = 0$.

We can then proceed as follows:

$$\overline{P}_l = \begin{cases} \frac{1}{2}\left(1-\sqrt{\frac{S_{M,2l-1}^2\overline{\mathsf{snr}}}{S_{M,2l-1}^2\overline{\mathsf{snr}}+1}}\right) - A_1(S_{M,2l+1}^2\overline{\mathsf{snr}}, \vartheta_{2l+1,2l-1}^{+}) & \\ \quad -A_1(S_{M,2l-1}^2\overline{\mathsf{snr}}, \vartheta_{2l-1,2l+1}^{+}), & \text{if } l = 1,\ldots,\frac{M}{2}-1 \\ 2A_1(S_{M,1}^2\overline{\mathsf{snr}}, \vartheta_{1,1}^{-}), & \text{if } l = \frac{M}{2} \\ \sqrt{\frac{S_{M,1}^2\overline{\mathsf{snr}}}{S_{M,1}^2\overline{\mathsf{snr}}+1}} + 2A_1(S_{M,1}^2\overline{\mathsf{snr}}, \vartheta_{1,1}^{-}), & \text{if } l = 0 \end{cases}, \quad (6.118)$$

where $\vartheta_{l,t}^{+} \triangleq \vartheta(S_{M,l}, S_{M,t}, \cos 2\pi/M)$ *and* $\vartheta_{l,t}^{-} \triangleq \vartheta(S_{M,l}, S_{M,t}, -\cos 2\pi/M)$.

The bounds are much simpler to calculate. In particular, taking expectation over (6.92) *and* (6.96) *we obtain*

$$\overline{\mathrm{UB}}_l^{\mathrm{PEP}} = \frac{1}{2}\left(1-\sqrt{\frac{S_{M,l}^2\overline{\mathsf{snr}}}{S_{M,l}^2\overline{\mathsf{snr}}+1}}\right) \quad (6.119)$$

and

$$\overline{\mathrm{UB}}_l^{\mathrm{1D}} = \begin{cases} \frac{1}{2}\left(1-\sqrt{\frac{S_{M,2l-1}^2\overline{\mathsf{snr}}}{S_{M,2l-1}^2\overline{\mathsf{snr}}+1}}\right), & \text{if } 0 < l \le M/4 \\ \frac{1}{2}\left(1-\sqrt{\frac{\overline{\mathsf{snr}}}{\overline{\mathsf{snr}}+1}}\right), & \text{if } M/4 < l \le M/2 \end{cases}. \quad (6.120)$$

The results are compared in Fig. 6.10. The simplicity of the bounding techniques is the price to pay for the gap in the results: while the bounding techniques provide a tight estimate of the TP for high SNR in nonfading channels (cf. Example 6.8), expectations over these bounds are affected by the differences appearing for small values of SNR (see Fig. 6.9).

Example 6.12 (8PSK **Labeled by the BRGC in Rayleigh Fading Channel)** *The results in Example 6.11 may seem discouraging when looking at the TP. However, including the effect of the labeling may significantly change the picture. In particular, considering the case treated in Example 6.6, i.e., assuming an* 8PSK *constellation labeled by the BRGC, we immediately see that, owing to the form of the decision regions* $\mathcal{Y}_{1,b}$ *and* $\mathcal{Y}_{2,b}$ *(see Fig. 6.6) we can exactly evaluate the BEP for* $k = 1, 2$ *using* (6.65) *and* (6.113)*, i.e.,*

$$\overline{P}_{\mathrm{b},k} = \overline{\mathrm{UB}}_{\mathrm{b},k}^{\mathrm{1D}} = \frac{1}{4}\left(2-\sqrt{\frac{S_{8,1}^2\overline{\mathsf{snr}}}{S_{8,1}^2\overline{\mathsf{snr}}+1}}-\sqrt{\frac{S_{8,3}^2\overline{\mathsf{snr}}}{S_{8,3}^2\overline{\mathsf{snr}}+1}}\right), \quad k = 1, 2. \quad (6.121)$$

This is possible because, instead of adding the bounds on the TPs P_l*, we first defined the bits' decision regions over which the integration turns out to be easy to deal with. This approach is known as "expurgation:" because the bounding implies integration over many decision regions, it is much more accurate to combine many regions to be described by a single inequality; many of the individual bounds are then removed. We will develop this idea further in Section 6.2.2.*

For the bit's position $k = 3$*, using Fig. 6.6, we can write*

$$P_{\mathrm{b},3} = \Pr\{\boldsymbol{Y} \in \mathcal{Y}_2 \cup \mathcal{Y}_3 \cup \mathcal{Y}_6 \cup \mathcal{Y}_7 | \boldsymbol{X} = \boldsymbol{x}_1\} \quad (6.122)$$

$$= \Pr\{\boldsymbol{Y} \in \mathcal{Y}_2 \cup \mathcal{Y}_3 | \boldsymbol{X} = \boldsymbol{x}_1\} + \Pr\{\boldsymbol{Y} \in \mathcal{Y}_6 \cup \mathcal{Y}_7 | \boldsymbol{X} = \boldsymbol{x}_1\}, \quad (6.123)$$

and because we easily see in Fig. 6.6 that

$$\mathcal{Y}_2 \cup \mathcal{Y}_3 = \{\boldsymbol{y} : \xi_{2,1}(\boldsymbol{y}) \le 0 \wedge \xi_{3,4}(\boldsymbol{y}) \le 0\}, \quad (6.124)$$

$$\mathcal{Y}_6 \cup \mathcal{Y}_7 = \{\boldsymbol{y} : \xi_{6,5}(\boldsymbol{y}) \le 0 \wedge \xi_{7,8}(\boldsymbol{y}) \le 0\}, \quad (6.125)$$

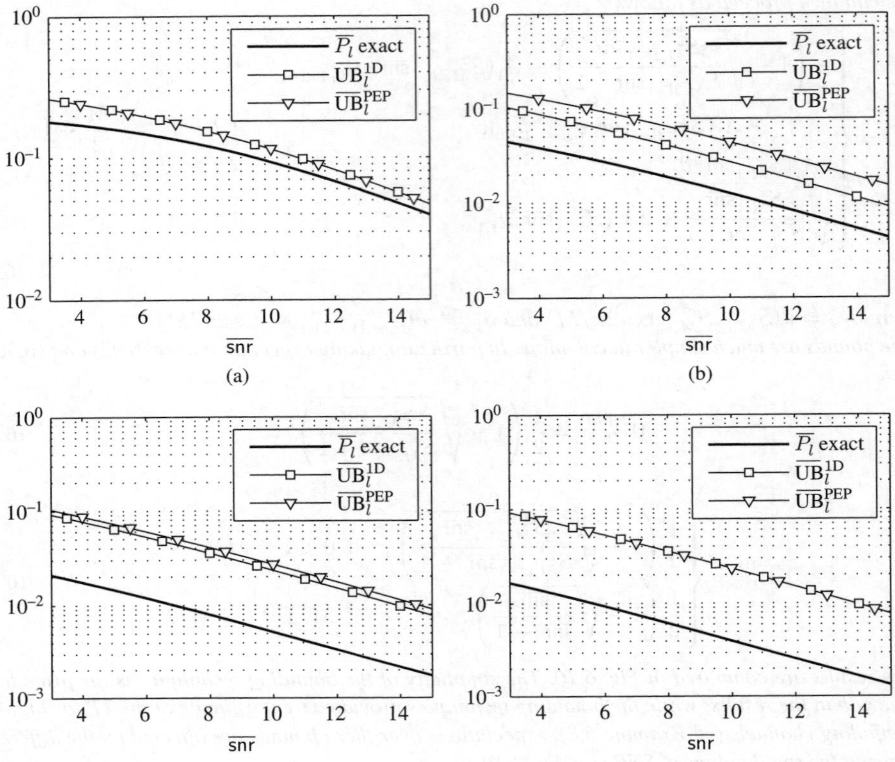

Figure 6.10 Evaluation of TPs for 8PSK in Rayleigh fading channel

we can use the 1D bound to obtain

$$P_{\mathrm{b},3} \leq \mathrm{UB}_{\mathrm{b},3}^{\mathrm{1D}} = \Pr\{\xi_{2,1}(\boldsymbol{Y}) \leq 0 | \boldsymbol{X} = \boldsymbol{x}_1\} + \Pr\{\xi_{7,8}(\boldsymbol{Y}) \leq 0 | \boldsymbol{X} = \boldsymbol{x}_1\} \tag{6.126}$$

$$= Q(\sqrt{2\,\mathsf{snr}}\, S_{8,1}) + Q(\sqrt{2\,\mathsf{snr}}\, S_{8,3}). \tag{6.127}$$

Consequently, we obtain

$$\overline{\mathrm{UB}}_{\mathrm{b},3}^{\mathrm{1D}} = \frac{1}{2}\left(2 - \sqrt{\frac{S_{8,1}^2 \overline{\mathsf{snr}}}{S_{8,1}^2 \overline{\mathsf{snr}} + 1}} - \sqrt{\frac{S_{8,3}^2 \overline{\mathsf{snr}}}{S_{8,3}^2 \overline{\mathsf{snr}} + 1}}\right), \tag{6.128}$$

which we show in Fig. 6.11.

6.2 Coded Transmission

When trying to analyze the performance of the decoder we face issues similar to those appearing in the uncoded case. However, because geometric considerations would require high-dimensional analysis,

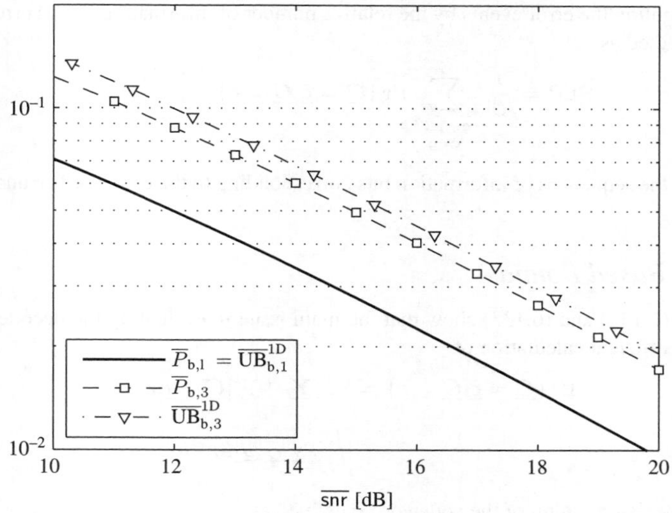

Figure 6.11 BEP $\overline{P}_{b,k}$ and the corresponding bounds $\overline{\text{UB}}^{1D}_{b,k}$ for $k = 1, 3$ and an 8PSK constellation labeled by the BRGC in a Rayleigh fading channel

exact solutions are very difficult to find. Instead, we resort to bounding techniques based on the evaluation of PEPs, which are similar in spirit to those presented in Section 6.1.3.

The decisions made by the *maximum likelihood* (ML) and the BICM decoders can both be expressed as (see (3.9) and (3.22))

$$\hat{\boldsymbol{c}} = \underset{\boldsymbol{c}\in\mathcal{C}}{\text{argmax}}\{\mathbb{M}(\underline{\boldsymbol{c}}, \boldsymbol{y})\}, \tag{6.129}$$

where $\mathbb{M}(\underline{\boldsymbol{c}}, \boldsymbol{y})$ is a decoding metric that depends on the considered decoder. Note that with a slight abuse of notation, throughout this chapter we also use the notation $\mathbb{M}(\underline{\boldsymbol{x}}, \boldsymbol{y})$ and $\mathbb{M}(\underline{\boldsymbol{c}}, \boldsymbol{\lambda})$ to denote this metric.

An error occurs when the decoded codeword $\underline{\hat{\boldsymbol{c}}}$ is different from the transmitted one $\underline{\boldsymbol{c}}$. The probability of detecting an incorrect codeword is the so-called WEP and is defined as

$$\mathsf{WEP} \triangleq \Pr\{\underline{\hat{\boldsymbol{C}}} \neq \underline{\boldsymbol{C}}\}, \tag{6.130}$$

where $\underline{\boldsymbol{C}}$ and $\underline{\hat{\boldsymbol{C}}}$ are the random variables representing, respectively, the transmitted and detected codewords.

The WEP can be expressed as

$$\mathsf{WEP} = \sum_{\underline{\boldsymbol{c}}\in\mathcal{C}} \Pr\{\underline{\hat{\boldsymbol{C}}} \neq \underline{\boldsymbol{C}} | \underline{\boldsymbol{C}} = \underline{\boldsymbol{c}}\} \Pr\{\underline{\boldsymbol{C}} = \underline{\boldsymbol{c}}\} \tag{6.131}$$

$$= \frac{1}{|\mathcal{C}|} \sum_{\substack{(\underline{\boldsymbol{c}}, \underline{\hat{\boldsymbol{c}}})\in\mathcal{C}^2 \\ \underline{\hat{\boldsymbol{c}}}\neq\underline{\boldsymbol{c}}}} \Pr\{\underline{\hat{\boldsymbol{C}}} = \underline{\hat{\boldsymbol{c}}} | \underline{\boldsymbol{C}} = \underline{\boldsymbol{c}}\}, \tag{6.132}$$

where $(\underline{\boldsymbol{c}}, \underline{\hat{\boldsymbol{c}}}) \in \mathcal{C}^2$ denotes the pairs $(\underline{\boldsymbol{c}}, \underline{\hat{\boldsymbol{c}}})$ taken from $\mathcal{C}^2 = \mathcal{C} \times \mathcal{C}$ and to obtain (6.132) we assumed equiprobable codewords, i.e., $\Pr\{\underline{\boldsymbol{C}} = \underline{\boldsymbol{c}}\} = \frac{1}{|\mathcal{C}|} = 2^{-N_b}$.

Similarly, weighting the error events by the relative number of information bits in error, we obtain the average BEP, defined as

$$\mathsf{BEP} \triangleq \frac{1}{|\mathcal{C}|} \sum_{\substack{(\underline{\boldsymbol{c}},\hat{\underline{\boldsymbol{c}}})\in\mathcal{C}^2\\ \hat{\underline{\boldsymbol{c}}}\neq\underline{\boldsymbol{c}}}} \Pr\{\underline{\hat{\boldsymbol{C}}} = \hat{\underline{\boldsymbol{c}}} | \underline{\boldsymbol{C}} = \underline{\boldsymbol{c}}\} \frac{d_{\mathrm{H}}(\underline{i},\underline{\hat{\imath}})}{N_{\mathrm{b}}}, \tag{6.133}$$

where $\underline{i}$ and $\underline{\hat{\imath}}$ are the sequences of information bits corresponding to the codewords $\underline{\boldsymbol{c}}$ and $\hat{\underline{\boldsymbol{c}}}$, respectively.

6.2.1 PEP-Based Bounds

The expressions (6.132) and (6.133) show that the main issue in evaluating the decoder's performance boils down to an efficient calculation of

$$\Pr\{\underline{\hat{\boldsymbol{C}}} = \hat{\underline{\boldsymbol{c}}} | \underline{\boldsymbol{C}} = \underline{\boldsymbol{c}}\} = \Pr\{\underline{\boldsymbol{Y}} \in \mathcal{Y}_{\hat{\underline{\boldsymbol{c}}}} | \underline{\boldsymbol{C}} = \underline{\boldsymbol{c}}\} \tag{6.134}$$

$$= \int_{\mathcal{Y}_{\hat{\underline{\boldsymbol{c}}}}} p_{\underline{\boldsymbol{Y}}|\underline{\boldsymbol{C}}}(\underline{\boldsymbol{y}}|\underline{\boldsymbol{c}})\, \mathrm{d}\underline{\boldsymbol{y}}, \tag{6.135}$$

where $\mathcal{Y}_{\hat{\underline{\boldsymbol{c}}}}$ is the decision region of the codeword $\hat{\underline{\boldsymbol{c}}}$ defined as

$$\mathcal{Y}_{\hat{\underline{\boldsymbol{c}}}} \triangleq \bigcap_{\substack{\underline{\boldsymbol{c}}'\in\mathcal{C}\\ \underline{\boldsymbol{c}}'\neq\hat{\underline{\boldsymbol{c}}}}} \mathcal{Y}^{\mathrm{PEP}}_{\underline{\boldsymbol{c}}'\to\hat{\underline{\boldsymbol{c}}}}, \tag{6.136}$$

where

$$\mathcal{Y}^{\mathrm{PEP}}_{\underline{\boldsymbol{c}}'\to\hat{\underline{\boldsymbol{c}}}} \triangleq \{\underline{\boldsymbol{Y}} \in \mathbb{R}^{N\times N_s} : \mathbb{M}(\hat{\underline{\boldsymbol{c}}}, \underline{\boldsymbol{y}}) > \mathbb{M}(\underline{\boldsymbol{c}}', \underline{\boldsymbol{y}}), \underline{\boldsymbol{c}}' \neq \hat{\underline{\boldsymbol{c}}}\}. \tag{6.137}$$

As in the case of uncoded transmission, the evaluation of $\Pr\{\underline{\hat{\boldsymbol{C}}} = \hat{\underline{\boldsymbol{c}}} | \underline{\boldsymbol{C}} = \underline{\boldsymbol{c}}\}$ in (6.134) boils down to finding the decision region $\mathcal{Y}_{\hat{\underline{\boldsymbol{c}}}}$ and, more importantly, to evaluating the multidimensional integral in (6.135). Since usually N_{s} is very large, the exact calculation of this NN_{s}-dimensional integral is most often considered infeasible. One of the most popular simplifications to tackle this problem is based on a PEP-based bounding technique. Let the PEP be defined as

$$\mathrm{PEP}(\underline{\boldsymbol{c}}, \hat{\underline{\boldsymbol{c}}}) \triangleq \Pr\{\underline{\boldsymbol{Y}} \in \mathcal{Y}^{\mathrm{PEP}}_{\underline{\boldsymbol{c}}\to\hat{\underline{\boldsymbol{c}}}} | \underline{\boldsymbol{C}} = \underline{\boldsymbol{c}}\} \tag{6.138}$$

$$= \Pr\{\mathbb{M}(\hat{\underline{\boldsymbol{c}}}, \underline{\boldsymbol{Y}}) > \mathbb{M}(\underline{\boldsymbol{c}}, \underline{\boldsymbol{Y}}) \mid \underline{\boldsymbol{C}} = \underline{\boldsymbol{c}}\}. \tag{6.139}$$

By knowing that $\mathcal{Y}_{\hat{\underline{\boldsymbol{c}}}} \subset \mathcal{Y}^{\mathrm{PEP}}_{\underline{\boldsymbol{c}}'\to\hat{\underline{\boldsymbol{c}}}}$ and using $\underline{\boldsymbol{c}}' = \underline{\boldsymbol{c}}$ we can then upper bound $\Pr\{\underline{\hat{\boldsymbol{C}}} = \hat{\underline{\boldsymbol{c}}} | \underline{\boldsymbol{C}} = \underline{\boldsymbol{c}}\}$ in (6.134) as

$$\Pr\{\underline{\hat{\boldsymbol{C}}} = \hat{\underline{\boldsymbol{c}}} | \underline{\boldsymbol{C}} = \underline{\boldsymbol{c}}\} \leq \mathrm{PEP}(\underline{\boldsymbol{c}}, \hat{\underline{\boldsymbol{c}}}). \tag{6.140}$$

The WEP in (6.132) is then bounded as

$$\mathsf{WEP} \leq \frac{1}{|\mathcal{C}|} \sum_{\substack{(\underline{\boldsymbol{c}},\hat{\underline{\boldsymbol{c}}})\in\mathcal{C}^2\\ \hat{\underline{\boldsymbol{c}}}\neq\underline{\boldsymbol{c}}}} \mathrm{PEP}(\underline{\boldsymbol{c}}, \hat{\underline{\boldsymbol{c}}}) \tag{6.141}$$

$$= \frac{1}{|\mathcal{X}|} \sum_{\substack{(\underline{\boldsymbol{x}},\hat{\underline{\boldsymbol{x}}})\in\mathcal{X}^2\\ \hat{\underline{\boldsymbol{x}}}\neq\underline{\boldsymbol{x}}}} \mathrm{PEP}(\underline{\boldsymbol{x}}, \hat{\underline{\boldsymbol{x}}}) \tag{6.142}$$

$$= \frac{1}{|\mathcal{B}|} \sum_{\substack{(\underline{\boldsymbol{b}},\hat{\underline{\boldsymbol{b}}})\in\mathcal{B}^2\\ \hat{\underline{\boldsymbol{b}}}\neq\underline{\boldsymbol{b}}}} \mathrm{PEP}(\underline{\boldsymbol{b}}, \hat{\underline{\boldsymbol{b}}}), \tag{6.143}$$

where to pass from (6.141) to (6.142) or (6.143) we use the fact that the mapping between the binary codewords $\underline{\boldsymbol{c}}$ and the codewords $\underline{\boldsymbol{x}}$ or $\underline{\boldsymbol{b}}$ is bijective. In a similar way, we can bound the BEP in (6.133) as

$$\mathsf{BEP} \leq \frac{1}{|\mathcal{C}|} \sum_{\substack{(\underline{\boldsymbol{c}},\hat{\underline{\boldsymbol{c}}})\in\mathcal{C}^2 \\ \hat{\underline{\boldsymbol{c}}}\neq\underline{\boldsymbol{c}}}} \mathrm{PEP}(\underline{\boldsymbol{c}},\hat{\underline{\boldsymbol{c}}}) \frac{d_{\mathrm{H}}(\underline{i},\hat{\underline{i}})}{N_{\mathrm{b}}} \tag{6.144}$$

$$= \frac{1}{|\mathcal{X}|} \sum_{\substack{(\underline{\boldsymbol{x}},\hat{\underline{\boldsymbol{x}}})\in\mathcal{X}^2 \\ \hat{\underline{\boldsymbol{x}}}\neq\underline{\boldsymbol{x}}}} \mathrm{PEP}(\underline{\boldsymbol{x}},\hat{\underline{\boldsymbol{x}}}) \frac{d_{\mathrm{H}}(\underline{i},\hat{\underline{i}})}{N_{\mathrm{b}}}, \tag{6.145}$$

where $\underline{i}$ and $\hat{\underline{i}}$ are the sequences of information bits corresponding to the codewords $\underline{\boldsymbol{x}}$ and $\hat{\underline{\boldsymbol{x}}}$, respectively.

We emphasize that while the equations for the coded and uncoded cases are very similar, the PEP depends on the metrics used for decoding. In the following sections we show how these metrics affect the calculation of the PEP.

6.2.2 *Expurgated Bounds*

For uncoded transmission, we have already seen in Example 6.12 that PEP-based bounds such as those in (6.141) and (6.144) may be quite loose and to tighten them it is possible to remove or *expurgate* some terms from the bound. This expurgation strategy can be extended to the case of coded transmission, which we show below.

We start by re-deriving (6.141) in a more convenient form. To this end, we write (6.132) as

$$\mathsf{WEP} = \frac{1}{|\mathcal{C}|} \sum_{\underline{\boldsymbol{c}}\in\mathcal{C}} \Pr\left\{ \underline{\boldsymbol{Y}} \in \bigcup_{\hat{\underline{\boldsymbol{c}}}\in\mathcal{C},\hat{\underline{\boldsymbol{c}}}\neq\underline{\boldsymbol{c}}} \mathcal{Y}_{\hat{\underline{\boldsymbol{c}}}} \,\middle|\, \underline{\boldsymbol{C}} = \underline{\boldsymbol{c}} \right\}. \tag{6.146}$$

We note that

$$\bigcup_{\hat{\underline{\boldsymbol{c}}}\in\mathcal{C},\hat{\underline{\boldsymbol{c}}}\neq\underline{\boldsymbol{c}}} \mathcal{Y}_{\hat{\underline{\boldsymbol{c}}}} = \bigcup_{\hat{\underline{\boldsymbol{c}}}\in\mathcal{C},\hat{\underline{\boldsymbol{c}}}\neq\underline{\boldsymbol{c}}} \mathcal{Y}^{\mathrm{PEP}}_{\underline{\boldsymbol{c}}\to\hat{\underline{\boldsymbol{c}}}}, \tag{6.147}$$

and thus, applying a union bound to (6.146) we obtain

$$\mathsf{WEP} = \frac{1}{|\mathcal{C}|} \sum_{\underline{\boldsymbol{c}}\in\mathcal{C}} \Pr\left\{ \underline{\boldsymbol{Y}} \in \bigcup_{\hat{\underline{\boldsymbol{c}}}\in\mathcal{C},\hat{\underline{\boldsymbol{c}}}\neq\underline{\boldsymbol{c}}} \mathcal{Y}^{\mathrm{PEP}}_{\underline{\boldsymbol{c}}\to\hat{\underline{\boldsymbol{c}}}} \,\middle|\, \underline{\boldsymbol{C}} = \underline{\boldsymbol{c}} \right\}$$

$$\leq \frac{1}{|\mathcal{C}|} \sum_{\substack{(\underline{\boldsymbol{c}},\hat{\underline{\boldsymbol{c}}})\in\mathcal{C}^2 \\ \hat{\underline{\boldsymbol{c}}}\neq\underline{\boldsymbol{c}}}} \Pr\left\{ \underline{\boldsymbol{Y}} \in \mathcal{Y}^{\mathrm{PEP}}_{\underline{\boldsymbol{c}}\to\hat{\underline{\boldsymbol{c}}}} \,\middle|\, \underline{\boldsymbol{C}} = \underline{\boldsymbol{c}} \right\}, \tag{6.148}$$

where the inequality in (6.148) follows from the fact that, in general, the sets $\mathcal{Y}^{\mathrm{PEP}}_{\underline{\boldsymbol{c}}\to\hat{\underline{\boldsymbol{c}}}}$ in the *right-hand side* (r.h.s.) of (6.147) are not disjoint. The general idea of expurgating the bound consists then in eliminating redundant sets in the r.h.s. of (6.147), while maintaining an inequality in (6.148). In other words, we aim at reducing the number of terms in the r.h.s. of (6.148), and by doing so, we tighten the bound on the WEP. We consider here decoders for which the decoding metric $\mathbb{M}(\underline{\boldsymbol{c}},\underline{\boldsymbol{y}})$ can be expressed as

$$\mathbb{M}(\underline{\boldsymbol{c}},\underline{\boldsymbol{y}}) = \sum_{n=1}^{N_{\mathrm{s}}} \mathbbm{m}(\boldsymbol{c}[n],\boldsymbol{y}[n]). \tag{6.149}$$

In fact, in view of (3.7) and (3.23), we can conclude that the decoding metrics of the ML and BICM decoders can both be expressed as in (6.149).

For decoders with decoding metric in the form of (6.149), the sets $\mathcal{Y}^{\mathrm{PEP}}_{\underline{\boldsymbol{c}}\to\hat{\underline{\boldsymbol{c}}}}$ can be expressed as

$$\mathcal{Y}^{\mathrm{PEP}}_{\underline{\boldsymbol{c}}\to\hat{\underline{\boldsymbol{c}}}} = \left\{ \underline{\boldsymbol{y}} : L^{\Sigma}_{\underline{\boldsymbol{c}},\hat{\underline{\boldsymbol{c}}}} > 0 \right\}, \tag{6.150}$$

where

$$L^{\Sigma}_{\underline{\mathbf{c}},\hat{\underline{\mathbf{c}}}} \triangleq \sum_{n=1}^{N_s} \mathfrak{m}(\hat{\mathbf{c}}[n], \boldsymbol{y}[n]) - \mathfrak{m}(\mathbf{c}[n], \boldsymbol{y}[n]) \tag{6.151}$$

$$= \sum_{\substack{n=1\\ \mathbf{c}[n]\neq\hat{\mathbf{c}}[n]}}^{N_s} \mathfrak{m}(\hat{\mathbf{c}}[n], \boldsymbol{y}[n]) - \mathfrak{m}(\mathbf{c}[n], \boldsymbol{y}[n]). \tag{6.152}$$

We assume that the codeword $\hat{\underline{\mathbf{c}}}$ can be expressed as $\hat{\underline{\mathbf{c}}} = \underline{\mathbf{c}} \oplus \underline{\mathbf{e}}_1 \oplus \underline{\mathbf{e}}_2$, where the binary error codewords $\underline{\mathbf{e}}_1$ and $\underline{\mathbf{e}}_2$ are orthogonal (i.e., $\sum_{n=1}^{N_s} \langle \mathbf{e}_1[n], \mathbf{e}_2[n]\rangle = 0$). We also assume that $\hat{\underline{\mathbf{c}}}_1 = \underline{\mathbf{c}} \oplus \underline{\mathbf{e}}_1$ and $\hat{\underline{\mathbf{c}}}_2 = \underline{\mathbf{c}} \oplus \underline{\mathbf{e}}_2$ are codewords. Because of the orthogonality of $\underline{\mathbf{e}}_1$ and $\underline{\mathbf{e}}_2$, the codewords $\hat{\underline{\mathbf{c}}}_1$ and $\hat{\underline{\mathbf{c}}}_2$ differ with $\underline{\mathbf{c}}$ at different time instants, and thus, we can decompose (6.152) as

$$L^{\Sigma}_{\underline{\mathbf{c}},\hat{\underline{\mathbf{c}}}} = L^{\Sigma}_{\underline{\mathbf{c}},\hat{\underline{\mathbf{c}}}_1} + L^{\Sigma}_{\underline{\mathbf{c}},\hat{\underline{\mathbf{c}}}_2}. \tag{6.153}$$

This allows us to conclude that if $L^{\Sigma}_{\underline{\mathbf{c}},\hat{\underline{\mathbf{c}}}} > 0$, then either $L^{\Sigma}_{\underline{\mathbf{c}},\hat{\underline{\mathbf{c}}}_1} > 0$ or $L^{\Sigma}_{\underline{\mathbf{c}},\hat{\underline{\mathbf{c}}}_2} > 0$, which also means that if the condition $\boldsymbol{y} \in \mathcal{Y}^{\mathrm{PEP}}_{\underline{\mathbf{c}}\to\hat{\underline{\mathbf{c}}}}$ is satisfied then either $\boldsymbol{y} \in \mathcal{Y}^{\mathrm{PEP}}_{\underline{\mathbf{c}}\to\hat{\underline{\mathbf{c}}}_1}$ or $\boldsymbol{y} \in \mathcal{Y}^{\mathrm{PEP}}_{\underline{\mathbf{c}}\to\hat{\underline{\mathbf{c}}}_1}$ is satisfied. We then immediately obtain

$$\mathcal{Y}^{\mathrm{PEP}}_{\underline{\mathbf{c}}\to\hat{\underline{\mathbf{c}}}} \subset \mathcal{Y}^{\mathrm{PEP}}_{\underline{\mathbf{c}}\to\hat{\underline{\mathbf{c}}}_1} \cup \mathcal{Y}^{\mathrm{PEP}}_{\underline{\mathbf{c}}\to\hat{\underline{\mathbf{c}}}_2}. \tag{6.154}$$

From (6.154), we conclude that if the sets $\mathcal{Y}^{\mathrm{PEP}}_{\underline{\mathbf{c}}\to\hat{\underline{\mathbf{c}}}}$, $\mathcal{Y}^{\mathrm{PEP}}_{\underline{\mathbf{c}}\to\hat{\underline{\mathbf{c}}}_1}$, and $\mathcal{Y}^{\mathrm{PEP}}_{\underline{\mathbf{c}}\to\hat{\underline{\mathbf{c}}}_2}$ all appear in the r.h.s. of (6.147), the set $\mathcal{Y}^{\mathrm{PEP}}_{\underline{\mathbf{c}}\to\hat{\underline{\mathbf{c}}}}$ is redundant and can be removed from the union. Therefore, the term $\mathrm{PEP}(\underline{\mathbf{c}}, \hat{\underline{\mathbf{c}}})$ can be removed from (6.148), and thus, the bound is tightened.

The above considerations were made using only two codewords $\hat{\underline{\mathbf{c}}}_1$ and $\hat{\underline{\mathbf{c}}}_2$; however, they straightforwardly generalize to the case where the codewords are defined via K orthogonal error codewords $\underline{\mathbf{e}}_k$, $k = 1, \ldots, K$. In such a case, if we can write $\hat{\mathbf{c}} = \underline{\mathbf{c}} \oplus \underline{\mathbf{e}}_1 \oplus \cdots \oplus \underline{\mathbf{e}}_K$, and $\underline{\mathbf{c}} \oplus \underline{\mathbf{e}}_k \in \mathcal{C}$, $k = 1, \ldots, K$, then the contribution of $\hat{\underline{\mathbf{c}}}$ can be expurgated.

6.2.3 *ML Decoder*

The ML decoder chooses the codeword via (3.9), and thus, the metric in (6.129) is

$$\mathbb{M}(\underline{\boldsymbol{x}}, \underline{\boldsymbol{y}}) = \log(p_{\underline{\boldsymbol{Y}}|\underline{\boldsymbol{X}}}(\underline{\boldsymbol{y}}|\underline{\boldsymbol{x}})) \tag{6.155}$$

$$= -\frac{1}{\mathsf{N}_0} \sum_{n=1}^{N_s} \|\boldsymbol{y}[n] - h[n]\boldsymbol{x}[n]\|^2, \tag{6.156}$$

where (6.156) follows from (2.28). The PEP is then calculated as

$$\mathrm{PEP}(\underline{\boldsymbol{x}}, \hat{\underline{\boldsymbol{x}}}) = \Pr\{\mathbb{M}(\hat{\underline{\boldsymbol{x}}}, \underline{\boldsymbol{Y}}) > \mathbb{M}(\underline{\boldsymbol{x}}, \underline{\boldsymbol{Y}}) | \underline{\boldsymbol{X}} = \underline{\boldsymbol{x}}\} \tag{6.157}$$

$$= \Pr\left\{-\frac{1}{\mathsf{N}_0} \sum_{n=1}^{N_s} |h[n]|^2(\|\hat{\boldsymbol{x}}[n]\|^2 - \|\boldsymbol{x}[n]\|^2) - 2h[n](\hat{\boldsymbol{x}}[n] - \boldsymbol{x}[n])^{\mathrm{T}}\boldsymbol{Y}[n] > 0\right\} \tag{6.158}$$

$$= \Pr\left\{-\frac{1}{\mathsf{N}_0} \sum_{n=1}^{N_s} \|h[n](\hat{\boldsymbol{x}}[n] - \boldsymbol{x}[n]) + \boldsymbol{Z}[n]\|^2 > -\frac{1}{\mathsf{N}_0} \sum_{n=1}^{N_s} \|\boldsymbol{Z}[n]\|^2\right\} \tag{6.159}$$

$$= \Pr\left\{L^{\Sigma}_{\underline{\boldsymbol{x}},\hat{\underline{\boldsymbol{x}}}} > 0\right\}, \tag{6.160}$$

where

$$L^{\Sigma}_{\underline{\boldsymbol{x}},\underline{\hat{\boldsymbol{x}}}} = \sum_{n=1}^{N_s} L[n], \tag{6.161}$$

and

$$L[n] \triangleq -\frac{h^2[n]}{\mathsf{N}_0}\|\boldsymbol{x}[n]-\hat{\boldsymbol{x}}[n]\|^2 - 2\frac{h[n]}{\mathsf{N}_0}(\boldsymbol{x}[n]-\hat{\boldsymbol{x}}[n])^{\mathrm{T}}\boldsymbol{Z}[n] \tag{6.162}$$

$$= -\frac{|h[n]|^2}{\mathsf{N}_0}(\|\hat{\boldsymbol{x}}[n]\|^2 - \|\boldsymbol{x}[n]\|^2) - \frac{2h[n]}{\mathsf{N}_0}(\hat{\boldsymbol{x}}[n]-\boldsymbol{x}[n])^{\mathrm{T}}\boldsymbol{Y}[n]. \tag{6.163}$$

For known $h[1],\ldots,h[N_{\mathrm{s}}]$, $\underline{\boldsymbol{x}}$ and $\underline{\hat{\boldsymbol{x}}}$, we see from (6.162) that $L[1],\ldots,L[N_{\mathrm{s}}]$ are independent Gaussian random variables with PDF given by

$$p_{L[n]}(l) = \Psi(l, -\mathsf{snr}(h[n])\|\hat{\boldsymbol{x}}[n]-\boldsymbol{x}[n]\|^2, 2\,\mathsf{snr}(h[n])\|\hat{\boldsymbol{x}}[n]-\boldsymbol{x}[n]\|^2). \tag{6.164}$$

This explains the notation $L[n]$, i.e., $L[n]$ has the same distribution as an L-value conditioned on transmitting a bit $b=0$ using binary modulation with symbols $\Phi(1)=\boldsymbol{x}[n]$ and $\Phi(0)=\hat{\boldsymbol{x}}[n]$; see (3.63) and (3.64).

In the case of nonfading channels, i.e., when $\mathsf{snr}(h[n]) = \mathsf{snr}$, we easily see that $L^{\Sigma}_{\underline{\boldsymbol{x}},\underline{\hat{\boldsymbol{x}}}}$ is also a Gaussian random variable with PDF

$$p_{L^{\Sigma}_{\underline{\boldsymbol{x}},\underline{\hat{\boldsymbol{x}}}}}(l) = \Psi\left(l, -\mathsf{snr}\sum_{n=1}^{N_{\mathrm{s}}}\|\hat{\boldsymbol{x}}[n]-\boldsymbol{x}[n]\|^2, 2\,\mathsf{snr}\sum_{n=1}^{N_{\mathrm{s}}}\|\hat{\boldsymbol{x}}[n]-\boldsymbol{x}[n]\|^2\right). \tag{6.165}$$

Thus, similarly to (6.79),

$$\mathsf{PEP}(\underline{\boldsymbol{x}},\underline{\hat{\boldsymbol{x}}}) = \mathsf{PEP}^{\mathrm{ED}}(\|\underline{\boldsymbol{x}}-\underline{\hat{\boldsymbol{x}}}\|) \triangleq Q\left(\sqrt{2\,\mathsf{snr}}\frac{\|\underline{\boldsymbol{x}}-\underline{\hat{\boldsymbol{x}}}\|}{2}\right). \tag{6.166}$$

In fading channels, i.e., when $\mathsf{snr}(h[n])$ is modeled as a random variable SNR, the analysis is slightly more involved so we postpone it to Section 6.3.4.

Since the PEP in (6.166) is a function of the *Euclidean distance* (ED) between the codewords only, in what follows, we define the *Euclidean distance spectrum* (EDS) of the code $\mathcal{X}$ and the *input-dependent Euclidean distance spectrum* (IEDS) of the respective *coded modulation* (CM) encoder.

Definition 6.13 (EDS of the Code $\mathcal{X}$) *The EDS of a code $\mathcal{X}$, denoted by $C^{\mathcal{X}}_{d}$, is defined as the average number of pair of codewords in $\mathcal{X}$ at ED equal to d, i.e.,*

$$C^{\mathcal{X}}_{d} \triangleq \frac{1}{|\mathcal{X}|}|\{(\underline{\boldsymbol{x}},\underline{\boldsymbol{x}}')\in\mathcal{X}^2 : \|\underline{\boldsymbol{x}}-\underline{\boldsymbol{x}}'\| = d\}|, \quad d\in\mathcal{D}, \tag{6.167}$$

where $\mathcal{D}$ is the set of all possible EDs between codewords in the code $\mathcal{X}$.

Definition 6.14 (IEDS of the CM Encoder) *The IEDS of a CM encoder, denoted by $B^{\mathcal{X}}_{d}$, is defined as the average Hamming distance between all input sequences corresponding to the pairs of codewords $(\underline{\boldsymbol{x}},\underline{\hat{\boldsymbol{x}}})\in\mathcal{X}^2$ at ED d, i.e.,*

$$B^{\mathcal{X}}_{d} \triangleq \frac{1}{|\mathcal{X}|}\sum_{\substack{(\underline{\boldsymbol{x}},\underline{\hat{\boldsymbol{x}}})\in\mathcal{X}^2\\ \|\underline{\boldsymbol{x}}-\underline{\hat{\boldsymbol{x}}}\|=d}}\frac{d_{\mathrm{H}}(\underline{i},\underline{\hat{i}})}{N_{\mathrm{b}}}, \tag{6.168}$$

where $\underline{i}$ and $\underline{\hat{i}}$ are the sequences of information bits corresponding to the codewords $\underline{\boldsymbol{x}}$ and $\underline{\hat{\boldsymbol{x}}}$, respectively.

We note here the analogy between the EDS $C_d^{\mathcal{X}}$ in (6.167) and the *distance spectrum* (DS) $C_d^{\mathcal{B}}$ of the code $\mathcal{B}$ in (2.105) (cf. also $B_d^{\mathcal{X}}$ in (6.168) and $B_d^{\mathcal{B}}$ in (2.111)). The difference is that the index d in $C_d^{\mathcal{X}}$ has a meaning of ED, while in $C_d^{\mathcal{B}}$ it denotes HD. We also emphasize that the IEDS is a property of the CM encoder (i.e., it depends on how the information sequences are mapped to the symbol sequences) while the EDS is solely a property of the code $\mathcal{X}$.

Now, combining (6.142) with (6.166), we obtain the following upper bound on the WEP for ML decoders:

$$\mathsf{WEP} \leq \sum_{d\in\mathcal{D}} C_d^{\mathcal{X}} \mathsf{PEP}^{\mathrm{ED}}(d) \tag{6.169}$$

$$= \sum_{d\in\mathcal{D}} C_d^{\mathcal{X}} Q\left(\sqrt{2\,\mathsf{snr}}\frac{d}{2}\right). \tag{6.170}$$

Similarly, using (6.145), the BEP is bounded as

$$\mathsf{BEP} \leq \sum_{d\in\mathcal{D}} B_d^{\mathcal{X}} \mathsf{PEP}^{\mathrm{ED}}(d) \tag{6.171}$$

$$= \sum_{d\in\mathcal{D}} B_d^{\mathcal{X}} Q\left(\sqrt{2\,\mathsf{snr}}\frac{d}{2}\right). \tag{6.172}$$

Example 6.15 (Trellis Encoders and ML Decoding) *We consider here a trellis encoder (see Fig. 2.4) and its corresponding ML (e.g., Viterbi) decoder. In this case, the trellis representation of the code simplifies the analysis: any codeword $\underline{\boldsymbol{x}}$ can be associated with a path in the trellis and the visited trellis states uniquely determine the codeword.*

Let $\hat{\mathcal{S}}_\ell(\underline{\boldsymbol{x}})$ be the set of length-ℓ codewords that start at time t_0 at the same trellis state as $\underline{\boldsymbol{x}}$ and then, up to time $t_0+\ell-1$, pass through states that are different from the states visited by $\underline{\boldsymbol{x}}$. Finally, at time $t_0+\ell$, these codewords pass through the same trellis state as $\underline{\boldsymbol{x}}$. The codewords in $\hat{\mathcal{S}}_\ell(\underline{\boldsymbol{x}})$ thus correspond to paths in the trellis that diverged from the path defined by the codeword $\underline{\boldsymbol{x}}$ and remerged after ℓ trellis stages. Detecting $\hat{\underline{\boldsymbol{x}}}$ instead of $\underline{\boldsymbol{x}}$ is sometimes called "first-error event" and its probability can be upper bounded by

$$P_{1,\mathrm{e}} \leq \sum_{\ell=1}^{\infty} \sum_{\underline{\boldsymbol{x}}\in\mathcal{S}_\ell} \frac{1}{L_\ell} \sum_{\hat{\underline{\boldsymbol{x}}}\in\hat{\mathcal{S}}_\ell(\underline{\boldsymbol{x}})} \mathrm{PEP}(\underline{\boldsymbol{x}}, \hat{\underline{\boldsymbol{x}}}), \tag{6.173}$$

where $\mathcal{S}_\ell$ is the set of codewords $\underline{\boldsymbol{x}}$ corresponding to all different length-ℓ paths in the trellis, and $L_\ell = |\mathcal{S}_\ell|$ in (6.173) *denotes the number of different length-ℓ paths in the trellis that start at time t_0 and terminate at time $t_0+\ell$. Assuming independent and identically distributed (i.i.d.) information bits, all such paths are equiprobable. Since there are S states and there are 2^{k_c} branches leaving each state of the trellis at each time instant, we obtain*

$$L_\ell = S 2^{k_c \ell}. \tag{6.174}$$

As we showed in (6.166)*, for nonfading channels the PEP depends only on the ED between $\underline{\boldsymbol{x}}$ and $\hat{\underline{\boldsymbol{x}}}$, and thus, we can express* (6.173) *as*

$$P_{1,\mathrm{e}} \leq \sum_{\ell=1}^{\infty} \frac{1}{S}\frac{1}{2^{k_c\ell}} \sum_{d\in\mathcal{D}} F_{d,\ell}^{\mathcal{X}} Q\left(\frac{d}{2}\sqrt{2\,\mathsf{snr}}\right) \tag{6.175}$$

$$= \sum_{d\in\mathcal{D}} F_d^{\mathcal{X}} Q\left(\frac{d}{2}\sqrt{2\,\mathsf{snr}}\right), \tag{6.176}$$

where

$$F_d^{\mathcal{X}} \triangleq \sum_{\ell=1}^{\infty} \frac{1}{S}\frac{1}{2^{k_c\ell}} \sum_{v=1}^{\infty} F_{v,d,\ell}^{\mathcal{X}} \tag{6.177}$$

gives the average number of pairs of codewords at ED d, $F^{\mathcal{X}}_{v,d,\ell}$ is the number of pairs at ED d generated by input sequences at HD v, and $\mathcal{D}$ is the set of all possible EDs between any two codewords corresponding to diverging paths in the trellis. To obtain a bound on the WEP, we use the bound

$$\mathsf{WEP} \leq \sum_{d\in\mathcal{D}} N_{\mathrm{s}} F^{\mathcal{X}}_{d} Q\left(\frac{d}{2}\sqrt{2\,\mathsf{snr}}\right), \tag{6.178}$$

which is obtained as a generalization of the bound used for convolutional encoders (CENCs) (more details are given in Section 6.5).

A similar bound can be obtained for the BEP. In this case, each error event must be weighted by the number of information bits in error (v out of k_{c}), i.e.,

$$\begin{aligned}\mathsf{BEP} &\leq \sum_{d\in\mathcal{D}} \sum_{\ell=1}^{\infty} \frac{1}{S}\frac{1}{2^{k_{\mathrm{c}}\ell}} \sum_{v=1}^{\infty} \frac{v}{k_{\mathrm{c}}} F^{\mathcal{X}}_{v,d,\ell} Q\left(\frac{d}{2}\sqrt{2\,\mathsf{snr}}\right)\\ &= \sum_{d\in\mathcal{D}} G^{\mathcal{X}}_{d} Q\left(\frac{d}{2}\sqrt{2\,\mathsf{snr}}\right),\end{aligned} \tag{6.179}$$

where

$$G^{\mathcal{X}}_{d} \triangleq \sum_{\ell=1}^{\infty} \frac{1}{S}\frac{1}{2^{k_{\mathrm{c}}\ell}} \sum_{v=1}^{\infty} \frac{v}{k} F^{\mathcal{X}}_{v,d,\ell} \tag{6.180}$$

may be interpreted as the average input HD between input sequences $\underline{\imath}$ and $\hat{\underline{\imath}}$ corresponding to pairs of codewords $\underline{\boldsymbol{x}}$ and $\hat{\underline{\boldsymbol{x}}}$ at ED d.

At this point, it is interesting to compare the WEP bound in (6.170) with the one in (6.178) (cf. also (6.172) and (6.179)). In particular, we note that $F^{\mathcal{X}}_{d}$ in (6.170) does not enumerate all the codewords in the code $\mathcal{X}$ (but $C^{\mathcal{X}}_{d}$ does). Specifically, the codewords $\hat{\underline{\boldsymbol{x}}}$ corresponding to a path in the trellis that diverged from $\underline{\boldsymbol{x}}$, converged, and then diverged and converged again, are not taken into account in $F^{\mathcal{X}}_{d}$ (while they are counted in $C^{\mathcal{X}}_{d}$). However, because $\hat{\underline{\boldsymbol{x}}} = \underline{\boldsymbol{x}} + \underline{\mathbf{e}}_1 + \underline{\mathbf{e}}_2$ (where $\underline{\mathbf{e}}_1$ and $\underline{\mathbf{e}}_2$ are orthogonal and correspond to two diverging terms), $\hat{\underline{\boldsymbol{x}}}_1 = \underline{\boldsymbol{x}} + \underline{\mathbf{e}}_1$ and $\hat{\underline{\boldsymbol{x}}}_1 = \underline{\boldsymbol{x}} + \underline{\mathbf{e}}_2$ are all codewords, from the analysis outlined in Section 6.2.2 we know that removing the contribution of $\hat{\underline{\boldsymbol{x}}}$ only tightens the bound.[8] Consequently, (6.178) and (6.179) may be treated as expurgated bounds and we may write

$$C^{\mathcal{X}}_{d} \approx N_{\mathrm{s}} F^{\mathcal{X}}_{d}, \tag{6.181}$$

$$B^{\mathcal{X}}_{d} \approx G^{\mathcal{X}}_{d}, \tag{6.182}$$

where the approximation is due to the expurgation.

Example 6.16 (TCM over the AWGN Channel) *We consider a trellis encoder formed by a rate $R_{\mathrm{c}} = 1/2$ CENC and a* 4PAM *constellation labeled by the natural binary code (NBC), which gives a rate $R = 1$ bit/symbol. We consider $\nu = 4$ and $\nu = 6$ (i.e., $S = 16$ and $S = 64$ trellis states) and two CENCs for each case: the original encoder proposed by Ungerboeck and a CENC chosen to be optimal in terms of $G^{\mathcal{X}}_{d}$ in* (6.180).[9] *The encoders are $\mathbf{G} = [23, 4]$ (Ungerboeck) and $\mathbf{G} = [23, 10]$ for $\nu = 4$ and $\mathbf{G} = [103, 24]$ (Ungerboeck) and $\mathrm{G} = [107, 32]$ for $\nu = 6$. The results obtained are shown in Fig. 6.12, where*

[8] Note that in Section 6.2.2 we considered binary codewords $\underline{\mathbf{c}}$, which implied addition in GF(2). In the case of codewords $\underline{\boldsymbol{x}}$, the addition is in the domain of real numbers.

[9] These CENCs are chosen in analogy to how the *optimal distance spectrum* (ODS) CENCs in Table 2.1 are obtained. The key difference is that here the code is chosen to be optimal in terms of $G^{\mathcal{X}}_{d}$ in (6.180) and not in terms of its *weight distribution* (WD) $F^{\mathcal{C}}_{d}$.

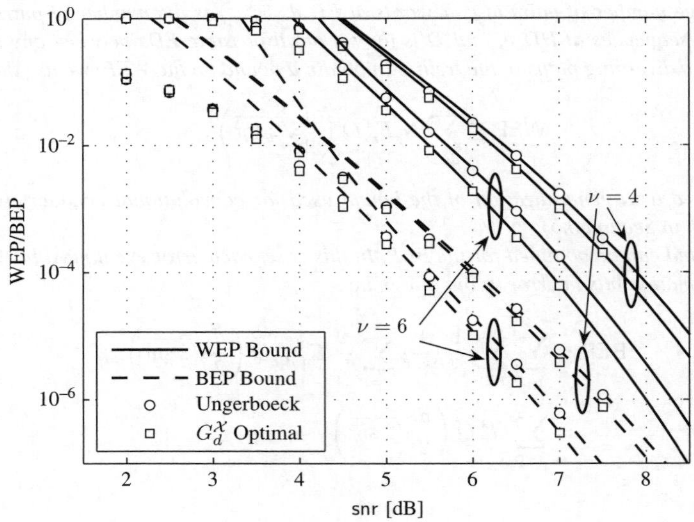

Figure 6.12 WEP and BEP bounds (lines) in (6.178) and (6.179) and simulations (markers) for Ungerboeck's encoders and CENCs optimal in terms of $G_d^{\mathcal{X}}$ for $N_s = 1000$, 4PAM, $R = 1/2$, and $\nu = 4, 6$

we compare simulation results with the bound in (6.178) *and* (6.179) *(computed using 20 terms in the bound). These results show a very good match between the simulations and the bound as well as the suboptimality of the original design by Ungerboeck.*

6.2.4 BICM Decoder

The decoding metric in (6.129) for the BICM decoder is defined in (3.22), and thus,

$$\mathbb{M}(\underline{\boldsymbol{c}}, \underline{\boldsymbol{\lambda}}) = \sum_{n=1}^{N_s} \sum_{q=1}^{n_c} c_q[n]\lambda_q[n], \tag{6.183}$$

where

$$\underline{\boldsymbol{\lambda}} = [\boldsymbol{\lambda}[1], \ldots, \boldsymbol{\lambda}[N_q]], \tag{6.184}$$

$$\boldsymbol{\lambda}[n] = [\lambda_1[n], \ldots, \lambda_{n_c}[n]]^T. \tag{6.185}$$

The PEP in (6.139) is then calculated as

$$\mathrm{PEP}(\underline{\boldsymbol{c}}, \hat{\underline{\boldsymbol{c}}}) = \Pr\{\mathbb{M}(\hat{\underline{\boldsymbol{c}}}, \underline{\boldsymbol{\Lambda}}) > \mathbb{M}(\underline{\boldsymbol{c}}, \underline{\boldsymbol{\Lambda}}) |\ \underline{\boldsymbol{C}} = \underline{\boldsymbol{c}}\} \tag{6.186}$$

$$= \Pr\left\{ \sum_{n=1}^{N_s} \sum_{q=1}^{n_c} (\hat{c}_q[n] - c_q[n])\Lambda_q[n] > 0 \,\middle|\, \underline{\boldsymbol{C}} = \underline{\boldsymbol{c}} \right\} \tag{6.187}$$

$$= \Pr\left\{ \sum_{n=1}^{N_s} \sum_{q=1}^{n_c} (-1)^{c_q[n]} e_q[n]\Lambda_q[n] > 0 \,\middle|\, \underline{\boldsymbol{C}} = \underline{\boldsymbol{c}} \right\}, \tag{6.188}$$

where $\Lambda_q[n]$ are the random variables modeling the L-values $\lambda_q[n]$, $\hat{\underline{\boldsymbol{c}}}$ was expressed as $\hat{\underline{\boldsymbol{c}}} = \underline{\boldsymbol{c}} \oplus \underline{\boldsymbol{e}}$, and where, because of the linearity of the code, the *error codeword* $\underline{\boldsymbol{e}}$ satisfies $\underline{\boldsymbol{e}} \in \mathcal{C}$.

For any n, q such that $c_q[n] = \hat{c}_q[n]$ (i.e., when $e_q[n] = 0$), the L-values $\Lambda_q[n]$ do not affect the PEP calculation. Therefore, there are only $d = d_{\text{H}}(\underline{\mathbf{c}}, \hat{\underline{\mathbf{c}}}) = w_{\text{H}}(\underline{\mathbf{e}})$ pairs n, q that are relevant for the PEP calculation in (6.187), and the PEP depends solely on $\underline{\mathbf{c}}$ and $\underline{\mathbf{e}}$. We can thus write

$$\text{PEP}(\underline{\mathbf{c}}, \hat{\underline{\mathbf{c}}}) = \text{PEP}(\underline{\mathbf{c}}, \underline{\mathbf{c}} \oplus \underline{\mathbf{e}}) \tag{6.189}$$

$$= \Pr\left\{ \Lambda^{\Sigma}_{\underline{\mathbf{c}},\underline{\mathbf{c}}\oplus\underline{\mathbf{e}}} > 0 \middle| \underline{\boldsymbol{C}} = \underline{\mathbf{c}} \right\}, \tag{6.190}$$

where

$$\Lambda^{\Sigma}_{\underline{\mathbf{c}},\underline{\mathbf{c}}\oplus\underline{\mathbf{e}}} \triangleq \sum_{t=1}^{d} (-1)^{c_{(t)}} \Lambda_{(t)}, \tag{6.191}$$

and where $\Lambda_{(t)}$ and $c_{(t)}$, $t = 1, \ldots, d$ are reindexed versions of $\Lambda_q[n]$, and $c_q[n]$, respectively. The reindexing enumerates only the elements with indices n, q for which $c_q[n] \neq \hat{c}_q[n]$, i.e., when $e_q[n] = 1$.

Moreover, knowing that the bits $b_k[n]$ are an interleaved version of the bits $c_q[n]$ and the L-values $\lambda_q[n]$ are obtained via deinterleaving of $l_k[n]$, we can write

$$\Lambda^{\Sigma}_{\underline{\mathbf{c}},\underline{\mathbf{c}}\oplus\underline{\mathbf{e}}} = \sum_{t=1}^{d} (-1)^{b_{(t)}} L_{(t)}, \tag{6.192}$$

where $L_{(t)}$ and $b_{(t)}$ are the interleaved/deinterleaved versions of $\Lambda_{(t)}$ and $c_{(t)}$, respectively.

The main reason to use (6.192) instead of (6.191) is that we already know how to model the L-values L_k (see Chapter 5). We now have to address the more delicate issue of the *joint* probabilistic model of the L-values $L_{(t)}$. The situation which is the simplest to analyze occurs when the L-values $L_{(t)}$ are independent, which we illustrate in Fig. 6.13 (a). In this case, the L-values $L_{(t)}$ are obtained from d different channel outcomes $\boldsymbol{Y}[n]$. A more complicated situation to deal with arises when we can find at least two indices $t_2 \neq t_1$ such that $L_{(t_1)}$ and $L_{(t_2)}$ correspond to $L_{k_1}[n]$ and $L_{k_2}[n]$, i.e., when $L_{(t_1)}$ and $L_{(t_2)}$ are L-values calculated at the same time n (and thus, they are dependent). The case when the L-values are independent as well as the requirements on the interleaver for this assumption to hold are discussed below. The case where the L-values are not independent is discussed in Chapter 9.

In what follows we provide a lemma that will simplify the discussions thereafter.

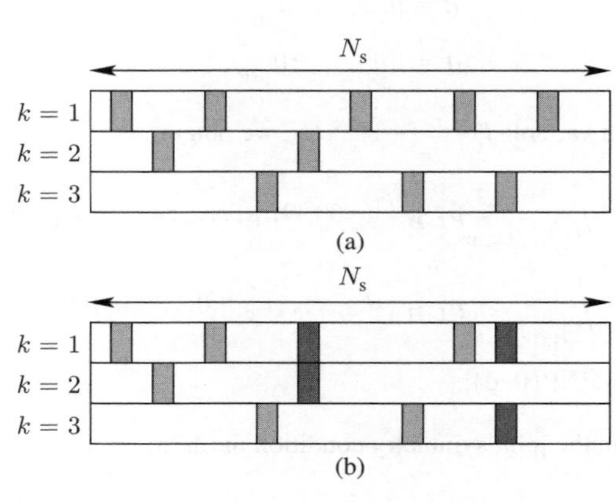

Figure 6.13 Codewords $\underline{\boldsymbol{b}}$ corresponding to different interleaved error codewords $\underline{\mathbf{e}}$. The L-values $L_{(t)}$ in (6.192) (indicated by rectangles) are (a) independent or (b) dependent random variables. The L-values, which are calculated using the same $\boldsymbol{Y}[n]$, i.e., at the same time n, are indicated with dark-gray rectangles

Lemma 6.17 *For all error codewords $\underline{\mathbf{e}} \in \mathcal{C}(d)$*

$$\mathrm{PEP}(\underline{\mathbf{c}}, \underline{\mathbf{c}} \oplus \underline{\mathbf{e}}) = \mathrm{PEP}(\underline{\mathbf{0}}, \underline{\mathbf{e}}). \tag{6.193}$$

Proof: For a given interleaving vector $\boldsymbol{\pi}$, the codewords $\underline{\mathbf{e}} \in \mathcal{B}(d)$ can be classified into two groups: the ones attaining their maximum diversity d and the ones not attaining their maximum diversity d. For clarity, we deal separately with these essentially different cases. First, we consider $\underline{\mathbf{e}} \in \mathcal{C}^{\mathrm{div}}(d)$, and next $\underline{\mathbf{e}} \notin \mathcal{C}^{\mathrm{div}}(d)$. In the first case, by definition, $\underline{\mathbf{e}} \in \mathcal{C}^{\mathrm{div}}(d)$ has all its d nonzero bits interleaved into distinct labels $\boldsymbol{b}[n]$, which means that the L-values $L_{(t)}$ are independent. Thus, we may find the PDF of $\Lambda^{\Sigma}_{\underline{\mathbf{c}},\underline{\mathbf{c}}\oplus\underline{\mathbf{e}}}$ in (6.192) as follows:

$$p_{\Lambda^{\Sigma}_{\underline{\mathbf{c}},\underline{\mathbf{c}}\oplus\underline{\mathbf{e}}}}(\lambda) = p_{L_{(1)}|B_{(1)}}((-1)^{b_{(1)}}\lambda|b_{(1)}) * \ldots * p_{L_{(d)}|B_{(d)}}((-1)^{b_{(d)}}\lambda|b_{(d)}) \tag{6.194}$$

$$= p_{L_{(1)}|B_{(1)}}(\lambda|0) * \ldots * p_{L_{(d)}|B_{(d)}}(\lambda|0), \tag{6.195}$$

where (6.195) follows from the symmetry condition in (3.74). The expression in (6.195) shows that $p_{\Lambda^{\Sigma}_{\underline{\mathbf{c}},\underline{\mathbf{c}}\oplus\underline{\mathbf{e}}}}(\lambda)$ does not depend on the codeword $\underline{\mathbf{c}}$, but only on the nonzero bits in $\underline{\mathbf{e}}$ (which determine $L_{(1)}, L_{(2)}, \ldots, L_{(d)}$), i.e.,

$$p_{\Lambda^{\Sigma}_{\underline{\mathbf{c}},\underline{\mathbf{c}}\oplus\underline{\mathbf{e}}}}(\lambda) = p_{\Lambda^{\Sigma}_{\underline{\mathbf{0}},\underline{\mathbf{e}}}}(\lambda), \tag{6.196}$$

which together with (6.190) proves (6.193) when $\underline{\mathbf{e}} \in \mathcal{C}^{\mathrm{div}}(d)$.

In the case when $\underline{\mathbf{e}} \in \mathcal{C}(d)$ but $\underline{\mathbf{e}} \notin \mathcal{C}^{\mathrm{div}}(d)$, the L-values are not i.i.d., and we cannot express the PDF $p_{\Lambda^{\Sigma}_{\underline{\mathbf{c}},\underline{\mathbf{c}}\oplus\underline{\mathbf{e}}}}(\lambda)$ as a convolution. Instead, we express the PEP using the joint PDF of all L-values as

$$\mathrm{PEP}(\underline{\mathbf{c}}, \underline{\mathbf{c}} \oplus \underline{\mathbf{e}}) = \int_{\sum_{t=1}^{d}(-1)^{b(t)}l_{(t)}>0} p_{\boldsymbol{L}|\boldsymbol{B}}([l_{(1)}, \ldots, l_{(d)}]|\boldsymbol{b})\, \mathrm{d}\boldsymbol{l}, \tag{6.197}$$

where we abbreviate the notation we use

$$\boldsymbol{L} \triangleq [L_{(1)}, \ldots, L_{(d)}], \tag{6.198}$$

$$\boldsymbol{B} \triangleq [B_{(1)}, \ldots, B_{(d)}], \tag{6.199}$$

$$\boldsymbol{b} \triangleq [b_{(1)}, \ldots, b_{(d)}], \tag{6.200}$$

$$\mathrm{d}\boldsymbol{l} \triangleq \mathrm{d}l_{(1)} \cdots \mathrm{d}l_{(d)}. \tag{6.201}$$

Then, with the change of variable $l'_{(t)} = (-1)^{b(t)} l_{(t)}$, we obtain

$$\mathrm{PEP}(\underline{\mathbf{c}}, \underline{\mathbf{c}} \oplus \underline{\mathbf{e}}) = \int_{\sum_{t=1}^{d} l_{(t)}>0} p_{\boldsymbol{L}|\boldsymbol{B}}\left([(-1)^{b_{(1)}}l_{(1)}, \ldots, (-1)^{b_{(d)}}l_{(d)}]|\boldsymbol{b}\right) \mathrm{d}\boldsymbol{l} \tag{6.202}$$

$$= \int_{\sum_{t=1}^{d} l_{(t)}>0} p_{\boldsymbol{L}|\boldsymbol{B}}\left([l_{(1)}, \ldots, l_{(d)}]|[0, \ldots, 0]\right) \mathrm{d}\boldsymbol{l} \tag{6.203}$$

$$= \mathrm{PEP}(\underline{\mathbf{0}}, \underline{\mathbf{e}}), \tag{6.204}$$

where (6.203) follows from the joint symmetry condition in (3.75). □

Lemma 6.17 allows us to simplify the WEP expression in (6.141) as follows:

$$\mathsf{WEP} \leq \frac{1}{|\mathcal{C}|} \sum_{\underline{\boldsymbol{c}} \in \mathcal{C}} \sum_{\substack{\underline{\hat{\boldsymbol{c}}} \in \mathcal{C} \\ \underline{\hat{\boldsymbol{c}}} \neq \underline{\boldsymbol{c}}}} \mathrm{PEP}(\underline{\boldsymbol{c}}, \underline{\hat{\boldsymbol{c}}}) \tag{6.205}$$

$$= \frac{1}{|\mathcal{C}|} \sum_{\underline{\boldsymbol{c}} \in \mathcal{C}} \sum_{\substack{\underline{\boldsymbol{e}} \in \mathcal{C} \\ \underline{\boldsymbol{e}} \neq \underline{\boldsymbol{0}}}} \mathrm{PEP}(\underline{\boldsymbol{c}}, \underline{\boldsymbol{c}} \oplus \underline{\boldsymbol{e}}) \tag{6.206}$$

$$= \sum_{d=d_{\mathrm{free}}}^{N_{\mathrm{c}}} \sum_{\underline{\boldsymbol{e}} \in \mathcal{C}(d)} \mathrm{PEP}(\underline{\boldsymbol{0}}, \underline{\boldsymbol{e}}) \tag{6.207}$$

$$\approx \sum_{d=d_{\mathrm{free}}}^{d_{\max}} \sum_{\underline{\boldsymbol{e}} \in \mathcal{C}(d)} \mathrm{PEP}(\underline{\boldsymbol{0}}, \underline{\boldsymbol{e}}), \tag{6.208}$$

where the approximation in (6.208) follows from the truncation we make in the number of *Hamming weights* (HWs) to consider ($d_{\mathrm{free}} \leq d \leq d_{\max}$). To transform (6.208) into a more suitable form we use a simple observation given by the following lemma.

Lemma 6.18 *For all error codewords $\underline{\boldsymbol{e}} \in \mathcal{C}^{\mathsf{div}}(d)$, the PEP in* (6.193) *depends only on the generalized Hamming weight (GHW) of the interleaved codeword $\boldsymbol{w} = W_{\mathrm{H}}(\Pi[\underline{\boldsymbol{e}}])$, i.e.,*

$$\mathrm{PEP}(\underline{\boldsymbol{0}}, \underline{\boldsymbol{e}}) = \mathsf{PEP}(\boldsymbol{w}), \tag{6.209}$$

where

$$\mathsf{PEP}(\boldsymbol{w}) \triangleq \int_0^\infty p_{\Lambda^{\Sigma}_{\underline{\boldsymbol{0}},\underline{\boldsymbol{e}}}}(\lambda)\, \mathrm{d}\lambda \tag{6.210}$$

and

$$p_{\Lambda^{\Sigma}_{\underline{\boldsymbol{0}},\underline{\boldsymbol{e}}}}(\lambda) = \{p_{L_1|B_1}(\lambda|0)\}^{*w_1} * \cdots * \{p_{L_m|B_m}(\lambda|0)\}^{*w_m}. \tag{6.211}$$

Proof: Since w_k denotes the number of nonzero elements in the error codeword $\underline{\boldsymbol{e}}$ mapped to the positions k in the label $\boldsymbol{b}[n]$, there will be w_k PDFs $p_{L_k|B_k}(l|b)$ in (6.195), and thus, we transform the latter into (6.211), which depends solely on the GHW $\boldsymbol{w}$. □

To illustrate the result of Lemma 6.18 we show in Fig. 6.14 two error codewords $\underline{\boldsymbol{e}}$ with the same HW d interleaved into codewords with different GHWs $\boldsymbol{w} = [w_1, \ldots, w_n]^{\mathrm{T}}$.

We now develop (6.208) as

$$\mathsf{WEP} \approx \sum_{d=d_{\mathrm{free}}}^{d_{\max}} \left[\sum_{\underline{\boldsymbol{e}} \in \mathcal{C}^{\mathsf{div}}(d)} \mathrm{PEP}(\underline{\boldsymbol{0}}, \underline{\boldsymbol{e}}) + \sum_{\substack{w_{\mathrm{H}}(\underline{\boldsymbol{e}})=d \\ \underline{\boldsymbol{e}} \notin \mathcal{C}^{\mathsf{div}}(d)}} \mathrm{PEP}(\underline{\boldsymbol{0}}, \underline{\boldsymbol{e}}) \right] \tag{6.212}$$

$$\approx \sum_{d=d_{\mathrm{free}}}^{d_{\max}} \sum_{\substack{\boldsymbol{w} \in \mathbb{N}^m \\ \|\boldsymbol{w}\|_1 = d}} \sum_{\substack{\underline{\boldsymbol{e}} \in \mathcal{C}^{\mathsf{div}}(d) \\ w_{\mathrm{H}}(\underline{\boldsymbol{e}}) = \boldsymbol{w}}} \mathrm{PEP}(\underline{\boldsymbol{0}}, \underline{\boldsymbol{e}}) + A \tag{6.213}$$

$$= \sum_{d=d_{\mathrm{free}}}^{d_{\max}} \sum_{\substack{\boldsymbol{w} \in \mathbb{N}^m \\ \|\boldsymbol{w}\|_1 = d}} C^{\mathcal{B}}_{\boldsymbol{w}} D^{\pi}_{d} \mathsf{PEP}(\boldsymbol{w}) + A, \tag{6.214}$$

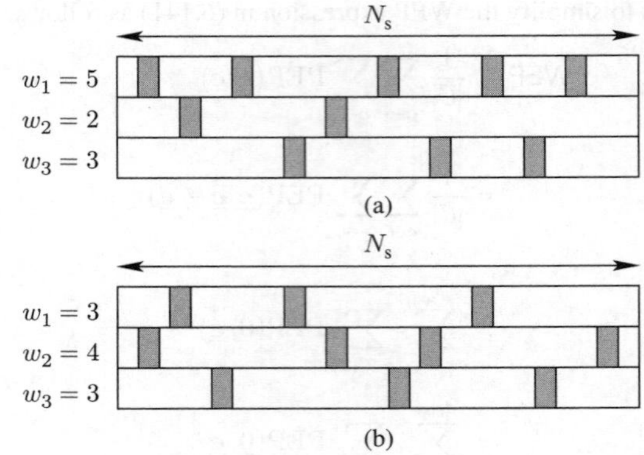

Figure 6.14 The codewords $\underline{e}$ with the same HW $d = w_{\mathrm{H}}(\underline{e}) = 10$ are interleaved into codewords with different GHWs $\boldsymbol{w} = W_{\mathrm{H}}(\Pi(\underline{e}))$: (a) GHW $\boldsymbol{w} = [5, 2, 3]^{\mathrm{T}}$ and (b) GHW $\boldsymbol{w} = [3, 4, 3]^{\mathrm{T}}$

where (6.214) follows from (6.209) and the third condition of the quasirandom interleaver in Definition 2.46, and where

$$A \triangleq \sum_{d=d_{\mathrm{free}}}^{d_{\max}} \sum_{\substack{w_{\mathrm{H}}(\underline{e})=d \\ \underline{e} \notin \mathcal{C}^{\mathrm{div}}(d)}} \mathrm{PEP}(\underline{\mathbf{0}}, \underline{e}). \tag{6.215}$$

Lemma 6.19 *For codes that satisfy* $\lim_{N_c \to \infty} C_d^{\mathcal{C}} < \infty$ *and any interleaver satisfying the first condition of quasirandomness in Definition 2.46, we have*

$$\lim_{N_c \to \infty} A = 0. \tag{6.216}$$

Proof: We bound (6.215) as

$$A \leq \sum_{d=d_{\mathrm{free}}}^{d_{\max}} \sum_{\substack{w_{\mathrm{H}}(\underline{e})=d \\ \underline{e} \notin \mathcal{C}^{\mathrm{div}}(d)}} 1 = \sum_{d=d_{\mathrm{free}}}^{d_{\max}} C_d^{\mathcal{C}}(1 - D_d^{\pi}), \tag{6.217}$$

which tends to zero for any finite $C_d^{\mathcal{C}}$ owing to the first condition of quasirandomness (see (2.141)). □

The condition of having the spectrum $C_d^{\mathcal{C}}$ bounded for given d and growing N_c is satisfied for *turbo codes* (TCs) (in fact, for TCs $C_d^{\mathcal{C}}$ actually decreases with N_c). On the other hand, for *convolutional codes* (CCs), the spectrum $C_d^{\mathcal{C}}$ may increase with N_c, but as discussed in Section 6.2.5, we may still neglect the effect of A for high SNR and large N_c. Thus, to streamline further discussions, from now on we do not consider A in (6.214). Furthermore, if we assume the first condition of quasirandomness in Definition 2.46 to be satisfied, we can express (6.214) as

$$\mathsf{WEP} \approx \sum_{d=d_{\mathrm{free}}}^{d_{\max}} \sum_{\substack{\boldsymbol{w} \in \mathbb{N}^m \\ \|\boldsymbol{w}\|_1 = d}} C_{\boldsymbol{w}}^{\mathcal{B}} \mathsf{PEP}(\boldsymbol{w}). \tag{6.218}$$

Using steps similar to those that lead to (6.208) and then to (6.214), we can develop (6.144) as follows:

$$\mathsf{BEP} \leq \frac{1}{|\mathcal{B}|} \sum_{\underline{\boldsymbol{b}} \in \mathcal{B}} \sum_{\substack{\hat{\underline{\boldsymbol{b}}} \in \mathcal{B} \\ \hat{\underline{\boldsymbol{b}}} \neq \underline{\boldsymbol{b}}}} \mathrm{PEP}(\underline{\boldsymbol{b}}, \hat{\underline{\boldsymbol{b}}}) \frac{d_{\mathrm{H}}(\underline{\boldsymbol{i}}, \hat{\underline{\boldsymbol{i}}})}{N_{\mathrm{b}}} \tag{6.219}$$

$$= \frac{1}{N_{\mathrm{b}}} \sum_{\substack{\underline{\boldsymbol{e}} \in \mathcal{B} \\ \underline{\boldsymbol{e}} \neq \underline{\boldsymbol{0}}}} \mathrm{PEP}(\underline{\boldsymbol{0}}, \underline{\boldsymbol{e}}) w_{\mathrm{H}}(\underline{\boldsymbol{i}}_{\underline{\boldsymbol{e}}}) \tag{6.220}$$

$$= \frac{1}{N_{\mathrm{b}}} \sum_{v=1}^{N_{\mathrm{b}}} \sum_{d=d_{\mathrm{free}}}^{N_{\mathrm{c}}} \sum_{\substack{\boldsymbol{w} \in \mathbb{N}^m \\ \|\boldsymbol{w}\|_1 = d}} \sum_{\underline{\boldsymbol{e}} \in \mathcal{B}(v, \boldsymbol{w})} \mathrm{PEP}(\underline{\boldsymbol{0}}, \underline{\boldsymbol{e}}) w_{\mathrm{H}}(\underline{\boldsymbol{i}}_{\underline{\boldsymbol{e}}}) \tag{6.221}$$

$$\approx \frac{1}{N_{\mathrm{b}}} \sum_{d=d_{\mathrm{free}}}^{d_{\max}} \sum_{\substack{\boldsymbol{w} \in \mathbb{N}^m \\ \|\boldsymbol{w}\|_1 = d}} \sum_{v=1}^{N_{\mathrm{b}}} v C^{\mathcal{B}}_{v,\boldsymbol{w}} \mathsf{PEP}(\boldsymbol{w}) \tag{6.222}$$

$$= \sum_{d=d_{\mathrm{free}}}^{d_{\max}} \sum_{\substack{\boldsymbol{w} \in \mathbb{N}^m \\ \|\boldsymbol{w}\|_1 = d}} B^{\mathcal{B}}_{\boldsymbol{w}} \mathsf{PEP}(\boldsymbol{w}), \tag{6.223}$$

where $\underline{\boldsymbol{i}}_{\underline{\boldsymbol{e}}}$ in (6.220) is the input sequence corresponding to the codeword $\underline{\boldsymbol{e}}$, $C^{\mathcal{B}}_{v,\boldsymbol{w}}$ is the *generalized input–output distance spectrum* (GIODS) (see Definition 2.37), and to pass from (6.222) to (6.223) we used (2.114).

We note that (6.218) and (6.223) hold for any interleaver that satisfies the first and the third conditions in Definition 2.46. The second condition of quasirandomness is not exploited yet and, as we show in the following theorem, it simplifies the analysis further.

Theorem 6.20 (BEP and WEP in BICM with Quasirandom Interleavers) *The WEP and BEP in BICM with quasirandom interleavers can be approximated using the DS* $C^{\mathcal{B}}_d$ *and the input-dependent distance spectrum (IDS)* $B^{\mathcal{B}}_d$ *as*

$$\mathsf{WEP} \approx \sum_{d=d_{\mathrm{free}}}^{d_{\max}} C^{\mathcal{B}}_d \mathsf{PEP}(d), \tag{6.224}$$

$$\mathsf{BEP} \approx \sum_{d=d_{\mathrm{free}}}^{d_{\max}} B^{\mathcal{B}}_d \mathsf{PEP}(d), \tag{6.225}$$

where

$$\mathsf{PEP}(d) \triangleq \int_0^{\infty} \{p_{\Lambda}(\lambda)\}^{*d} \mathrm{d}\lambda \tag{6.226}$$

and Λ *is a random variable with PDF*

$$p_{\Lambda}(\lambda) \triangleq \frac{1}{m} \sum_{k=1}^{m} p_{L_k|B_k}(\lambda|0). \tag{6.227}$$

Proof: Using the second condition of quasirandomness in Definition 2.46 lets us express (6.218) as

$$\mathsf{WEP} \approx \sum_{d=d_{\mathrm{free}}}^{d_{\max}} C^{\mathcal{B}}_d \sum_{\substack{\boldsymbol{w} \in \mathbb{N}^m \\ \|\boldsymbol{w}\|_1 = d}} \frac{1}{m^d} \binom{d}{\boldsymbol{w}} \mathsf{PEP}(\boldsymbol{w}). \tag{6.228}$$

The innermost sum in (6.228) can be expressed using (6.211) and (6.210) as

$$\sum_{\substack{\boldsymbol{w}\in\mathbb{N}^m\\ \|\boldsymbol{w}\|_1=d}} \frac{1}{m^d}\binom{d}{\boldsymbol{w}} \mathsf{PEP}(\boldsymbol{w}) = \int_0^\infty t(\lambda)\,\mathrm{d}\lambda, \tag{6.229}$$

where

$$t(\lambda) \triangleq \sum_{\substack{\boldsymbol{w}\in\mathbb{N}^m\\ \|\boldsymbol{w}\|_1=d}} \frac{1}{m^d}\binom{d}{\boldsymbol{w}} \{p_{L_1|B_1}(\lambda|0)\}^{*w_1} * \ldots * \{p_{L_m|B_m}(\lambda|0)\}^{*w_m}\,\mathrm{d}\lambda. \tag{6.230}$$

The function $t(\lambda)$ can be interpreted as the PDF of a random variable T with MGF

$$\mathsf{P}_T(s) = \sum_{\substack{\boldsymbol{w}\in\mathbb{N}^m\\ \|\boldsymbol{w}\|_1=d}} \frac{1}{m^d}\binom{d}{\boldsymbol{w}} \left[\mathsf{P}_{L_1}(s)\right]^{w_1}\cdot \ldots \cdot \left[\mathsf{P}_{L_m}(s)\right]^{w_m} \tag{6.231}$$

$$= \left(\frac{1}{m}\sum_{k=1}^{m}\mathsf{P}_{L_k}(s)\right)^d, \tag{6.232}$$

where (6.232) follows from (2.18) and $\mathsf{P}_{L_k}(s) = \mathbb{E}_{L_k|B_k=0}[\mathrm{e}^{sL_k}]$ is the MGF of L_k conditioned on $B_k = 0$. It is seen immediately that $t(\lambda) = \{p_\Lambda(\lambda)\}^{*d}$, where $p_\Lambda(\lambda)$ is given by (6.227). We therefore conclude that

$$\sum_{\substack{\boldsymbol{w}\in\mathbb{N}^m\\ \|\boldsymbol{w}\|_1=d}} \frac{1}{m^d}\binom{d}{\boldsymbol{w}} \mathsf{PEP}(\boldsymbol{w}) = \mathsf{PEP}(d), \tag{6.233}$$

which used in (6.228) gives (6.224).

The proof to obtain (6.225) follows similar steps to the ones above, with the difference that we have to include the weights of the input sequences, and thus, use the IDS $B_d^{\mathcal{B}}$ instead of the DS $C_d^{\mathcal{B}}$. □

The key difference between the bounds for the ML decoder in (6.169) and (6.171) with those for the BICM decoder in (6.224) and (6.225), is that the performance of BICM depends on the DS of the binary code $\mathcal{B}$, and not on the DS of the code $\mathcal{X}$. Further, because of the numerous approximations we made, the expressions (6.224) and (6.225) are not true bounds, but only approximations. Nevertheless, we will later show that these approximations indeed give accurate predictions of the WEP and BEP for BICM.

The somewhat lengthy derivations above may be simplified using the model of independent and *random* assignment of the bits $c_q[n]$ to the positions of the mapper in Fig. 2.24, and assuming that all codewords achieve their maximum diversity. Then, $p_\Lambda(\lambda)$ is immediately obtained as the average of the PDF of the L-values over the different bit positions. Our objective here was to show that, in order to use such a random-assignment model, the interleaver (which is a deterministic element of the transceiver) must comply with all conditions of the quasirandomness in Definition 2.46.

The quasirandomness condition is not automatically satisfied by all interleavers. For example, we showed in Example 2.47 that for a *convolutional code* (CC), a rectangular interleaver does not satisfy the second condition of quasirandomness, which allowed us to replace $\frac{C_{\boldsymbol{w}}^{\mathcal{B}}}{C_d^{\mathcal{B}}}$ by $\frac{1}{m^d}\binom{d}{\boldsymbol{w}}$ in (6.228). Therefore, while the first and the third condition of the quasirandomness hold for the rectangular interleaver and we may apply (6.218),[10] we cannot use Theorem 6.20 in this case. The importance of the second condition of the quasirandomness is illustrated in the following example.

[10] To do so, we would need to calculate the *generalized distance spectrum* (GDS) of the code after using the rectangular interleaver, i.e, the GDS $C_{\boldsymbol{w}}^{\mathcal{B}}$ of the code $\mathcal{B}$ (see Definitions 2.35 and 2.38).

Example 6.21 (Interleaver Effects) *We assume $m = 2$ where each of the bits $b_k[n]$ is transmitted over an independent channel using a 2PAM constellation, i.e., $\boldsymbol{x}[n] = [x_1[n], x_2[n]]^{\mathrm{T}}$, where $x_k[n] = \Phi(b_k[n])$. The SNR in each of the subchannels is given by snr_1 and snr_2 respectively, with $\mathsf{snr}_2 = 5\ \mathsf{snr}_1$. The code bits $\underline{\mathbf{c}}$ obtained at the output of the rate $R_{\mathrm{c}} = 1/2$ CENC with constraint length $\nu = 2$ and generator polynomials $\mathbf{G} = [5, 7]$ are interleaved using two different interleavers: a pseudorandom interleaver (see Definition 2.34) and a rectangular interleaver with period $T = 447$ (see Definition 2.32). In the simulations we encode $N_{\mathrm{b}} = 10^5$ information bits and add two dummy zeros in order to bring the encoder to the zero state. Thus, the effective coding rate is given by $\frac{N_{\mathrm{b}}}{N_{\mathrm{b}}/R_{\mathrm{c}}+2} \approx R_{\mathrm{c}}$.*

From the PDF of the L-values $p_{L_k}(l|0) = \Psi(l, -4\,\mathsf{snr}_k, 8\,\mathsf{snr}_k)$ (see Example 3.7), we obtain

$$p_{\Lambda}(\lambda) = \frac{1}{2}\left(p_{L_1}(\lambda|0) + p_{L_2}(\lambda|0)\right) \tag{6.234}$$

$$= \frac{1}{2}\left(\Psi(l, -4\,\mathsf{snr}_1, 8\,\mathsf{snr}_1) + \Psi(l, -4\,\mathsf{snr}_2, 8\,\mathsf{snr}_2)\right), \tag{6.235}$$

and thus,

$$\mathsf{PEP}(d) = \sum_{k=0}^{d} \frac{1}{2^d}\binom{d}{k} Q\left(\sqrt{2(k\,\mathsf{snr}_1 + (d-k)\,\mathsf{snr}_2)}\right). \tag{6.236}$$

In Fig. 6.15 we compare the simulation results and the analytical bound (6.225) with $d_{\max} = 25$ and where the PEP is given by (6.236). These results clearly show that because the second condition of quasi-randomness is not fulfilled by the rectangular interleaver, we cannot apply the performance evaluation approach based on the "average" PDF in (6.227).

For the remainder of this chapter, we use only pseudorandom interleavers, and thus, assume the conditions of quasirandomness in Definition 2.46 are fulfilled. In analogy to Example 6.15, the following example shows the performance of BICM with CENCs.

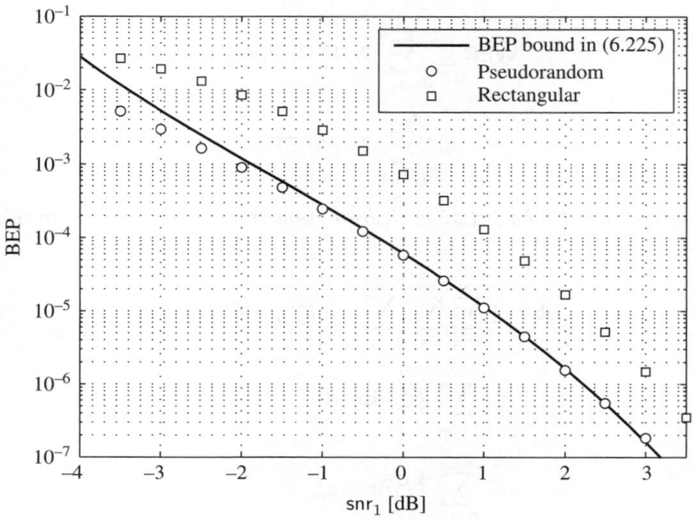

Figure 6.15 The analytical bound obtained using (6.225) and (6.236) (solid line) and the simulation results (markers) for pseudorandom and rectangular interleavers

Example 6.22 (Convolutional Encoders and BICM Decoding) *Similarly to the bound we derived in* (6.173)*, we now consider a CENC-based BICM (note that in Example 6.15 we used a trellis encoder, i.e., the interleaver was not present). As in Example 6.15 we consider the first-error events corresponding to the diverging paths in the trellis. Here, because of the linearity of the code and the assumption of quasirandomness we may arbitrarily set the reference path corresponding to the transmitted codeword* $\underline{\mathbf{c}}$ *and analyze diverging paths* $\underline{\hat{\mathbf{c}}} = \underline{\mathbf{c}} \oplus \underline{\mathbf{e}}$*. Thus, we fix* $\underline{\mathbf{c}} = \underline{\mathbf{0}}$ *and consider error codewords* $\underline{\hat{\mathbf{c}}} = \underline{\mathbf{e}}$ *knowing that* $\underline{\mathbf{c}} = \underline{\mathbf{0}}$ *corresponds to a path in the trellis that passes only through zero states. The first-error event can be then upper bounded as*

$$P_{1,\mathrm{e}} \leq \sum_{\ell=1}^{\infty} \sum_{\underline{\hat{\mathbf{e}}} \in \hat{\mathcal{C}}_{\ell}(\underline{\mathbf{0}})} \mathsf{PEP}(\underline{\mathbf{0}}, \underline{\hat{\mathbf{e}}}) \tag{6.237}$$

$$= \sum_{\ell=1}^{\infty} \sum_{\underline{\hat{\mathbf{e}}} \in \hat{\mathcal{C}}_{\ell}(\underline{\mathbf{0}})} \mathsf{PEP}(w_{\mathrm{H}}(\underline{\hat{\mathbf{e}}})), \tag{6.238}$$

where $\hat{\mathcal{C}}_{\ell}(\underline{\mathbf{0}})$ *is the set of codewords corresponding to the paths that start at the zero state of the trellis at a given time* t_0 *and end at the zero state of the trellis* $k_{\mathrm{c}}\ell$ *times instants later. Denoting by* $F^{\mathcal{C}}_{d,\ell}$ *the number of codewords* $\underline{\hat{\mathbf{e}}} \in \hat{\mathcal{C}}_{\ell}(\underline{\mathbf{0}})$ *with HW d and by* $F^{\mathcal{C}}_{v,d,\ell}$ *the number of codewords* $\underline{\hat{\mathbf{e}}} \in \hat{\mathcal{C}}_{\ell}(\underline{\mathbf{0}})$ *with HW d generated by input sequences with HW v, the bound in* (6.237) *can be expressed as*

$$P_{1,\mathrm{e}} \leq \sum_{\ell=1}^{\infty} \sum_{d=d_{\mathrm{free}}}^{\infty} F^{\mathcal{C}}_{d,\ell} \mathsf{PEP}(d) = \sum_{d=d_{\mathrm{free}}}^{\infty} F^{\mathcal{C}}_{d} \mathsf{PEP}(d), \tag{6.239}$$

where $F^{\mathcal{C}}_{d}$ *is the WD of the CENC defined as*[11]

$$F^{\mathcal{C}}_{d} \triangleq \sum_{\ell=1}^{\infty} F^{\mathcal{C}}_{d,\ell} = \sum_{\ell=1}^{\infty} \sum_{v=1}^{\infty} F^{\mathcal{C}}_{v,d,\ell}. \tag{6.240}$$

In analogy to (6.178)*, the WEP for BICM can be bounded as*

$$\mathsf{WEP} \leq \sum_{d=d_{\mathrm{free}}}^{\infty} N_{\mathrm{q}} F^{\mathcal{C}}_{d} \mathsf{PEP}(d) \tag{6.241}$$

$$\approx \sum_{d=d_{\mathrm{free}}}^{d_{\max}} N_{\mathrm{q}} F^{\mathcal{C}}_{d} \mathsf{PEP}(d). \tag{6.242}$$

To obtain bounds on the BEP for BICM, each error event in (6.239) *is weighted by the number of bits in error (v out of* k_{c}*), i.e.,*

$$\mathsf{BEP} \leq \sum_{\ell=1}^{\infty} \sum_{v=1}^{\infty} \sum_{d=d_{\mathrm{free}}}^{\infty} \frac{v}{k_{\mathrm{c}}} F^{\mathcal{C}}_{v,d,\ell} \mathsf{PEP}(d)$$

$$\leq \sum_{d=d_{\mathrm{free}}}^{\infty} G^{\mathcal{C}}_{d} \mathsf{PEP}(d) \tag{6.243}$$

$$\approx \sum_{d=d_{\mathrm{free}}}^{d_{\max}} G^{\mathcal{C}}_{d} \mathsf{PEP}(d), \tag{6.244}$$

[11] We note that the WD ignores how the input sequences are mapped into the codewords, and thus, the WD is in fact a property of the code only.

where $G^{\mathcal{C}}_d$ is the input-dependent weight distribution (IWD) of the CENC defined as

$$G^{\mathcal{C}}_d \triangleq \sum_{\ell=1}^{\infty}\sum_{v=1}^{\infty} \frac{v}{k_c} F^{\mathcal{C}}_{v,d,\ell}. \tag{6.245}$$

The IWD of the CENC $G^{\mathcal{C}}_d$ in (6.245) is the total HW of all input sequences that generate error events with HW d, where an error event is defined as a path in the trellis representation of the code that leaves the zero state and remerges with it after an arbitrary number of trellis stages. In the following example, we show how to calculate this IWD.

Example 6.23 (WD and IWD of the Encoder $\mathbf{G} = [5, 7]$) *Consider the rate $R_c = 1/2$ CENC with constraint length $\nu = 2$ and polynomial generators $\mathbf{G} = [5, 7]$ (see Table 2.1). The encoder for this code is shown in Fig. 6.16 (a). Fig. 6.16 (b) shows the state machine associated to the encoder, where the four states represent the states of the two memories. We label each state transition with dummy variables V and D, where the exponents of V and D represent the HW of the input and output of the encoder, respectively. The presence of a V means that the input bit was one (marked as a solid line) and the absence of a V means that the input bit was a zero (marked as a dashed lines). In addition, each state is labeled by a dummy variable $\{\xi_a, \xi_b, \xi_c, \xi_d, \xi_e\}$, which can be used to write the state equations:*

$$\xi_b = VD^2\xi_a + V\xi_c, \tag{6.246}$$
$$\xi_c = D\xi_d + D\xi_b, \tag{6.247}$$
$$\xi_d = VD\xi_b + VD\xi_d, \tag{6.248}$$
$$\xi_e = D^2\xi_c. \tag{6.249}$$

The equations (6.246)–(6.249) can be used to find a transfer function of the code $T(V, D)$ as

$$T(V, D) = \frac{\xi_e}{\xi_a} = \frac{VD^5}{1-2VD} = D^5V + 2D^6V^2 + 4D^7V^3 + 8D^8V^4 + \cdots.$$

The WD of the code can be obtained as

$$F^{\mathcal{C}}_d = \frac{1}{d!}\frac{\partial^d}{\partial D^d} T(1, D)|_{D=0} \tag{6.250}$$
$$= \begin{cases} 2^{d-5}, & \text{if } d \geq 5 \\ 0, & \text{otherwise} \end{cases}. \tag{6.251}$$

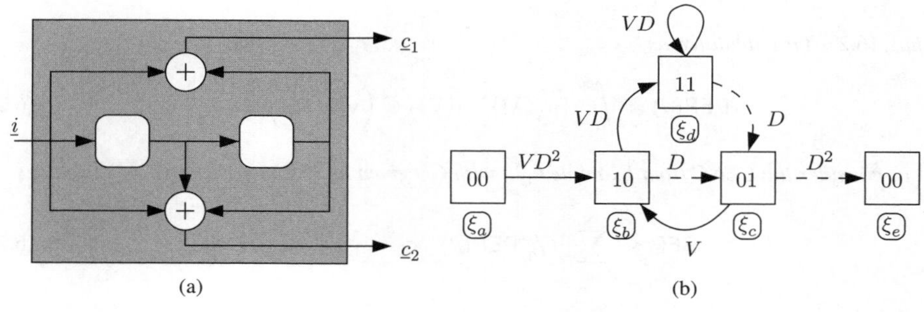

Figure 6.16 CENC $\mathbf{G} = [5, 7]$: (a) encoder and (b) state machine

The IWD of the code can be finally obtained as

$$G_d^{\mathcal{C}} = \frac{1}{d!}\frac{\partial^d}{\partial D^d}\frac{\partial}{\partial V}T(V,D)|_{D=0,V=1} \tag{6.252}$$

$$= \begin{cases}(d-4)2^{d-5}, & \text{if } d \geq 5\\ 0, & \text{otherwise}\end{cases}. \tag{6.253}$$

We note that the enumeration of the diverging and merging paths in the trellis we analyzed in Example 6.23 does not consider all the codewords with HW d because we do not take into account paths that diverged/merged multiple times. Assigning a codeword $\underline{\mathbf{e}}_k$ to the kth diverging/merging path, we see that all error codewords are mutually orthogonal and their sum $\underline{\mathbf{e}} = \underline{\mathbf{e}}_1 \oplus \cdots \oplus \underline{\mathbf{e}}_k$ is also a valid codeword. Thus, they comply with the conditions we defined in Section 6.2.2, which lets us expurgate the codeword $\underline{\mathbf{e}}$ from the bound. In other words, the enumeration of the uniquely-diverging codewords provides us with an expurgated bound. Therefore, for CENCs we have

$$C_d^{\mathcal{B}} \approx N_{\text{q}} F_d^{\mathcal{C}}, \tag{6.254}$$

$$B_d^{\mathcal{B}} \approx G_d^{\mathcal{C}}, \tag{6.255}$$

where the approximation is due to the expurgation.

Since the performance of the receiver is directly affected by the characteristics of the CENC (via $F_d^{\mathcal{C}}$ and $G_d^{\mathcal{C}}$), efforts should be made to optimize the CENC. In fact, the CENCs we show in Table 2.1 are the encoders optimal in terms of the IWD $G_d^{\mathcal{C}}$. To find them, an exhaustive search over the CENC universe $\mathcal{G}_{k_c,n_c,\nu}$ is performed: first, the *free Hamming distance* (FHD) d_{free} is maximized and next, the corresponding term $G_{d_{\text{free}}}^{\mathcal{C}}$ is minimized. If multiple encoders with the same d_{free} and the same $G_{d_{\text{free}}}^{\mathcal{C}}$ are found, those that minimize $G_{d_{\text{free}}+1}^{\mathcal{C}}$ are chosen, then $G_{d_{\text{free}}+2}^{\mathcal{C}}$ is analyzed, and so on. We refer to these optimized CENCs shown in Table 2.1 as "ODS CENCs" even if the literature uses the name "ODS codes." We use the name ODS CENCs because $G_d^{\mathcal{C}}$ depends on the encoder and not only on the code.

Example 6.24 (BEP for 2PAM **in the AWGN Channel)** *Arguably the simplest case to deal with is that of* 2PAM *transmission over the AWGN channel we considered in Example 3.7. Since the L-value L is a Gaussian random variable with PDF given by (see* (3.66)*)*

$$p_L(l|0) = \Psi\left(l, -4\,\mathsf{snr}, 8\,\mathsf{snr}\right), \tag{6.256}$$

from (6.211) *we obtain*

$$p_{\Lambda^{\Sigma}_{\underline{\mathbf{0}},\underline{\mathbf{e}}}}(\lambda) = \{p_L(\lambda|0)\}^{*d} = \Psi\left(\lambda, -4d\,\mathsf{snr}, 8d\,\mathsf{snr}\right), \tag{6.257}$$

and thus, (6.226) *is calculated as*

$$\mathsf{PEP}(d) = \int_0^{\infty}\{p_{\Lambda}(\lambda)\}^{*d}\,\mathrm{d}\lambda = Q\left(\sqrt{2d\,\mathsf{snr}}\right). \tag{6.258}$$

Let us consider the CENC from Example 6.23 with $G_d^{\mathcal{C}}$ given by (6.253)*. Then,* (6.243) *becomes*

$$\mathsf{BEP} \leq \sum_{d=d_{\text{free}}}^{\infty} G_d^{\mathcal{C}}\,\mathsf{PEP}(d) \tag{6.259}$$

$$= \sum_{d=5}^{\infty}(d-4)2^{d-5}Q\left(\sqrt{2d\,\mathsf{snr}}\right) \tag{6.260}$$

$$\leq \frac{Q\left(\sqrt{8\,\mathsf{snr}}\right)}{2} \sum_{d=1}^{\infty} d2^d \exp(-d\,\mathsf{snr}) \tag{6.261}$$

$$= \frac{Q\left(\sqrt{8\,\mathsf{snr}}\right)}{2} \sum_{d=1}^{\infty} dr^d \tag{6.262}$$

$$= Q\left(\sqrt{8\,\mathsf{snr}}\right) \frac{\mathrm{e}^{-\mathsf{snr}}}{(1-2\mathrm{e}^{-\mathsf{snr}})^2}, \tag{6.263}$$

where $d_{\text{free}} = 5$ *follows from* (2.18) *and in* (6.261) *we bounded* (6.258) *using* $Q(\sqrt{a+b}) \leq Q(\sqrt{a})\exp(-b^2/2)$. *To obtain* (6.262) *we used* $r = 2\mathrm{e}^{-\mathsf{snr}}$ *and to obtain* (6.263) *we used the closed-form expression for the arithmetic-geometric series* $\sum_{d=1}^{\infty} dr^d = \frac{r}{(1-r)^2}$, *which converges for* $r < 1$, *i.e., for* $\mathsf{snr} > \log(2) = -1.59\,\text{dB}$. *In a similar way we can bound the WEP using* (6.241) *and* (6.251)*:*

$$\mathsf{WEP} \leq \sum_{d=d_{\text{free}}}^{\infty} N_{\text{q}} F_d^{\mathcal{C}} \mathsf{PEP}(d) \tag{6.264}$$

$$= N_{\text{b}} \sum_{d=5}^{\infty} 2^{d-5} Q\left(\sqrt{2d\,\mathsf{snr}}\right) \tag{6.265}$$

$$= Q\left(\sqrt{10\,\mathsf{snr}}\right) \frac{N_{\text{b}}}{1-2\mathrm{e}^{-\mathsf{snr}}}. \tag{6.266}$$

Using this simple encoder, we can now illustrate some of the assumptions we made in the previous section concerning the convergence of the upper bounds. In Fig. 6.17 we show the simulation results obtained and compare them with the bound in (6.263) *and the approximation in* (6.244) *for* $d_{\text{max}} = d_{\text{free}} + 10$ *and* $d_{\text{max}} = d_{\text{free}} + 500$. *This figure shows that* $d_{\text{max}} = d_{\text{free}} + 10$ *is sufficient to approximate the BEP or WEP using* (6.244) *or* (6.242) *for the range of SNR where the bound is actually useful. In other words, only a few terms in the approximation* (6.244) *(here* $d_{\text{max}} - d_{\text{free}} + 1 = 11$*) are needed to approximate well the BEP or SEP.*

6.2.5 *BICM Decoder Revisited*

To obtain the results in Theorem 6.20, we removed from (6.214) the contributions of the codewords that do not achieve their maximum diversity. Here, we want to have another look at this issue, but instead of proofs, we outline a few assumptions that hold indeed in practice and are sufficient to explain why we can neglect the factor A in (6.214).

To explicitly show the dependence of the PEP on the SNR, we use $\mathsf{PEP}(d|\mathsf{snr})$ to denote the PEP in (6.226), $\mathrm{PEP}(\underline{\mathbf{0}}, \underline{\mathbf{e}}|\mathsf{snr})$ to denote the PEP in (6.193), and $\mathsf{PEP}(\boldsymbol{w}|\mathsf{snr})$ to denote (6.210).

We then assume that $\mathsf{PEP}(d|\mathsf{snr})$ is monotonically decreasing in both arguments, i.e.,

$$\mathsf{PEP}(d|\mathsf{snr}) \leq \mathsf{PEP}(d'|\mathsf{snr}), \quad \forall d' \leq d, \tag{6.267}$$

$$\mathsf{PEP}(d|\mathsf{snr}) \leq \mathsf{PEP}(d|\mathsf{snr}'), \quad \forall \mathsf{snr}' \leq \mathsf{snr}. \tag{6.268}$$

In other words, we assert that increasing the HD between the codewords or increasing the SNR decreases the PEP.

Now we are able to relate (6.268) to the PEP expression $\mathsf{PEP}(\boldsymbol{w}|\mathsf{snr})$. In general, $\mathsf{PEP}(\boldsymbol{w}|\mathsf{snr}) \neq \mathsf{PEP}(d|\mathsf{snr})$, but because $\mathsf{PEP}(\boldsymbol{w}|\mathsf{snr})$ is finite, for any snr and $\boldsymbol{w} \in \mathbb{N}^m$, and thanks to (6.268), we can always find snr' such that

$$\mathsf{PEP}(\boldsymbol{w}|\mathsf{snr}) \leq \mathsf{PEP}(d|\mathsf{snr}'), \tag{6.269}$$

where $d = \|\boldsymbol{w}\|_1$.

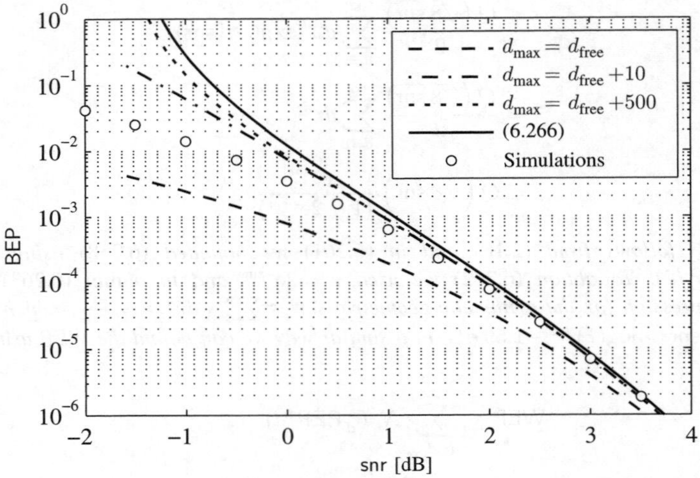

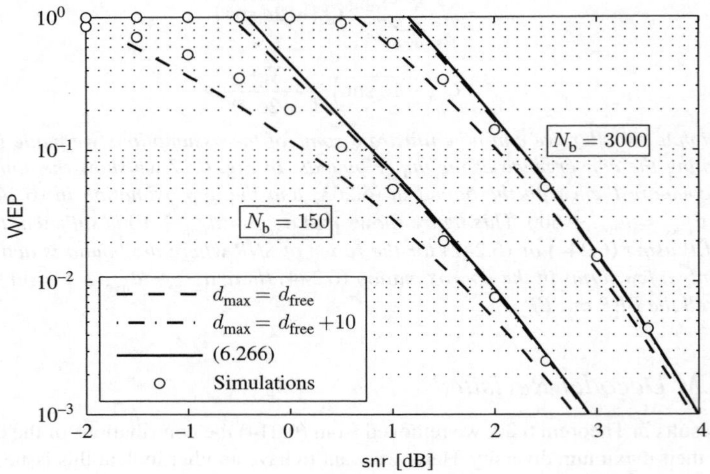

Figure 6.17 Simulation results (circles) are compared to the bounds (6.263) and (6.266) (solid lines) and the approximations (6.244) and (6.242) for the ODS CENC with $\mathbf{G} = [5, 7]$; the WEP results are shown for two values of N_b. The bounds and approximations are only shown in the range of SNR where (6.259) is guaranteed to converge, i.e., for $\mathsf{snr} > -1.59$ dB

Further, we assume that if $\underline{\mathbf{e}}^{\text{div}} \in \mathcal{C}(d')$ is a "subcodeword" of $\underline{\mathbf{e}} \in \mathcal{C}(d)$,[12] i.e., $e_q^{\text{div}}[n] = 1 \Rightarrow e_q[n] = 1$, and $\underline{\mathbf{e}}^{\text{div}}$ is such that it achieves its maximum diversity, i.e., $\underline{\mathbf{e}}^{\text{div}} \in \mathcal{C}^{\text{div}}(d')$, then

$$\text{PEP}(\underline{\mathbf{0}}, \underline{\mathbf{e}}|\mathsf{snr}) \leq \text{PEP}(\underline{\mathbf{0}}, \underline{\mathbf{e}}^{\text{div}}|\mathsf{snr}), \tag{6.270}$$

where $d' \leq d$ and where (6.270) holds for any $\underline{\mathbf{e}}$ and any $\underline{\mathbf{e}}^{\text{div}}$.

Since $\text{PEP}(\underline{\mathbf{0}}, \underline{\mathbf{e}}|\mathsf{snr}) = \Pr\{L^{\Sigma} > 0|\mathsf{snr}\}$, where $L^{\Sigma} \triangleq \sum_{t=1}^{d} L_{(t)}$ is the sum of L-values corresponding to the d nonzero bits in $\underline{\mathbf{e}}$, to obtain $\text{PEP}(\underline{\mathbf{0}}, \underline{\mathbf{e}}^{\text{div}}|\mathsf{snr})$ we may need to remove some elements from $\underline{\mathbf{e}}$. Thus, the expression in (6.270) shows that removing one (or more) L-values from the sum $L^{\Sigma} = \sum_{t=1}^{d} L_{(t)}$ yields a larger PEP. The intuition behind such an assertion is the following: each

[12] The codeword $\underline{\mathbf{e}}$ we consider here may or may not achieve its maximum diversity.

L-value contributes to improve the reliability of the detection so removing any element should produce a larger PEP. When the sum L^{Σ} is composed of i.i.d. L-vales $L_{(t)}$, this can be better understood from Sections 6.3.3 and 6.3.4, where we showed that an upper bound on the PEP is determined by the exponential of the sum of the *cumulant-generating functions* (CGFs) corresponding to each of the L-values. As the CGFs are negative, removing one L-value corresponds to increasing the bound. On the other hand, if some of the L-values in L^{Σ} are not i.i.d., our assertion is slightly less obvious. We treat such a case in more detail in Chapter 9, where we show that adding dependent L-values decreases the PEP, and thus, removing L-values from the sum must increase the PEP.

Combining (6.270) and (6.269) we have that for any $\underline{\boldsymbol{e}} \in \mathcal{C}(d)$ and snr, we can find $\underline{\boldsymbol{e}}^{\mathrm{div}} \in \mathcal{C}^{\mathsf{div}}(d')$ such that

$$\mathrm{PEP}(\underline{\boldsymbol{0}}, \underline{\boldsymbol{e}}|\mathsf{snr}) \leq \mathrm{PEP}(\underline{\boldsymbol{0}}, \underline{\boldsymbol{e}}^{\mathrm{div}}|\mathsf{snr}) \tag{6.271}$$

$$= \mathsf{PEP}(W_{\mathrm{H}}(\underline{\boldsymbol{e}}^{\mathrm{div}})|\mathsf{snr}) \tag{6.272}$$

$$\leq \mathsf{PEP}(d'|\mathsf{snr}') \tag{6.273}$$

$$\leq \mathsf{PEP}(\lceil d/m \rceil|\mathsf{snr}'), \tag{6.274}$$

where $\mathsf{snr}' \leq \mathsf{snr}$ and where we used the fact that we can always find a codeword $\underline{\boldsymbol{e}}^{\mathrm{div}}$ with interleaving diversity d' satisfying $d' \geq \lceil d/m \rceil$ (see (2.116)).

We further make a simple extension to the conditions of quasirandomness we introduced in Section 2.7. Namely, we assume that if the interleaver is quasirandom, then the diversity efficiency $D_d^{\boldsymbol{\pi}}$ decreases monotonically with d, i.e.,

$$D_{d'}^{\boldsymbol{\pi}} \geq D_d^{\boldsymbol{\pi}}, \quad \forall d' \leq d. \tag{6.275}$$

This property holds for random interleavers as we showed in (2.127). Here we extend it to the case of quasirandom (but fixed) interleavers, as we already did with other properties of the interleavers in Definition 2.46.

Let us recall now the formula for the WEP given by (6.224), where we reconsider the term A in (6.215). Then, the WEP can be expressed as

$$\mathsf{WEP} \approx \overline{\mathsf{WEP}}(\mathsf{snr}) + \sum_{d=d_{\mathrm{free}}}^{d_{\max}} \sum_{\substack{\underline{\boldsymbol{e}} \notin \mathcal{C}^{\mathsf{div}}(d) \\ w_{\mathrm{H}}(\underline{\boldsymbol{e}})=d}} \mathrm{PEP}(\underline{\boldsymbol{0}}, \underline{\boldsymbol{e}}|\mathsf{snr}), \tag{6.276}$$

where

$$\overline{\mathsf{WEP}}(\mathsf{snr}) \triangleq \sum_{d=d_{\mathrm{free}}}^{d_{\max}} C_d^{\mathcal{B}} \mathsf{PEP}(d|\mathsf{snr}) \tag{6.277}$$

denotes the WEP approximation in (6.224), which ignores the codewords that do not achieve maximum diversity.

The WEP in (6.276) can be expressed as

$$\mathsf{WEP} \leq \overline{\mathsf{WEP}}(\mathsf{snr}) + \sum_{d=d_{\mathrm{free}}}^{d_{\max}} (C_d^{\mathcal{B}} - |\mathcal{C}^{\mathsf{div}}(d)|) \cdot \mathsf{PEP}(\lceil d/m \rceil|\mathsf{snr}') \tag{6.278}$$

$$= \overline{\mathsf{WEP}}(\mathsf{snr}) + \sum_{d=d_{\mathrm{free}}}^{d_{\max}} C_d^{\mathcal{B}}(1 - D_d^{\boldsymbol{\pi}}) \cdot \mathsf{PEP}(\lceil d/m \rceil|\mathsf{snr}') \tag{6.279}$$

$$\leq \overline{\mathsf{WEP}}(\mathsf{snr}) + \sum_{d=d_{\mathrm{free}}}^{d_{\max}} C_d^{\mathcal{B}}(1 - D_{d_{\max}}^{\boldsymbol{\pi}}) \cdot \mathsf{PEP}(d|\mathsf{snr}'') \tag{6.280}$$

$$= \overline{\mathsf{WEP}}(\mathsf{snr}) + (1 - D_{d_{\max}}^{\boldsymbol{\pi}})\overline{\mathsf{WEP}}(\mathsf{snr}''), \tag{6.281}$$

where $\mathsf{snr} \geq \mathsf{snr}' \geq \mathsf{snr}''$. In order to obtain (6.278) we used (6.271) and to obtain (6.279) we used (2.117). To obtain (6.280) we used (6.267) and (6.268), where $\mathsf{snr}'' \leq \mathsf{snr}'$ should be sufficiently small so as to compensate for the increase of the first argument of the PEP (from $\lceil d/m \rceil$ to d). Here we also used (6.275), which implies that $(1 - D_d^{\pi}) \leq (1 - D_{d_{\max}}^{\pi})$.

From (6.281) we conclude that for sufficiently high SNR, we can neglect the term $(1 - D_{d_{\max}}^{\pi})\, \overline{\mathsf{WEP}}(\mathsf{snr}'')$ provided that we increase N_{c} so as to make the term $(1 - D_{d_{\max}}^{\pi})$ sufficiently small. This condition is fulfilled when using quasirandom interleavers. An alternative interpretation of this result is that the true WEP for a finite N_{c} is formed by two terms, where the second term $(1 - D_{d_{\max}}^{\pi})\overline{\mathsf{WEP}}(\mathsf{snr}'')$ corresponds to a WEP obtained for an equivalent SNR $\mathsf{snr}'' \leq \mathsf{snr}$. This factor might be relevant for the evaluation of the WEP for small values of N_{c}.

It is worth noting that all the considerations in this section may be repeated in the case of a BEP analysis, as well as for fading channels. For the former, the PEP should be weighted by the number of bits in error, and for the latter, $\mathsf{PEP}(d|\mathsf{snr})$ should be replaced by $\mathsf{PEP}(d|\overline{\mathsf{snr}})$.

6.3 PEP Evaluation

From the results in Sections 6.2.4 and 6.2.5, we know that finding the PEP in (6.166) and (6.226) is instrumental to evaluate the performance of the ML and BICM decoders, respectively. In the case of the BICM decoder we need to evaluate

$$\mathsf{PEP}(d) = \int_0^{\infty} p_{\Lambda^{\Sigma}}(\lambda)\, \mathrm{d}\lambda, \tag{6.282}$$

where

$$p_{\Lambda^{\Sigma}}(\lambda) = \{p_{\Lambda}(\lambda)\}^{*d} \tag{6.283}$$

and $p_{\Lambda}(\lambda)$ is the average PDF given in (6.227). Alternatively, we might want to calculate

$$\mathsf{PEP}(\boldsymbol{w}) = \int_0^{\infty} p_{\Lambda^{\Sigma}}(\lambda)\, \mathrm{d}\lambda, \tag{6.284}$$

where

$$p_{\Lambda^{\Sigma}}(\lambda) = \{p_{L_1}(\lambda|0)\}^{*w_1} * \ldots * \{p_{L_m}(\lambda|0)\}^{*w_m}, \tag{6.285}$$

and carry out the summation (6.233) over $\boldsymbol{w} \in \mathbb{N}^m$. Note that, to simplify the notation we removed the subindexing with $\underline{\mathbf{0}}, \underline{\mathbf{e}}$ we used in (6.210).

Although we used the same notation in (6.283) and (6.285), the random variables Λ^{Σ} described by these PDFs are not the same. In the first case, to evaluate the PEP, we have to consider only i.i.d. random variables. In the case of (6.285) the random variables are non-i.i.d.; we treat this case in Section 6.3.4.

6.3.1 Numerical Integration

In general, we will have to deal with distributions different from the Gaussian case we showed in Example 6.24. The direct way to obtain the tail integral of the multiple convolution is via direct and inverse Laplace-type transforms, i.e., from the relationship linking the MGF (see (2.6)) of the L-values Λ and their PDF $p_{\Lambda}(\lambda)$ given in (6.227).[13] More specifically,

$$\mathsf{P}_{\Lambda}(s) = \mathbb{E}_{\Lambda}[\mathrm{e}^{s\Lambda}] \tag{6.286}$$

$$= \int_{-\infty}^{\infty} p_{\Lambda}(\lambda) \mathrm{e}^{\lambda s}\, \mathrm{d}\lambda \tag{6.287}$$

[13] Recall that, although not explicitly shown in the notation, the PEP calculation assumes that the all-zero codeword was transmitted.

and

$$p_\Lambda(\lambda) = \frac{1}{2\pi \mathrm{j}} \int_{\alpha-\mathrm{j}\infty}^{\alpha+\mathrm{j}\infty} \mathsf{P}_\Lambda(s) \mathrm{e}^{-s\lambda}\, \mathrm{d}s, \tag{6.288}$$

where $\alpha \in \mathbb{R}$ is taken from the domain of the MGF.

Expressing (6.282) as

$$\mathsf{PEP}(d) = u(\lambda) * p_{\Lambda^\Sigma}(\lambda)|_{\lambda=\lambda_0} \tag{6.289}$$

$$= u(\lambda) * \{p_\Lambda(\lambda)\}^{*d}|_{\lambda=\lambda_0}, \tag{6.290}$$

where $\lambda_0 = 0$ and $u(\lambda) = \mathbb{I}_{[\lambda<0]}$ is the inverted step function with MGF $\mathsf{U}(s) = \frac{1}{s}$. The PEP can be then calculated by inverting the MGF of Λ^Σ, i.e.,

$$\begin{aligned}\mathsf{PEP}(d) &= \frac{1}{2\pi \mathrm{j}} \int_{\alpha-\mathrm{j}\infty}^{\alpha+\mathrm{j}\infty} \mathsf{P}_{\Lambda^\Sigma}(s)\mathsf{U}(s)\mathrm{e}^{-s\lambda_0}\, \mathrm{d}s \\ &= \frac{1}{2\pi \mathrm{j}} \int_{\alpha-\mathrm{j}\infty}^{\alpha+\mathrm{j}\infty} \frac{[\mathsf{P}_\Lambda(s)]^d}{s}\, \mathrm{d}s.\end{aligned} \tag{6.291}$$

Finding exact analytical solutions of the integral (6.291) is often impossible so numerical integration is then used. Before studying this approach, we show a useful lemma.

Lemma 6.25 *If the PDF of* Λ *satisfies the symmetry-consistency condition* (3.85), *then* $\mathsf{P}_\Lambda\left(\frac{1}{2} + \mathrm{j}\omega\right)$ *is a real and even function of* $\omega \in \mathbb{R}$.

Proof: Throughout this proof, we use $p_{\Lambda|C}(\lambda|0)$ to denote the PDF $p_\Lambda(\lambda)$ in (6.227), where C can be interpreted as the binary random variable at the input of the BICM channel in Fig. 5.1. From the definition of the MGF we obtain

$$\mathsf{P}_\Lambda\left(\frac{1}{2} + \mathrm{j}\omega\right) = \int_{-\infty}^{\infty} p_{\Lambda|C}(\lambda|0)\mathrm{e}^{\frac{1}{2}\lambda}(\cos(\omega\lambda) + \mathrm{j}\sin(\omega\lambda))\, \mathrm{d}\lambda \tag{6.292}$$

$$= \int_{-\infty}^{\infty} p_{\Lambda|C}(\lambda|0)\mathrm{e}^{\frac{1}{2}\lambda}\cos(\omega\lambda)\, \mathrm{d}\lambda, \tag{6.293}$$

where we are able to pass from (6.292) to (6.293) because the product of $\sin(\omega\lambda)$ (odd function of λ) and $p_{\Lambda|C}(\lambda|0)\mathrm{e}^{\frac{\lambda}{2}}$ (even function of λ due to (3.85)) is an odd function of λ that integrates to zero. This proves that $\mathsf{P}_\Lambda\left(\frac{1}{2} + \mathrm{j}\omega\right)$ is real. To prove that $\mathsf{P}_\Lambda\left(\frac{1}{2} + \mathrm{j}\omega\right)$ is an even function of ω, we use the fact that the integrand of (6.293) is even. □

In what follows, we show how to numerically calculate (6.291). To this end, we use a *Gauss-Chebyshev* (GCh) quadrature, which states that for any function $g(x)$ that can be represented by polynomials for $x \in (-1, 1)$,

$$\int_{-1}^{1} \frac{g(x)}{\sqrt{1-x^2}}\, \mathrm{d}x \approx \frac{\pi}{J}\sum_{j=1}^{J} g(x_j), \tag{6.294}$$

where $x_j = \cos(\nu_j)$ and $\nu_j = \frac{\pi(2j-1)}{2J}$.

Using the substitution $s = \frac{1}{2} + \mathrm{j}\omega$, (6.291) becomes

$$\mathsf{PEP}(d) = \frac{1}{2\pi} \int_{-\infty}^{\infty} \frac{\left[\mathsf{P}_\Lambda\left(\frac{1}{2} + \mathrm{j}\omega\right)\right]^d}{\left(\frac{1}{2}\right)^2 + \omega^2} \left(\frac{1}{2} - \mathrm{j}\omega\right) \mathrm{d}\omega \tag{6.295}$$

$$= \frac{1}{2\pi}\int_{-\infty}^{\infty} \frac{\left[\mathsf{P}_{\Lambda}\left(\frac{1}{2}+\mathrm{j}\omega\right)\right]^{d}\frac{1}{2}}{\left(\frac{1}{2}\right)^2+\omega^2} - \mathrm{j}\frac{\left[\mathsf{P}_{\Lambda}\left(\frac{1}{2}+\mathrm{j}\omega\right)\right]^{d}}{\left(\frac{1}{2}\right)^2+\omega^2}\omega\, \mathrm{d}\omega \tag{6.296}$$

$$= \frac{1}{2\pi}\int_{-\infty}^{\infty} \frac{\left[\mathsf{P}_{\Lambda}\left(\frac{1}{2}+\mathrm{j}\omega\right)\right]^{d}\frac{1}{2}}{\left(\frac{1}{2}\right)^2+\omega^2}\, \mathrm{d}\omega, \tag{6.297}$$

where (6.297) follows from the fact that the second part of the integrand in (6.297) is odd. This is due to Lemma 6.25, which shows that $\mathsf{P}_{\Lambda}\left(\frac{1}{2}+\mathrm{j}\omega\right)$ is real and even.

After the change of variable $\omega = \frac{x}{2\sqrt{1-x^2}}$, (6.297) becomes

$$\mathsf{PEP}(d) = \frac{1}{2\pi}\int_{-1}^{1} \frac{g(x)}{\sqrt{1-x^2}}\, \mathrm{d}x, \tag{6.298}$$

where

$$g(x) = \left[\mathsf{P}_{\Lambda}\left(\frac{1}{2}+\mathrm{j}\frac{x}{2\sqrt{1-x^2}}\right)\right]^{d}. \tag{6.299}$$

Thus, using (6.294) in (6.298) yields

$$\mathsf{PEP}(d) \approx \frac{1}{2J}\sum_{j=1}^{J} g(\cos\nu_j). \tag{6.300}$$

We then use $\tau_j = \cos(\nu_j)/\sqrt{1-\cos^2(\nu_j)} = \cot(\nu_j)$ and take advantage of the symmetry $\tau_j = -\tau_{J-j+1}$ and $\mathsf{P}_{\Lambda}(s^*) = \mathsf{P}^*_{\Lambda}(s)$. This yields the final expression for the PEP

$$\mathsf{PEP}(d) \approx \frac{1}{J}\sum_{j=1}^{J/2}\left[\mathsf{P}_{\Lambda}\left(\frac{1}{2}(1+\mathrm{j}\tau_j)\right)\right]^{d}, \tag{6.301}$$

where $\tau_j \triangleq \cot\left(\frac{\pi(2j-1)}{2J}\right)$ and $\mathsf{P}_{\Lambda}(s)$ is the MGF of the random variable Λ with PDF given by (6.227).

To evaluate the PEP in (6.301), one would ideally use a large value of J. However, a good tradeoff between accuracy and implementation complexity is typically obtained for relatively small values of J. Most of the examples in this chapter were calculated using $J = 10$ or $J = 20$.

Example 6.26 (PEP for 2PAM in Rayleigh Fading Channel) *The MGF of the L-values Λ for* 2PAM *and the AWGN channel is*

$$\mathsf{P}_{\Lambda}(s;\mathsf{snr}) = \mathbb{E}_{\Lambda}[\exp(s\Lambda)] = \exp(-4\,\mathsf{snr}\,s + 4\,\mathsf{snr}\,s^2). \tag{6.302}$$

Averaging (6.302) *over the distribution of* SNR *in* (2.35) *(with* $\mathsf{m}=1$*), we obtain*

$$\overline{\mathsf{P}}_{\Lambda}(s) = \mathbb{E}_{\mathsf{SNR}}\left[\mathsf{P}_{\Lambda}(s;\mathsf{SNR})\right] \tag{6.303}$$

$$= \int_{0}^{\infty} \mathsf{P}_{\Lambda}(s;\mathsf{snr})p_{\mathsf{SNR}}(\mathsf{snr},1,\overline{\mathsf{snr}})\,\mathrm{d}\mathsf{snr} \tag{6.304}$$

$$= \frac{1}{\overline{\mathsf{snr}}}\int_{0}^{\infty} \exp\left(-4\,\mathsf{snr}\,s + 4\,\mathsf{snr}\,s^2 - \frac{\mathsf{snr}}{\overline{\mathsf{snr}}}\right)\mathrm{d}\mathsf{snr} \tag{6.305}$$

$$= \frac{1}{1+4\,\overline{\mathsf{snr}}\,s(1-s)}. \tag{6.306}$$

Then,

$$\overline{\mathsf{P}}_{\Lambda}\left(\frac{1}{2}(1+\mathrm{j}\tau_j)\right) = \frac{1}{1+\overline{\mathsf{snr}}(1+\tau_j^2)}, \tag{6.307}$$

which used in (6.301) *yields*

$$\mathsf{PEP}(d) \approx \frac{1}{J}\sum_{j=1}^{J/2} \frac{1}{[1+\overline{\mathsf{snr}}(1+\tau_j^2)]^d}. \tag{6.308}$$

Numerical results obtained by evaluating (6.308) are shown in Section 6.3.3. Here we show a graphical representation of the numerical integration. This is shown in Fig. 6.18, where we assume $d = 1$ *and we identify the integration line* $s = \frac{1}{2} + \mathrm{j}\omega$ *along which* $\overline{\mathsf{P}}_{\Lambda}(s)$ *in (6.303) is real as we have shown in Lemma 6.25. The quadrature nodes for* $J = 8$ *are shown with circles.*

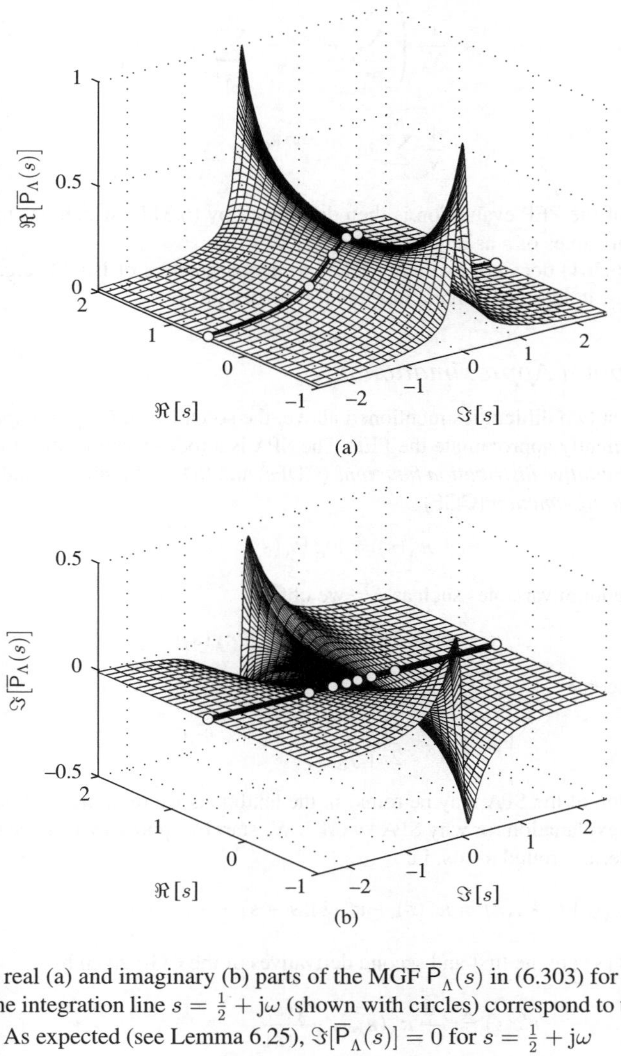

Figure 6.18 The real (a) and imaginary (b) parts of the MGF $\overline{\mathsf{P}}_{\Lambda}(s)$ in (6.303) for $d = 1$ and $\overline{\mathsf{snr}} = 2$. The points along the integration line $s = \frac{1}{2} + \mathrm{j}\omega$ (shown with circles) correspond to the GCh quadrature nodes with $J = 8$. As expected (see Lemma 6.25), $\Im[\overline{\mathsf{P}}_{\Lambda}(s)] = 0$ for $s = \frac{1}{2} + \mathrm{j}\omega$

We conclude this section by making some remarks about the PEP expression in (6.301):

- The numerical complexity grows with the number of terms required by (6.243), as the integration (6.301) must be carried out for each value of d. This brings implementation burden when we need to evaluate $\mathsf{PEP}(d)$ for many different values of d.
- The MGF $\mathsf{P}_\Lambda(s)$ must be known for $J/2$ values of s required in (6.301). While in some cases we may be able to analytically calculate the MGF from the analytical form of the PDF $p_\Lambda(\lambda)$, in others, it is too cumbersome or even impossible. In fact, it may be difficult to derive tractable expressions for the PDF $p_\Lambda(\lambda)$ in the first place, e.g., when the L-values depend on many random parameters.
- In the cases where the PDF is not known, we may replace (6.286) by Monte Carlo integration. To this end, we use a sum over realizations $\lambda[n]$ of Λ, conditioned on the inputs of the BICM channel in Fig. 5.1, i.e.,

$$\mathsf{P}_\Lambda(s) = \mathbb{E}_{\Lambda|C=0}[\mathrm{e}^{s\Lambda}] = \mathbb{E}_{\Lambda|C=1}[\mathrm{e}^{-s\Lambda}] \tag{6.309}$$

$$\approx \frac{1}{N_\mathrm{c}}\left(\sum_{\substack{n=1\\c[n]=0}}^{N_\mathrm{c}} \mathrm{e}^{s\lambda[n]} + \sum_{\substack{n=1\\c[n]=1}}^{N_\mathrm{c}} \mathrm{e}^{-s\lambda[n]}\right) \tag{6.310}$$

$$= \frac{1}{N_\mathrm{c}}\sum_{n=1}^{N_\mathrm{c}} \mathrm{e}^{(-1)^{c[n]}s\lambda[n]}. \tag{6.311}$$

  The complexity of the PEP evaluation is then dominated by (6.311), which has to be evaluated $J/2$ times for different values of s as required by (6.301).
- The expression (6.301) does not explicitly relate to the parameters of BICM (e.g., constellation and labeling), and thus, it does not provide insights for the design.

6.3.2 Saddlepoint Approximation

To go around the first two difficulties mentioned above, the so-called *saddlepoint approximation* (SPA) may be used to efficiently approximate the PEP. The SPA is a tool known in statistical analysis to efficiently calculate *cumulative distribution functions* (CDFs) and PDFs of a sum of random variables using the *cumulant generating function* (CGF)

$$\kappa_\Lambda(s) \triangleq \log \mathsf{P}_\Lambda(s). \tag{6.312}$$

For a sum of i.i.d. random variables such as Λ^Σ, we obtain

$$\kappa_{\Lambda^\Sigma}(s) = \log\,[\mathsf{P}_\Lambda(s)]^d = d\,\kappa_\Lambda(s), \tag{6.313}$$

and thus, (6.291) becomes

$$\mathsf{PEP}(d) = \frac{1}{2\pi\mathrm{j}}\int_{\alpha-\mathrm{j}\infty}^{\alpha+\mathrm{j}\infty} \frac{1}{s}\mathrm{e}^{d\kappa_\Lambda(s)}\,\mathrm{d}s. \tag{6.314}$$

A formal derivation of the SPA may be found in the textbooks we reference in Section 6.5. Here, we provide an intuitive explanation for why SPA "work". We start by approximating the CGF in (6.312) via a truncated Taylor series around $s = \hat{s}$, i.e.,

$$\kappa_\Lambda(s) \approx \tilde{\kappa}_\Lambda(s) = \kappa_\Lambda(\hat{s}) + \kappa'_\Lambda(\hat{s})(s-\hat{s}) + \frac{1}{2}\kappa''_\Lambda(\hat{s})(s-\hat{s})^2, \tag{6.315}$$

where $\kappa'_\Lambda(s)$ and $\kappa''_\Lambda(s)$ are the first and second derivatives of the CGF given by

$$\kappa'_\Lambda(s) \triangleq \frac{\mathrm{d}}{\mathrm{d}s}\kappa_\Lambda(s) = \frac{\mathsf{P}'_\Lambda(s)}{\mathsf{P}_\Lambda(s)}, \tag{6.316}$$

$$\kappa''_{\Lambda}(s) \triangleq \frac{\mathrm{d}^2}{\mathrm{d}s^2}\kappa_{\Lambda}(s) = \frac{\mathsf{P}''_{\Lambda}(s)}{\mathsf{P}_{\Lambda}(s)} - [\kappa'_{\Lambda}(s)]^2. \tag{6.317}$$

The Taylor expansion in (6.315) is done around the so-called saddlepoint $\hat{s} = \hat{s}^{\mathrm{spa}}$, chosen to satisfy $\kappa'_{\Lambda}(\hat{s}^{\mathrm{spa}}) = 0$, which used in (6.314) gives the SPA

$$\mathsf{PEP}(d) \approx \frac{1}{2\pi\mathrm{j}}\int_{\alpha-\mathrm{j}\infty}^{\alpha+\mathrm{j}\infty} \frac{1}{s}\mathrm{e}^{d\tilde{\kappa}_{\Lambda}(\hat{s}^{\mathrm{spa}})}\,\mathrm{d}s \tag{6.318}$$

$$= \frac{1}{2\pi\mathrm{j}}\int_{\alpha-\mathrm{j}\infty}^{\alpha+\mathrm{j}\infty} \frac{1}{s}\mathrm{e}^{d\kappa_{\Lambda}(\hat{s}^{\mathrm{spa}})+\frac{d}{2}\kappa''_{\Lambda}(\hat{s}^{\mathrm{spa}})(s-\hat{s}^{\mathrm{spa}})^2}\,\mathrm{d}s. \tag{6.319}$$

After the change of variables $s = \hat{s}^{\mathrm{spa}} + \mathrm{j}\omega$ we obtain

$$\begin{aligned}\mathsf{PEP}(d) &\approx \frac{\mathrm{e}^{d\kappa_{\Lambda}(\hat{s}^{\mathrm{spa}})}}{2\pi}\int_{-\infty}^{\infty} \mathrm{e}^{-\frac{d}{2}\kappa''_{\Lambda}(\hat{s}^{\mathrm{spa}})\omega^2}\frac{\hat{s}^{\mathrm{spa}} - j\omega}{(\hat{s}^{\mathrm{spa}})^2 + \omega^2}\mathrm{d}\omega \\ &= \frac{\hat{s}^{\mathrm{spa}}\mathrm{e}^{d\kappa_{\Lambda}(\hat{s}^{\mathrm{spa}})}}{2\pi}\int_{-\infty}^{\infty} \frac{\exp\left(-\frac{d}{2}\kappa''_{\Lambda}(\hat{s}^{\mathrm{spa}})\omega^2\right)}{(\hat{s}^{\mathrm{spa}})^2 + \omega^2}\mathrm{d}\omega \\ &= \mathsf{PEP}^{\mathrm{spa}}(d),\end{aligned} \tag{6.320}$$

where

$$\mathsf{PEP}^{\mathrm{spa}}(d) \triangleq \mathrm{e}^{d\kappa_{\Lambda}(\hat{s}^{\mathrm{spa}})}Q\left(\hat{s}^{\mathrm{spa}}\sqrt{d\kappa''_{\Lambda}(\hat{s}^{\mathrm{spa}})}\right)\exp\left(\frac{1}{2}(\hat{s}^{\mathrm{spa}})^2 d\kappa''_{\Lambda}(\hat{s}^{\mathrm{spa}})\right), \tag{6.321}$$

and where (6.320) results from $\int_{-\infty}^{\infty}\exp(-b^2x^2)\frac{1}{a^2+x^2}\,\mathrm{d}x = \frac{2\pi}{a}Q(ab\sqrt{2})\exp(a^2b^2)$.

Using $Q(x) \approx \frac{1}{2}\exp(-\frac{x^2}{2})$ transforms (6.321) into

$$\mathsf{PEP}^{\mathrm{spa}}(d) \approx \frac{1}{2}\mathrm{e}^{d\kappa_{\Lambda}(\hat{s}^{\mathrm{spa}})}. \tag{6.322}$$

Alternatively, if the (tighter) approximation $Q(x) \approx \frac{1}{x\sqrt{2\pi}}\exp(-\frac{x^2}{2})$ is used, we obtain

$$\mathsf{PEP}^{\mathrm{spa}}(d) \approx \frac{\mathrm{e}^{d\kappa_{\Lambda}(\hat{s}^{\mathrm{spa}})}}{\hat{s}^{\mathrm{spa}}\sqrt{2\pi d\kappa''_{\Lambda}(\hat{s}^{\mathrm{spa}})}}. \tag{6.323}$$

We can also transform the expression in (6.322) and (6.323) back into the MGF domain via (6.312). For example, (6.323) becomes

$$\mathsf{PEP}^{\mathrm{spa}}(d) = \frac{[\mathsf{P}_{\Lambda}(\hat{s}^{\mathrm{spa}})]^{d+\frac{1}{2}}}{\hat{s}^{\mathrm{spa}}\sqrt{2\pi d\mathsf{P}''_{\Lambda}(\hat{s}^{\mathrm{spa}})}}. \tag{6.324}$$

We have obtained expressions that depend solely on the MGF or CGF evaluated at $s = \hat{s}^{\mathrm{spa}}$. This is an important simplification when compared to (6.301), as once $\mathsf{P}_{\Lambda}(\hat{s}^{\mathrm{spa}})$ and $\mathsf{P}''_{\Lambda}(\hat{s}^{\mathrm{spa}})$ are known, we can evaluate $\mathsf{PEP}(d)$ for any d. The caveat is that now, $\hat{s}^{\mathrm{spa}}$ must be found. This is usually the most difficult part of the SPA method because solving nonlinear saddlepoint equation $\kappa'_{\Lambda}(\hat{s}^{\mathrm{spa}}) = 0$, or equivalently

$$\frac{\mathsf{P}'_{\Lambda}(\hat{s}^{\mathrm{spa}})}{\mathsf{P}_{\Lambda}(\hat{s}^{\mathrm{spa}})} = 0, \tag{6.325}$$

is, in general, not trivial. However, for the cases we study here, it is in fact quite simple.

We note that to solve (6.325) we need to choose $\hat{s}^{\mathsf{spa}}$ to satisfy

$$\mathsf{P}'_\Lambda(\hat{s}^{\mathsf{spa}}) = \int_{-\infty}^{\infty} p_\Lambda(\lambda)\lambda \mathrm{e}^{\hat{s}^{\mathsf{spa}}\lambda}\, \mathrm{d}\lambda = \int_{-\infty}^{\infty} p_\Lambda(\lambda)\mathrm{e}^{\frac{\lambda}{2}}\lambda \mathrm{e}^{(\hat{s}^{\mathsf{spa}}-\frac{1}{2})\lambda}\, \mathrm{d}\lambda = 0. \tag{6.326}$$

From (3.85) we know that the function $p_\Lambda(\lambda)\mathrm{e}^{\lambda/2}\lambda$ is odd, which allows us to conclude that for

$$\hat{s}^{\mathsf{spa}} = \frac{1}{2} \tag{6.327}$$

the integrand in (6.326) is also odd, and thus, (6.327) solves (6.325). To show that this solution is unique, we need to demonstrate that $\kappa_\Lambda(s)$ is convex, i.e.,

$$\kappa''_\Lambda(s) = \frac{\mathbb{E}_\Lambda[\Lambda^2\mathrm{e}^{s\Lambda}]\mathbb{E}_\Lambda[\mathrm{e}^{s\Lambda}] - \mathbb{E}^2_\Lambda[\Lambda\mathrm{e}^{s\Lambda}]}{\mathbb{E}^2_\Lambda[\mathrm{e}^{s\Lambda}]} \geq 0, \tag{6.328}$$

which can be recognized as Hölder's inequality, and therefore, holds independently of the distribution of the random variable Λ.

Example 6.27 (SPA for 2PAM**)** *Let us consider again the case of* 2PAM *transmission we already used in Example 6.24. From* (6.302) *we obtain the CGF as well as its first and second derivatives*

$$\kappa_\Lambda(s) = -4\,\mathsf{snr}\,s + 4\,\mathsf{snr}\,s^2 \tag{6.329}$$

$$\kappa'_\Lambda(s) = -4\,\mathsf{snr} + 8\,\mathsf{snr}\,s \tag{6.330}$$

$$\kappa''_\Lambda(s) = 8\,\mathsf{snr}, \tag{6.331}$$

where we can immediately validate (6.327)*, i.e.,* $\kappa'_\Lambda\left(\frac{1}{2}\right) = 0$*. Using* (6.329) *and* (6.330) *in* (6.320) *yields*

$$\mathsf{PEP}(d) = Q(\sqrt{2d\,\mathsf{snr}}), \tag{6.332}$$

where we use equality because the solution is exact (as already derived in (6.258)*). This is not surprising: the CGF we consider here is a quadratic function, so* $\tilde{\kappa}_\Lambda(s) = \kappa_\Lambda(s)$*. Thus,* (6.320) *is indeed the exact value of the PEP because the only approximation involved up to this point was due to the Taylor series truncation, which in the case of a quadratic function does not entail any loss. While this is not a particularly interesting solution, we show it as a sanity check. Further, if we use the simplified SPA solution* (6.323)*, we obtain*

$$\mathsf{PEP}^{\mathsf{spa}}(d) = \frac{\mathrm{e}^{-d\,\mathsf{snr}}}{2\sqrt{\pi d\,\mathsf{snr}}}. \tag{6.333}$$

Example 6.28 (SPA for 2PAM **in Fading Channels)** *We can now turn our attention to a less obvious calculation of the PEP, i.e., for transmission over fading channels as we did already in Example 6.26. From the MGF* $\overline{\mathsf{P}}_\Lambda(s)$ *shown in* (6.303)*, we obtain the CGF*

$$\kappa_\Lambda(s) = -\log(1 + 4\,\overline{\mathsf{snr}}\,s(1-s)). \tag{6.334}$$

Then $\kappa_\Lambda(1/2) = -\log(1+\overline{\mathsf{snr}})$ *and* $\kappa''_\Lambda(1/2) = \frac{8\,\overline{\mathsf{snr}}}{1+\overline{\mathsf{snr}}}$*, and* (6.320) *is given by*

$$\mathsf{PEP}^{\mathsf{spa}}(d) = \frac{(1+\overline{\mathsf{snr}})^{-d+\frac{1}{2}}}{2\sqrt{\pi d\,\overline{\mathsf{snr}}}}. \tag{6.335}$$

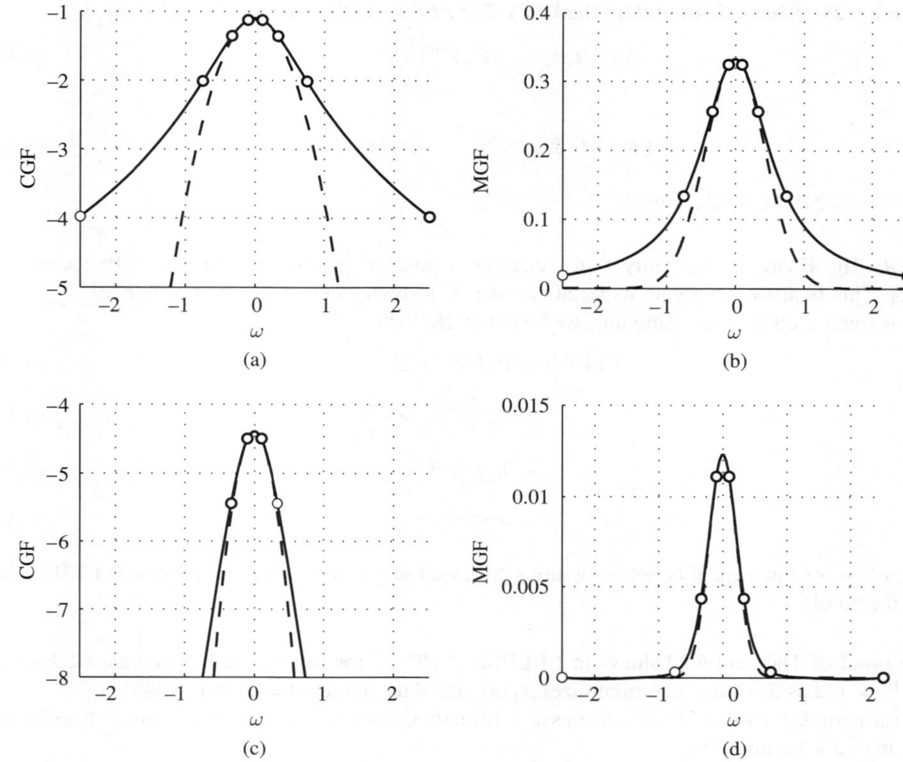

Figure 6.19 CGF $d\,\kappa_\Lambda(\frac{1}{2}+\mathrm{j}\omega)$ (solid lines) and its second-order approximation $d\,\tilde{\kappa}_\Lambda(\frac{1}{2}+\mathrm{j}\omega)$ (dashed lines) are shown together with the corresponding MGF $\mathrm{e}^{d\kappa_\Lambda(s)}$ (solid lines) and its approximation $\mathrm{e}^{d\tilde{\kappa}_\Lambda(s)}$ (dashed lines): (a) $d=1$, (b) $d=1$, (c) $d=4$, and (d) $d=4$. The circles indicate the position of the $J=8$ quadrature nodes necessary to implement the solution from Section 6.3.1. Here $\overline{\mathsf{snr}}=2$

In Fig. 6.19 (a) and (c) we compare the CGF $d\kappa_\Lambda(s)$ and its approximation $d\tilde{\kappa}_\Lambda(s)$ for $d=1$ and $d=4$, respectively, and for $s=\frac{1}{2}+\mathrm{j}\omega$. The corresponding values of the MGF $\mathsf{P}_\Lambda(s)$ and the MGF approximation $\mathrm{e}^{\tilde{\kappa}_\Lambda(s)}$ are shown in Fig. 6.19 (b) and d. The form of the MGF corresponds to the bell-like shape in Fig. 6.18 while the logarithmic scale of CGF accentuates the differences between the tails. When d increases, these differences become negligible, and thus, the accuracy of the PEP evaluation improves.

Thanks to the SPA, once we know $\mathsf{P}_\Lambda(s)$ and $\mathsf{P}''_\Lambda(s)$ for one real argument $\hat{s}^{\mathrm{spa}}=\frac{1}{2}$, we may calculate $\mathsf{PEP}(d)$ for any value of d. This stands in contrast to the numerical integration approach, where we have to evaluate the MGF for $J/2$ points (see (6.301)), and next carry out the summations for each d. Thus, not only does the SPA allow us to obtain analytical solutions in some cases, but it also offers a clear advantage over numerical integration when the MGF is difficult to acquire.

6.3.3 Chernoff Bound

Another useful approximation of the PEP relies on using upper bounding techniques, as shown in the following theorem.

Theorem 6.29 (Chernoff Bound on the PEP) *The PEP in* (6.282) *is upper bounded as*

$$\mathsf{PEP}(d) \leq \mathsf{PEP}^{\mathrm{ub}}(d), \tag{6.336}$$

where

$$\mathsf{PEP}^{\mathrm{ub}}(d) \triangleq \mathrm{e}^{d\kappa_\Lambda(\hat{s}^{\mathrm{spa}})} = [\mathsf{P}_\Lambda(\hat{s}^{\mathrm{spa}})]^d \tag{6.337}$$

and $\hat{s}^{\mathrm{spa}} = 1/2$ *is the saddlepoint.*

Proof: The Chernoff inequality states that for a positive random variable X, $x \Pr\{X > x\} \leq \mathbb{E}_X[X]$. This is proven applying expectation over X to both sides of the relationship $x\mathbb{I}_{[X>x]} \leq X$, which is true for all $X \geq 0$. Using this, we bound (6.282) via

$$\mathsf{PEP}(d) = \Pr\{\Lambda^\Sigma > 0\} \tag{6.338}$$

$$= \Pr\left\{\mathrm{e}^{s\Lambda^\Sigma} > 1\right\} \tag{6.339}$$

$$\leq \mathbb{E}_{\Lambda^\Sigma}\left[\mathrm{e}^{s\Lambda^\Sigma}\right] \tag{6.340}$$

$$= \mathrm{e}^{\kappa_{\Lambda^\Sigma}(s)}, \tag{6.341}$$

where $s > 0$. As the inequality holds for any $s > 0$, we use $s = \hat{s}^{\mathrm{spa}}$, which together with (6.313) completes the proof. □

The proof of Theorem 6.29 shows that $\mathsf{PEP}(d) \leq \mathrm{e}^{d\kappa_\Lambda(s)}$ for any $s > 0$. In Theorem 6.29 we use $s = \hat{s}^{\mathrm{spa}} = 1/2$ as this value of s minimizes $\kappa_\Lambda(s)$, and thus, tightens the bound (6.340).

For the particular case of 2PAM transmission, from (6.329) we have $\kappa_\Lambda(1/2) = -\mathsf{snr}$, so the Chernoff bound in (6.336) is given by

$$\mathsf{PEP}^{\mathrm{ub}}(d) = \mathrm{e}^{-d\,\mathsf{snr}}. \tag{6.342}$$

Using $Q(x) \leq \frac{1}{x\sqrt{2\pi}} \exp\left(-\frac{1}{2}x^2\right)$, we bound the true PEP in (6.258) as

$$\mathsf{PEP}(d) \leq \frac{1}{2\sqrt{\pi d\,\mathsf{snr}}}\mathrm{e}^{-d\,\mathsf{snr}}, \tag{6.343}$$

which shows that the Chernoff bound in (6.342) goes to zero (as $\mathsf{snr} \to \infty$) slower than the actual value of the PEP. The next example shows PEP calculation for 2PAM in fading channels.

Example 6.30 (PEP for 2PAM in Fading Channels) *For Rayleigh fading, from* (6.306) *we have* $\kappa_\Lambda\left(\frac{1}{2}\right) = -\log(1+\mathsf{snr})$ *and the upper bound becomes*

$$\mathsf{PEP}^{\mathrm{ub}}(d) = (1+\mathsf{snr})^{-d}, \tag{6.344}$$

while the SPA solution is given by (6.335)

$$\mathsf{PEP}^{\mathrm{spa}}(d) = \frac{(1+\mathsf{snr})^{-d+\frac{1}{2}}}{2\sqrt{\pi d\,\mathsf{snr}}}. \tag{6.345}$$

For high SNR we can write

$$\mathsf{PEP}^{\mathrm{ub}}(d) \approx \mathsf{snr}^{-d}, \tag{6.346}$$

$$\mathsf{PEP}^{\mathrm{spa}}(d) \approx \frac{\mathsf{snr}^{-d}}{2\sqrt{\pi d}}. \tag{6.347}$$

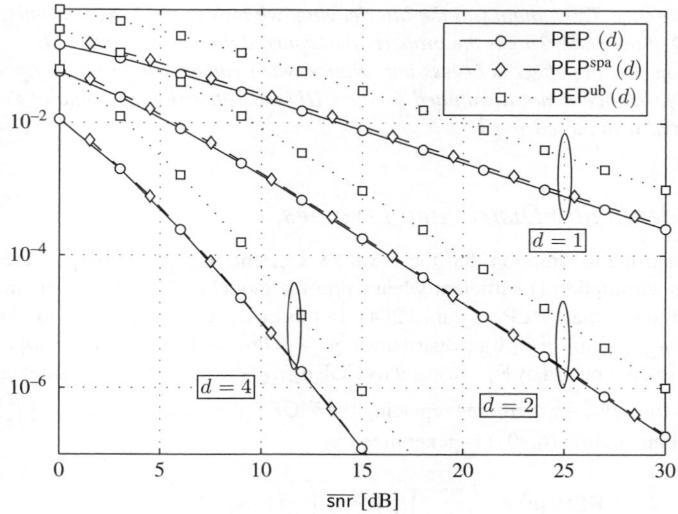

Figure 6.20 Comparison of the PEP estimation methods for a 2PAM constellation in Rayleigh fading channel. To obtain $\mathsf{PEP}(d)$ we used (6.308) with $J = 10$ quadrature points

Treating $\mathsf{PEP}^{\text{ub}}(d)$ *and* $\mathsf{PEP}^{\text{spa}}(d)$ *as a function of* $\overline{\mathsf{snr}}$ *we can conclude that, for high* $\overline{\mathsf{snr}}$*, there is a constant SNR gap between* $\mathsf{PEP}^{\text{ub}}(d)$ *and* $\mathsf{PEP}^{\text{spa}}(d)$ *of*

$$\Delta_{\overline{\mathsf{snr}}} = \frac{10}{d}\log_{10}\left(2\sqrt{\pi d}\right) \text{ [dB]}. \tag{6.348}$$

The solutions obtained are compared in Fig. 6.20, where we can appreciate that while for $d = 1$ *the SPA-based evaluation* $\mathsf{PEP}^{\text{spa}}(d)$ *is slightly larger than the exact solution* $\mathsf{PEP}(d)$*, it becomes very accurate already for* $d = 2$*. For* $d = 4$*, the differences between* $\mathsf{PEP}(d)$ *and* $\mathsf{PEP}^{\text{spa}}(d)$ *are practically indistinguishable. This is a consequence of the very good approximation of the MGF, whose accuracy improves with* d*, as shown in Fig. 6.19. Figure 6.20 also shows the bound* $\mathsf{PEP}^{\text{ub}}(d)$*, which considerably overestimates the exact solution, as quantified in* (6.348).

Using the asymptotic PEP expression in (6.346) *we can evaluate the so-called transmission diversity defined as*

$$\mathsf{div} \triangleq -\lim_{\overline{\mathsf{snr}}\to\infty} \frac{\log \mathsf{WEP}}{\log \overline{\mathsf{snr}}}. \tag{6.349}$$

Then, in the high-SNR regime, we obtain

$$\mathsf{WEP} \approx \sum_{d=d_{\text{free}}}^{d_{\max}} C_d^{\mathcal{C}} \mathsf{PEP}(d) \tag{6.350}$$

$$\approx C_{d_{\text{free}}}^{\mathcal{C}} \mathsf{PEP}(d_{\text{free}}) \tag{6.351}$$

$$\approx \frac{C_{d_{\text{free}}}^{\mathcal{C}}}{\overline{\mathsf{snr}}^{d_{\text{free}}}}, \tag{6.352}$$

so we conclude that $\mathsf{div} = d_{\text{free}}$*, and thus, the transmission diversity equals the FHD of the code* d_{free}*. We note that in order to obtain this result, the codewords* $\underline{\mathbf{c}} \in \mathcal{C}^{\mathsf{div}}(d)$ *must attain their maximum* *(interleaving) diversity. Thus, thanks to the interleaving, we may control the transmission diversity by varying the FHD of the code. This is the property that spurred the interest in BICM.*

This diversity-increasing effect is of less importance when capacity-approaching codes are used. In this case the performance is not dominated by the FHD but rather by the value of the DS C^{c}_{d}, which becomes very small with increasing N_c.

6.3.4 Nonidentically Distributed L-values

We can now revisit the assumption that the L-values $L_{(t)}$ entering the metric Λ^{Σ} are identically distributed. Such an assumption is sufficient when using the model (6.227), but in the most general case (6.218), we need to calculate $\mathsf{PEP}(\boldsymbol{w})$ in (6.284). In this case, Λ^{Σ} is a sum of d random variables, w_1 with distribution $p_{L_1|B_1}(l|b)$, w_2 with a distribution $p_{L_2|B_2}(l|b)$, and so on, where $\|\boldsymbol{w}\|_1 = d$. We denote the MGF of the kth distribution by $\mathsf{P}_{L_k}(s)$, and its CGF by $\kappa_k(s) = \log \mathsf{P}_{L_k}(s)$. The extension of the previously obtained formulas thus requires replacing the MGF $\mathsf{P}_{\Lambda}(s)$ with the product $\prod_{k=1}^{m} \left[\mathsf{P}_{L_k}(s)\right]^{w_k}$.

The numerical integration (6.301) is generalized as

$$\mathsf{PEP}(\boldsymbol{w}) = \frac{1}{J} \sum_{j=1}^{J/2} \prod_{k=1}^{m} \left(\mathsf{P}_{L_k}\left(\frac{1}{2}(1 + \mathrm{j}\tau_j)\right)\right)^{w_k}. \tag{6.353}$$

The SPA-based solution (6.323) becomes

$$\mathsf{PEP}^{\mathrm{spa}}(\boldsymbol{w}) = \frac{\exp(\sum_{k=1}^{m} w_k \kappa_k(\hat{s}))}{\hat{s}\sqrt{2\pi \sum_{k=1}^{m} w_k \kappa_k''(\hat{s})}} \tag{6.354}$$

and the upper bound is generalized as

$$\mathsf{PEP}^{\mathrm{ub}}(\boldsymbol{w}) = \exp\left(\sum_{k=1}^{m} w_k \kappa_k(\hat{s})\right) \tag{6.355}$$

$$= \prod_{k=1}^{m} \left[\mathsf{P}_{L_k}(\hat{s})\right]^{w_k}, \tag{6.356}$$

where

$$\hat{s} = \underset{s \geq 0}{\operatorname{argmin}} \left\{\sum_{k=1}^{m} w_k \kappa_k(s)\right\}, \tag{6.357}$$

and thus $\hat{s} = \frac{1}{2}$.

We will exploit the bound (6.355) in Lemma 7.13 and use it in the following example to find the PEP of *trellis-coded modulation* (TCM) receivers.

Example 6.31 (PEP for TCM in Fading Channels) *The case of nonidentically distributed L-values appear in the PEP calculation for TCM. In this case the difference in the distribution of the L-values is due to the varying distances between the symbols $\hat{\boldsymbol{x}}[n]$ and $\boldsymbol{x}[n]$. From* (6.160)–(6.164) *we can write*

$$\mathrm{PEP}(\underline{\boldsymbol{x}}, \hat{\underline{\boldsymbol{x}}}) = \Pr\left\{L^{\Sigma} > 0\right\} = \int_0^{\infty} p_{L^{\Sigma}}(l)\,\mathrm{d}l, \tag{6.358}$$

where

$$\mathsf{P}_{L^{\Sigma}}(s) = \prod_{n=1}^{N_s} \mathsf{P}_{L[n]}(s), \tag{6.359}$$

and where for a given $\boldsymbol{x}[n]$ and $\hat{\boldsymbol{x}}[n]$, $L[n]$ is a Gaussian random variable with MGF

$$\mathsf{P}_{L[n]}(s) = \exp(-\mathsf{snr}[n]\|\boldsymbol{x}[n] - \hat{\boldsymbol{x}}[n]\|^2 s - \mathsf{snr}[n]\|\boldsymbol{x}[n] - \hat{\boldsymbol{x}}[n]\|^2 s^2). \tag{6.360}$$

Since only the elements $\hat{\boldsymbol{x}}[n] \neq \boldsymbol{x}[n]$ matter in the PEP calculation, to simplify the notation, we reindex them as $\boldsymbol{x}_{(t)}, t = 1, \ldots, T$, where T is the number of different symbols between $\boldsymbol{x}[n]$ and $\hat{\boldsymbol{x}}[n]$ and $\|\boldsymbol{x}_{(t)} - \hat{\boldsymbol{x}}_{(t)}\| \neq 0$. Similarly, we also use $L_{(t)}$ to denote the corresponding reindexed version of $L[n]$.

As in (6.303)–(6.306), assuming Rayleigh fading, the average MGF is given by

$$\overline{\mathsf{P}}_{L_{(t)}}(s) = \frac{1}{1 + \mathsf{snr}\|\boldsymbol{x}_{(t)} - \hat{\boldsymbol{x}}_{(t)}\|^2 s(1-s)}, \quad t = 1, \ldots, T. \tag{6.361}$$

The numerical quadrature (6.301) then generalizes to

$$\mathrm{PEP}(\underline{\boldsymbol{x}}, \hat{\underline{\boldsymbol{x}}}) = \frac{1}{J}\sum_{j=1}^{J/2} \frac{1}{\prod_{t=1}^{T}\left(1 + \frac{1}{4}\mathsf{snr}\|\boldsymbol{x}_{(t)} - \hat{\boldsymbol{x}}_{(t)}\|^2(1+\tau_j^2)\right)} \tag{6.362}$$

$$\leq \frac{1}{J\mathsf{X}^2_{\underline{\boldsymbol{x}},\hat{\underline{\boldsymbol{x}}}}}\sum_{j=1}^{J/2}\left(\frac{4}{\mathsf{snr}(1+\tau_j^2)}\right)^T \tag{6.363}$$

$$= \left(\frac{4}{\mathsf{snr}}\right)^T \frac{1}{J\mathsf{X}^2_{\underline{\boldsymbol{x}},\hat{\underline{\boldsymbol{x}}}}}\sum_{j=1}^{J/2}(1+\tau_j^2)^{-T}, \tag{6.364}$$

where

$$\mathsf{X}_{\underline{\boldsymbol{x}},\hat{\underline{\boldsymbol{x}}}} = \prod_{t=1}^{T}\|\boldsymbol{x}_{(t)} - \hat{\boldsymbol{x}}_{(t)}\| \tag{6.365}$$

is the "product distance" between the symbols.

On the other hand, the upper bound is generalized using (6.356) *to*

$$\mathsf{PEP}^{\mathrm{ub}}(\underline{\boldsymbol{x}}, \hat{\underline{\boldsymbol{x}}}) = \frac{1}{\prod_{t=1}^{T}\left(1 + \frac{1}{4}\mathsf{snr}\|\boldsymbol{x}_{(t)} - \hat{\boldsymbol{x}}_{(t)}\|^2\right)} \tag{6.366}$$

$$\leq \left(\frac{4}{\mathsf{snr}}\right)^T \frac{1}{\mathsf{X}^2_{\underline{\boldsymbol{x}},\hat{\underline{\boldsymbol{x}}}}}. \tag{6.367}$$

The bounds in (6.363) *and* (6.367) *are good for high SNR and show that, while in nonfading channels the PEP depends on the Euclidean distance (ED) (see* (6.166)*), in fading channels it depends on the product distance. We can also appreciate that, at high SNR,* $\mathrm{PEP}(\underline{\boldsymbol{x}}, \hat{\underline{\boldsymbol{x}}})$ *and* $\mathsf{PEP}^{\mathrm{ub}}(\underline{\boldsymbol{x}}, \hat{\underline{\boldsymbol{x}}})$ *are the same up to a scaling factor.*

6.4 Performance Evaluation of BICM via Gaussian Approximations

Two major handicaps of the numerical integration we discussed at the end of Section 6.3.1 were removed in Sections 6.3.2 and 6.3.3. However, the problem of a direct connection between the PEP and the parameters of the constellation and/or labeling still remains. In order to solve this problem, in this section we use the simplified Gaussian forms for the PDF we derived in Section 5.5.

6.4.1 PEP Calculation using Gaussian Approximations

For MPAM and MPSK constellations labeled by the BRGC, we showed in Section 5.5 that we may use Gaussian functions to approximate the PDF of the L-values conditioned on the transmitted symbols. More specifically, for a given transmitted symbol $\boldsymbol{x}_r$, we have

$$p_{L_k|\boldsymbol{X}}(l|\boldsymbol{x}_r) \approx \tilde{p}_{L_k|\boldsymbol{X}}(l|\boldsymbol{x}_r) = \Psi\left(l, \mathsf{snr}\,\tilde{\mu}_k(\boldsymbol{x}_r), \mathsf{snr}\,\tilde{\sigma}_k^2(\boldsymbol{x}_r)\right), \tag{6.368}$$

where $r = 1, \dots, 2^{m-k}$, $\tilde{\mu}_k(\boldsymbol{x}_r)$ and $\tilde{\sigma}_k^2(\boldsymbol{x}_r)$ depend on the transmitted symbol $\boldsymbol{x}_r$, bit position k, constellation $\mathcal{S}$, labeling $\mathbf{Q}$, and adopted approximation model (*consistent model* (CoM) or *zero-crossing model* (ZcM)). The mean values and variances $\tilde{\mu}_k(\boldsymbol{x}_r)$ and $\tilde{\sigma}_k^2(\boldsymbol{x}_r)$ in (6.368) are given in Tables 5.1 and 5.2 for MPAM and MPSK, respectively.

The PDF of the L-values conditioned on the transmitted bit $B_k = 0$ can then be obtained via marginalization, i.e.,

$$\tilde{p}_{L_k|B_k}(l|0) = \sum_{\boldsymbol{x}_r \in \mathcal{S}_{k,0}} \tilde{p}_{L_k|\mathbf{X}}(l|\boldsymbol{x}_r) P_{\mathbf{X}|B_k}(\boldsymbol{x}_r|0) \tag{6.369}$$

$$= \frac{2}{M} 2^{k-1} \sum_{r=1}^{2^{m-k}} \Psi(l, \mathsf{snr}\, \tilde{\mu}_k(\boldsymbol{x}_r), \mathsf{snr}\, \tilde{\sigma}_k^2(\boldsymbol{x}_r)), \tag{6.370}$$

where (6.370) follows from (6.368) and (2.77). In (6.370) we use a factor 2^{k-1} to take into account that the same PDF approximation will be obtained in each of 2^{k-1} "groups" of symbols in the constellation labeled by the BRGC for a given k (see more details in Section 5.5.4).

By inspecting Tables 5.1–5.2, it is possible to see that for a given m, the possible values of $\tilde{\mu}_k(\boldsymbol{x}_r)$ and $\tilde{\sigma}_k^2(\boldsymbol{x}_r)$ obtained for $k = 1$ include those obtained for $k = 2$, which in turn include those obtained for $k = 3$, and so on. In other words, the set of mean and variances obtained for $k = 1$ covers the mean and variances for $k = 2, \dots, m$. Thus, for simplicity, and assuming there are at most G different Gaussian PDFs, we define μ_g and σ_g^2 as the mean and variance of the gth Gaussian PDF. Furthermore, we assume $0 < \mu_1 < \mu_2 < \dots < \mu_G$, and $\sigma_1^2 \le \sigma_2^2 \le \dots \le \sigma_G^2$.[14] We can then express the PDFs (6.370) as the following Gaussian mixture

$$\tilde{p}_{L_k|B_k}(l|0) = \sum_{g=1}^{G} \zeta_{k,g} \Psi(l, -\mathsf{snr}\, \mu_g, \mathsf{snr}\, \sigma_g^2), \tag{6.371}$$

where the proportion of gth Gaussian PDFs in the mixture, denoted by $\zeta_{k,g}$, can be interpreted as the probability that the L-value L_k is distributed according to the gth Gaussian PDF.

Example 6.32 (4PAM **and the BRGC)** *Consider the* 4PAM *constellation labeled by the BRGC in Example 5.16, where the CoM is used to model the L-values. In this case,* (5.183) *already shows the two Gaussian PDFs for $k = 1$: $\Psi(l, -16\gamma, 32\gamma)$ and $\Psi(l, -4\gamma, 8\gamma)$, and from* (5.184) *we see that the Gaussian PDF $\Psi(l, -4\gamma, 8\gamma)$ is used for $k = 2$. Using $\gamma = \mathsf{snr}/5$, we then conclude that $G = 2$, $\mu_1 = 4/5$, $\mu_2 = 16/5$, $\sigma_1^2 = 8/5$, $\sigma_2^2 = 32/5$, $\zeta_{1,1} = \zeta_{1,2} = 1/2$, $\zeta_{2,1} = 1$ and $\zeta_{2,2} = 0$. Thus,* (6.371) *becomes*

$$\tilde{p}_{L_1|B_1}(l|0) = \frac{1}{2}\left(\Psi(l, -\mathsf{snr}\, \mu_1, \mathsf{snr}\, \sigma_1^2) + \Psi(l, -\mathsf{snr}\, \mu_2, \mathsf{snr}\, \sigma_2^2)\right), \tag{6.372}$$

$$\tilde{p}_{L_2|B_2}(l|0) = \Psi(l, -\mathsf{snr}\, \mu_1, \mathsf{snr}\, \sigma_1^2). \tag{6.373}$$

We can gather the weighting factors $\zeta_{k,g}$ in a matrix

$$\mathbf{Z}_\Phi \triangleq \begin{bmatrix} \zeta_{1,1} & \zeta_{1,2} & \cdots & \zeta_{1,G} \\ \zeta_{2,1} & \zeta_{2,2} & \cdots & \zeta_{2,G} \\ \vdots & \vdots & \ddots & \vdots \\ \zeta_{m,1} & \zeta_{m,2} & \cdots & \zeta_{m,G} \end{bmatrix}. \tag{6.374}$$

[14] Note that we considered here a reordered (and sign changed) version of the mean values $\tilde{\mu}_k(\boldsymbol{x}_r)$, which are all negative according to Tables 5.1 and 5.2. We also use $\le$ (instead of $<$) when sorting the variances, because in the case of the ZcM, the variance is constant.

The parameters of the gth Gaussian PDF uniquely depend on the ED between the symbol $\boldsymbol{x}_g \in \mathcal{S}_{k,0}$ and the closest symbol in $\mathcal{S}_{k,1}$, and therefore, the elements of $\mathbf{Z}_\Phi$ may be seen as a generalization of the EDS $\boldsymbol{d}_\mathcal{S} = [d_1, d_2, \ldots, d_D]^\mathrm{T}$ defined in Section 2.5.1. We will thus refer to $\mathbf{Z}_\Phi$ as a (normalized) *constellation bit-wise Euclidean distance spectrum* (CBEDS).

Example 6.33 (**Z_Φ for** 8PAM **and the BRGC**) *The CBEDS for the* 8PAM *constellation labeled by the BRGC in Fig. 2.13 (a), which we reproduce in Fig. 6.21, is given by*

$$\mathbf{Z}_\Phi = \begin{bmatrix} 1/4 & 1/4 & 1/4 & 1/4 \\ 1/2 & 1/2 & 0 & 0 \\ 1 & 0 & 0 & 0 \end{bmatrix}. \tag{6.375}$$

The elements in $\mathbf{Z}_\Phi$ can be obtained from direct inspection of Fig. 6.21, where the relevant distances for the subconstellation $\mathcal{S}_{k,0}$ is also schematically shown. This figure also shows the "grouping" of symbols (2^{k-1} groups for each bit position) we discussed above.

Under the quasirandom interleaving assumption, the PDF $p_\Lambda(\lambda)$ in (6.227) can be expressed using (6.371) as

$$p_\Lambda(\lambda) \approx \frac{1}{m} \sum_{k=1}^{m} \sum_{g=1}^{G} \zeta_{k,g} \Psi(\lambda, -\mathsf{snr}\, \mu_g, \mathsf{snr}\, \sigma_g^2) \tag{6.376}$$

$$= \sum_{g=1}^{G} w_g \Psi(\lambda, -\mathsf{snr}\, \mu_g, \mathsf{snr}\, \sigma_g^2), \tag{6.377}$$

where

$$w_g \triangleq \frac{1}{m} \sum_{k=1}^{m} \zeta_{k,g} \tag{6.378}$$

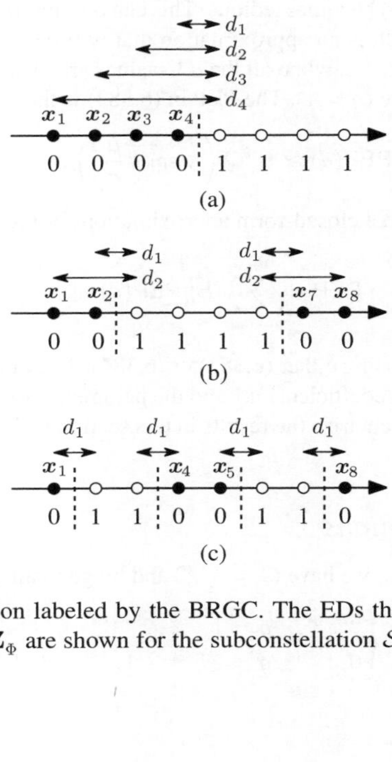

Figure 6.21 8PAM constellation labeled by the BRGC. The EDs that are relevant from the point of view of obtaining the CBEDS $\mathbf{Z}_\Phi$ are shown for the subconstellation $\mathcal{S}_{k,0}$ for (a) $k = 1$, (b) $k = 2$, and (c) $k = 3$

has a meaning of the probability that an L-value passed to the decoder is distributed according to the gth Gaussian PDF. Later in this section we compute w_g for MPAM and MPSK constellations.

With a closed-form approximation for the PDF of Λ in (6.377), we are ready to compute the PDF of Λ^{Σ} in (6.283). This is done as follows:

$$p_{\Lambda^{\Sigma}}(\lambda) \approx \sum_{g_1=1}^{G} w_{g_1} \Psi(\lambda, -\mathsf{snr}\, \mu_{g_1}, \mathsf{snr}\, \sigma^2_{g_d}) * \ldots * \sum_{g_d=1}^{G} w_{g_d} \Psi(\lambda, -\mathsf{snr}\, \mu_{g_d}, \mathsf{snr}\, \sigma^2_{g_d}) \tag{6.379}$$

$$= \sum_{g_1=1}^{G} \ldots \sum_{g_d=1}^{G} \left(\prod_{f=1}^{d} w_{g_f} \right) \Psi \left(\lambda, -\mathsf{snr} \sum_{f=1}^{d} \mu_{g_f}, \mathsf{snr} \sum_{f=1}^{d} \sigma^2_{g_f} \right) \tag{6.380}$$

$$= \sum_{\substack{\boldsymbol{r}\in\mathbb{N}^G \\ \|\boldsymbol{r}\|_1=d}} \binom{d}{\boldsymbol{r}} \Psi \left(\lambda, -\mathsf{snr} \sum_{g=1}^{G} r_g \mu_g, \mathsf{snr} \sum_{g=1}^{G} r_g \sigma^2_g \right) \prod_{g=1}^{G} w_g^{r_g}, \tag{6.381}$$

where (6.381) follows from reorganizing the terms in (6.380), and where r_g in $\boldsymbol{r} \triangleq [r_1, \ldots, r_G]$ denotes the number of L-values distributed according to the gth Gaussian PDF.

Using (6.381) in (6.282) we find

$$\mathsf{PEP}(d) \approx \sum_{\substack{\boldsymbol{r}\in\mathbb{N}^G \\ \|\boldsymbol{r}\|_1=d}} \binom{d}{\boldsymbol{r}} \prod_{g=1}^{G} w_g^{r_g} \int_0^{\infty} \Psi \left(\lambda, -\mathsf{snr} \sum_{g=1}^{G} r_g \mu_g, \mathsf{snr} \sum_{g=1}^{G} r_g \sigma^2_g \right) \mathrm{d}\lambda \tag{6.382}$$

$$= \sum_{\substack{\boldsymbol{r}\in\mathbb{N}^G \\ \|\boldsymbol{r}\|_1=d}} \binom{d}{\boldsymbol{r}} \prod_{g=1}^{G} w_g^{r_g} Q \left(\frac{\sqrt{\mathsf{snr}} \sum_{g=1}^{G} r_g \mu_g}{\sqrt{\sum_{g=1}^{G} r_g \sigma^2_g}} \right). \tag{6.383}$$

The PEP approximation in (6.383) is in closed-form; however, for large values of G and/or d, the enumeration of all the terms in (6.383) becomes tedious. This can be simplified by taking only the Q-function with the smallest argument, which is an approximation that will be tight for $\mathsf{snr} \to \infty$. This dominant Q-function is obtained for $r_1 = d$, i.e., when all the d L-values are distributed according to the Gaussian PDF with the smallest mean value ($g = 1$). The PEP in (6.383) is then approximated as

$$\mathsf{PEP}(d) \approx w_1^d Q \left(\sqrt{\mathsf{snr}\, d}\, \frac{\mu_1}{\sigma_1} \right). \tag{6.384}$$

Finally, from (6.225) we obtain a closed-form approximation for the BEP in BICM

$$\mathsf{BEP} \approx \sum_{d=d_{\text{free}}}^{d_{\max}} B_d^{\mathcal{B}} \mathsf{PEP}(d), \tag{6.385}$$

where $\mathsf{PEP}(d)$ can be evaluated using either (6.383) or (6.384). To evaluate (6.385), we need the IDS of the binary encoder, the weighting coefficients w_g and the parameters of the Gaussian approximations μ_g and σ^2_g. In the following, we particularize the results in this section to MPAM and MPSK constellations.

6.4.2 *M*PAM *Constellations*

For MPAM labeled by the BRGC, we have $G = M/2$ and by generalizing Example 6.33, we obtain

$$\zeta_{k,g} = \begin{cases} 2^{k-m}, & g = 1, \ldots, 2^{m-k} \\ 0, & g = 2^{m-k} + 1, \ldots, 2^{m-1} \end{cases}, \tag{6.386}$$

for $k = 1, \ldots, m$. The mean values and variances are obtained from Table 5.1 as

$$\mu_g = \begin{cases} 4\Delta^2 g^2, & \text{for CoM} \\ 4\Delta^2(2g-1), & \text{for ZcM} \end{cases} \tag{6.387}$$

$$\sigma_g^2 = \begin{cases} 8\Delta^2 g^2, & \text{for CoM} \\ 8\Delta^2, & \text{for ZcM,} \end{cases} \tag{6.388}$$

and w_g in (6.378) is

$$w_g = \frac{1}{m}\sum_{k=1}^{m} 2^{k-m}\mathbb{I}_{[g\leq 2^{m-k}]} \tag{6.389}$$

$$= \frac{2(2^{m-\lceil \log_2(g)\rceil}-1)}{mM} \tag{6.390}$$

$$= \frac{1}{m}\left(\frac{2}{2^{\lceil \log_2(g)\rceil}} - \frac{2}{M}\right), \tag{6.391}$$

where $g = 1, \ldots, 2^{m-1}$.

Using (6.391) we express (6.383) as

$$\mathsf{PEP}(d) \approx \sum_{\substack{\boldsymbol{r}\in\mathbb{N}^{M/2}\\ \|\boldsymbol{r}\|_1=d}} \binom{d}{\boldsymbol{r}} \prod_{g=1}^{M/2} \frac{1}{m^{r_g}}\left(\frac{2}{2^{\lceil \log_2(g)\rceil}} - \frac{2}{M}\right)^{r_g} Q\left(\sqrt{2\,\mathsf{snr}\,\Delta^2 \sum_{g=1}^{M/2} r_g g^2}\right) \tag{6.392}$$

for CoM and

$$\mathsf{PEP}(d) \approx \sum_{\substack{\boldsymbol{r}\in\mathbb{N}^{G}\\ \|\boldsymbol{r}\|_1=d}} \binom{d}{\boldsymbol{r}} \prod_{g=1}^{M/2} \frac{1}{m^{r_g}}\left(\frac{2}{2^{\lceil \log_2(g)\rceil}} - \frac{2}{M}\right)^{r_g} Q\left(\frac{\sqrt{2\,\mathsf{snr}\,\Delta^2}(-d+\sum_{g=1}^{M/2} 2gr_g)}{\sqrt{d}}\right) \tag{6.393}$$

for ZcM.

The simplified PEP approximation in (6.384) (the same result is obtained by applying CoM and ZcM) for MPAM is then given by

$$\mathsf{PEP}(d) \approx \left(\frac{2}{m}\right)^d \left(1 - \frac{1}{2^m}\right)^d Q\left(\sqrt{2\,\mathsf{snr}\,\Delta^2 d}\right) \tag{6.394}$$

$$= \left(\frac{2}{m}\right)^d \left(1 - \frac{1}{2^m}\right)^d Q\left(\sqrt{\frac{6\,\mathsf{snr}\,d}{2^{2m}-1}}\right), \tag{6.395}$$

where we used (2.47). We can thus conclude that an increase on the size of the constellation $M = 2^m$ reduces the multiplicative factor before the Q-function. More importantly, an increase of m by one is equivalent to decreasing the SNR by $10\log_{10}\frac{2^{2(m+1)}-1}{2^{2m}-1} \approx 6$ dB. This SNR shift will dominate the behavior of the BEP at high SNR.

The following example show the approximations for particular values of M.

Example 6.34 (BEP and PEP for MPAM and CENC in AWGN Channel) *As a sanity check, consider a* 2PAM *constellation, i.e.,* $m = 1$, $M = 2$, *and* $\Delta = 1$. *In this case, we have* $G = 1$ *so both* (6.392) *and* (6.393) *produce*

$$\mathsf{PEP}(d) = Q\left(\sqrt{2d\,\mathsf{snr}}\right), \tag{6.396}$$

which is consistent with the results in Example 6.24. In this case we can use the equality sign because the PEP in (6.396) is exact.

For a 4PAM *constellation we have $G = 2$, $r_1 + r_2 = d$, $w_1 = 3/4$ and $w_2 = 1/4$, which used in* (6.381) *yield*

$$p_{\Lambda\Sigma}(\lambda) \approx \sum_{r_1=0}^{d} \left(\frac{3}{4}\right)^{r_1} \left(\frac{1}{4}\right)^{d-r_1} \frac{d!}{r_1!(d-r_1)!} \Psi\left(l, -\mathsf{snr}(r_1\mu_1 + (d-r_1)\mu_2), \mathsf{snr}(r_1\sigma_1^2 + (d-r_1)\sigma_2^2)\right), \tag{6.397}$$

and thus, the PEP is given by

$$\mathsf{PEP}(d) \approx \sum_{r_1=0}^{d} \binom{d}{r_1} \frac{3^{r_1}}{4^d} Q\left(\frac{r_1\mu_1 + (d-r_1)\mu_2}{\sqrt{r_1\sigma_1^2 + (d-r_1)\sigma_2^2}}\right), \tag{6.398}$$

where μ_1, μ_2, σ_1^2, and σ_2^2 are given in (6.387) *and* (6.388).

If we consider the approximation in (6.394) *instead, i.e., if we neglect the terms that vanish for high SNR, we obtain*

$$\mathsf{PEP}(d) \approx \left(\frac{3}{4}\right)^{d} Q\left(\sqrt{2\,\mathsf{snr}\,\Delta^2 d}\right), \tag{6.399}$$

where from (2.47), $\Delta^2 = 1/5$.

For a 8PAM *constellation we have $G = 4$, $r_1 + r_2 + r_3 + r_4 = d$, and from* (6.375), $w_1 = 7/12$, $w_2 = 3/12$, *and* $w_3 = w_4 = 1/12$. *Although a closed-form expression similar to* (6.398) *can be written, we only show the approximation in* (6.394), *which gives*

$$\mathsf{PEP}(d) \approx \left(\frac{7}{12}\right)^{d} Q\left(\sqrt{2\,\mathsf{snr}\,\Delta^2 d}\right), \tag{6.400}$$

where from (2.47), $\Delta^2 = 1/21$. *Similarly, it can be shown that for* 16PAM

$$\mathsf{PEP}(d) \approx \left(\frac{15}{32}\right)^{d} Q\left(\sqrt{2\,\mathsf{snr}\,\Delta^2 d}\right), \tag{6.401}$$

where $\Delta^2 = 1/85$.

The results for 4PAM, 8PAM, *and* 16PAM *are compared in Fig. 6.22 for a CENC and the AWGN channel. We show the simulation results (markers), the CoM approximations from* (6.392), *the ZcM approximations from* (6.393), *and the single-term approximation from* (6.384) *specialized above in* (6.399)–(6.401). *We can see that, as expected, for high SNR, increasing m by one shifts the BEP curve by approximately* 6 dB.

6.4.3 *M*PSK *Constellations*

In the case of MPSK, we know that $G = 2^{m-2} = M/4$ and from Table 5.2 we read

$$\mu_g = \begin{cases} 4S_{M,g}^2, & \text{for CoM} \\ 4S_{M,2g-1}S_{M,1}, & \text{for ZcM} \end{cases} \tag{6.402}$$

$$\sigma_g^2 = \begin{cases} 8S_{M,g}^2, & \text{for CoM} \\ 8S_{M,1}^2, & \text{for ZcM,} \end{cases} \tag{6.403}$$

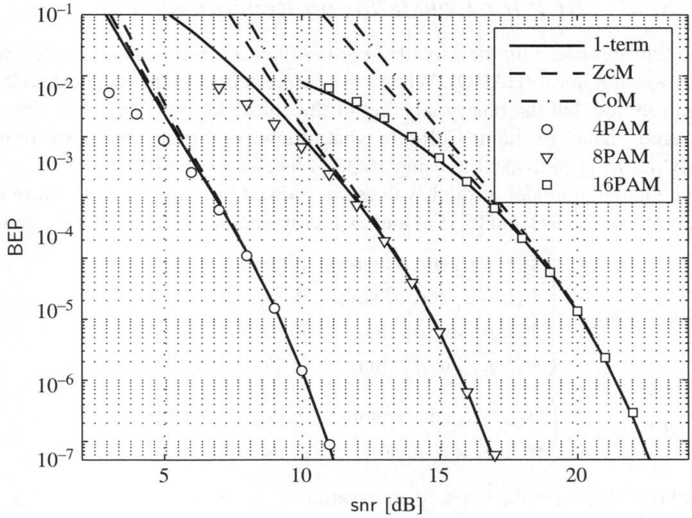

Figure 6.22 BEP approximations (lines) and simulations (markers) for a CENC with $\mathbf{G} = [5, 7]$ over the AWGN channel: 4PAM (circles), 8PAM (triangles) and 16PAM (squares). The dashed and dashed-dotted lines are the approximations obtained using (6.392) and (6.393) and the "1-term" approximation is obtained using the single-term expressions for the PEP shown in (6.399)–(6.401)

and knowing that the PDF of L_1 is the same as the PDF of L_2 (see also Example 5.7), we have

$$\zeta_{1,g} = \zeta_{2,g}, \tag{6.404}$$

and

$$\zeta_{k,g} = \begin{cases} 2^{k-m}, & g = 1, \dots, 2^{m-k} \\ 0, & g = 2^{m-k} + 1, \dots, 2^{m-2} \end{cases}, \tag{6.405}$$

for $k = 2, \dots, m$.

In analogy to (6.391) we can find w_g as

$$w_g = \frac{1}{m} \left(\zeta_{2,g} + \sum_{k=2}^{m} \zeta_{k,g} \right) \tag{6.406}$$

$$= \frac{1}{m} \left(2^{2-m} + \sum_{k=2}^{m} 2^{k-m} \mathbb{I}_{[g \le 2^{m-k}]} \right) \tag{6.407}$$

$$= \frac{2}{m} \frac{1}{2^{\lceil \log_2(g) \rceil}}, \tag{6.408}$$

where $g = 1, \dots, 2^{m-2}$.

The results above show that in the case of MPSK constellations, the L-values can again be approximated as a Gaussian mixture, where the parameters of the Gaussian PDFs as well as the weights are known in closed form. Using these closed-form expressions, it is possible to repeat the analysis we presented before, which we do not include here as it is mostly a repetition of the developments for MPAM in Section 6.4.2.

6.4.4 *Case Study: BEP for Constellation Rearrangement*

We consider here transmission with the so-called *constellation rearrangement* (CoRe), which is used in *hybrid automatic repeat request* (HARQ). When errors are detected in the received codeword, the same codeword is retransmitted, but the binary labeling of the constellation is changed. The constellation is therefore "rearranged", hence the name CoRe. In what follows we briefly outline the principles of CoRe in the case of 4PAM; this is equivalent to using 16QAM labeled by the BRGC.

We will use the Gaussian model of the L-values we showed in Example 5.19. More specifically, we reorganize the term from (5.197) and (5.198) and explicitly condition on the bits B_1 and B_2:

$$\tilde{p}_{L_1|B_1,B_2}(l|b_1,b_2) = \begin{cases} \Psi(l,-\mu_2\,\mathsf{snr},\sigma_2^2\,\mathsf{snr}), & \text{if } [b_1,b_2]=[0,0] \\ \Psi(l,-\mu_1\,\mathsf{snr},\sigma_1^2\,\mathsf{snr}), & \text{if } [b_1,b_2]=[0,1] \\ \Psi(l,\mu_1\,\mathsf{snr},\sigma_1^2\,\mathsf{snr}), & \text{if } [b_1,b_2]=[1,1] \\ \Psi(l,\mu_2\,\mathsf{snr},\sigma_2^2\,\mathsf{snr}), & \text{if } [b_1,b_2]=[1,0] \end{cases}, \tag{6.409}$$

$$\tilde{p}_{L_2|B_1,B_2}(l|b_1,b_2) = \begin{cases} \Psi(l,-\mu_1\,\mathsf{snr},\sigma_1^2\,\mathsf{snr}), & \text{if } [b_1,b_2]\in\{[0,0],[1,0]\} \\ \Psi(l,\mu_1\,\mathsf{snr},\sigma_1^2\,\mathsf{snr}), & \text{if } [b_1,b_2]\in\{[0,1],[1,1]\} \end{cases}. \tag{6.410}$$

Using (6.409) and (6.410), we make two key observations:

- The L-value L_1 has a "high-protection" distribution $\Psi(l,\pm\mu_2\,\mathsf{snr},\sigma_2^2\,\mathsf{snr})$ if $b_2=0$ and a "low-protection" distribution $\Psi(l,\pm\mu_1\,\mathsf{snr},\sigma_1^2\,\mathsf{snr})$ if $b_2=1$.
- The L-value L_2 has always a low-protection distribution, irrespective of the value of the transmitted bits.

CoRe can be then seen as a process that equalizes the "protection" experienced by the bits passing through the different bit positions in different transmissions. This is possible because in each transmission the same bits B_1 and B_2 are transmitted. More specifically, CoRe is based on two operations: negation of the bit at position $k=2$ (i.e., negation of the second row of the matrix $\mathbf{Q}$) and swapping the position of the labels of $k=1$ and $k=2$ (i.e., swapping the first and second row of $\mathbf{Q}$).

The negation operation is connected with the first observation we made above: as depending on the value of B_2 the L-value L_1 changes its distribution, using $B_2=b_2$ in the first transmission and $B_2=\overline{b}_2$ in the next one, we can guarantee that "high" protection is offered to the bit B_1 in one out of the two transmissions.

The swapping responds to the second observation we made above, and is meant to transmit the bits B_2 at position $k=1$ so that B_2 can take advantage of the "high" protection offered in that bit position.

At the receiver, the L-values for different transmissions are calculated, negated and/or swapped (if necessary), and then added to form aggregated CoRe L-values $L_{k,I}^{\text{CoRe}}$.[15] These L-values are then passed to the decoder. After I transmissions, we obtain the following distributions for L_1

$$\tilde{p}_{L_{1,I}^{\text{CoRe}}|B_1}(l|0) = \begin{cases} \frac{1}{2}\left(\Psi(l,-\mu_1\,\mathsf{snr},\sigma_1^2\,\mathsf{snr})+\Psi(l,-\mu_2\,\mathsf{snr},\sigma_2^2\,\mathsf{snr})\right), & \text{if } I=1 \\ \Psi(l,-(\mu_1+\mu_2)\,\mathsf{snr},(\sigma_1^2+\sigma_2^2)\,\mathsf{snr}), & \text{if } I=2 \\ \Psi(l,-(\mu_1(I-1)+\mu_2)\,\mathsf{snr},(\sigma_1^2(I-1)+\sigma_2^2)\,\mathsf{snr}), & \text{if } I=3,4 \end{cases}, \tag{6.411}$$

and for L_2

$$\tilde{p}_{L_{2,I}^{\text{CoRe}}|B_2}(l|0) = \begin{cases} \Psi(l,-\mu_1 I\,\mathsf{snr},\sigma_1^2 I\,\mathsf{snr}), & \text{if } I=1,2 \\ \frac{1}{2}(\Psi(l,-3\mu_1\,\mathsf{snr},3\sigma_1^2\,\mathsf{snr}) & \\ \quad +\Psi(l,-(2\mu_1+\mu_2)\,\mathsf{snr},(2\sigma_1^2+\sigma_2^2)\,\mathsf{snr})), & \text{if } I=3 \\ \Psi(l,-(3\mu_1+\mu_2)\,\mathsf{snr},(3\sigma_1^2+\sigma_2^2)\,\mathsf{snr}). & \text{if } I=4 \end{cases}. \tag{6.412}$$

15 Adding L-values obtained in different transmissions is suboptimal; however, it is often done for the simplicity of resulting operations.

The PDF of the L-values passed to the decoder is then given by

$$\tilde{p}_{L_I^{\text{CoRe}}}(l) = \frac{1}{2}\left(\tilde{p}_{L_{1,I}^{\text{CoRe}}|B_1}(l|0) + \tilde{p}_{L_{2,I}^{\text{CoRe}}|B_2}(l|0)\right) \tag{6.413}$$

$$= \tilde{w}_{1,I}\Psi(l, -\tilde{\mu}_{1,I}\,\mathsf{snr}, \tilde{\sigma}_{1,I}^2\,\mathsf{snr}) + \tilde{w}_{2,I}\Psi(l, -\tilde{\mu}_{2,I}\,\mathsf{snr}, \tilde{\sigma}_{2,I}^2\,\mathsf{snr}), \tag{6.414}$$

where

$$\tilde{\mu}_{1,I} = I\mu_1, \qquad \tilde{\mu}_{2,I} = \mu_2 + (I-1)\mu_1,$$

$$\tilde{\sigma}_{1,I}^2 = I\sigma_1^2, \qquad \tilde{\sigma}_{2,I}^2 = \sigma_2^2 + (I-1)\sigma_1^2,$$

$$\tilde{w}_{1,I} = 1 - \frac{I}{4}, \qquad \tilde{w}_{2,I} = \frac{I}{4}.$$

Since the PDF in (6.414) is again a Gaussian mixture, an approximation similar to the one in (6.398) may be used

$$\mathsf{PEP}(d) \approx \sum_{r_1=0}^{d} \tilde{w}_{1,I}^{r_1}\tilde{w}_{2,I}^{d-r_1}\binom{d}{r_1} Q\left(\frac{r_1\tilde{\mu}_{1,I} + (d-r_1)\tilde{\mu}_{2,I}}{\sqrt{r_1\tilde{\sigma}_{1,I}^2 + (d-r_1)\tilde{\sigma}_{2,I}^2}}\right). \tag{6.415}$$

In Fig. 6.23 the results obtained via numerical simulations are contrasted against the approximations of the PEP in (6.392) and (6.393). Unlike in Fig. 6.22, the differences between ZcM and CoM are now very clear. The ZcM provides a tight approximation on the coded BEP, especially when the number of transmissions increases. In particular, for $I = 4$, L_1 and L_2 have the same PDF and μ_2 and σ_2^2 have to be used in the model. Thus, HARQ accentuates the importance of the adequate modeling of the "high-protection" effect. Note that without HARQ and CoRe, the effect of "high-protection" is less pronounced and can be even neglected, e.g., using the one-term "low-protection" approximation (6.399). In the presence of CoRe we cannot do this because for $I = 4$ we have $\tilde{w}_{1,4} = 0$, i.e., the "low-protection" is entirely removed.

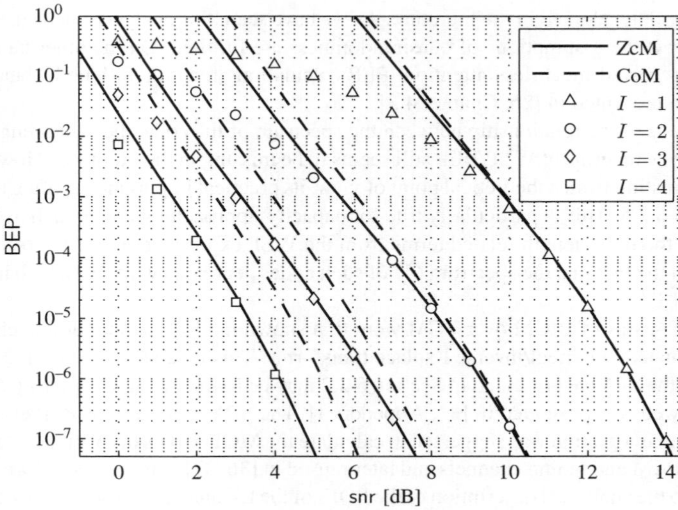

Figure 6.23 BEP approximations using the ZcM (solid lines) and the CoM (dashed lines) for a CENC with $\mathbf{G} = [5, 7]$ over an AWGN channel with CoRe and I retransmissions. The simulation results are shown with markers: $I = 1$ (triangles), $I = 2$ (circles), $I = 3$ (diamonds), and $I = 4$ (squares)

6.5 Bibliographical Notes

Performance evaluation in uncoded transmission has been the focus of research for many decades. The initial approximations of the SEP [1] were later replaced by calculations for regularly spaced constellations MQAM and/or MPSK [2, 3]. In this chapter we showed expression for $N = 1, 2$. The calculation of the BEP for 3D constellations was considered in [4].

The BEP for uncoded transmission in (6.28) has been studied in detail in [5, 6], where the asymptotic optimality of the BRGC for regular constellations is proved. Significant efforts have been made to evaluate the BEP in fading channels, i.e., to average the expressions for the AWGN channel over the fading distribution. This was considered, e.g., in [7–10]. Some of the formulas we presented in this chapter were shown, for integer m in [11, eq. (6)], for half-integer m in [12, eq. (15)], and for arbitrary m (using hypergeometric functions) in [13]. The literature is abundant in this area, so it is in fact quite difficult to recognize all the contributions. For a hopefully more complete list, we refer the reader to [14].

An alternative representation of the bivariate Q-function (6.99) was shown in [15] and the identity (6.102), simplified with respect to [16, eq. (11)], via the use of Q-functions. The form (6.99) is called *Craig's form* after the author of [17], who derived first a simple alternative form of Q-function we showed in (6.108).

The error event appearing in (6.173) is sometimes called "first-error event" [18, Section 6.2], [19, Section I], [20, Section 12.2] or error probability per node [21, Section 4.3]. The upper bound on the WEP in the case of the TCM transmissions can be found in [18, eq. (6.6)], [20, eq. (12.20)], [22, eq. (4.1)]. The expressions for the WEP in (6.178) and (6.241) are straightforward generalizations of the bound presented in [23] for CCs. TCM encoders with optimal distance spectrum (similar to the ones we used in Example 6.16) were presented in [24].

The expression in (6.243) is the most common expression for the upper bound for BICM, cf. [25, eq. (26)], [26, eq. (4.12)]. The upper bound in (6.243) can be found in almost any existing book on digital communications or coding (see, e.g., [20, eq. (12.28)], [27, eq. (7.9)], [28, eq. (8.2–19)], and it was originally defined for channels in which the metrics for the code bits passed to the decoder are i.i.d., e.g., 2PAM over the AWGN channel, in which the conditional L-values follow a Gaussian distribution as shown in (3.63).

The performance analysis of BICM transmission under random infinite-length interleaving proposed in [25] has been widely adopted in the literature. As we have seen, this analysis can be used in the case of a fixed interleaver if the assumptions of quasirandomness are fulfilled. On the other hand, finite-length interleaving has received much less attention. PEP calculations for infinite-length (but random) interleaving have been presented in [26, Chapter 4.3].

A formal analysis of the relationship between the spectrum of the code, the finite-length interleaving, and the performance in terms of WEP/BEP still seems to be missing in the literature. However, while this issue may be interesting from a theoretical point of view, its practical importance is often negligible as we argued in Section 6.2.5. This is particularly true for capacity-approaching codes such as TCs, for which we can eliminate the finite-length related terms from the WEP expression in (6.214) (see Lemma 6.19). This follows from the fact that the spectrum $C_d^{\mathcal{B}}$ of such codes decreases with $N_{\rm c}$, which has been shown, e.g., in [29, Fig. 10].

Insights into the gains of BICM over TCM were first shown in [30] via bounding techniques. In [31] the PEP was evaluated via direct/inverse Laplace transforms. This idea was refined in [32, 33]. The formal derivation of the SPA we presented in Section 6.3.2 may be found in [34, Chapter 2], and the intuitive approach we showed was presented in [33, Appendix I]. The SPA was used for PEP evaluation in [33], where Monte Carlo integration was suggested to calculate the MGF and its derivatives. The SPA was then used in [35] for 2PAM and fading channels and later reused in [36, 37], where closed-form formulas were obtained thanks to the analytical description of the PDFs of the L-values. The use of (zero-crossing or consistent) Gaussian approximations to simplify the integration was made popular in [38]. The zero-crossing approximation is due to [39], where it was first applied to analyze uncoded HARQ transmission based on the CoRe.

The CBEDS we used in this chapter was first introduced in [40, 41] where all the binary labelings for 8PSK having a different CBEDS were classified. The same concept was also presented in [42, Chapter 4]. The CBEDS in fact corresponds to a generalization of the ED spectrum of [43] in the sense that it considers the bit positions separately.

Mapping diversity has been studied, e.g., in [44–47]. CoRe was recommended by the *third-generation partnership project* (3GPP) working group because of its simplicity [48] and is only slightly suboptimal when compared to metrics calculation based on the outcomes of all transmissions (as required in other mapping diversity schemes). More details about CoRe may be found in [49]. Various mapping diversity schemes are analyzed from an information-theoretic point of view in [50].

The WD and the IWD we used for CENCs in Example 6.22 can be extended to *turbo encoders* (TENCs) using the concept of uniform and random interleaver introduced in [51, 52]. For the numerical results in this chapter, we used a breadth-first search algorithm [53]. Alternatively, a transfer function approach could be used, which works well for small values of memories ν. For large values of ν the *Bayesian evolutionary analysis by sampling trees* (BEAST) algorithm recently introduced in [54] (see also [55]) is more appropriate.

References

[1] Foschini, G. J., Gitlin, R. D., and Weinstein, S. B. (1974) Optimization of two-dimensional signal constellations in the presence of Gaussian noise. *IEEE Trans. Commun.*, **22** (1), 28–38.

[2] Lassing, J., Ström, E. G., Agrell, E., and Ottosson, T. (2003) Computation of the exact bit error rate of coherent M-ary PSK with Gray code bit mapping. *IEEE Trans. Commun.*, **51** (11), 1758–1760.

[3] Lassing, J. (2005) On the labeling of signal constellations. PhD dissertation, Chalmers University of Technology, Göteborg, Sweden.

[4] Khabbazian, M., Hossain, M. J., Alouini, M. S., and Bhargava, V. K. (2009) Exact method for the error probability calculation of three-dimensional signal constellations. *IEEE Trans. Commun.*, **57** (4), 922–925.

[5] Agrell, E., Lassing, J., Ström, E. G., and Ottosson, T. (2004) On the optimality of the binary reflected Gray code. *IEEE Trans. Inf. Theory*, **50** (12), 3170–3182.

[6] Agrell, E., Lassing, J., Ström, E. G., and Ottosson, T. (2007) Gray coding for multilevel constellations in Gaussian noise. *IEEE Trans. Inf. Theory*, **53** (1), 224–235.

[7] Simon, M. and Alouini, M. S. (1998) A simple single integral representation of the bivariate Rayleigh distribution. *IEEE Commun. Lett.*, **2** (5), 128–130.

[8] Alouini, M. S. and Goldsmith, A. J. (1999) A unified approach for calculating error rates of linearly modulated signals over generalized fading channels. *IEEE Trans. Commun.*, **47** (9), 1324–1334.

[9] Dong, X., Beaulieu, N. C., and Wittke, P. H. (1999) Error probabilities of two-dimensional m-ary signalling in fading. *IEEE Trans. Commun.*, **47** (3), 352–355.

[10] Annamalai, A., Tellambura, C., and Bhargava, V. K. (2005) A general method for calculating error probabilities over fading channels. *IEEE Trans. Commun.*, **53** (5), 841–852.

[11] Annamalai, A. and Tellambura, C. (2001) Error rates for Nakagami-m fading multichannel reception of binary and m-ary signals. *IEEE Trans. Commun.*, **49** (1), 58–68.

[12] Xu, H., Benjillali, M., and Szczecinski, L. (2008) Closed-form expression for the bit error rate in rectangular QAM with arbitrary constellation mapping in transmissions over Nakagami-m fading channel. *Wiley J. Wireless Commun. Mob. Comput.*, **8** (1), 93–99.

[13] Shin, H. and Lee, J. H. (2004) On the error probability of binary and M-ary signals in Nakagami-m fading channels. *IEEE Trans. Commun.*, **52** (4), 536–539.

[14] Simon, M. K. and Alouini, M.-S. (2000) *Digital Communications Over Fading Channels: A Unified Approach to Performance Analysis*, 1st edn, John Wiley & Sons, Inc., New York.

[15] Simon, M. K. (2002) A simpler form of the Craig representation for the two-dimensional joint Gaussian Q-function. *IEEE Commun. Lett.*, **6** (2), 49–51.

[16] Zhong, L., Alajaji, F., and Takahara, G. (2005) Error analysis for nonuniform signaling over Rayleigh fading channels. *IEEE Trans. Commun.*, **53** (1), 39–43.

[17] Craig, J. W. (1991) A new, simple and exact result for calculating the probability of error for two-dimensional signal constellations. Military Communications Conference (MILCOM), November 1991, McLean, VA.

[18] Schlegel, C. B. and Perez, L. C. (2004) *Trellis and Turbo Coding*, 1st edn, John Wiley & Sons.

[19] Rouanne, M. and Costello, D. J. Jr. (1989) An algorithm for computing the distance spectrum of trellis codes. *IEEE J. Sel. Areas Commun.*, **7** (6), 929–940.

[20] Lin, S. and Costello, D. J. Jr. (2004) *Error Control Coding*, 2nd edn, Prentice Hall, Englewood Cliffs, NJ.

[21] Viterbi, A. J. and Omura, J. K. (1979) *Principles of Digital Communications and Coding*, McGraw-Hill.

[22] Benedetto, S. and Biglieri, E. (1999) *Principles of Digital Transmission with Wireless Applications*, Kluwer Academic.

[23] Caire, G. and Viterbo, E. (1998) Upper bound on the frame error probability of terminated trellis codes. *IEEE Commun. Lett.*, **2** (1), 2–4.

[24] Alvarado, A., Graell i Amat, A., Brännström, F., and Agrell, E. (2013) On optimal TCM encoders. *IEEE Trans. Commun.*, **61** (6), 2178–2189.

[25] Caire, G., Taricco, G., and Biglieri, E. (1998) Bit-interleaved coded modulation. *IEEE Trans. Inf. Theory*, **44** (3), 927–946.

[26] Guillén i Fàbregas, A., Martinez, A., and Caire, G. (2008) Bit-interleaved coded modulation. *Found. Trends Commun. Inf. Theory*, **5** (1–2), 1–153.

[27] Burr, A. (2001) *Modulation and Coding for Wireless Communications*, Prentice Hall.

[28] Proakis, J. G. and Salehi, M. (2008) *Digital Communications*, 5th edn, McGraw-Hill.

[29] Perez, L. C., Seghers, J., and Costello, D. J. Jr. (1996) A distance spectrum interpretation of turbo codes. *IEEE Trans. Inf. Theory*, **42** (16), 1698–1709.

[30] Zehavi, E. (1992) 8-PSK trellis codes for a Rayleigh channel. *IEEE Trans. Commun.*, **40** (3), 873–884.

[31] Biglieri, E., Caire, G., Taricco, G., and Ventura-Traveset, J. (1996) Simple method for evaluating error probabilities. *Electron. Lett.*, **32** (2), 191–192.

[32] Biglieri, E., Caire, G., Taricco, G., and Ventura-Traveset, J. (1998) Computing error probabilities over fading channels: A unified approach. *Eur. Trans. Telecommun.*, **9** (1), 15–25.

[33] Martinez, A., Guillén i Fàbregas, A., and Caire, G. (2006) Error probability analysis of bit-interleaved coded modulation. *IEEE Trans. Inf. Theory*, **52** (1), 262–271.

[34] Butler, R. W. (2007) *Saddlepoint Approximation with Applications*, Cambridge University Press.

[35] Martinez, A., Guillén i Fàbregas, A., and Caire, G. (2007) A closed-form approximation for the error probability of BPSK fading channels. *IEEE Trans. Wireless Commun.*, **6** (6), 2051–2054.

[36] Szczecinski, L., Alvarado, A., and Feick, R. (2008) Closed-form approximation of coded BER in QAM-based BICM faded transmission. IEEE Sarnoff Symposium, April 2008, Princeton, NJ.

[37] Kenarsari-Anhari, A. and Lampe, L. (2010) An analytical approach for performance evaluation of BICM over Nakagami-m fading channels. *IEEE Trans. Commun.*, **58** (4), 1090–1101.

[38] Alvarado, A., Szczecinski, L., Feick, R., and Ahumada, L. (2009) Distribution of L-values in Gray-mapped M^2-QAM: closed-form approximations and applications. *IEEE Trans. Commun.*, **57** (7), 2071–2079.

[39] Benjillali, M., Szczecinski, L., Aissa, S., and Gonzalez, C. (2008) Evaluation of bit error rate for packet combining with constellation rearrangement. *Wiley J. Wireless Commun. Mob. Comput.*, **8**, 831–844.

[40] Brännström, F. (2004) Convergence analysis and design of multiple concatenated codes. PhD dissertation, Chalmers University of Technology, Göteborg, Sweden.

[41] Brännström, F. and Rasmussen, L. K. (2009) Classification of unique mappings for 8PSK based on bit-wise distance spectra. *IEEE Trans. Inf. Theory*, **55** (3), 1131–1145.

[42] Schreckenbach, F. (2007) Iterative decoding of bit-interleaved coded modulation. PhD dissertation, Technische Universität München, Munich, Germany.

[43] Schreckenbach, F., Görtz, N., Hagenauer, J., and Bauch, G. (2003) Optimized symbol mappings for bit-interleaved coded modulation with iterative decoding. IEEE Global Telecommunications Conference (GLOBECOM), December 2003, San Francisco, CA.

[44] Metzner, J. (1977) Improved sequential signaling and decision techniques for nonbinary block codes. *IEEE Trans. Commun.*, **25** (5), 561–563.

[45] Benelli, G. (1992) A new method for integration of modulation and channel coding in an ARQ protocol. *IEEE Trans. Commun.*, **40** (10), 1594–1606.

[46] Szczecinski, L. and Bacic, M. (2005) Constellations design for multiple transmissions: Maximizing the minimum squared Euclidean distance. IEEE Wireless Communications and Networking Conference (WCNC), March 2005, New Orleans, LA.

[47] Samra, H., Ding, Z., and Hahn, P. M. (2005) Symbol mapping diversity design for multiple packet transmissions. *IEEE Trans. Commun.*, **53**, 810–817.
[48] Panasonic (2001) Enhaced HARQ method with signal constellation rearrangement. Technical Report, 3GPP TSG RAN WG1.
[49] Wengerter, C., von Elbwart, A., Seidel, E., Velev, G., and Schmitt, M. (2002) Advanced hybrid ARQ technique employing a signal constellation rearrangement. IEEE Vehicular Technology Conference (VTC-Fall), September 2002, Vancouver, BC, Canada.
[50] Szczecinski, L., Diop, F.-K., and Benjillali, M. (2008) On the performance of BICM with mapping diversity in hybrid ARQ. *Wiley J. Wireless Commun. Mob. Comput.*, **8** (7), 963–972.
[51] Benedetto, S. and Montorsi, G. (1995) Average performance of parallel concatenated block codes. *Electron. Lett.*, **31** (3), 156–158.
[52] Benedetto, S. and Montorsi, G. (1996) Unveiling turbo codes: Some results on parallel concatenated coding schemes. *IEEE Trans. Inf. Theory*, **42** (2), 409–428.
[53] Belzile, J. and Haccoun, D. (1993) Bidirectional breadth-first algorithms for the decoding of convolutional codes. *IEEE Trans. Commun.*, **41** (2), 370–380.
[54] Bocharova, I. E., Handlery, M., Johannesson, R., and Kudryashov, B. D. (2004) A BEAST for prowling in trees. *IEEE Trans. Inf. Theory*, **50** (6), 1295–1302.
[55] Hug, F. (2012) Codes on graphs and more. PhD dissertation, Lund University, Lund, Sweden.

7

Correction of L-values

The L-values are the signals/messages exchanged in *bit-interleaved coded modulation* (BICM) receivers. We saw in Chapter 3 that, thanks to the formulation of the L-values in the logarithm domain, multiplication of probabilities/likelihoods transforms into addition of the corresponding L-values. This makes the L-values well suited for numerical implementation. Furthermore, the L-values can be processed in abstraction of how they were calculated. The designer can then connect well-defined processing blocks, which is done under the assumption that the L-values can be transformed into probabilities/likelihoods. This assumption does not hold in all cases, as the L-values might have been incorrectly calculated, resulting in *mismatched* L-values. In this chapter, we take a closer look at the problem of how to deal with this mismatch.

We will discuss the motivation for L-values correction in Section 7.1 and explain optimal processing rules in Section 7.2. Suboptimal (linear) correction strategies will be studied in Section 7.3.

7.1 Mismatched Decoding and Correction of L-values

As discussed in Section 3.2, the BICM decoding rule

$$\hat{\underline{\boldsymbol{b}}}^{\text{bi}} = \operatorname*{argmax}_{\underline{\boldsymbol{b}}\in\mathcal{B}} \left\{ \sum_{n=1}^{N_s} \sum_{k=1}^{m} b_k[n] l_k[n] \right\} \tag{7.1}$$

is solely based on the L-values $l_k[n]$ calculated as

$$l_k[n] = \theta_k(\boldsymbol{y}[n]) = \log \frac{p_{\boldsymbol{Y}|B_k}(\boldsymbol{y}[n]|1)}{p_{\boldsymbol{Y}|B_k}(\boldsymbol{y}[n]|0)}. \tag{7.2}$$

Thus, decoding is carried out in abstraction of the channel model, i.e., the L-values are assumed to carry all the information necessary for decoding.

We also know from Section 3.2 that the decoding rule (7.1) is suboptimal. More specifically, the decoding metric in (7.1) implicitly uses the following approximation:

$$p_{\boldsymbol{Y}|\boldsymbol{B}}(\boldsymbol{y}[n]|\boldsymbol{b}[n]) \approx \prod_{k=1}^{m} p_{\boldsymbol{Y}|B_k}(\boldsymbol{y}[n]|b_k[n]), \tag{7.3}$$

Bit-Interleaved Coded Modulation: Fundamentals, Analysis, and Design, First Edition.
Leszek Szczecinski and Alex Alvarado.

where, using (3.38), we have

$$p_{\boldsymbol{Y}|B_k}(\boldsymbol{y}[n]|b_k[n]) = \exp\left(b_k[n]l_k[n] + C\right). \tag{7.4}$$

We showed in Example 3.3 that (7.3) may hold with equality in some particular cases, but in general, the decoding metric used by (7.1) is mismatched. As we explained in Section 3.2, an achievable rate for the (mismatched) BICM decoder is then given by the *generalized mutual information* (GMI).

In previous chapters, the analysis of the BICM decoder in (7.1) was done assuming (7.2) is implemented. In this chapter, however, we are interested in cases when (7.2) is no applied *exactly*, i.e., when the L-values are incorrectly calculated, or–in other words–when they are *mismatched*. We emphasize that this mismatch comes on the top of the mismatch caused by the suboptimal decoding rule (7.1).

The main reasons for the L-values to be mismatched are the following:

- **Modeling Errors**: The model used to derive $p_{\boldsymbol{Y}|B_k}(\boldsymbol{y}[n]|b_k[n])$ is not accurate. For example, if the actual distribution of the additive noise $\boldsymbol{Z}$ is ignored (e.g., when $\boldsymbol{Z}$ is not Gaussian). Similarly, the parameters of the distributions might have been incorrectly estimated. This is, in fact, almost always the case in practice, as estimation implies unavoidable estimation errors.
- **Simplifications**: The L-values are calculated using simplifications introduced to diminish the computational load. This includes the popular max-log approximation as well as other methods discussed in Section 3.3.3. Approximations are in fact, unavoidable in interference-limited transmission scenarios, where the effective constellation size grows exponentially with the number of data streams (which corresponds, e.g., to the number of transmitting antennas in *multiple-input multiple-output* (MIMO) systems).
- **L-values Processing**: After the L-values are calculated (possibly with modeling errors and simplifications), some extra operations may be necessary to render them suitable for decoding. This may include quantization, truncation, scaling, or any other operation directly applied on the L-values.

In the first two cases above, the likelihoods $p_{\boldsymbol{Y}|B_k}(\boldsymbol{y}[n]|b_k[n])$ are incorrectly calculated. Formally, we assume that the decoder uses $\tilde{p}_{\boldsymbol{Y}|B_k}(\boldsymbol{y}[n]|b_k[n])$ instead of $p_{\boldsymbol{Y}|B_k}(\boldsymbol{y}[n]|b_k[n])$. The binary decoder remains unaware of the mismatch and will blindly use the former. The mismatched metric is then $\tilde{\mathbb{q}}_k(b,\boldsymbol{y}) \propto \tilde{p}_{\boldsymbol{Y}|B_k}(\boldsymbol{y}[n]|b_k[n])$. In analogy to (4.30)–(4.33), we have that the BICM decoding rule in this case is

$$\hat{\underline{\boldsymbol{b}}}^{\mathrm{bi}} = \underset{\underline{\boldsymbol{b}}\in\mathcal{B}}{\operatorname{argmax}} \left\{ \sum_{n=1}^{N_{\mathrm{s}}} \sum_{k=1}^{m} \log\left[\tilde{\mathbb{q}}_k(b_k[n], \boldsymbol{y}[n])\right] \right\}. \tag{7.5}$$

We can now define mismatched L-values as

$$\tilde{l}_k[n] = \tilde{\theta}_k(\boldsymbol{y}[n]) = \log\frac{\tilde{\mathbb{q}}_k(1,\boldsymbol{y}[n])}{\tilde{\mathbb{q}}_k(0,\boldsymbol{y}[n])}, \tag{7.6}$$

where $\tilde{\theta}_k(\boldsymbol{y})$ is a mismatched demapper.

Transforming (7.6) into

$$\hat{\underline{\boldsymbol{b}}}^{\mathrm{bi}} = \underset{\underline{\boldsymbol{b}}\in\mathcal{B}}{\operatorname{argmax}} \left\{ \sum_{n=1}^{N_{\mathrm{s}}} \sum_{k=1}^{m} \log\frac{\tilde{\mathbb{q}}_k(b_k[n],\boldsymbol{y}[n])}{\tilde{\mathbb{q}}_k(0,\boldsymbol{y}[n])} \right\} \tag{7.7}$$

$$= \underset{\underline{\boldsymbol{b}}\in\mathcal{B}}{\operatorname{argmax}} \left\{ \sum_{n=1}^{N_{\mathrm{s}}} \sum_{k=1}^{m} b_k[n]\tilde{l}_k[n] \right\}, \tag{7.8}$$

we obtain again a form similar to (7.1), i.e., the decoder remains unaware of the mismatch and blindly uses the mismatched L-values $\tilde{l}_k[n]$. Moreover, because (7.5) and (7.8) are equivalent, we know now that the mismatch in the likelihoods translates into an equivalent mismatch in the L-values. Thus, from now on, we do not analyze the source of the mismatch, and we uniquely consider the mismatched L-values calculated by the mismatched demapper $\tilde{\theta}_k(\cdot)$ defined in (7.6). This is also convenient because our considerations will be done in the domain of the L-values, so from now on we will avoid references to the observations $\boldsymbol{y}[n]$. Instead, we consider the mismatched L-values as the outputs of an equivalent channel, shown in Fig. 7.1, which encompasses the actual transmission and the mismatched demapping.

Owing to the memoryless property of the channel, the BICM decoding rule based on the mismatched L-values $\tilde{\boldsymbol{l}}[n] = [\tilde{l}_1[n], \ldots, \tilde{l}_m[n]]$ can be again written as

$$\hat{\underline{\boldsymbol{b}}}^{\text{bi}} = \underset{\underline{\boldsymbol{b}} \in \mathcal{B}}{\operatorname{argmax}} \left\{ \prod_{n=1}^{N_s} \breve{\mathfrak{q}}(\boldsymbol{b}[n], \tilde{\boldsymbol{l}}[n]) \right\}, \tag{7.9}$$

where $\breve{\mathfrak{q}}(\boldsymbol{b}, \tilde{\boldsymbol{l}})$ is used to denote the metric defined for the equivalent channel outputs, i.e., a metric based on the mismatched L-values rather than on the channel observations. We know from Section 3.1 that the metric $\breve{\mathfrak{q}}(\boldsymbol{b}, \tilde{\boldsymbol{l}}) = p_{\tilde{\boldsymbol{L}}|\boldsymbol{B}}(\tilde{\boldsymbol{l}}, \boldsymbol{b})$ is optimal in the sense of minimizing the probability of decoding error. However, we want to preserve the bitwise operation, so we will prefer the suboptimal decoding

$$\hat{\underline{\boldsymbol{b}}}^{\text{bi}} = \underset{\underline{\boldsymbol{b}} \in \mathcal{B}}{\operatorname{argmax}} \left\{ \prod_{n=1}^{N_s} \prod_{k=1}^{m} \breve{\mathfrak{q}}_k(b_k[n], \tilde{\boldsymbol{l}}[n]) \right\}, \tag{7.10}$$

where we decomposed the metric $\breve{\mathfrak{q}}(\boldsymbol{b}, \tilde{\boldsymbol{l}})$ into a product of bitwise metrics $(b_k, \tilde{\boldsymbol{l}})$, i.e.,

$$\breve{\mathfrak{q}}(\boldsymbol{b}, \tilde{\boldsymbol{l}}) = \prod_{k=1}^{m} \breve{\mathfrak{q}}_k(b_k, \tilde{\boldsymbol{l}}), \tag{7.11}$$

where $\boldsymbol{b} = [b_1, \ldots, b_m]$.

We can now develop (7.10) along the lines of (7.7) and (7.8); namely,

$$\hat{\underline{\boldsymbol{b}}}^{\text{bi}} = \underset{\underline{\boldsymbol{b}} \in \mathcal{B}}{\operatorname{argmax}} \left\{ \sum_{n=1}^{N_s} \sum_{k=1}^{m} \log \frac{\breve{\mathfrak{q}}_k(b_k[n], \tilde{\boldsymbol{l}}[n])}{\breve{\mathfrak{q}}_k(0, \tilde{\boldsymbol{l}}[n])} \right\} \tag{7.12}$$

$$= \underset{\underline{\boldsymbol{b}} \in \mathcal{B}}{\operatorname{argmax}} \left\{ \sum_{n=1}^{N_s} \sum_{k=1}^{m} b_k[n] \tilde{l}_k^{\text{c}}[n] \right\}, \tag{7.13}$$

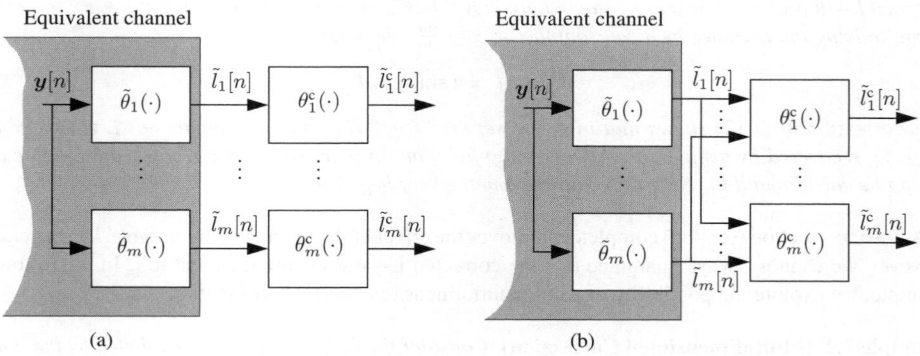

Figure 7.1 The mismatched L-values $\tilde{l}_k[n]$ are the result of a mismatched demapping and can be corrected using (a) scalar or (b) multidimensional functions $\theta_k^{\text{c}}(\cdot)$. These functions act as demappers for the equivalent channel (shown as a shaded rectangle), which concatenates the actual transmission channel (with output $\boldsymbol{y}[n]$) and the mismatched demapper $\tilde{\theta}_k(\cdot)$

where

$$\tilde{l}_k^{\rm c}[n] = \theta_k^{\rm c}(\tilde{\boldsymbol{l}}[n]) = \log \frac{\check{\mathbb{q}}_k(1, \tilde{\boldsymbol{l}}[n])}{\check{\mathbb{q}}_k(0, \tilde{\boldsymbol{l}}[n])}. \tag{7.14}$$

Thus, treating $\tilde{\boldsymbol{l}}[n]$ as the outcome of the equivalent channel, $\tilde{l}_k^{\rm c}[n]$ has a meaning of a new L-value calculated on the basis of the metrics $\check{\mathbb{q}}_k(b, \tilde{\boldsymbol{l}}[n])$.

The decoding results will clearly change depending on the choice of the decoding metric $\check{\mathbb{q}}_k(b, \tilde{\boldsymbol{l}})$. In particular, using $\check{\mathbb{q}}_k(b, \tilde{\boldsymbol{l}}) = \check{\mathbb{q}}_k^0(b, \tilde{\boldsymbol{l}})$ with

$$\check{\mathbb{q}}_k^0(b, \tilde{\boldsymbol{l}}) \triangleq \exp(b \cdot \tilde{l}_k)\, t(\tilde{l}_k), \tag{7.15}$$

and $t(l)$ being an arbitrary nonzero function independent of b, we obtain the relationship $\tilde{l}_k^{\rm c}[n] = \tilde{l}_k[n]$ from (7.14). In other words, by using $\check{\mathbb{q}}_k(b, \tilde{\boldsymbol{l}}) = \check{\mathbb{q}}_k^0(b, \tilde{\boldsymbol{l}})$, we guarantee that (7.13) and (7.8) are equivalent. This means that the decoder using the observations $\boldsymbol{y}[n]$ and the metrics $\mathbb{q}_k(b, \boldsymbol{y})$ carries out the same operation as the decoder using the observations $\tilde{l}_k[n]$ and the metrics $\check{\mathbb{q}}_k^0(b, \tilde{\boldsymbol{l}})$. Using such a metric, we actually assume that the relationship (7.4) holds,[1] and thus, we ignore the fact that $\tilde{l}_k[n]$ is mismatched.

On the other hand, if we are aware of the presence of the mismatch, i.e., if we know that (7.4) does not hold, we should use a different metric $\check{\mathbb{q}}_k(b, \tilde{\boldsymbol{l}})$. This implies a different form of the function $\theta_k^{\rm c}(\cdot)$, which is meant to "correct" the effect of the mismatch. This is the focus of this chapter. From this perspective, the metric $\check{\mathbb{q}}_k^0(b, \tilde{l})$ in (7.15) is "neutral" as it does lead to the trivial correction $\theta_k^{\rm c}(\tilde{\boldsymbol{l}}) = \tilde{l}$.

We say that the correction is *multidimensional* if it follows the general formulation in (7.14) and depends on a vector $\tilde{\boldsymbol{l}}$. We say a correction is *scalar* if the kth corrected L-value is obtained using only the corresponding mismatched L-value, i.e., $\check{\mathbb{q}}_k(b, \tilde{\boldsymbol{l}}) = \check{\mathbb{q}}_k(b, \tilde{l}_k)$. Then, the correction function takes the form

$$\tilde{l}_k^{\rm c}[n] = \theta_k^{\rm c}(\tilde{l}_k[n]) = \log \frac{\check{\mathbb{q}}_k(1, \tilde{l}_k[n])}{\check{\mathbb{q}}_k(0, \tilde{l}_k[n])}. \tag{7.16}$$

These two correction strategies are illustrated in Fig. 7.1.

Example 7.1 (Mismatched L-values in 2PAM**)** *Let us consider again the case of 2-ary pulse amplitude modulation (2PAM) transmission from Example 3.7, i.e.,* $\boldsymbol{x}_1 = -1$ *and* $\boldsymbol{x}_2 = 1$. *From the received signal* $\boldsymbol{y} = \Phi(b) + \boldsymbol{z}$, *the L-values are calculated as* $l = \theta(\boldsymbol{y}) = 4\,\mathsf{snr}\,\boldsymbol{y}$. *We further assume that, instead of the exact signal-to-noise ratio (SNR) value* snr, *the demapper uses its estimate* $\hat{\mathsf{snr}}$, *and thus, mismatched L-values* $\tilde{l} = 4\,\hat{\mathsf{snr}}\,\boldsymbol{y}$ *are obtained, as shown in Example 3.15. In this case, it is easy to see that by multiplying the L-values by a constant factor* $\alpha = \frac{\mathsf{snr}}{\hat{\mathsf{snr}}}$, *we obtain*

$$\tilde{l}^{\rm c} = \alpha \tilde{l} = 4\,\mathsf{snr}\,\boldsymbol{y} = l. \tag{7.17}$$

The expression in (7.17) *shows that in this case, the "true" (i.e., exactly calculated) L-values can be perfectly recovered by using the right correcting function. In other words, we recover the L-value as it should be calculated if the SNR were known from the very beginning.*

A linear correction (scaling) completely removes the effect of the mismatch in Example 7.1. In general, however, we cannot always guarantee that the corrected L-values $\tilde{l}^{\rm c}$ are identical to l. In the following example, we explore the possibility of using multidimensional correction functions.

Example 7.2 (Multidimensional Correction) *Consider the* 4PAM *constellation labeled by the binary reflected Gray code (BRGC) shown in Fig. 5.3 and the L-values calculated using the max-log simplification as shown in Example 5.5. From Fig. 5.10, we immediately see that the function* $\tilde{\theta}_1(\boldsymbol{y})$ *is bijective,*[2] *and thus, knowing* $\tilde{l}_1$ *we can recover* $\boldsymbol{y} = \tilde{\theta}_1^{-1}(\tilde{l}_1)$. *The exact L-values are then obtained using* (5.6)

[1] The constant factor C in (7.4) is irrelevant to the decoding and metrics' manipulation so it may be omitted.
[2] We use the notation $\tilde{\theta}_1(\boldsymbol{y})$ to emphasize that the L-value obtained via the max-log simplification is mismatched.

and (5.15)*, i.e.,* $l_1 = \theta_1(\boldsymbol{y})$ *and* $l_2 = \theta_2(\boldsymbol{y})$*. Note that knowing solely* $\tilde{l}_2$ *gives us information only on the absolute value of* $\boldsymbol{y}$*, which is not enough to recalculate* l_1*.*

Moreover, we note that the function $\tilde{\theta}_1(\boldsymbol{y})$ *is bijective for any* MPAM *constellation labeled by the BRGC or the natural binary code (NBC). Therefore, the above considerations may be immediately generalized so that the exact L-values can be recovered as*

$$\tilde{l}_k^{\mathrm{c}} = l_k = \theta_k^{\mathrm{c}}(\tilde{\boldsymbol{l}}) = \theta_k(\tilde{\theta}_1^{-1}(\tilde{l}_1)), \qquad k = 1, \ldots, m. \tag{7.18}$$

As $\tilde{l}_k^{\mathrm{c}}$ *depends on* $\tilde{l}_1$*, the correction* (7.18) *is multidimensional.*

While multidimensional correction functions can offer advantages (as shown in Example 7.2), their main drawback with respect to their scalar counterpart is that they are, in general, more complex to implement and to design. This is important when the mismatch is a result of a simplified processing introduced for complexity reduction. In this situation, the correction of the mismatch should also have low complexity. This is why scalar correction is an interesting alternative. Such a scalar correction, which neglects the relationship between the L-values $\tilde{l}_1, \ldots, \tilde{l}_m$ is, in fact, well aligned with the spirit of BICM where all the operations are done at a bit level.

In this chapter, we take a general view on the problem of correcting L-values and focus on finding suitable scalar correction strategies for the mismatched L-values. There are two issues to consider. The first one is of a fundamental nature: what is an optimal correction strategy? The second relates to practical implementation aspects: what are good simplified (and thus suboptimal) correction strategies? As we will see in this chapter, an optimal correction function $\theta_k^{\mathrm{c}}(\cdot)$ can be defined but is in general nonlinear, and thus, its implementation may be cumbersome. This will lead us to mainly focus on (suboptimal) linear corrections.

7.2 Optimal Correction of L-values

The assumption that the corrected L-values $\tilde{l}_k[n]$ will be used by the decoder defined in (7.13) leads to the obvious question: how is the decoder affected by the mismatch? Once this question is answered, we may design correction strategies aiming at improving the decoder's performance.

7.2.1 GMI-Optimal Correction

From an information-theoretic point of view, considering the effect of the mismatched L-values on the performance of a BICM decoder in (7.13) is similar to the problem of analyzing the mismatched decoding metric in a BICM decoder, as we did in Section 4.3.1. The main difference now is that the decoder in (7.10) uses the metrics $\check{\mathfrak{q}}_k(b, \tilde{l}_k)$ calculated for the outputs of the equivalent channel $\tilde{l}_k[n]$. As the decoding (7.10) has the form of the BICM decoder, an achievable rate, in this case, can also be characterized by the GMI, i.e.,

$$I_{\check{\mathfrak{q}}}^{\mathrm{gmi}}(\boldsymbol{B}; \tilde{\boldsymbol{L}}) = \max_{s \geq 0} \left\{ \sum_{k=1}^{m} I_{\check{\mathfrak{q}}_k, s}^{\mathrm{gmi}}(B_k; \tilde{\boldsymbol{L}}) \right\}, \tag{7.19}$$

where, thanks to (7.11), we used Theorem 4.11 to express the GMI as the sum of bitwise GMIs, which are defined in the same way as in (4.47), i.e.,

$$I_{\check{\mathfrak{q}}_k, s}^{\mathrm{gmi}}(B_k; \tilde{\boldsymbol{L}}) = \mathbb{E}_{B_k, \tilde{\boldsymbol{L}}} \left[\log_2 \frac{[\check{\mathfrak{q}}_k(B_k, \tilde{\boldsymbol{L}})]^s}{\sum_{b \in \mathbb{B}} P_{B_k}(b) [\check{\mathfrak{q}}_k(b, \tilde{\boldsymbol{L}})]^s} \right] \tag{7.20}$$

$$= \mathbb{E}_{B_k, \tilde{L}_k} \left[\log_2 \frac{[\check{\mathfrak{q}}_k(B_k, \tilde{L}_k)]^s}{\sum_{b \in \mathbb{B}} P_{B_k}(b) [\check{\mathfrak{q}}_k(b, \tilde{L}_k)]^s} \right] \tag{7.21}$$

$$= I_{\check{\mathfrak{q}}_k, s}^{\mathrm{gmi}}(B_k; \tilde{L}_k). \tag{7.22}$$

To pass from (7.20) to (7.21), we used the scalar correction principle, i.e., $\check{\mathfrak{q}}_k(b, \tilde{\boldsymbol{L}}) = \check{\mathfrak{q}}_k(b, \tilde{L}_k)$. Using (7.22) in (7.19) yields

$$I^{\text{gmi}}_{\check{\mathfrak{q}}}(\boldsymbol{B}; \tilde{\boldsymbol{L}}) = \max_{s\geq 0}\left\{\sum_{k=1}^{m} I^{\text{gmi}}_{\check{\mathfrak{q}}_k,s}(B_k; \tilde{L}_k)\right\}. \tag{7.23}$$

The GMI in (7.23) is an achievable rate for the decoder in (7.13), which uses the observations $\tilde{l}_k[n]$ and applies scalar correction functions. The scalar correction assumption remains implicit throughout the rest of this chapter.

Theorem 7.3 (Equivalence of GMI) *For any mismatched metric $\tilde{\mathfrak{q}}_k(b, \boldsymbol{y})$, the GMI in* (4.34) *is equal to the GMI in* (7.23) *if $\check{\mathfrak{q}}_k(b, \tilde{l}_k) = \check{\mathfrak{q}}^0_k(b, \tilde{l}_k)$, i.e.,*

$$I^{\text{gmi}}_{\tilde{\mathfrak{q}}}(\boldsymbol{B}; \boldsymbol{Y}) = I^{\text{gmi}}_{\check{\mathfrak{q}}^0}(\boldsymbol{B}; \tilde{\boldsymbol{L}}). \tag{7.24}$$

Proof: Using (4.34), (4.46), and (7.23), we express (7.24) as

$$\max_{s\geq 0}\left\{\sum_{k=1}^{m} I^{\text{gmi}}_{\tilde{\mathfrak{q}}_k,s}(B_k; \boldsymbol{Y})\right\} = \max_{s\geq 0}\left\{\sum_{k=1}^{m} I^{\text{gmi}}_{\check{\mathfrak{q}}^0_k,s}(B_k; \tilde{L}_k)\right\}. \tag{7.25}$$

Next, by using similar steps to those in the proof of Theorem 4.20, we express (7.21) as

$$I^{\text{gmi}}_{\check{\mathfrak{q}}^0_k,s}(B_k; \tilde{L}_k) = \mathcal{H}(B_k) - \sum_{b\in\mathbb{B}} P_{B_k}(b)\mathbb{E}_{\tilde{L}_k|B_k=b}\left[\log_2\left(1 + \frac{P_{B_k}(\bar{b})}{P_{B_k}(b)}\frac{[\check{\mathfrak{q}}^0_k(\bar{b}, \tilde{L}_k)]^s}{[\check{\mathfrak{q}}^0_k(b, \tilde{L}_k)]^s}\right)\right] \tag{7.26}$$

$$= \mathcal{H}(B_k) - \sum_{b\in\mathbb{B}} P_{B_k}(b)\mathbb{E}_{\tilde{L}_k|B_k=b}\left[\log_2(1 + \mathrm{e}^{(-1)^b[s\tilde{L}_k + l^{\mathrm{a}}_k]})\right] \tag{7.27}$$

$$= I^{\text{gmi}}_{\tilde{\mathfrak{q}}_k,s}(B_k; \boldsymbol{Y}), \tag{7.28}$$

where to pass from (7.26) to (7.27), we used (7.15), and (7.28) is due to (4.75). As (7.28) holds for any k and s, it guarantees that (7.25) holds too, which completes the proof. □

Clearly, using the "neutral" metric $\check{\mathfrak{q}}^0_k(b, \tilde{l}_k)$ in (7.15) corresponds to feeding the decoder with the mismatched, uncorrected L-values $\tilde{l}_k[n]$, so the achievable rates are not changed when compared to those obtained by decoding on the basis of $\boldsymbol{y}[n]$ and the metrics $\tilde{\mathfrak{q}}_k(b, \boldsymbol{y})$. Theorem 7.3 is provided here to connect the analysis based on the metrics $\tilde{\mathfrak{q}}_k(b, \boldsymbol{y})$ and the analysis of the performance based on the metrics $\check{\mathfrak{q}}_k(b, \tilde{l}_k)$.

As the GMI is suitable to assess the performance of BICM receivers, we would like to perform a correction that improves this very criterion; this motivates the following definition of optimality.

Definition 7.4 (GMI-Optimal Correction of the L-values) *The correction functions $\theta^{\mathrm{c}}_k(\tilde{l})$ are said to be GMI-*optimal *if:*

(i) they maximize the GMI $I^{\text{gmi}}_{\check{\mathfrak{q}}}(\boldsymbol{B}; \tilde{\boldsymbol{L}})$ in (7.23)*, and*
(ii) $\theta^{\mathrm{c}}_k(\tilde{l}) = \tilde{l}$ when $\tilde{L}_k = L_k$.

To explain the origin of the condition (ii) in Definition 7.4, we note that using $\alpha\theta^{\mathrm{c}}_k(l)$ with any $\alpha \neq 0$ does not change the GMI in (7.23). Thus, to avoid this nonuniqueness of the solutions, we postulate that the trivial correction $\theta^{\mathrm{c}}_k(\tilde{l}_k) = \tilde{l}_k$ should be applied when the L-values are not mismatched, i.e., when $\tilde{L}_k = L_k$.

Theorem 7.5 (GMI-Optimal Metric) *The maximum value of the bitwise GMI* $I^{\text{gmi}}_{\breve{q}_k,s}(B_k; \tilde{L}_k)$ *is given by the bitwise mutual information (MI)* $I(B_k; \tilde{L}_k)$ *and is attained using* $\breve{q}_k(b, \tilde{l}) \propto p_{\tilde{L}_k|B_k}(\tilde{l}|b)$. *These metrics also maximize the GMI in* (7.23).

Proof: The proof of the first part of the theorem follows by using Theorem 4.10 with $\boldsymbol{Y} = \tilde{L}_k$ and $\boldsymbol{B} = B_k$. Theorem 4.10 also shows that the maximum of the bitwise GMI is obtained for $s = 1$. The proof of the second part of theorem then follows from the fact that each of the terms in the sum in (7.23) is maximized for the same value of s. □

Theorem 7.5 shows that, not surprisingly, the maximum GMI is obtained when the bit metrics are matched to the channel outcomes, in this case: to the mismatched L-values $\tilde{L}_k$. The next corollary gives a similar result in terms of the optimal correction functions.

Corollary 7.6 (GMI-Optimal Correction Function) *The GMI-optimal correction functions are*

$$\theta^{\text{c}}_k(\tilde{l}) = \log \frac{p_{\tilde{L}_k|B_k}(\tilde{l}|1)}{p_{\tilde{L}_k|B_k}(\tilde{l}|0)}, \qquad k = 1, \dots, m. \tag{7.29}$$

Proof: The proof follows by applying the results of Theorem 7.5 to (7.16). □

The results in Corollary 7.6 show the GMI-optimal correction functions, i.e., it explicitly shows the best correction strategy (in terms of GMI). An intuitive explanation of Corollary 7.6 may be provided as follows: knowing the conditional *probability density function* (PDF) $p_{\tilde{L}_k|B_k}(\tilde{l}|b)$, and treating $\tilde{l}_k[n]$ as an observation of the equivalent channel, we "recalculate" an L-value using the definition (3.29), which leads to (7.29).

As the GMI-optimal correction means that the corrected L-values $\tilde{L}^{\text{c}}_k$ are calculated using the likelihoods $p_{\tilde{L}_k|B_k}(\tilde{l}|b)$, they should have properties similar to those of the exact L-values L_k. This is shown in the next theorem.

Theorem 7.7 *The PDF of the L-values* $\tilde{L}^{\text{c}}_k = \theta^{\text{c}}_k(\tilde{L}_k)$ *obtained using the correction function* (7.29) *satisfies the consistency condition* (3.67).

Proof: The proof follows by using Theorem 3.10 and using $\boldsymbol{Y} = \tilde{L}_k$. □

Observe that if the PDF of the L-values $\tilde{L}_k$ already satisfies the consistency condition (3.67), the correction in (7.29) is useless, as on combining (3.67) with (7.16), we obtain a trivial correction function $\theta^{\text{c}}_k(l) = l$, i.e., $\tilde{l}^{\text{c}}_k[n] = \tilde{l}_k[n]$. We emphasize that this does not mean that L-values $\tilde{l}_k[n]$ obtained using the metrics $\breve{q}_k(b, \boldsymbol{y})$ are the same as the L-values $l_k[n]$ obtained using the true likelihoods $p_{\boldsymbol{Y}|B_k}(\boldsymbol{y}|b)$, i.e., the consistency condition does not imply "global" optimality. It only means that the L-values may be used by subsequent processing blocks (the decoder, in particular) and may be correctly interpreted in the domain of the likelihoods $p_{\tilde{L}_k|B_k}(\tilde{l}_k[n]|b) \propto \exp(b \cdot \tilde{l}_k[n])$. With this reasoning, Theorem 7.7 tells us that the L-values corrected via (7.29) are consistent and thus, cannot be corrected any further. This points to a rather obvious fact: if a further correction was possible, it would mean that the correction defined via (7.29) was not optimal.

Corollary 7.8 (GMI-Optimal Correction of Gaussian L-values) *If the L-values are distributed according to a symmetric Gaussian density* (sgd)

$$p_{\tilde{L}|B}(\tilde{l}|1) = p_{\tilde{L}|B}(-\tilde{l}|0) = \Psi(\tilde{l}, \mu_{\tilde{L}}, \sigma^2_{\tilde{L}}), \tag{7.30}$$

where $\mu_{\tilde{L}} = \mathbb{E}_{\tilde{L}|B=1}[\tilde{L}] = -\mathbb{E}_{\tilde{L}|B=0}[\tilde{L}]$ *and* $\sigma^2_{\tilde{L}} = \mathbb{V}\mathrm{ar}_{\tilde{L}|B=1}[\tilde{L}]$, *the optimal correction function is linear, that is,*

$$\theta^{\mathrm{c}}(\tilde{l}) = \alpha^{\mathrm{sgd}} \cdot \tilde{l}, \tag{7.31}$$

and the correction factor is

$$\alpha^{\mathrm{sgd}} = \frac{2\mu_{\tilde{L}}}{\sigma^2_{\tilde{L}}}. \tag{7.32}$$

Proof: The proof follows from (7.30) and Corollary 7.6. □

Example 7.9 (GMI-Optimal Correction in 2PAM with SNR mismatch) *A Gaussian form of the PDF of the L-values was indeed obtained in the case we analyzed in Example 7.1, where the L-values were calculated using an incorrectly estimated SNR value. The parameters of the distribution can be shown to be*

$$\mu_{\tilde{L}} = 4\,\hat{\mathsf{snr}}, \qquad \sigma^2_{\tilde{L}} = 2\mu_{\tilde{L}}\frac{\hat{\mathsf{snr}}}{\mathsf{snr}}, \tag{7.33}$$

and therefore, from (7.31), *we obtain the scaling factor*

$$\alpha^{\mathrm{sgd}} = \frac{2\mu_{\tilde{L}}}{\sigma^2_{\tilde{L}}} = \frac{\mathsf{snr}}{\hat{\mathsf{snr}}}. \tag{7.34}$$

The scaling factor in (7.34) is the same as the optimal correction factor we found in Example 7.1, where, to remove the mismatch, we exploited the knowledge of relationship between $\tilde{l}$ and $\boldsymbol{y}$. On the other hand, in Example 7.9, we do not need to know how the mismatched demapper $\tilde{\theta}_k(\boldsymbol{y})$ works and the mismatch is corrected knowing solely the distribution $p_{\tilde{L}|B}(\tilde{l}|b)$. In fact, because of the Gaussian assumption, only two parameters (mean and variance) of this distribution are necessary. This illustrates well the fact that, to correct the L-values, we do not need to know how they were calculated, but instead, we rely only on the knowledge of their distribution or of some of their parameters.

Example 7.10 (Mismatched L-values in 4PAM**)** *The PDF of the max-log L-values L_1 and L_2 for* 4PAM *were found in Example 5.5. Although we did not use the explicit notation $\tilde{L}_1$ and $\tilde{L}_2$, because of the max-log approximation, these L-values are mismatched. The correction function $\theta^{\mathrm{c}}(\tilde{l})$ in (7.29), found using the densities in Fig. 5.12 (dashed lines), is shown in Fig. 7.2. Note that for high SNR, the correction function is nearly linear confirming the fact that the max-log L-values are close to the optimal ones.*

7.2.2 *PEP-Optimal Correction*

If instead of information-theoretic considerations, we want to analyze the performance of a decoder in terms of decoding error probability, we may use the analysis and tools described in Section 6.2.4. The *word-error probability* (WEP) can be then approximated as

$$\mathsf{WEP} \approx \sum_{d=d_{\mathrm{free}}}^{d_{\max}} \sum_{\substack{\boldsymbol{w}\in\mathbb{N}^m \\ \|\boldsymbol{w}\|_1=d}} C^{\mathcal{B}}_{\boldsymbol{w}} \mathsf{PEP}^{\mathrm{c}}(\boldsymbol{w}), \tag{7.35}$$

where

$$\mathsf{PEP}^{\mathrm{c}}(\boldsymbol{w}) \triangleq \Pr\left\{\tilde{\Lambda}^{\Sigma,\mathrm{c}}_{\underline{\mathbf{0}},\underline{\mathbf{e}}} > 0\right\}, \tag{7.36}$$

$$\tilde{\Lambda}^{\Sigma,\mathrm{c}}_{\underline{\mathbf{0}},\underline{\mathbf{e}}} \triangleq \sum_{t=1}^{d} \tilde{L}^{\mathrm{c}}_{(t)}, \tag{7.37}$$

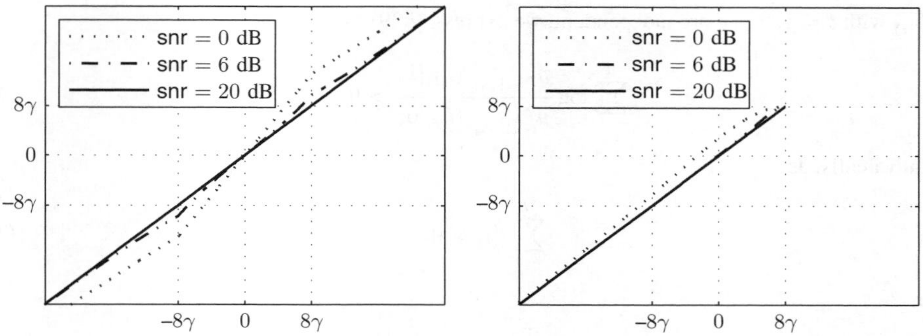

(a) (b)

Figure 7.2 The correction function $\theta_k^{\text{c}}(\tilde{l})$ from (7.29) obtained treating the max-log L-values L_k for 4PAM labeled by the BRGC as mismatched L-values $\tilde{L}_k$, $\gamma = \mathsf{snr}/5$ (see (5.7)), and for (a) $k = 1$ and (b) $k = 2$. Note that the PDF of the L-values $\tilde{L}_2$ is zero for the arguments in $[8\gamma, \infty)$, cf. (5.17), and thus, the correction function $\theta_2^{\text{c}}(\tilde{l})$ is undefined in this interval

$d = \|\boldsymbol{w}\|_1$, and $\tilde{L}_{(t)}^{\text{c}}$ are the corrected L-values corresponding to the d nonzero bits in the error codeword $\underline{\mathbf{e}}$.

Here, as in Section 6.2.4, we consider only the error codewords $\underline{\mathbf{e}}$ whose nonzero bits are mapped to different time instants n. This assumption (already discussed in Section 6.2.4) allows us to treat all the L-values $\tilde{L}_{(t)}^{\text{c}}$ as independent.

Definition 7.11 (PEP-Optimal Correction of the L-values) *The correction functions $\theta_k^{\text{c}}(\cdot)$ are said to be pairwise-error probability (PEP)-*optimal *if (i) they minimize* $\mathsf{PEP}^{\text{c}}(\boldsymbol{w})$ *for all generalized Hamming weight (GHW) $\boldsymbol{w}$, and (ii) are "neutral" ($\theta_k^{\text{c}}(l) = l$) if the L-values are not mismatched, i.e., if $\tilde{L}_k = L_k$.*

The condition (ii) in Definition 7.11 is useful again to ensure uniqueness of the solutions, as we note that multiplication of all the L-values by a common factor $\alpha \neq 0$ does not change the PEP.

Theorem 7.12 (PEP-Optimal Correction Function) *The PEP-optimal correction functions are*

$$\theta_k^{\text{c}}(\tilde{l}) = \log \frac{p_{\tilde{L}_k|B_k}(\tilde{l}|1)}{p_{\tilde{L}_k|B_k}(\tilde{l}|0)}, \quad k = 1, \ldots, m. \tag{7.38}$$

Proof: The PEP is, by definition, the probability of error in the following binary hypothesis testing problem. One hypothesis is that the codeword $\underline{\boldsymbol{E}} = \underline{\mathbf{e}}$ was transmitted, where $\underline{\mathbf{e}}$ has d nonzero bits and GHW, $\boldsymbol{w}$ (i.e., $\|\boldsymbol{w}\|_1 = d$). The second hypothesis is that the codeword $\underline{\boldsymbol{E}} = \underline{\mathbf{0}}$ was transmitted. The decision is made using the L-values $\tilde{l}_{(1)}, \ldots, \tilde{l}_{(d)}$, and assuming equally likely codewords, a decision in favor of $\underline{\mathbf{e}}$ which minimizes the probability of error (and thus also the PEP) will be made if

$$\log \frac{p_{\tilde{L}_{(1)},\ldots,\tilde{L}_{(d)}|B_{(1)},\ldots,B_{(d)}}(\tilde{l}_{(1)},\ldots,\tilde{l}_{(d)}|1,\ldots,1)}{p_{\tilde{L}_{(1)},\ldots,\tilde{L}_{(d)}|B_{(1)},\ldots,B_{(d)}}(\tilde{l}_{(1)},\ldots,\tilde{l}_{(d)}|0,\ldots,0)} > 0. \tag{7.39}$$

As $\tilde{L}_{(t)}$ with $t = 1, \dots, d$ are independent, we express (7.39) as

$$\sum_{t=1}^{d} \log \frac{p_{\tilde{L}_{(t)}|B_{(t)}}(\tilde{l}_{(t)}|1)}{p_{\tilde{L}_{(t)}|B_{(t)}}(\tilde{l}_{(t)}|0)} > 0, \tag{7.40}$$

or equivalently, as

$$\sum_{t=1}^{d} \tilde{l}^{\mathrm{c}}_{(t)} > 0, \tag{7.41}$$

where

$$\tilde{l}^{\mathrm{c}}_{(t)} \triangleq \log \frac{p_{\tilde{L}_{(t)}|B_{(t)}}(\tilde{l}_{(t)}|1)}{p_{\tilde{L}_{(t)}|B_{(t)}}(\tilde{l}_{(t)}|0)}. \tag{7.42}$$

Thus, deciding in favor of the hypothesis $\underline{\boldsymbol{E}} = \underline{\boldsymbol{e}}$ on the basis of the sum of $\tilde{l}^{\mathrm{c}}_{(t)}$ in (7.41) minimizes the probability of decision error. As this is equivalent to minimizing the PEP, and the optimal decision variable in the *left-hand side* (l.h.s.) of (7.41) is the sum of $\tilde{l}^{\mathrm{c}}_{(t)}$ given by (7.42) with the same form as (7.38), using corrected L-values in (7.38) minimizes the PEP. □

The PEP-optimal correction indeed minimizes the PEP independently of the GHW $\boldsymbol{w}$. This is a consequence of the assumption of the L-values being independent. Moreover, we note that the PEP-optimal and GMI-optimal correction functions are the same.

7.3 Suboptimal Correction of L-values

In Section 7.2, we defined the optimal correction function $\theta^{\mathrm{c}}_k(\cdot)$. However, as finding or implementing $\theta^{\mathrm{c}}_k(\cdot)$ is not necessarily simple, we use it as a reference rather than as a practical approach. Moreover, its form may provide us with guidelines to design suboptimal but simple-to-implement correction functions. This is the objective of this section.

Our objective is to find a suitable approximate correction function

$$\hat{\theta}^{\mathrm{c}}_k(l; \boldsymbol{\alpha}_k) \approx \theta^{\mathrm{c}}_k(l), \tag{7.43}$$

where $\boldsymbol{\alpha}_k$ represents parameters of the approximate correction function. Because of the approximation in (7.43), the corrected L-values are still mismatched, i.e., we lose the relationship with their interpretation in terms of a posteriori probabilities or likelihoods.

7.3.1 Adaptation Strategies

To find a correction function $\hat{\theta}^{\mathrm{c}}_k(l; \boldsymbol{\alpha}_k)$, first we need to decide on the form of the approximate function, and then, we have to find the parameters $\boldsymbol{\alpha}_k$. There are essentially two different ways to "adapt" the correction parameters $\boldsymbol{\alpha}_k$, which we present schematically in Fig. 7.3:

- **Feedback-Based Adaptation** consists in adjusting the correction parameters using the decoding results. Such an approach requires off-line simulations in the targeted conditions of operation. The advantage is that we explicitly optimize the criterion of interest (e.g., *bit-error probability* (BEP) after decoding). The disadvantage, however, is that, in most cases, we need to carry out an exhaustive search over the entire domain of the correction parameters. While this is acceptable if only a few parameters need to be found, it is unfeasible when the number of parameters is large or when the operating conditions (such as the SNR or the interference level) are variable.

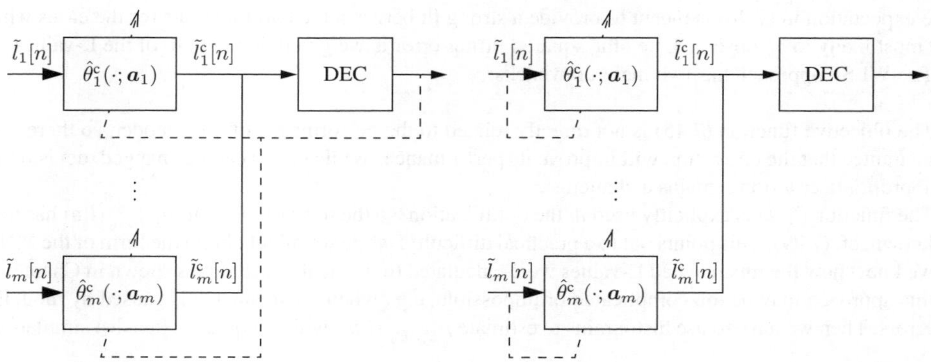

Figure 7.3 Finding the parameters $\boldsymbol{a}_k$ of the correction function via (a) a feedback-based strategy where the parameters are adjusted using the statistics of the output (e.g., the decoding results), and (b) a predictive strategy where the parameters are adjusted, before decoding, using the PDF of the mismatched L-values only

- **Predictive Adaptation** consists in finding the parameters of the correction function from the PDF of the mismatched L-values and from the model of the decoder. The postulate, inspired by the optimal correction in Section 7.2, is that the correction should be done for each of the mismatched L-values independently of the others. The caveat is that a simple strategy to deal with the mismatch needs to be devised.

While the feedback-based adaptation is robust and may be interesting to use from a practical point of view, it has little theoretical interest and provides no insights into the correction principles. In this chapter, we focus on predictive adaptation strategies.

Probably the simplest correction is done via a linear function

$$\hat{\theta}_k^{c}(l; \alpha_k) = \alpha_k l, \tag{7.44}$$

which has an appealing simplicity. Furthermore, in many cases, $\theta_k^{c}(l)$ is observed to be relatively "well" approximated by a linear function, see, e.g., Example 7.10.

In the following, we will discuss different ways of determining the optimal values of α_k. The first one is based on a heuristic approach based on function fitting, the second one aims at maximizing achievable rates of the resulting mismatched decoder, and the third one aims at minimizing the decoding error probability. It is important to note that the maximization of achievable rates and the minimization of the error probability are not the same for the (suboptimal) linear correction we consider here.

7.3.2 Function Fitting

A heuristic adaptation strategy aiming at fulfilling the condition (7.43) is based on finding the parameter α_k which approximates "well" $\theta_k^{c}(l)$ (which is PEP- and GMI-optimal). As an example, we consider here the *weighted least-squares fit* (WLSF)

$$\hat{\alpha}_k^{\text{wlsf}} \triangleq \underset{\alpha \geq 0}{\operatorname{argmin}} \left\{ \phi(\theta_k^{c}(\tilde{l}_k), \alpha \tilde{l}_k) \right\}, \tag{7.45}$$

where the fitting criterion is defined as

$$\phi(a(l), b(l)) \triangleq \mathbb{E}_{\tilde{L}_k}\left[(a(\tilde{L}_k) - b(\tilde{L}_k))^2 \right]. \tag{7.46}$$

The expectation in (7.46) is meant to provide a strong fit between the two functions for the cases which are most likely to be observed, i.e., the squared fitting-error is weighted by the PDF of the L-values.

The WLSF approach has two main drawbacks:

- The objective function (7.45) is not directly related to the performance of the decoder, so there is no guarantee that the correction will improve its performance. While (7.45) can be changed, devising the appropriate criterion remains a challenge.
- The function $\theta_k^{\mathrm{c}}(l)$ is explicitly used in the optimization, so the form of the PDF $p_{\tilde{L}_k|B_k}(l|b)$ has to be known, cf. (7.29). This points out to a practical difficulty: while we might obtain the form of the PDF if we knew how the mismatched L-values were calculated (e.g., via the techniques shown in Chapter 5), this approach may be too complicated or impossible, e.g., when the mismatch is caused by modeling errors. Then we have to use histograms to estimate $p_{\tilde{L}_k|B_k}(l|b)$, which requires extensive simulations.

Before proceeding further, we introduce a lemma that we will use in the following subsections.

Lemma 7.13 *For any concave functions $f_k(x), k = 1, \ldots, m,$ each with global maximum $\hat{x}_k = \operatorname{argmax}_{x\in\mathbb{R}} f_k(x)$ and any real m weights $w_k, k = 1, \ldots, m$, the function*

$$f(x, \boldsymbol{\alpha}) = \sum_{k=1}^{m} w_k\, f_k(x\alpha_k) \tag{7.47}$$

with $\boldsymbol{\alpha} = [\alpha_1, \ldots, \alpha_m]$ satisfies

$$\max_{x\in\mathbb{R}, \boldsymbol{\alpha}\in\mathbb{R}^m} f(x, \boldsymbol{\alpha}) = f(x_0, \hat{\boldsymbol{\alpha}}), \tag{7.48}$$

where $x_0 \neq 0$ is an arbitrary constant and $\hat{\boldsymbol{\alpha}} = [\hat{\alpha}_1, \ldots, \hat{\alpha}_m]$ and $\hat{\alpha}_k = \hat{x}_k/x_0, k = 1, \ldots, m.$

Proof: To prove (7.48), it is enough to prove that $f(x, \boldsymbol{\alpha}) \leq f(x_0, \hat{\boldsymbol{\alpha}}), \forall x \in \mathbb{R}, \boldsymbol{\alpha} \in \mathbb{R}^m$. To this end, we note that the function $f(x, \boldsymbol{\alpha})$ in (7.47) can be bounded as

$$f(x, \boldsymbol{\alpha}) \leq \max_{x\in\mathbb{R}, \boldsymbol{\alpha}\in\mathbb{R}^m} \sum_{k=1}^{m} w_k\, f_k(x\alpha_k) \leq \sum_{k=1}^{m} w_k\, f_k(\hat{x}_k), \tag{7.49}$$

which combined with

$$f(x_0, \hat{\boldsymbol{\alpha}}) = \sum_{k=1}^{m} w_k\, f_k(x_0\hat{\alpha}_k) = \sum_{k=1}^{m} w_k\, f_k(\hat{x}_k) \tag{7.50}$$

completes the proof. □

7.3.3 GMI-Optimal Linear Correction

The GMI-optimality principle in Section 7.2 has to be revisited taking into account the fact that we constrain the correction function to have a linear form. To this end, we rewrite (7.23) as

$$I_{\tilde{\mathsf{q}}}^{\mathrm{gmi}}(\boldsymbol{B}; \tilde{\boldsymbol{L}}) = \max_{s\geq 0} \left\{ \sum_{k=1}^{m} \tilde{I}_k^{\mathrm{gmi}}(s) \right\} \tag{7.51}$$

$$= \sum_{k=1}^{m} \tilde{I}_k^{\mathrm{gmi}}(\hat{s}^{\mathrm{gmi}}), \tag{7.52}$$

where, for notation simplicity, in (7.51) we use $\tilde{I}_k^{\text{gmi}}(s) \triangleq I_{\check{q}_k,s}^{\text{gmi}}(B_k; \tilde{L}_k)$, and $\hat{s}^{\text{gmi}}$ is the solution of the optimization problem in the *right-hand side* (r.h.s.) of (7.51). Owing to Theorem 4.22, this solution is known to be unique.

To analyze the effect of the linear correction of the L-values $\tilde{l}_k^{\text{c}}[n] = \alpha_k \tilde{l}_k[n]$ on the GMI, we move the correction effect from the L-values' domain to the domain of the metrics $\check{q}_k(b, \tilde{l})$. From (7.16), we obtain the relationship between the corrected and the uncorrected metrics, i.e.,

$$\tilde{l}_k^{\text{c}}[n] = \alpha_k \tilde{l}_k[n] \tag{7.53}$$

$$= \alpha_k \log \frac{\check{q}_k^0(1, \tilde{l}_k[n])}{\check{q}_k^0(0, \tilde{l}_k[n])} \tag{7.54}$$

$$= \log \frac{\left[\check{q}_k^0(1, \tilde{l}_k[n])\right]^{\alpha_k}}{\left[\check{q}_k^0(0, \tilde{l}_k[n])\right]^{\alpha_k}}. \tag{7.55}$$

By comparing (7.16) with (7.55), we conclude that the linear correction implies that the corrected decoding metric depends on the "neutral" metric via

$$\check{q}_k(b, \tilde{l}_k[n]) = \left[\check{q}_k^0(b, \tilde{l}_k[n])\right]^{\alpha_k}, \tag{7.56}$$

Which, when used in (7.21) yields

$$I_{\check{q}_k,s}^{\text{gmi}}(B_k; \tilde{L}_k) = I_{\check{q}_k^0, s\alpha_k}^{\text{gmi}}(B_k; \tilde{L}_k) = \tilde{I}_k^{\text{gmi}}(s\alpha_k). \tag{7.57}$$

Therefore, in order to maximize the GMI in (7.52) using linear correction functions, we have to solve the following optimization problem:

$$\max_{\boldsymbol{\alpha} \in \mathbb{R}_+^m} \left\{ I_{\check{q}}^{\text{gmi}}(\boldsymbol{B}; \tilde{\boldsymbol{L}}) \right\} = \max_{s \in \mathbb{R}_+, \boldsymbol{\alpha} \in \mathbb{R}_+^m} \left\{ \sum_{k=1}^{m} \tilde{I}_k^{\text{gmi}}(s\alpha_k) \right\}, \tag{7.58}$$

where $\boldsymbol{\alpha} \triangleq [\alpha_1, \ldots, \alpha_m]$. The solution to the problem in (7.58) is given in the following theorem.

Theorem 7.14 (GMI Maximization) *The linear correction factors that maximize the GMI* $I_{\check{q}}^{\text{gmi}}(\boldsymbol{B}; \tilde{\boldsymbol{L}})$ *are given by*

$$\hat{\alpha}_k^{\text{gmi}} = \hat{s}_k^{\text{gmi}}, \quad k = 1, \ldots, m, \tag{7.59}$$

where

$$\hat{s}_k^{\text{gmi}} \triangleq \underset{s \geq 0}{\text{argmax}} \left\{ \tilde{I}_k^{\text{gmi}}(s) \right\}. \tag{7.60}$$

Proof: Combining (7.28) and (7.57), we obtain $\tilde{I}_k^{\text{gmi}}(s) = I_{\check{q}_k,s}^{\text{gmi}}(B_k; \boldsymbol{Y})$. Then, by Theorem 4.22 (which is valid for any q_k), $\tilde{I}_k^{\text{gmi}}(s)$ is concave in s with global maxima $\hat{s}_k^{\text{gmi}}$ given by (7.60), and thus, we use Lemma 7.13 to obtain the solution $\hat{\alpha}_k^{\text{gmi}} = \hat{s}_k^{\text{gmi}}/s_0$, where $s_0 \neq 0$ is an arbitrary constant. To resolve the ambiguity with respect to s_0, we recall that when the L-values are not mismatched, the condition (ii) in Definition 7.4 requires that $\hat{\alpha}_k^{\text{gmi}} = 1$, i.e., no correction shall be made. This, combined with the fact that the GMI for matched metrics is maximized for $\hat{s}^{\text{gmi}} = 1$, gives $s_0 = 1$, which results in (7.59). □

It is important to note that the correction via $\hat{\alpha}_k^{\text{gmi}}$ does not change the *maximum value* of the bitwise GMI, i.e.,

$$\max_{s \geq 0} \left\{ \tilde{I}_k^{\text{gmi}}\left(s\hat{\alpha}_k^{\text{gmi}}\right) \right\} = \max_{s \geq 0} \left\{ \tilde{I}_k^{\text{gmi}}(s) \right\}. \tag{7.61}$$

Instead, the objective of the correction is to "align" the maxima for each $\tilde{I}_k^{\text{gmi}}(s\hat{\alpha}_k^{\text{gmi}})$, i.e., to make them occur for the same value of $s = 1$, and thus, maximize the sum.

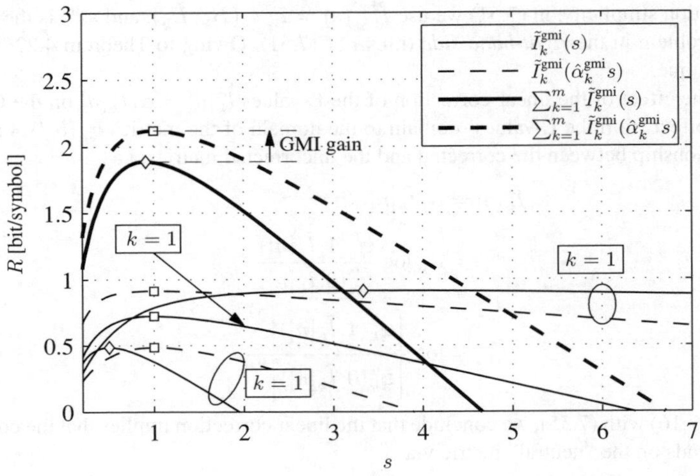

Figure 7.4 Linear correction of the L-values in the presence of an SNR mismatch. Thanks to the correction of the L-values $\tilde{L}_k$ by the optimal factors $\hat{\alpha}_k^{\text{gmi}}$ in (7.62), the maxima of all the functions $\tilde{I}_k^{\text{gmi}}(\hat{\alpha}_k^{\text{gmi}}s)$ are aligned at $s = 1$, and their sum is then maximized. In this example, $\mathsf{snr}_1 = \hat{\mathsf{snr}}_1$ so $\hat{\alpha}_1^{\text{gmi}} = 1$, and thus, the bitwise GMI functions before and after correction are the same. The gain offered by the correction in terms of GMI is also shown

Example 7.15 (Alignment of the GMI Maxima) *Suppose we transmit $m = 3$ bits $\boldsymbol{b} = [b_1, b_2, b_3]^{\text{T}}$ over three independent channels using a 2PAM constellation in each of them. The received signal is given by $\boldsymbol{y} = [y_1, \ldots, y_m]^{\text{T}}$, where $y_k = \Phi(b_k) + z_k$, where $k = 1, \ldots, m$. The SNR in each channel is defined by snr_k, and to calculate the L-values $\tilde{L}_k$, we use the SNR estimates $\hat{\mathsf{snr}}_k$. As we have seen in Example 3.15, the PDF of each L-value $\tilde{L}_k, k = 1, \ldots, m$ is given by $p_{\tilde{L}_k|B_k}(l|b) = \Psi(l, (-1)^{\bar{b}} 4\,\hat{\mathsf{snr}}_k, 8\,\hat{\mathsf{snr}}_k^2/\mathsf{snr}_k)$.*

In this case, the L-values have a Gaussian distribution, and thus, from Corollary 7.8, the GMI-optimal correction is linear and the correction factor for each of the bit's positions is given by (7.32), i.e.,

$$\hat{\alpha}_k^{\text{gmi}} = \hat{s}^{\text{gmi}} = \frac{\mathsf{snr}_k}{\hat{\mathsf{snr}}_k}. \tag{7.62}$$

In Fig. 7.4, we show the functions $\tilde{I}_k^{\text{gmi}}(\hat{\alpha}_k^{\text{gmi}}s)$ with $\hat{\alpha}_k^{\text{gmi}}$ given by (7.62) for $\mathsf{snr}_1 = 1$, $\hat{\mathsf{snr}}_1 = \mathsf{snr}_1$, $\mathsf{snr}_2 = 0.5$, $\hat{\mathsf{snr}}_2 = 1$, $\mathsf{snr}_3 = 2$, and $\hat{\mathsf{snr}}_3 = 0.6$.

The maxima of the functions $\tilde{I}_k^{\text{gmi}}(s)$ occurring at $s = \hat{\alpha}_k^{\text{gmi}}$ are shown with diamonds and the maxima of the function $\tilde{I}_k^{\text{gmi}}(\hat{\alpha}_k^{\text{gmi}}s)$ with squares. We can appreciate that the maximum of $\tilde{I}_{\tilde{\mathbb{q}}^0}^{\text{gmi}}(\boldsymbol{B}; \tilde{\boldsymbol{L}}) = \sum_{k=1}^m \tilde{I}_k^{\text{gmi}}(s)$ (i.e., before the correction) is attained for $s \approx 1$, which does not yet indicate that the underlying L-values do not need any scaling. In fact, the maximum of $\tilde{I}_{\tilde{\mathbb{q}}}^{\text{gmi}}(\boldsymbol{B}; \tilde{\boldsymbol{L}}) = \sum_{k=1}^m \tilde{I}_k^{\text{gmi}}(s\hat{\alpha}_k^{\text{gmi}})$ is attained exactly for $s = 1$ and gains are obtained, thanks to scaling of the mismatched L-values individually, so as to align their maxima.

7.3.4 PEP-Optimal Linear Correction

Considering the linear correction, (7.37) becomes

$$\tilde{\Lambda}^{\Sigma,\text{c}} = \sum_{t=1}^{d} \alpha_t \tilde{L}_{(t)} \tag{7.63}$$

and the PEP is given by

$$\mathsf{PEP}^{\mathrm{c}}(\boldsymbol{w};\boldsymbol{\alpha}) = \int_0^{\infty} p_{\tilde{\Lambda}^{\Sigma,\mathrm{c}}}(l)\,\mathrm{d}l, \tag{7.64}$$

where the PDF of the random variable $\tilde{\Lambda}^{\Sigma,\mathrm{c}}$ is given by

$$p_{\tilde{\Lambda}^{\Sigma,\mathrm{c}}}(l) = \{p_{\tilde{L}_1^{\mathrm{c}}|B_1}(l|0)\}^{*w_1} * \ldots * \{p_{\tilde{L}_m^{\mathrm{c}}|B_m}(l|0)\}^{*w_m} \tag{7.65}$$

$$= \left\{\frac{1}{\alpha_1} p_{\tilde{L}_1|B_1}(l/\alpha_1|0)\right\}^{*w_1} * \ldots * \left\{\frac{1}{\alpha_m} p_{\tilde{L}_m|B_m}(l/\alpha_m|0)\right\}^{*w_m}, \tag{7.66}$$

where (7.66) follows from (7.63).

In order to minimize the upper bound on the performance of the decoder in (7.35), we should find $\boldsymbol{\alpha}$ that minimizes $\mathsf{PEP}^{\mathrm{c}}(\boldsymbol{w};\boldsymbol{\alpha})$ for all $\boldsymbol{w}$, i.e.,

$$\hat{\boldsymbol{\alpha}}^{\mathrm{pep}} \triangleq \operatorname*{argmin}_{\boldsymbol{\alpha}\in\mathbb{R}_+^m} \{\mathsf{PEP}^{\mathrm{c}}(\boldsymbol{w};\boldsymbol{\alpha})\}, \quad \forall \boldsymbol{w} \in \mathbb{N}_+^m. \tag{7.67}$$

At first sight, the PEP minimization problem in (7.67) may appear intractable because of the dependence on the unknown $\boldsymbol{w}$ affecting the convolution of the PDFs in (7.65). It is thus instructive to have a look at a very simple case which we show in the following example.

Example 7.16 (PEP Minimization for 2PAM with SNR Mismatch) *Assume again the scenario of Example 7.15, where we transmit m bits $b_k[n]$, $n = 1, \ldots, N_{\mathrm{s}}$ using a 2PAM constellation over the channel with SNR $\hat{\mathsf{snr}}_k$. Assume also that the SNR is estimated (with errors) at the receiver as $\hat{\mathsf{snr}}_k \neq \mathsf{snr}_k$, and thus, the L-values $\tilde{l}_k[n] = 4y_k[n]\hat{\mathsf{snr}}_k$, $n = 1, \ldots, N_{\mathrm{s}}$ are mismatched. From Example 3.7 we know that the PDF of the mismatched L-values is given by $p_{\tilde{L}_k|B}(l|0) = \Psi(l, -4\,\hat{\mathsf{snr}}_k, 8\,\hat{\mathsf{snr}}_k/\mathsf{snr}_k)$ (see also Examples 7.15 and 7.8), and thus, the PDF of the corrected L-values $\tilde{L}_k^{\mathrm{c}} = \alpha_k \tilde{L}_k$ is given by*

$$p_{\tilde{L}_k^{\mathrm{c}}|B_k}(l|0) = \Psi\left(l, -4\alpha_k\hat{\mathsf{snr}}_k, 8\alpha_k^2\hat{\mathsf{snr}}_k^2/\mathsf{snr}_k\right). \tag{7.68}$$

As all the L-values in the sum affecting the PEP are independent and identically distributed (i.i.d.) Gaussian random variables, their sum $\tilde{\Lambda}^{\Sigma,\mathrm{c}}$ in (7.64) is a Gaussian random variable too with PDF

$$p_{\tilde{\Lambda}}^{\Sigma,\mathrm{c}}(l) = \Psi\left(l, -\sum_{k=1}^{m} 4\alpha_k w_k \hat{\mathsf{snr}}_k, \sum_{k=1}^{m} 8\alpha_k^2 w_k^2 \frac{\hat{\mathsf{snr}}_k^2}{\mathsf{snr}_k}\right). \tag{7.69}$$

Therefore, we can write (7.64) as

$$\mathsf{PEP}^{\mathrm{c}}(\boldsymbol{w};\boldsymbol{\alpha}) = Q\left(\sqrt{2}\frac{\sum_{k=1}^{m}\alpha_k w_k \hat{\mathsf{snr}}_k}{\sqrt{\sum_{k=1}^{m}\alpha_k^2 w_k \hat{\mathsf{snr}}_k^2/\mathsf{snr}_k}}\right). \tag{7.70}$$

The minimization of the PEP in (7.67) is obtained by setting the derivatives of $\mathsf{PEP}^{\mathrm{c}}(\boldsymbol{w};\boldsymbol{\alpha})$ with respect to α_k to zero, which gives

$$\hat{\alpha}_k^{\mathrm{pep}} = \frac{\mathsf{snr}_k}{\hat{\mathsf{snr}}_k}, \quad k = 1, \ldots m. \tag{7.71}$$

To show that $\hat{\boldsymbol{\alpha}}^{\mathrm{pep}} = [\hat{\alpha}_1^{\mathrm{pep}}, \ldots, \hat{\alpha}_m^{\mathrm{pep}}]^{\mathrm{T}}$ is the global minimum of the problem (7.67), we will demonstrate that

$$\mathsf{PEP}^{\mathrm{c}}(\boldsymbol{w};\hat{\boldsymbol{\alpha}}^{\mathrm{pep}}) \leq \mathsf{PEP}^{\mathrm{c}}(\boldsymbol{w};\boldsymbol{\alpha}), \quad \forall \boldsymbol{\alpha}. \tag{7.72}$$

To this end, we use the fact that the Q-function is monotonically decreasing, and thus, it is sufficient to show that the argument of the Q-function in (7.70) *for any* $\boldsymbol{\alpha}$ *is smaller than the corresponding argument for* $\hat{\boldsymbol{\alpha}}^{\text{pep}}$, *i.e., we have to demonstrate that the following holds*

$$\frac{\sum_{k=1}^{m} \alpha_k w_k \hat{\mathsf{snr}}_k}{\sqrt{\sum_{k=1}^{m} \alpha_k^2 w_k \hat{\mathsf{snr}}_k^2 / \mathsf{snr}_k}} \leq \frac{\sum_{k=1}^{m} \hat{\alpha}_k^{\text{pep}} w_k \hat{\mathsf{snr}}_k}{\sqrt{\sum_{k=1}^{m} (\hat{\alpha}_k^{\text{pep}})^2 w_k \hat{\mathsf{snr}}_k^2 / \mathsf{snr}_k}} = \sqrt{\sum_{k=1}^{m} w_k \mathsf{snr}_k}, \tag{7.73}$$

where the equality on the r.h.s. of (7.73) *is obtained using* (7.71). *Rearranging the terms further, we obtain*

$$\left(\sum_{k=1}^{m} \alpha_k w_k \hat{\mathsf{snr}}_k\right)^2 \leq \left(\sum_{k=1}^{m} w_k \mathsf{snr}_k\right)\left(\sum_{k=1}^{m} \alpha_k^2 w_k \frac{\hat{\mathsf{snr}}_k^2}{\mathsf{snr}_k}\right), \tag{7.74}$$

which we recognize as the Cauchy–Schwarz inequality, so indeed, (7.74) *holds and the solutions in* (7.71) *provide the global minimum of the PEP.*

The results in Example 7.16 show that the correction factor in this case is independent of $\boldsymbol{w}$. While this is an encouraging conclusion, recall that we solved the PEP-minimization problem (7.67), thanks to the Gaussian form of the PDF. As we cannot do this for arbitrary distributions, we now turn our attention to approximations.

Consider the upper (Chernoff) bound we introduced in Theorem 6.29 and generalized to the case of nonidentically distributed L-values in (6.355). We upper bound (7.64) as

$$\mathsf{PEP}^{\text{c}}(\boldsymbol{w};\boldsymbol{\alpha}) \leq \mathsf{PEP}^{\text{ub}}(\boldsymbol{w};\boldsymbol{\alpha}) = \min_{s\geq 0} \exp\left(\sum_{k=1}^{m} w_k \tilde{\kappa}_k^{\text{c}}(s)\right), \tag{7.75}$$

where $\tilde{\kappa}_k^{\text{c}}(s) \triangleq \log \mathbb{E}_{\tilde{L}_k^{\text{c}}|B_k=0}[\exp(\tilde{L}_k^{\text{c}} s)]$ is the *cumulant-generating function* (CGF) of the corrected L-value $\tilde{L}_k^{\text{c}}$ conditioned on $B_k = 0$.

Theorem 7.17 *The upper bound* $\mathsf{PEP}^{\text{ub}}(\boldsymbol{w};\boldsymbol{\alpha})$ *in* (7.75) *is minimized by the linear correction factors*

$$\hat{\alpha}_k^{\text{pep}} = 2\hat{s}_k^{\text{spa}}, \quad k = 1, \ldots, m, \tag{7.76}$$

where

$$\hat{s}_k^{\text{spa}} = \underset{s\geq 0}{\operatorname{argmin}} \{\tilde{\kappa}_k(s)\} \tag{7.77}$$

is the saddlepoint of the CGF $\tilde{\kappa}_k(s) = \mathbb{E}_{\tilde{L}_k|B_k=0}[\exp(s\tilde{L}_k)]$.

Proof: The relationship

$$\tilde{\kappa}_k^{\text{c}}(s) = \mathbb{E}_{\tilde{L}_k|B_k=0}[\exp(s\alpha_k \tilde{L}_k)] = \tilde{\kappa}_k(\alpha_k s) \tag{7.78}$$

used in (7.75) allows us to write

$$\min_{\boldsymbol{\alpha}\in\mathbb{R}_+^m} \{\mathsf{PEP}^{\text{ub}}(\boldsymbol{w};\boldsymbol{\alpha})\} = \min_{\boldsymbol{\alpha}\in\mathbb{R}_+^m, s\in\mathbb{R}_+} \exp\left(\sum_{k=1}^{m} w_k \tilde{\kappa}_k(\alpha_k s)\right) \tag{7.79}$$

$$= \exp \min_{\boldsymbol{\alpha}\in\mathbb{R}_+^m, s\in\mathbb{R}_+} \left(\sum_{k=1}^{m} w_k \tilde{\kappa}_k(\alpha_k s)\right). \tag{7.80}$$

As $\tilde{\kappa}_k(s)$ are convex (see (6.328)), with minima $\hat{s}_k^{\text{spa}}$ given by (7.77), we use Lemma 7.13 to obtain the solution $\hat{\alpha}_k^{\text{pep}} = \hat{s}_k^{\text{spa}}/s_0$. To resolve the ambiguity with respect to $s_0 \neq 0$, we use the condition (ii) of Definition 7.11. We know that for exact L-values the saddlepoint is given by $\hat{s}_k^{\text{spa}} = \frac{1}{2}$ (cf. (6.327)), and thus, we use $s_0 = \frac{1}{2}$ to guarantee that, in that case, $\hat{\alpha}_k^{\text{pep}} = 2\hat{s}_k^{\text{spa}} = 1$, and thus, the exact L-values are not corrected. This yields (7.76), which completes the proof. □

It may be argued that the upper bound on the PEP may not be tight and more accurate approximations or even the exact numerical expressions we discussed in Section 6.3.3 should be used. While this might be done, more involved expressions would not allow us to find simple (and independent of $\boldsymbol{w}$) analytical rules for correction. Moreover, the numerical examples (see, e.g., Example 6.30) indicate that the upper bound follows the actual value of the PEP very closely (up to a multiplicative factor), so more involved expressions are unlikely to offer significant gains.

With this cautionary statement, we will refer to the rule defined in Theorem 7.76 as the *PEP-optimal linear correction*.

7.3.5 GMI- and PEP-Optimal Linear Corrections: A Comparison

To gain a quick insight into the differences between the GMI- and PEP-optimal corrections based on linear correction strategies, in what follows, we rewrite the optimality conditions in both cases, assuming the bits are uniformly distributed (i.e., $P_{B_k}(b) = \frac{1}{2}$) and the symmetry condition (3.74) is satisfied.

As the GMI is concave on s (see Theorem 4.22), a necessary and sufficient condition for the maximization of the bitwise GMI is given by

$$\frac{\mathrm{d}}{\mathrm{d}s}\tilde{I}_k^{\text{gmi}}(s)\bigg|_{s=\hat{\alpha}_k^{\text{gmi}}} = 0. \tag{7.81}$$

Using Theorem 7.3 and (4.94), we then obtain a condition which has to be satisfied by the GMI-optimal correction factor $\hat{\alpha}^{\text{gmi}}$

$$\int_{-\infty}^{\infty} p_{\tilde{L}_k|B_k}(l|0)\mathrm{e}^{\frac{\hat{\alpha}_k^{\text{gmi}}}{2}l}\frac{l}{\cosh\left(\frac{\hat{\alpha}_k^{\text{gmi}}}{2}l\right)}\,\mathrm{d}l = 0 \tag{7.82}$$

$$\int_{-\infty}^{\infty} p_{\tilde{L}_k^{\text{c}}|B_k}(l|0)\mathrm{e}^{\frac{l}{2}}\frac{l}{\cosh\left(\frac{l}{2}\right)}\,\mathrm{d}l = 0, \tag{7.83}$$

where, to pass from (7.82) to (7.83), we used $p_{\tilde{L}_k^{\text{c}}|B_k}(l|b) = \frac{1}{\hat{\alpha}_k^{\text{gmi}}}p_{\tilde{L}_k|B_k}(l/\hat{\alpha}_k^{\text{gmi}}|b)$.

Similarly, the PEP-minimization condition in (7.76) states that the saddlepoint should be related to the correction factor via $\hat{s}^{\text{spa}} = \hat{\alpha}^{\text{pep}}/2$, which may be written as follows:

$$\frac{\mathrm{d}}{\mathrm{d}s}\tilde{\kappa}_k(s)\bigg|_{s=\hat{\alpha}_k^{\text{pep}}/2} = \mathbb{E}_{\tilde{L}_k|B_k=0}\left[\tilde{L}_k\mathrm{e}^{s\tilde{L}_k}\right]\bigg|_{s=\hat{\alpha}^{\text{pep}}/2} = 0, \tag{7.84}$$

which implies

$$\int_{-\infty}^{\infty} p_{\tilde{L}_k^{\text{c}}|B_k}(l|0)\mathrm{e}^{\frac{l}{2}}l\,\mathrm{d}l = 0. \tag{7.85}$$

Comparing (7.83) and (7.85) we see that, in general, GMI-optimal and PEP-optimal linear correction factors will not be the same. This may come as a surprise because we know from Section 7.2 that the GMI-optimal and PEP-optimal correction functions are the same. Here, however, we are restricting our analysis to linear correction functions, while the optimal function are nonlinear in general. This constraint produces different results.

As for the practical aspects of using both correction principles, the main difference lies in the complexity of the search for the optimal scaling factor. In most cases, finding the maximum GMI will require numerical quadratures as the nonlinear functions will resist analytical integration, e.g., the hyperbolic cosine in (7.83). On the other hand, the PEP-optimal correction relies on the knowledge of the *moment-generating function* (MGF) of the PDF of the L-values, which, in some cases, may be calculated analytically. Consequently, finding the correction factor will be simplified, as we will illustrate in Section 7.3.6.

In both cases, if $p_{\tilde{L}_k|B_k}(l|0)$ is not known, Monte Carlo integration may be used to calculate the integrals (7.83) or (7.85), cf. (6.311) and (4.89), and then, the complexity of adaptation is similar for both approaches.

7.3.6 Case Study: Correcting Interference Effects

To illustrate the previous findings, we consider transmission using a 2PAM constellation, where the transmitted symbols $x[n] = 2b[n] - 1$ pass through a channel corrupted with *additive white Gaussian noise* (AWGN) and a 2PAM-modulated interference, i.e.,

$$\boldsymbol{y}[n] = h[n]\boldsymbol{x}[n] + \boldsymbol{z}[n] + g[n]\boldsymbol{d}[n], \tag{7.86}$$

where $h[n]$ is the channel gain, $\boldsymbol{z}[n]$ is Gaussian noise with variance $\mathsf{N}_0/2$, and $d[n] \in \{-1, 1\}$ is the interference signal received with gain $g[n]$. We assume that the channel gains $h[n]$ and $g[n]$ are known (estimated) and $g[n] < h[n]$, i.e., the interference is weaker than the desired signal.

The instantaneous SNR is defined as $\mathsf{snr}[n] = h[n]/\mathsf{N}_0$, the instantaneous *signal-to-interference ratio* (SIR) as $\mathsf{sir}[n] = h^2[n]/g^2[n]$, and the average SIR is given by $\overline{\mathsf{sir}} = 1/\mathbb{E}_G[G^2]$. We assume that the channel is memoryless, and thus, the time index n is dropped. The model (7.86) may be written using random variables as

$$\boldsymbol{Y} = h\boldsymbol{X} + \boldsymbol{Z} + g\boldsymbol{D}. \tag{7.87}$$

From (7.2), it is simple to calculate exactly the L-values in this case as

$$l = \log \frac{p_{\boldsymbol{Y}|\boldsymbol{X}}(\boldsymbol{y}|+1)}{p_{\boldsymbol{Y}|\boldsymbol{X}}(\boldsymbol{y}|-1)}, \tag{7.88}$$

where

$$\begin{aligned} p_{\boldsymbol{Y}|\boldsymbol{X}}(\boldsymbol{y}|\boldsymbol{x}) &= \frac{1}{2} \sum_{\boldsymbol{d}\in\{-1,1\}} p_{\boldsymbol{Y}|\boldsymbol{X},\boldsymbol{D}}(\boldsymbol{y}|\boldsymbol{x},\boldsymbol{d}) \\ &= \frac{1}{2\sqrt{\pi \mathsf{N}_0}} \left[\exp\left(-\overline{\mathsf{snr}}(\boldsymbol{y}-h-g)^2\right) + \exp\left(-\overline{\mathsf{snr}}(\boldsymbol{y}-h+g)^2\right)\right]. \end{aligned} \tag{7.89}$$

We assume now that the receiver ignores the presence of the interference (which is equivalent to using $g = 0$ in (7.89)). The L-values are obtained as in (3.65), i.e.,

$$\tilde{l} = 4\,\mathsf{snr}\,\boldsymbol{y} = \frac{2h\boldsymbol{y}}{\mathsf{N}_0/2}, \tag{7.90}$$

and are—because of the assumed absence of interference—mismatched.

To find the correction factors in the case of the PEP-optimal correction, we need to calculate the CGF of $\tilde{L} = 4\,\mathsf{snr}\,\boldsymbol{Y}$, conditioned on a transmitted bit $B = 0$, or, equivalently, on $\boldsymbol{X} = -1$. As $\tilde{L} = 4\ \mathsf{snr}(-h + \boldsymbol{Z} + g\boldsymbol{D})$ is a sum of independent random variables $\boldsymbol{Z}$, and $\boldsymbol{D}$, we obtain its CGF as follows:

$$\tilde{\kappa}(s) = \log \mathbb{E}_{\tilde{L}|B=0}\left[\mathrm{e}^{s\tilde{L}}\right] \tag{7.91}$$

$$= -4\,\mathsf{snr}\,sh + \log \mathbb{E}_{\boldsymbol{Z}}\left[\mathrm{e}^{4\,\mathsf{snr}\,s\boldsymbol{Z}}\right] + \log \mathbb{E}_{\boldsymbol{D}}\left[\mathrm{e}^{4\,\mathsf{snr}\,g\boldsymbol{D}s}\right] \tag{7.92}$$

$$= -4\,\mathsf{snr}\,sh + 4\,\mathsf{snr}^2 \frac{1}{2} s^2 \frac{\mathsf{N}_0}{2} + \log\left(\tfrac{1}{2}\mathrm{e}^{4\,\mathsf{snr}\,sg} + \tfrac{1}{2}\mathrm{e}^{-4\,\mathsf{snr}\,sg}\right) \tag{7.93}$$

$$= -4\,\mathsf{snr}\,sh + 4\,\mathsf{snr}\,hs^2 + \log\cosh\left(4\,sg\,\mathsf{snr}\right). \tag{7.94}$$

Setting the derivative of (7.94) with respect to s to zero and using $\alpha = 2s$, we obtain the nonlinear equation

$$h(1-\alpha) = g\tanh\left(2\alpha g\mathsf{snr}\right), \tag{7.95}$$

solved by $\alpha = \hat{\alpha}^{\mathrm{pep}}$, which we interpret graphically in Fig. 7.5 as the intersection of the r.h.s. and l.h.s. of (7.95).

In general, we may solve (7.95) numerically; however, using approximations, closed-form approximations may be obtained in particular cases, Namely, for $\mathsf{snr} \to 0$, using the linearization $\tanh(x) \approx x$ (shown as a dashed line in Fig. 7.5) in (7.95), we obtain

$$\hat{\alpha}^{\mathrm{pep}}_{\mathrm{low}} = \frac{h}{2g^2\mathsf{snr} + h} = \frac{\mathsf{N}_0}{2g^2 + \mathsf{N}_0}, \tag{7.96}$$

and then for a given h, we see that

$$\lim_{\mathsf{snr}\to 0} \hat{\alpha}^{\mathrm{pep}}_{\mathrm{low}} = 1. \tag{7.97}$$

The corrected L-values are next calculated as

$$\tilde{l}^{\mathrm{c}}_{\mathrm{low}} = \hat{\alpha}^{\mathrm{pep}}_{\mathrm{low}} \cdot 4h\boldsymbol{y}\,\mathsf{snr} = \frac{2h\boldsymbol{y}}{\mathsf{N}_0/2 + g^2}, \tag{7.98}$$

which is similar to (7.90) but with the denominator now having the meaning of the noise and interference power. This means that the effect of the noise and the interference is modeled as a Gaussian random variable with variance $\mathsf{N}_0/2 + g^2$. Such a model is, indeed, appropriate for low SNR, when the noise "dominates" the interference. We also note that, using

$$\mu_{\tilde{L}} = \mathbb{E}_{\tilde{L}|B=0}[\tilde{L}] = \frac{4h^2}{\mathsf{N}_0} \tag{7.99}$$

and

$$\sigma^2_{\tilde{L}} = \mathrm{Var}_{\tilde{L}|B=0}[\tilde{L}] = \frac{4h^2(\mathsf{N}_0/2 + g^2)}{\mathsf{N}_0^2/4} \tag{7.100}$$

in (7.31), yields exactly the same results $\alpha^{\mathrm{sgd}} = \hat{\alpha}^{\mathrm{pep}}_{\mathrm{low}}$.

We note that we always have $\hat{\alpha}^{\mathrm{pep}} \leq 1$ (see Fig. 7.5), i.e., we have to scale down the L-values to decrease their reliability (expressed by their amplitude); in other words, they are too "optimistic" when calculated ignoring the interference. On the other hand, as $\hat{\alpha}^{\mathrm{pep}}_{\mathrm{low}} \leq \hat{\alpha}^{\mathrm{pep}}$, we conclude that assuming a Gaussian interference is too pessimistic.

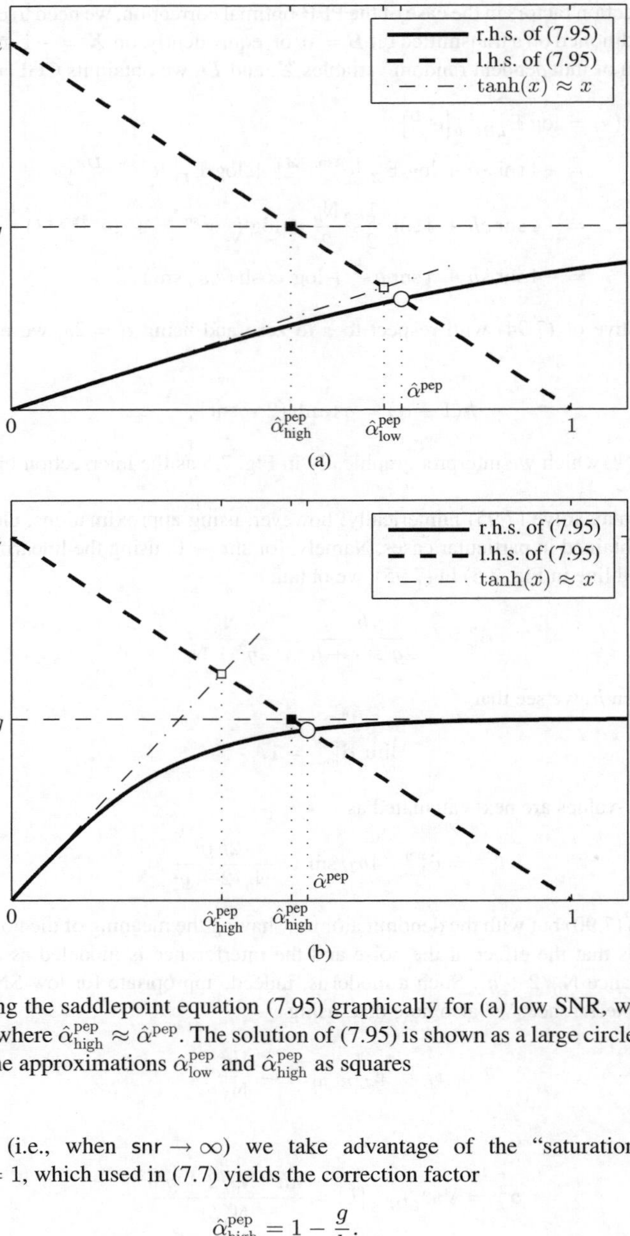

Figure 7.5 Solving the saddlepoint equation (7.95) graphically for (a) low SNR, where $\hat{\alpha}^{\text{pep}}_{\text{low}} \approx \hat{\alpha}^{\text{pep}}$, and (b) high SNR, where $\hat{\alpha}^{\text{pep}}_{\text{high}} \approx \hat{\alpha}^{\text{pep}}$. The solution of (7.95) is shown as a large circle and the solutions corresponding to the approximations $\hat{\alpha}^{\text{pep}}_{\text{low}}$ and $\hat{\alpha}^{\text{pep}}_{\text{high}}$ as squares

For high SNR (i.e., when $\mathsf{snr} \to \infty$) we take advantage of the "saturation" approximation $\lim_{x\to\infty} \tanh(x) = 1$, which used in (7.7) yields the correction factor

$$\hat{\alpha}^{\text{pep}}_{\text{high}} = 1 - \frac{g}{h}. \tag{7.101}$$

The corrected L-values in this case are calculated as

$$\tilde{l}^{\text{c}}_{\text{high}} = \frac{2\,(h-g)\,\boldsymbol{y}}{\mathsf{N}_0/2}, \tag{7.102}$$

which is, again, similar to (7.90), but now the gain of the desired signal h is reduced by the interference-related term g. This can be explained as follows: for high SNR, the interference can be "distinguished" from the noise and becomes part of the transmitted constellation, i.e., by sending bit $b = 1$, we will effectively be able to differentiate between $h - g$ and $h + g$. Moreover, for high SNR, the symbol that is the most likely to provoke the error is the one closest to the origin, i.e., $h - g$. This leads to assuming that 2PAM symbols are sent over a channel with gain $h - g$.

To obtain the GMI-optimal factor $\hat{\alpha}^{\rm gmi}$ we numerically solve (7.82)

$$\int_{-\infty}^{\infty} p_{\tilde{L}|B}(l|0) \frac{l \exp\left(\frac{\hat{\alpha}^{\rm gmi}}{2} l\right)}{\cosh(\frac{\hat{\alpha}^{\rm gmi}}{2} l)} \, {\rm d}l = 0, \tag{7.103}$$

using

$$p_{\tilde{L}|B}(l|0) = \frac{1}{4\,\mathsf{snr}} p_{\boldsymbol{Y}|\boldsymbol{X}}(l/(4\mathsf{snr})| - 1) \tag{7.104}$$

with $p_{\boldsymbol{Y}|\boldsymbol{X}}(\boldsymbol{y}|\boldsymbol{x})$ given by (7.89).

In Fig. 7.6, we show the values of the optimal correction factors as a function of SNR for different values SIR, where we can appreciate that GMI-optimal correction factors $\hat{\alpha}^{\rm gmi}$ in (7.59) are not identical but close to $\hat{\alpha}^{\rm pep}$. Both factors increase with SIR, as the case $\mathsf{sir} \to \infty$ corresponds to the assumed absence of interference, i.e., $\hat{\alpha}^{\rm pep} \to 1$. For comparison, in this figure, we also show the results of the ad hoc WLSF correction defined in (7.45) and (7.46). We can expect it to provide adequate results when $\theta^{\rm c}(l)$ is almost linear, i.e., when the PDF $p_{\tilde{L}|B}(l|0)$ is close to the Gaussian form (see Corollary 7.8). This happens when the interference is dominated by the noise, i.e., for low SNR and high SIR and then, as we can see in Fig. 7.6, $\hat{\alpha}^{\rm wlsf}$ is close to the GMI- and PEP-optimal correction. The results obtained are very different when the SNR increases.

To verify how the correction affects the performance of a practical decoder, we consider a block of $N_{\rm b} = 1000$ bits encoded using a *convolutional encoder* (CENC) with rate $R_{\rm c} = \frac{1}{2}$ and generating polynomials $\mathbf{G} = [15, 17]$ (see Table 2.1) as well as by the *turbo encoder* (TENC) with rate $R_{\rm c} = \frac{3}{4}$ from Example 2.31. We assume that H and G are Rayleigh random variables with $\mathbb{E}_H[H^2] = 1$ and $\mathbb{E}_G[G^2] = 1/\overline{\mathsf{sir}}$. The *average* SNR is $\overline{\mathsf{snr}} = 1/\mathsf{N}_0$. Thus, we may observe the effect of interference level

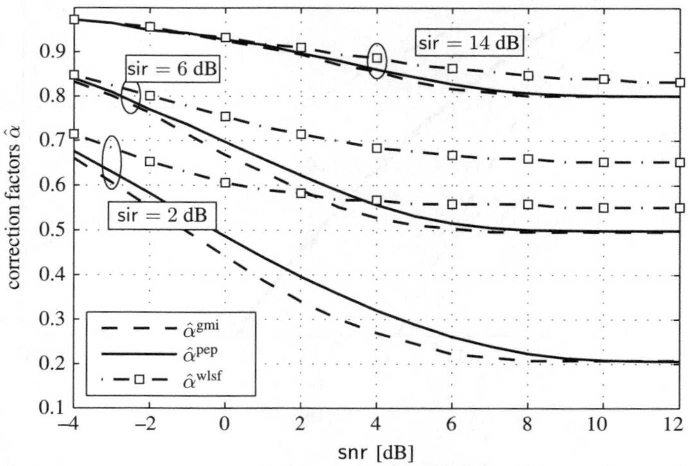

Figure 7.6 Linear correction factors $\hat{\alpha}^{\rm gmi}$ solving (7.103), $\hat{\alpha}^{\rm pep}$ solving (7.95), and $\hat{\alpha}^{\rm wlsf}$ in (7.45) as a function of SNR for various values of SIR. We may appreciate that for low SNR, the correction factors tend to 1 (see (7.97)), and for high SNR, they tend to $\hat{\alpha}^{\rm pep}_{\rm high}$ (see (7.101))

by changing the value of $\overline{\mathsf{sir}}$. The correction factor has to be found for each value of h and g, which we assumed were perfectly known at the receiver. The fact that we know the interference gain g but we do not use it in (7.89) allows us to highlight the functioning of the correction principle and shows the eventual gain of more complex processing in (7.89).

The average SIR is set to $\overline{\mathsf{sir}} = 6$ dB for the CENC and $\overline{\mathsf{sir}} = 8$ dB for the TENC. The results of the decoding in terms of BEP are shown in Fig. 7.7. For the CENC we used a *maximum likelihood* (ML) (Viterbi) decoder and for the TENC a turbo decoder with five iterations. Figure 7.7 shows the BEP for different correction strategies, where, for comparison, we also show the results of the decoding using exact L-values obtained via (7.88).

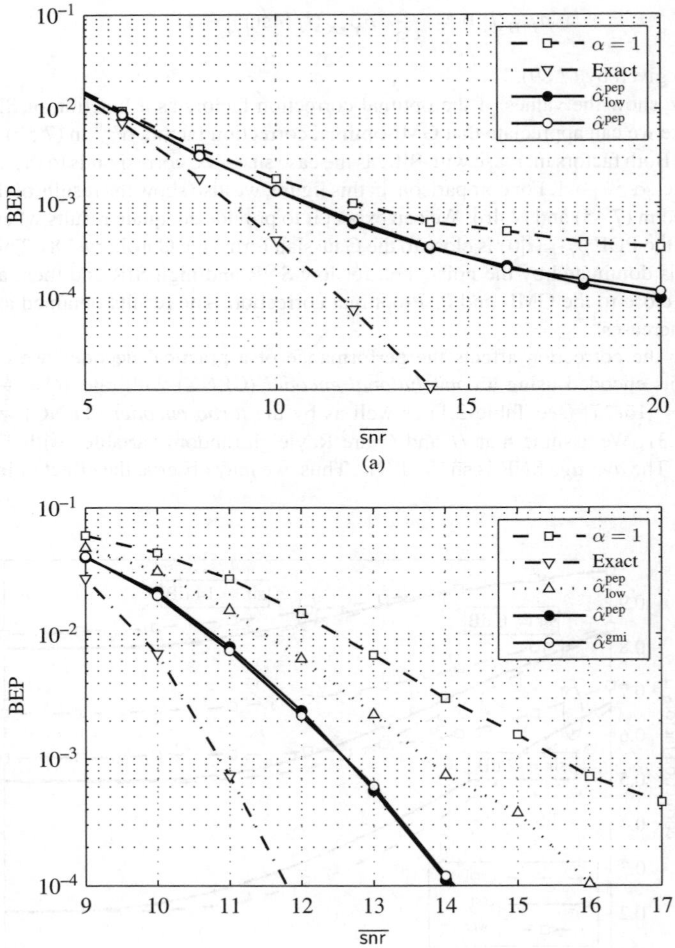

Figure 7.7 BEP results for (a) a CENC with $R_c = \frac{1}{2}$ and (b) a TENC with $R_c = \frac{3}{4}$ obtained using L-values without correction (7.90) ($\alpha = 1$), L-values with PEP-optimal correction (calculated solving (7.95)), exact L-values (7.88), and L-values corrected using $\hat{\alpha}^{\text{gmi}}$. For the TENC, the results obtained using a Gaussian model for the L-values are also shown ($\hat{\alpha}^{\text{pep}}_{\text{low}}$)

The results in Fig. 7.7 show that the correction results based on the PEP- and the GMI-optimal approaches are similar and manage to partially bridge the gap to the results based on exact L-values. The performance improvement is particularly notable for high average SNR, which is consistent with the results of Fig. 7.6, where the most notable correction (small values of $\hat{\alpha}^{\text{pep}}$) are also obtained for high SNR.

In Fig. 7.7, we also show the results of the correction derived assuming that the interference is Gaussian, yielding the correction factor $\hat{\alpha}^{\text{pep}}_{\text{low}}$. This coefficient is independent of the channel gains h, and thus, common to all the L-values and irrelevant to the performance of an ML decoder. For this reason, the results obtained with $\hat{\alpha}^{\text{pep}}_{\text{low}}$ and with $\hat{\alpha}^{\text{pep}}$ are identical for the CENC, where the ML (Viterbi) decoder is used, and thus, not shown in Fig. 7.7 (a). On the other hand, the turbo decoder, based on the iterative exchange of information between the constituent decoders, depends on an accurate representation of the a posteriori probabilities via the L-values. It is, therefore, sensitive to the scaling, which is why the correction with $\hat{\alpha}^{\text{pep}}_{\text{low}}$ improves the decoding results.

As might be expected from the results shown in Fig. 7.6, the BEP results do not change significantly when using GMI- or PEP-optimal linear correction. However, this example also clearly demonstrates that the PEP provides a much simpler approach to find the correction factor, which we obtained analytically, thanks to the adopted approximations.

7.4 Bibliographical Notes

Dealing with incorrectly evaluated or mismatched likelihoods can be traced back to early works in the context of iterative decoders, where simplified operations in one iteration affect the performance of the subsequent ones. This has been observed in the context of the decoding of *turbo codes* (TCs) [1–6] or *low-density parity-check* (LDPC) codes [7, 8]. The problem of mismatched L-values in the demapper, which was the focus of this chapter, was analyzed in [9–12]. The main difference between the analysis of the decoder and the demapper is that, in the latter, we may characterize analytically the distribution of the L-values. When the L-values follow a Gaussian distribution, a linear correction is optimal. As the Gaussian model is simple to deal with, the linear correction was very often assumed. This was also motivated by the simplicity of the resulting implementation and the simplified search of the scalar correction parameter [1–3]. This linear scaling, in fact, evolved as a heuristic compensation, and was studied in various scenarios [4, 13–16]. Multiparameter correction functions have also been studied, e.g., in [17, 18].

The feedback-based adaptation approach can be applied in the absence of simple models relating the parameters of the L-values to the decoding outcome (e.g., BEP). This is necessary when the results are obtained via actual implementation of the decoding algorithm on many different realizations of the channel outcome [2, 3], or when they are predicted using semi-analytical tools such density evolution (in the case of LDPC codes [17]) or extrinsic transfer charts (in the case of TCs [8]). In such cases, the correction factor is found through a brute-force search over the entire space of parameters, i.e., among the results obtained for different correction factors, the one ensuring the best performance is deemed optimal [2, 3, 6, 7, 19]. While this is a pragmatic approach when searching for one [7] or two [18] correction factors, it becomes very tedious when many [17] correction factors have to be found, because of the increased dimensionality of the search space.

The formalism of the decoding based on the mismatched metrics (the mismatched decoding perspective for BICM) was introduced in [20]. The idea of the optimal correction function from Section 7.2 was shown in [3] and formally developed as GMI-optimal in [21]. This inspired the idea of the GMI-optimal linear correction via the so-called "harmonization" of the GMI curves introduced in [10] and described in Section 7.3.3. Examples of applications can be found in [10–12, 22]. The PEP-optimal linear correction we presented in Section 7.3.4 was introduced in [12].

As explained in Section 7.3, the predictive adaptation approach exploits the knowledge of the probabilistic model of the L-values. Some works focused on ensuring that the corrected L-values are consistent [8, 23, 24]. This was achieved by using a heuristically adopted criterion, e.g., (7.45), and by estimating the PDF via histograms. As this approach is quite tedious, parameterization of the PDF may be used. In particular, a Gaussian model was used in [1] and a Gaussian mixture model in [18]. The drawback of these approaches is that the PDF must be estimated. More importantly, a good fit between the linear form and the optimal function does not necessarily translate into the decoding gains. On the other hand, the GMI-optimal and PEP-optimal methods formally address the problem of improving the performance of the decoder. These predictive approaches optimize the GMI and minimize the CGF, respectively. In the cases where we do not know the PDF, both the GMI and the CGF can be found via Monte Carlo simulations, which is still simpler than estimating the PDF itself.

References

[1] Papke, L., Robertson, P., and Villebrun, E. (1996) Improved decoding with the SOVA in a parallel concatenated (turbo-code) scheme. IEEE International Conference on Communications (ICC), June 1996, Dallas, TX.

[2] Vogt, J. and Finger, A. (2000) Improving the max-log-map turbo decoder. *IEEE Electron. Lett.*, **36** (23), 1937–1939.

[3] van Dijk, M., Janssen, A., and Koppelaar, A. (2003) Correcting systematic mismatches in computed log-likelihood ratios. *Eur. Trans. Telecommun.*, **14** (3), 227–224.

[4] Ould-Cheikh-Mouhamedou, Y., Guinand, P., and Kabal, P. (2003) Enhanced Max-Log-APP and enhanced Log-APP decoding for DVB-RCS. International Symposium on Turbo Codes and Related Topics, May 2003, Brest, France.

[5] Huang, C. X. and Ghrayeb, A. (2006) A simple remedy for the exaggerated extrinsic information produced by the SOVA algorithm. *IEEE Trans. Wireless Commun.*, **5** (5), 996–1002.

[6] Alvarado, A., Núñez, V., Szczecinski, L., and Agrell, E. (2009) Correcting suboptimal metrics in iterative decoders. IEEE International Conference on Communications (ICC), June 2009, Dresden, Germany.

[7] Chen, J. and Fossorier, M. (2002) Near optimum universal belief propagation based decoding of low-density parity check codes. *IEEE Trans. Commun.*, **50** (3), 406–414.

[8] Lechner, G. (2007) Efficient decoding techniques for LDPC codes. PhD dissertation, Vienna University of Technology, Vienna, Austria.

[9] Classon, B., Blankenship, K., and Desai, V. (2002) Channel coding for 4G systems with adaptive modulation and coding. *IEEE Wireless Commun. Mag.*, **9** (2), 8–13.

[10] Nguyen, T. and Lampe, L. (2011) Bit-interleaved coded modulation with mismatched decoding metrics. *IEEE Trans. Commun.*, **59** (2), 437–447.

[11] Yazdani, R. and Ardakani, M. (2011) Efficient LLR calculation for non-binary modulations over fading channels. *IEEE Trans. Commun.*, **59** (5), 1236–1241.

[12] Szczecinski, L. (2012) Correction of mismatched L-values in BICM receivers. *IEEE Trans. Commun.*, **60** (11), 3198–3208.

[13] Pyndiah, R. M. (1998) Near-optimum decoding of product codes: block turbo codes. *IEEE Trans. Commun.*, **46** (8), 1003–1010.

[14] Crozier, S., Gracie, K., and Hunt, A. (1999) Efficient turbo decoding techniques. 11th International Conference on Wireless Communications, July 1999, Calgary, AB, Canada.

[15] Gracie, K., Crozier, S., and Guinand, P. (2004) Performance of an mlse-based early stopping technique for turbo codes. IEEE Vehicular Technology Conference (VTC-Fall), Los Angeles, CA.

[16] Gracie, K., Hunt, A., and Crozier, S. (2006) Performance of turbo codes using MLSE-based early stopping and path ambiguity checking for input quatized to 4 bits. International Symposium on Turbo Codes and Related Topics, April 2006, Munich, Germany.

[17] Zhang, J., Fossorier, M., Gu, D., and Zhang, J. (2006) Two-dimensional correction for Min-Sum decoding of irregular LDPC codes. *IEEE Commun. Lett.*, **10** (3), 180–182.

[18] Jia, Q., Kim, Y., Seol, C., and Cheun, K. (2009) Improving the performance of SM-MIMO/BICM-ID systems with LLR distribution matching. *IEEE Trans. Commun.*, **57** (11), 3239–3243.

[19] Heo, J. and Chugg, K. (2005) Optimization of scaling soft information in iterative decoding via density evolution methods. *IEEE Trans. Commun.*, **53** (6), 957–961.

[20] Martinez, A., Guillén i Fàbregas, A., Caire, G., and Willems, F. M. J. (2009) Bit-interleaved coded modulation revisited: a mismatched decoding perspective. *IEEE Trans. Inf. Theory*, **55** (6), 2756–2765.

[21] Jaldén, J., Fertl, P., and Matz, G. (2010) On the generalized mutual information of BICM systems with approximate demodulation. IEEE Information Theory Workshop (ITW), January 2010, Cairo, Egypt.
[22] Nguyen, T. and Lampe, L. (2011) Mismatched bit-interleaved coded noncoherent orthogonal modulation. *IEEE Commun. Lett.*, **15** (5), 563–565.
[23] Lechner, G. and Sayir, J. (2004) Improved sum-min decoding of LDPC codes. International Symposium on Information Theory and its Applications (ISITIA), October 2004, Parma, Italy.
[24] Lechner, G. and Sayir, J. (2006) Improved sum-min decoding for irregular LDPC codes. International Symposium on Turbo Codes and Related Topics, April 2006, Munich, Germany.

8

Interleaver Design

In binary modulation all the bits receive the same "treatment" when transmitted over the channel, i.e., the L-values have the same conditional *probability density function* (PDF). The error probability is also the same for all bits, i.e., each bit is equally "protected" against errors. High-order modulations, on the other hand, introduce *unequal error protection* (UEP). This is an inherent property of *bit-interleaved coded modulation* (BICM) and depends on the labeling and the form of the constellation. While previously we considered the use of quasi-random interleavers (which average out the UEP), in this chapter we want to take advantage of the UEP. To this end, we propose to design interleavers for BICM, which requires the analytical tools developed in the previous chapters to be refined.

This chapter is organized as follows. In Section 8.1 we introduce the idea of UEP and study a particular interleaver that takes into account the presence of the UEP: the so-called *multiple-input interleaver* (M-interleaver). In Section 8.2 we study the performance of BICM with M-interleavers generalizing the results from Chapter 6. In Section 8.2 we also study the problem of a joint interleaver and code design for such BICM transceivers.

8.1 UEP in BICM and M-interleavers

UEP may be easily explained using the communication-theoretic tools of Chapter 6 or the information-theoretic tools of Chapter 4. Let us revisit these concepts through a simple example.

Example 8.1 *Consider the* 8PAM *constellation labeled by the binary reflected Gray code (BRGC) we used in Example 6.33. In Fig. 8.1 (a), we show the bit-error probability (BEP) for uncoded transmission* $P_{\mathrm{b},k}$ *in* (6.24)*, which clearly depends on the bit position: the "protection" is higher for* $k = 1$ *than for* $k = 2$*, which is, in turn, also higher than for* $k = 3$.[1] *Similar results are obtained when studying the BEP per bit position for* MPSK *and* MQAM *constellations labeled by the BRGC.*[2]

Using now an information-theoretic approach, we consider the mutual information (MI) of the m *parallel binary-input continuous-output (BICO) channels* $I(B_k; \boldsymbol{Y})$ *defining the BICM generalized mutual information (BICM-GMI) in* (4.54)*. In Fig. 8.1 (b), we show these MIs and we again appreciate*

[1] On purpose, we do not formalize the concept of "protection"; however, it may be interpreted here as the inverse of the BEP.

[2] For MPSK, however, not all the bit positions offer different protection. For MPSK constellations labeled by the BRGC, the protections for $k = 1$ and $k = 2$ are in fact equivalent (see Example 6.6).

Bit-Interleaved Coded Modulation: Fundamentals, Analysis, and Design, First Edition.
Leszek Szczecinski and Alex Alvarado.

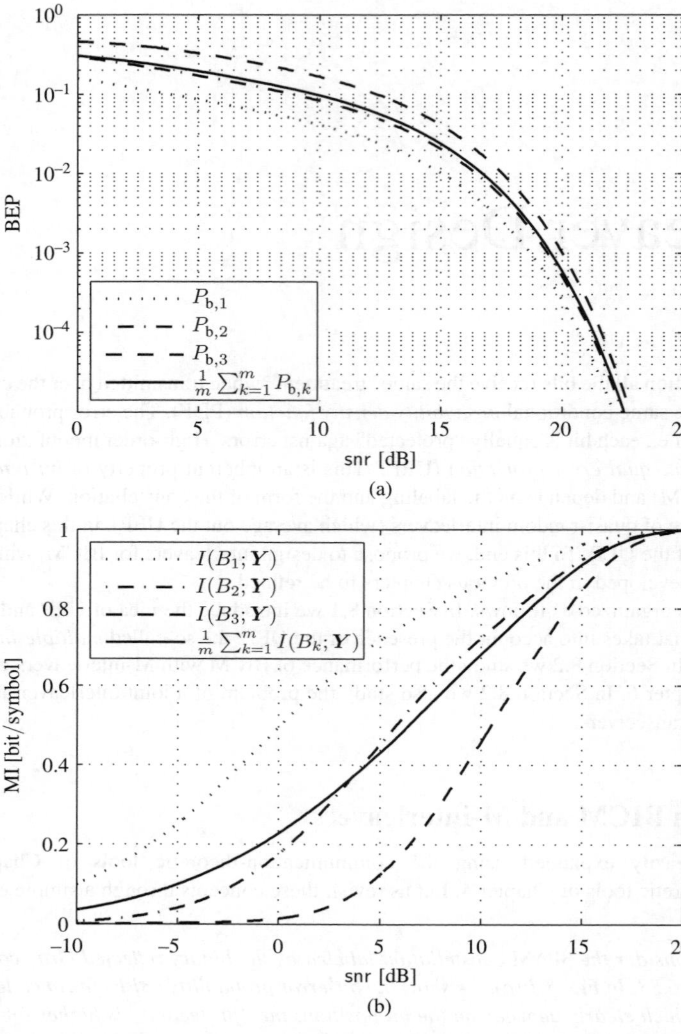

Figure 8.1 UEP for 8PAM constellation labeled by the BRGC over the AWGN channel: (a) BEP for uncoded transmission in (6.26) and (b) normalized BICM-GMI and the bit-wise MIs $I(B_k;\boldsymbol{Y})$

that $I(B_1;\boldsymbol{Y}) \geq I(B_2;\boldsymbol{Y}) \geq I(B_3;\boldsymbol{Y})$, *i.e., the first bit position provides higher protection*[3] *than the second position, which in turn provides higher protection than the third bit position. Figure 8.1 (b) is in fact the same as Fig. 4.15 (a).*

Example 8.1 illustrates the concept of UEP caused by the binary labeling. In what follows we focus on its communication-theoretic analysis. Namely, we characterize the performance of the BICM decoder by explicitly taking into account the fact that the L-values calculated by the demapper for different bit positions have different distributions.

[3] The term *protection* may be now directly interpreted as the MI between the bits B_k and the observations $\boldsymbol{Y}$.

8.1.1 Preserving UEP in BICM

Let us consider a model for the BICM transmission, which generalizes the BICM channel in Fig. 5.1. This is done in Fig. 8.2 where all the elements appearing after the encoder and before the decoder are treated as a BICM channel with a vectorial input $\boldsymbol{c}[n]$ and a vectorial output $\boldsymbol{\lambda}[n]$. In this way, the code bits from the n_c different classes are explicitly shown in Fig. 8.2 as the inputs to the BICM channel.

Example 8.2 (Zehavi's 1992 Interleaving) *When Zehavi introduced BICM in his 1992 paper, he considered a convolutional encoder (CENC) with* $n_c = 3$ *and an 8PSK constellation* ($m = 3$)*. Further, Zehavi proposed to use* $n_c = m$ *interleavers and send all the code bits from the first/second/third encoder's output through the first/second/third interleaver as we show in Fig. 8.3 (a).*

As discussed in Section 6.2.4, if the interleaver complies with the conditions of quasirandomness (see Definition 2.46), the L-values $\Lambda_1, \ldots, \Lambda_{n_c}$ may be considered independent. Using similar considerations, for $n_c = m$, we can replace all the elements of the generalized BICM channel by a set of parallel channels. This is shown in Fig. 8.3 (b), where each channel yields L-values having distribution $p_{\Lambda_q|C_q}(\lambda|c), q = 1, \ldots, n_c$.

Let us now compare the conventional *single-input interleaver* (S-interleaver) Π with the M-interleaver shown in Fig. 8.3. The S-interleaver is the one we studied in the previous chapters and operates without

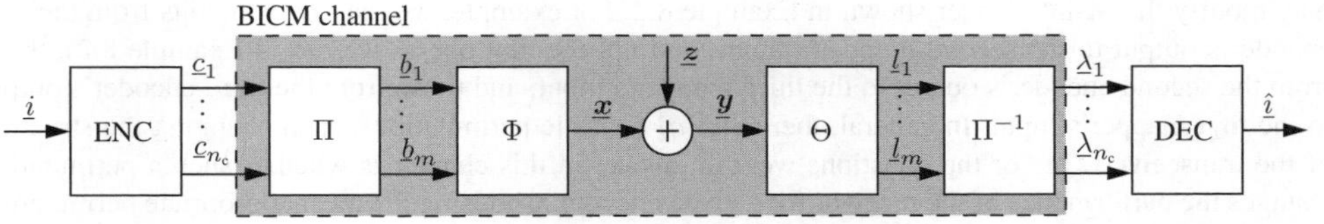

Figure 8.2 Refined model of the BICM channel: the n_c classes of code bits are identified at the inputs of the BICM channel

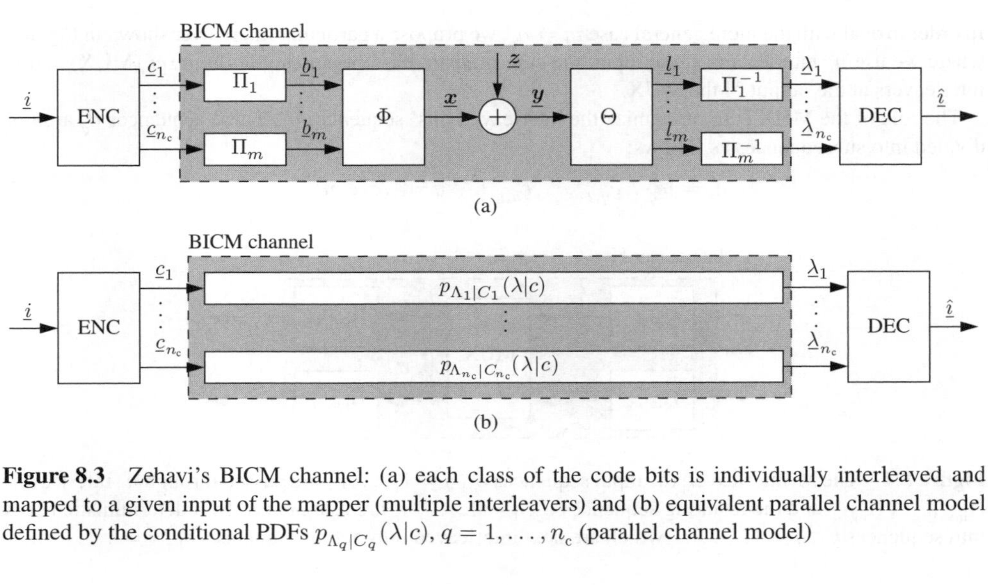

Figure 8.3 Zehavi's BICM channel: (a) each class of the code bits is individually interleaved and mapped to a given input of the mapper (multiple interleavers) and (b) equivalent parallel channel model defined by the conditional PDFs $p_{\Lambda_q|C_q}(\lambda|c), q = 1, \ldots, n_c$ (parallel channel model)

differentiating the classes of code bits. For the S-interleaver, according to the considerations in Section 6.2 leading to (6.227), we have

$$p_{\Lambda_q|C_q}(\lambda|c) = \frac{1}{m}\sum_{k=1}^{m} p_{L_k|B_k}(\lambda|c), \quad q = 1, \ldots, n_{\mathrm{c}}. \tag{8.1}$$

For the M-interleaver with $m = n_{\mathrm{c}}$, the qth L-values Λ_q is the deinterleaved version of L_q, so we obtain

$$p_{\Lambda_q|C_q}(\lambda|c) = p_{L_q|B_q}(\lambda|c), \quad q = 1, \ldots, n_{\mathrm{c}}. \tag{8.2}$$

Thus, using the S-interleaver we average the PDFs of the L-values and remove the effect of UEP. Eliminating UEP was, in fact, considered useful in some works on BICM (as it simplifies the analysis). On the other hand, the M-interleaver considered above allows us to preserve the UEP introduced by the binary labeling. While the idea of using M-interleavers was already present in the early works on BICM, this concept was not thoroughly studied in the literature.

We will show that the M-interleaver we propose are a generalization of S-interleavers, and thus, with an appropriate design, we can only obtain performance gains. This is shown in Section 8.1.4. From now, we refer to BICM transmission with S-interleavers as BICM-S, and BICM with M-interleaver will be called BICM-M. Further, in order to exploit the presence of UEP, we will generalize the M-interleaver to deal with the case of $m \neq n_{\mathrm{c}}$ and derive performance metrics that will allow us to design such M-interleavers.

The order in which we enumerate the classes of bits is arbitrary and so is the numbering of the mapper's inputs. Instead of changing the order of the encoder's output or the order of the mapper's inputs, we may modify the M-interleaver shown in Example 8.2. For example, we may pass the bits from the first encoder's output to the second mapper's input (and not the first one as we did in Example 8.2), those from the second encoder's output to the third mapper's input, and those from the third encoder's output to the first mapper's input. In general, there are $m!$ possible permutations, each changing the structure of the transceiver. One of the questions we will answer in this chapter is whether such a permutation changes the performance of the receiver for a given encoder. Choosing the most appropriate permutation becomes a part of *interleaver design*.

8.1.2 The M-Interleaver

In order to deal with the more general case $m \neq n_{\mathrm{c}}$, we propose a particular interleaver shown in Fig. 8.4, where we use n_{c} interleavers at the input, the deterministic bit-reorganizing *multiplexer* (MUX) and m interleavers at the output of the MUX.

The role of the MUX is to reorganize the interleaved bits' sequences $\underline{c}'_q$. These sequences $\underline{c}'_q$ are first divided into subsequences as follows:

$$\underline{c}'_q = [\underline{c}'_{q,1}, \underline{c}'_{q,2}, \ldots, \underline{c}'_{q,m}], \quad q = 1, \ldots, n_{\mathrm{c}}, \tag{8.3}$$

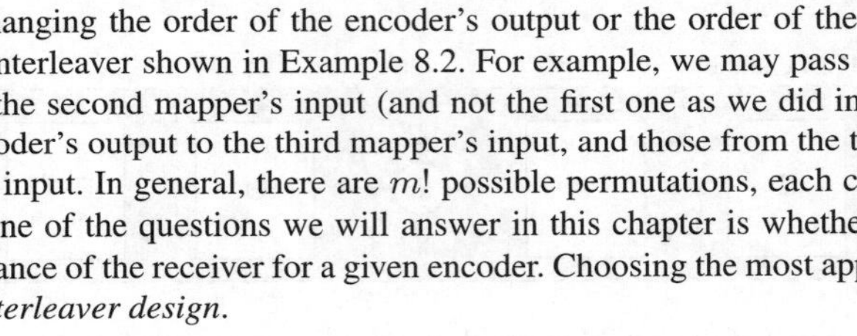

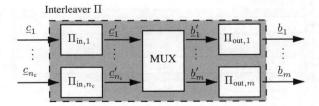

Figure 8.4 The M-interleaver: the input sequences $\underline{c}_q, q = 1, \ldots, n_{\mathrm{c}}$ are passed through n_{c} interleavers $\Pi_{\mathrm{in},1}, \ldots, \Pi_{\mathrm{in},n_{\mathrm{c}}}$ and the resulting sequences $\underline{c}'_q, q = 1, \ldots, n_{\mathrm{c}}$ are reorganized by a deterministic MUX into sequences $\underline{b}'_k, q = 1, \ldots, m$, which are next interleaved into output sequences $\underline{b}_k, k = 1, \ldots, m$

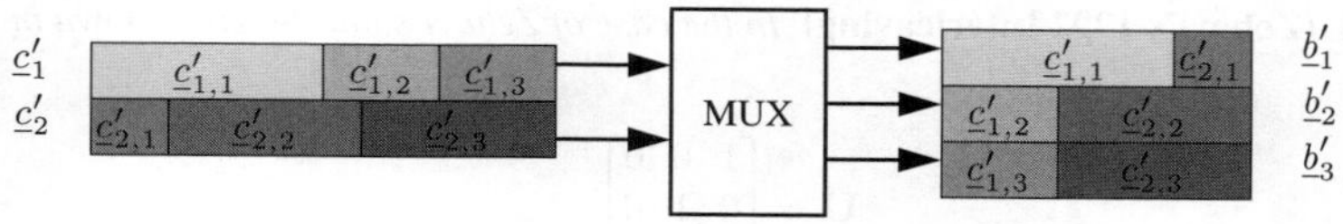

Figure 8.5 Example of interleaving via the M-interleaver with $n_c = 2$, $m = 3$. The values defining the MUX are $u_{1,1} = 1/2$, $u_{1,2} = 1/4$, $u_{1,3} = 1/4$, $u_{2,1} = 1/6$, and $u_{2,2} = u_{2,3} = 5/12$

where each subsequence $\underline{c}'_{q,k}$ contains $N_q u_{q,k}$ bits and $u_{q,k}$ is constrained to be such that $N_q u_{q,k} \in \mathbb{N}$. Next, the subsequences are concatenated into output sequences

$$\underline{b}'_k = [\underline{c}'_{1,k}, \underline{c}'_{2,k}, \ldots, \underline{c}'_{n_c,k}], \quad k = 1, \ldots, m, \tag{8.4}$$

each of them composed of $\sum_{q=1}^{n_c} u_{q,k} N_q = \frac{n_c}{m} N_q = \frac{N_c}{m} = N_s$ bits, as required to generate N_s labels $\boldsymbol{b}[n], n = 1, \ldots, N_s$.

Example 8.3 *To clarify the definitions above, in Fig. 8.5 we show an example for $n_c = 2$ and $m = 3$. The MUX is defined via $u_{1,1} = 1/2$, $u_{1,2} = 1/4$, $u_{1,3} = 1/4$, $u_{2,1} = 1/6$, and $u_{2,2} = u_{2,3} = 5/12$.*

The M-interleaver is then defined by a matrix $\mathbf{U} \in \mathbb{R}^{n_c \times m}$ with entries $u_{q,k}$, which have to satisfy two obvious constraints. The first one is that all the bits $c_q[n]$ must be assigned to one of the mapper's input, i.e.,

$$\sum_{k=1}^{m} u_{q,k} N_q = N_q, \tag{8.5}$$

which translates into

$$\sum_{k=1}^{m} u_{q,k} = 1. \tag{8.6}$$

The second constraint is that each mapper's input is equally "loaded" with bits from the encoder's output

$$\sum_{q=1}^{n_c} u_{q,k} N_q = N_s, \tag{8.7}$$

or equivalently,

$$\sum_{q=1}^{n_c} u_{q,k} = \frac{n_c}{m}. \tag{8.8}$$

The matrix $\mathbf{U}$ can then be written as

$$\mathbf{U} \triangleq \begin{bmatrix} u_{1,1} & \cdots & u_{1,m-1} & 1 - \sum_{k=1}^{m-1} u_{1,k} \\ u_{2,1} & \cdots & u_{2,m-1} & 1 - \sum_{k=1}^{m-1} u_{2,k} \\ \vdots & \ddots & \vdots & \vdots \\ u_{n_c-1,1} & \cdots & u_{n_c-1,m-1} & 1 - \sum_{k=1}^{m-1} u_{n_c-1,k} \\ \frac{n_c}{m} - \sum_{q=1}^{n_c-1} u_{q,1} & \cdots & \frac{n_c}{m} - \sum_{q=1}^{n_c-1} u_{q,m-1} & \frac{n_c}{m} - n_c + 1 + \sum_{q=1}^{n_c-1} \sum_{k=1}^{m-1} u_{q,k} \end{bmatrix}, \tag{8.9}$$

where the last row and the last column of $\mathbf{U}$ take into account the constraints imposed on $u_{q,k}$. Consequently, when designing $\mathbf{U}$, only $u_{q,k}$ for $q = 1, \ldots, n_c - 1$ and $k = 1, \ldots, m - 1$ may be freely set (considering also $0 \leq u_{q,k} \leq 1 \; \forall q, k$).

Example 8.4 (Zehavi's 1992 Interleaving) *In the case of Zehavi's interleaving shown in Example 8.2 we have*

$$\mathbf{U} = \begin{bmatrix} 1 & 0 & 0 \\ 0 & 1 & 0 \\ 0 & 0 & 1 \end{bmatrix}. \tag{8.10}$$

A reorganization of the connectivity between the encoders' output and the mapper's inputs boils down to a permutation of the columns of $\mathbf{U}$. *For example, to pass the bits from the first encoder's output to the second mapper's input, those from the second encoder's output to the third mapper's input, and those from the third encoder's output to the first mapper's input, we would use the following matrix*

$$\mathbf{U} = \begin{bmatrix} 0 & 1 & 0 \\ 0 & 0 & 1 \\ 1 & 0 & 0 \end{bmatrix}. \tag{8.11}$$

The motivation behind the structure of the M-interleaver should be now clear: we would like to assign the code bits to the protected positions so as to improve the performance of the receiver. However, we cannot do it arbitrarily. For example, we cannot assign all the bits $c_q[n]$ to the most protected positions (in Example 8.1 this would be the position $k = 1$) as this will violate the constraints defined in (8.8).

We note here that the particular case of $n_c = m$, i.e., when the matrix $\mathbf{U}$ is square and each row/column contains only one nonzero element, the double set of interleavers—input ($\Pi_{\text{in},q}$, $q = 1, \ldots, n_c$) and output ($\Pi_{\text{out},k}$, $k = 1, \ldots, m$)—is not necessary. When $n_c = m$ we may use just one set of interleavers as is shown already in Fig. 8.3. Having two sets of interleavers, however, is necessary for the more general case $n_c \neq m$.

8.1.3 Modeling the M-Interleaver

We recall that the S-interleaver in Fig. 2.24 is modeled as a random MUX assigning the code bits belonging to the class q to the kth position of the mapper. The randomness comes from the fact that the interleavers are generated randomly, and then, the position k is random as well, which we denote by K. For S-interleavers we considered up to now, this variable was uniformly distributed.

To analyze the M-interleaver and take advantage of the considerations in Section 2.7 and Section 6.2, we model the interleaving vectors $\boldsymbol{\pi}_{\text{in},q}$ and $\boldsymbol{\pi}_{\text{out},k}$ as random variables drawn with equal probability from the set of all possible permutations. Then,

$$\Pr\{c_q[n] \in \underline{c}'_{q,k}\} = u_{q,k}, \tag{8.12}$$

where $c_q[n] \in \underline{c}'_{q,k}$ means that $c_q[n]$ belongs to the subsequence $\underline{c}'_{q,k}$, i.e., it is assigned to the mapper's output k. Similarly,

$$\Pr\{b_k[n] \in \underline{c}'_{q,k}\} = u_{q,k}, \tag{8.13}$$

i.e., for any position n, the bit $b_k[n]$ is obtained from the qth code bits' class with probability $u_{q,k}$.

To adapt the model Fig. 2.24 to the case of the M-interleaver, we first need to consider a random switch indexed by q, i.e., we use K_q instead of K. We also replace the values of the assignment probabilities as shown in Fig. 8.6. Now $u_{q,k}$ has a meaning of the probability that the bit belonging to the class q is mapped to the position k of the mapper, i.e., $u_{q,k} = \Pr\{K_q = k\}$.

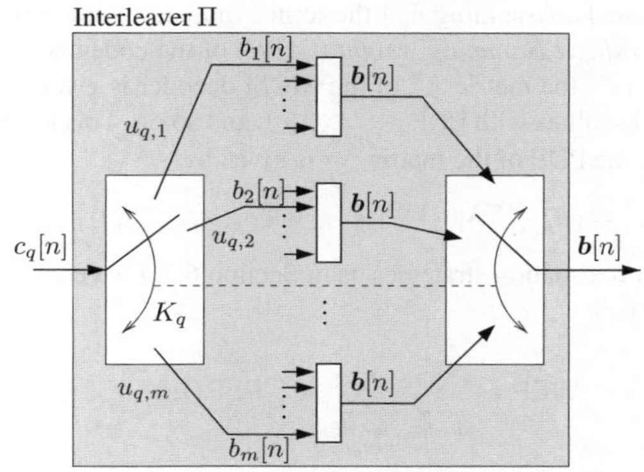

Figure 8.6 Probabilistic model of the M-interleaver defined in Section 8.1.2: the bits $c_q[n]$ are multiplexed to the position K_q within the label $\boldsymbol{b}[n]$, where K_q is a random variable with distribution $\Pr\{K_q = k\} = u_{q,k}, k = 1, \ldots, m$

The immediate consequence of the random assignment of the bits $c_q[n]$ to the positions K_q is that the corresponding L-value Λ_q has a distribution that changes with K_q. Thus the PDFs for Λ_q are given by

$$p_{\Lambda_q|C_q}(\lambda|c) = \sum_{k=1}^{m} p_{L_k|B_k}(\lambda|c) \Pr\{K_q = k\} \tag{8.14}$$

$$= \sum_{k=1}^{m} p_{L_k|B_k}(\lambda|c) u_{q,k}. \tag{8.15}$$

Using the PDFs of the L-values in (8.15) we transform the BICM channel in Fig. 8.2 to a set of parallel BICO channels with channel transition probabilities $p_{\Lambda_q|C_q}(\lambda|c)$, which is shown in Fig. 8.3 (b).

In the case of Zehavi's interleaving in Example 8.2, using (8.10) and (8.15) we recover the formulas we already had in (8.2). In the case of the S-interleaver, the matrix $\mathbf{U}$ should be chosen as

$$\mathbf{U} = \frac{1}{m} \begin{bmatrix} 1 & \ldots & 1 & 1 \\ 1 & \ldots & 1 & 1 \\ \vdots & \ddots & \vdots & \vdots \\ 1 & \ldots & 1 & 1 \end{bmatrix}, \tag{8.16}$$

so using (8.15) we obtain the PDF

$$p_{\Lambda_q|C_q}(\lambda|c) = \frac{1}{m} \sum_{k=1}^{m} p_{L_k|B_k}(\lambda|c), \tag{8.17}$$

which coincides, as expected, with (8.1).

We emphasize that in order to obtain the formula (8.15) and use it further to evaluate the performance of the decoder, we need to satisfy the conditions of quasi-randomness, similar to those we explained in Section 2.7. We do not dwell on this anymore and simply assume that such conditions are fulfilled.

8.1.4 Performance Evaluation

We have established a parallel channel model (see Fig. 8.3 (b)), which relies solely on the knowledge of the PDF of the L-values $\Lambda_q, q = 1, \ldots, n_c$ associated with the classes of code bits. To study the performance of the decoders, we should distinguish amongst the codewords according to the input they provide to the parallel channels. Assuming that the joint symmetry condition in (3.75) is satisfied, this input depends on the *generalized Hamming weight* (GHW) of the codeword $\underline{\mathbf{c}}$. Namely, for each codeword $\underline{\mathbf{c}} \in \mathcal{C}(\boldsymbol{w})$ with $\boldsymbol{w} \in \mathbb{N}^{n_c}$ the metric Λ^{Σ} of the BICM decoder is given by the sum of w_1 L-values with PDF $p_{\Lambda_1|C_1}(\lambda|c)$, w_2 L-values with PDF $p_{\Lambda_2|C_2}(\lambda|c)$, and so on. This is what we showed already in (6.211), and consequently, the PDF of the metric Λ^{Σ} is given by

$$p_{\Lambda^{\Sigma}}(\lambda) = \{p_{\Lambda_1|C_1}(\lambda|0)\}^{*w_1} * \ldots * \{p_{\Lambda_{n_c}|C_{n_c}}(\lambda|0)\}^{*w_{n_c}}. \tag{8.18}$$

Then using the same approximation strategies as in Section 6.2.4 we obtain an approximation for the *word-error probability* (WEP):

$$\mathsf{WEP} \approx \sum_{d=d_{\text{free}}}^{d_{\max}} \sum_{\substack{\boldsymbol{w}\in\mathbb{N}^{n_c}\\ \|\boldsymbol{w}\|_1=d}} C_{\boldsymbol{w}}^{\mathcal{C}} \mathsf{PEP}(\boldsymbol{w}), \tag{8.19}$$

where $C_{\boldsymbol{w}}^{\mathcal{C}}$ is the *generalized distance spectrum* (GDS) of the code $\mathcal{C}$ in Definition 2.23, and

$$\mathsf{PEP}(\boldsymbol{w}) = \int_0^{\infty} p_{\Lambda^{\Sigma}}(\lambda)\, d\lambda. \tag{8.20}$$

In analogy to Example 6.22, the following example shows the performance of BICM with *convolutional encoders* (CENCs), but now using M-interleaver.

Example 8.5 (Convolutional Encoders, M-interleavers and BICM Decoding) *The bounds obtained for the CENC in Example 6.22 can be generalized to take into account the weights for each of the classes of code bits. In particular, the BEP in* (6.243) *becomes*

$$\mathsf{BEP} \le \sum_{\ell=1}^{\infty}\sum_{v=1}^{\infty}\sum_{d=d_{\text{free}}}^{\infty} \sum_{\substack{\boldsymbol{w}\in\mathbb{N}^{n_c}\\ \|\boldsymbol{w}\|_1=d}} \frac{v}{k_c} F_{v,\boldsymbol{w},\ell}^{\mathcal{C}} \mathsf{PEP}(\boldsymbol{w}), \tag{8.21}$$

where $F_{v,\boldsymbol{w},\ell}^{\mathcal{C}}$ *is the number of sequences* $\underline{\hat{\mathbf{c}}} \in \hat{\mathcal{C}}_{\ell}(\underline{\mathbf{0}})$ *with GHW* $\boldsymbol{w}$ *generated by input sequences with Hamming weight (HW)* v.

The BEP in (8.21) *can be then approximated as*

$$\mathsf{BEP} \approx \sum_{d=d_{\text{free}}}^{d_{\max}} \sum_{\substack{\boldsymbol{w}\in\mathbb{N}^{n_c}\\ \|\boldsymbol{w}\|_1=d}} G_{\boldsymbol{w}}^{\mathcal{C}} \mathsf{PEP}(\boldsymbol{w}), \tag{8.22}$$

where $G_{\boldsymbol{w}}^{\mathcal{C}}$ *is the generalized input-dependent weight distribution (GIWD) of the CENC defined as*

$$G_{\boldsymbol{w}}^{\mathcal{C}} \triangleq \sum_{\ell=1}^{\infty}\sum_{v=1}^{\infty} \frac{v}{k_c} F_{v,\boldsymbol{w},\ell}^{\mathcal{C}}. \tag{8.23}$$

The GIWD in (8.23) *considers not only the total HW of* $\underline{\hat{\mathbf{e}}}$ *(see Example 6.22) but also the HW of each row of* $\underline{\hat{\mathbf{e}}}$. *Moreover, the GIWD* $G_{\boldsymbol{w}}^{\mathcal{C}}$ *in* (8.23) *and the input-dependent weight distribution (IWD)* $G_d^{\mathcal{C}}$ *defined in* (6.245) *are related via*

$$G_d^{\mathcal{C}} = \sum_{\substack{\boldsymbol{w}\in\mathbb{N}^{n_c}\\ \|\boldsymbol{w}\|_1=d}} G_{\boldsymbol{w}}^{\mathcal{C}}. \tag{8.24}$$

The WEP in (6.241) *can also be generalized as*

$$\mathsf{WEP} \approx \sum_{d=d_{\text{free}}}^{d_{\max}} \sum_{\substack{\boldsymbol{w}\in\mathbb{N}^{n_c}\\ \|\boldsymbol{w}\|_1=d}} N_q F_{\boldsymbol{w}}^{\mathcal{C}} \mathsf{PEP}(\boldsymbol{w}), \tag{8.25}$$

where

$$F^{\mathcal{C}}_{\boldsymbol{w}} \triangleq \sum_{\ell=1}^{\infty}\sum_{v=1}^{\infty} F^{\mathcal{C}}_{v,\boldsymbol{w},\ell} \tag{8.26}$$

is a generalization of (6.240), *which we therefore call* generalized weight distribution *(GWD).*

In the following example we show how to compute the GIWD $G^{\mathcal{C}}_{\boldsymbol{w}}$ and the GWD $F^{\mathcal{C}}_{\boldsymbol{w}}$ for a CENC, which follow as a generalization of the procedure in Example 6.23.

Example 8.6 (GWD and GIWD of the Encoder $\mathrm{G} = [5,7]$**)** *To calculate the GWD and GIWD for the CENC analyzed in Example 6.23 we use a generalized transfer function with $n_c + 1 = 3$ dummy variables instead of two, which was sufficient to characterize the CENC with S-interleavers. The generalized state machine in this case is shown in Fig. 8.7.*

The state equations are now given by

$$\xi_b = VW_1W_2\xi_a + V\xi_c, \tag{8.27}$$
$$\xi_c = W_1\xi_d + W_2\xi_b, \tag{8.28}$$
$$\xi_d = VW_1\xi_b + VW_2\xi_d, \tag{8.29}$$
$$\xi_e = W_1W_2\xi_c, \tag{8.30}$$

and we obtain the generalized transfer function of the encoder $T(V, W_1, W_2)$ as

$$T(V, W_1, W_2) = \frac{\xi_e}{\xi_a} = \frac{VW_1^2W_2^2(V(W_1^2 - W_2^2) + W_2)}{1 - V(V(W_1^2 - W_2^2) + 2W_2)}.$$

The GWD and GIWD of the encoder can be finally calculated using a generalization of (6.250) *and* (6.252), *i.e.,*

$$F^{\mathcal{C}}_{\boldsymbol{w}} = \frac{1}{d_1!d_2!}\frac{\partial^{d_1}}{\partial W_1^{d_1}}\frac{\partial^{d_2}}{\partial W_2^{d_2}}T(1, W_1, W_2)|_{W_1=0,W_2=0}, \tag{8.31}$$

$$G^{\mathcal{C}}_{\boldsymbol{w}} = \frac{1}{d_1!d_2!}\frac{\partial^{d_1}}{\partial W_1^{d_1}}\frac{\partial^{d_2}}{\partial W_2^{d_2}}\frac{\partial}{\partial V}T(V, W_1, W_2)|_{W_1=0,W_2=0,V=1}, \tag{8.32}$$

The first terms of the IWD (see (6.253)*) are given by*

$$[G^{\mathcal{C}}_1\ G^{\mathcal{C}}_2\ G^{\mathcal{C}}_3\ G^{\mathcal{C}}_4\ G^{\mathcal{C}}_5\ G^{\mathcal{C}}_6\ G^{\mathcal{C}}_7\ G^{\mathcal{C}}_8\ \ldots] = [0\ 0\ 0\ 0\ 1\ 4\ \mathbf{12}\ 32\ \ldots], \tag{8.33}$$

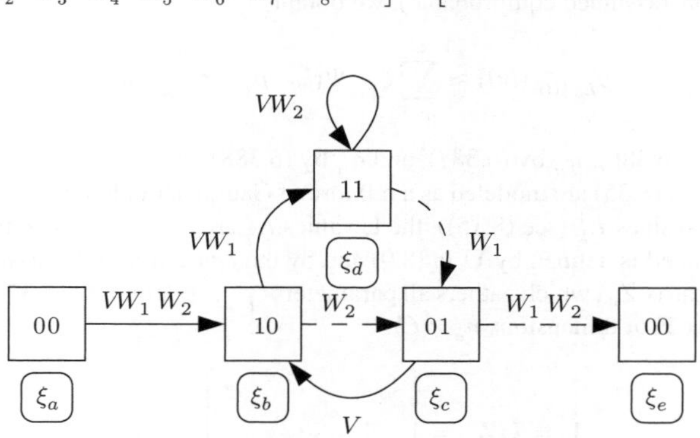

Figure 8.7 Generalized state machine for $\mathbf{G} = [5,7]$ used in the analysis of BICM-M

which can be compared to the first terms in the GIWD in (8.32)

$$\begin{bmatrix} G^{\mathcal{C}}_{[1,1]} & \cdots & G^{\mathcal{C}}_{[1,7]} & \cdots \\ \vdots & \ddots & \vdots & \\ G^{\mathcal{C}}_{[7,1]} & \cdots & G^{\mathcal{C}}_{[7,7]} & \cdots \\ \vdots & & \vdots & \ddots \end{bmatrix} = \begin{bmatrix} 0 & 0 & 0 & 0 & 0 & \mathbf{0} & 0 & \cdots \\ 0 & 0 & 1 & 2 & \mathbf{3} & 4 & 5 & \cdots \\ 0 & 0 & 0 & \mathbf{0} & 0 & 0 & 0 & \cdots \\ 0 & 2 & \mathbf{9} & 24 & 50 & 90 & 147 & \cdots \\ 0 & \mathbf{0} & 0 & 0 & 0 & 0 & 0 & \cdots \\ \mathbf{0} & 4 & 25 & 90 & 245 & 560 & 1134 & \cdots \\ 0 & 0 & 0 & 0 & 0 & 0 & 0 & \cdots \\ \vdots & \vdots & \vdots & \vdots & \vdots & \vdots & \vdots & \ddots \end{bmatrix}. \tag{8.34}$$

From (8.24)*, we know that the sum of the elements $G^{\mathcal{C}}_{\boldsymbol{w}}$ at the dth antidiagonal is equal to d. For the particular case of $d = 7$, the element $G^{\mathcal{C}}_7$ and the elements $G^{\mathcal{C}}_{\boldsymbol{w}}$ for $\|\boldsymbol{w}\|_1 = d$ are highlighted in* (8.33) *and* (8.34).

By using a vector of dummy variables $\boldsymbol{W} = [W_1, \ldots, W_{n_c}]^{\mathrm{T}}$, the procedure shown in Example 8.6 ($n_c = 2$) can be extended to any value of n_c, and in general, the GWD and GIWD can be expressed in terms of an n_c dimensional vector $\boldsymbol{w}$. Once the GIWD of the encoder is calculated, in order to predict the BEP performance of BICM-M via (8.22), we only need to compute $\mathsf{PEP}(\boldsymbol{w})$ in (8.20). This is done in the following section by exploiting the Gaussian simplifications we introduced in Section 6.4.

8.2 Exploiting UEP in *M*PAM Constellations

In Section 8.1 we defined the models to characterize the performance of BICM-M. In order to efficiently optimize the transceiver, we need closed-form expressions that relate the performance of the decoder to the multiplexing matrix $\mathbf{U}$ and the channel model. In this section we turn our attention to the relatively simple case of MPAM constellations for which we obtained simplified forms for the PDFs of the L-values (see Section 5.5).

8.2.1 The Generalized BICM Channel

For MPAM constellations labeled by the BRGC, the results in Section 6.4.2 tell us that the conditional PDF of the max-log L-values can be approximated via a Gaussian function. After marginalization over all transmitted symbols (assumed equiprobable), we obtain

$$p_{L_k|B_k}(l|0) \approx \sum_{g=1}^{G} \zeta_{k,g} \Psi(l, -\mu_g \mathsf{snr}, \sigma_g^2 \mathsf{snr}), \tag{8.35}$$

where $\zeta_{k,g}$ is given by (6.386), μ_g by (6.387), and σ_g^2 by (6.388).

As the L-values L_k in (8.35) are modeled as a mixture of Gaussian random variables, and the L-values Λ_q are a mixture of L-values L_k (see (8.15)), the L-values Λ_q are also a Gaussian mixture. The way the Gaussian PDFs are mixed is defined by $\mathbf{U}$ in (8.9) and by the *constellation bit-wise Euclidean distance spectrum* (CBEDS) matrix $\mathbf{Z}_\Phi$, which gathers all parameters $\zeta_{k,g}$, as defined in (6.374). More specifically, if we define the matrix $\mathbf{E}$ of dimensions $n_c \times G$

$$\mathbf{E} \triangleq \mathbf{U}\mathbf{Z}_\Phi = \begin{bmatrix} \xi_{1,1} & \cdots & \xi_{1,G} \\ \vdots & \ddots & \vdots \\ \xi_{n_c,1} & \cdots & \xi_{n_c,G} \end{bmatrix}, \tag{8.36}$$

the PDF of the L-values at the qth decoder's input is given by

$$p_{\Lambda_q|C_q}(\lambda|0) \approx \sum_{g=1}^{G} \xi_{q,g} \Psi(\lambda, -\mathsf{snr}\,\mu_g, \mathsf{snr}\,\sigma_g^2). \tag{8.37}$$

From (8.37), $\xi_{q,g}$ can be interpreted as the probability of observing the gth Gaussian PDF at the qth parallel channel in Fig. 8.3 (b).

Example 8.7 (BICM-S vs. BICM-M for 8PAM**)** *Consider an interleaver similar to the one given in Example 8.2, i.e., an MUX matrix*

$$\mathbf{U} = \begin{bmatrix} 0 & 1 & 0 \\ 1 & 0 & 0 \\ 0 & 0 & 1 \end{bmatrix}. \tag{8.38}$$

Using (6.375) *and* (8.36) *we obtain*

$$\mathrm{E} = \begin{bmatrix} 1/2 & 1/2 & 0 & 0 \\ 1/4 & 1/4 & 1/4 & 1/4 \\ 1 & 0 & 0 & 0 \end{bmatrix}, \tag{8.39}$$

which applied to (8.37) *yields*

$$p_{\Lambda_1|C_1}(\lambda|0) = \frac{1}{2}(\Psi(\lambda, -\mu_1\,\mathsf{snr}, \sigma_2^2\,\mathsf{snr}) + \Psi(\lambda, -\mu_2\,\mathsf{snr}, \sigma_2^2\,\mathsf{snr})), \tag{8.40}$$

$$p_{\Lambda_2|C_2}(\lambda|0) = \frac{1}{4}(\Psi(\lambda, -\mu_1\,\mathsf{snr}, \sigma_2^2\,\mathsf{snr}) + \Psi(\lambda, -\mu_2\,\mathsf{snr}, \sigma_2^2\,\mathsf{snr})$$
$$+ \Psi(\lambda, -\mu_3\,\mathsf{snr}, \sigma_3^2\,\mathsf{snr}) + \Psi(\lambda, -\mu_4\,\mathsf{snr}, \sigma_4^2\,\mathsf{snr})), \tag{8.41}$$

$$p_{\Lambda_3|C_3}(\lambda|0) = \Psi(\lambda, -\mu_1\,\mathsf{snr}, \sigma_2^2\,\mathsf{snr}). \tag{8.42}$$

The model introduced for BICM-M can also be used for BICM-S. This can be achieved using **U** *in* (8.16)*, which yields*

$$\mathbf{E} = \begin{bmatrix} 7/12 & 3/12 & 1/12 & 1/12 \\ 7/12 & 3/12 & 1/12 & 1/12 \\ 7/12 & 3/12 & 1/12 & 1/12 \end{bmatrix}, \tag{8.43}$$

which used in (8.37) *gives*

$$p_{\Lambda_q|C_q}(\lambda|0) = \frac{7}{12}\Psi(\lambda, -\mu_1\,\mathsf{snr}, \sigma_1^2\,\mathsf{snr}) + \frac{1}{4}\Psi(\lambda, -\mu_2\,\mathsf{snr}, \sigma_2^2\,\mathsf{snr})$$
$$+ \frac{1}{12}\Psi(\lambda, -\mu_3\,\mathsf{snr}, \sigma_3^2\,\mathsf{snr}) + \frac{1}{12}\Psi(\lambda, -\mu_4\,\mathsf{snr}, \sigma_4^2\,\mathsf{snr}), \tag{8.44}$$

for $q = 1, \ldots, m$*. As expected, for* BICM-S*, the PDF* $p_{\Lambda_q|C_q}(\lambda|0)$ *is independent of the code bit class* q*.*

8.2.2 *Performance Evaluation*

The expression in (8.37) gives an analytical approximation for the PDFs of the L-values for BICM-M. In this section we study the performance of such BICM-M transceivers. From now on, we limit our considerations to the parameters defined by the *zero-crossing model* (ZcM) approximation (see Section 6.4.2), which proved to yield more accurate results than those obtained from the *consistent model* (CoM) (see Section 6.4.4). The next theorem shows how the *pairwise-error probability* (PEP) in (8.20) can be approximated.

Theorem 8.8 *Using the approximate PDF in (8.37), the PEP in (8.20) can be approximated as*

$$\mathsf{PEP}(\boldsymbol{w}) \approx \sum_{\substack{\boldsymbol{r}_1 \in \mathbb{N}^G \\ \|\boldsymbol{r}_1\|_1 = d_1}} \cdots \sum_{\substack{\boldsymbol{r}_{n_c} \in \mathbb{N}^G \\ \|\boldsymbol{r}_{n_c}\|_1 = d_{n_c}}} g(\boldsymbol{r}_1, \ldots, \boldsymbol{r}_{n_c}) \cdot Q(\sqrt{\mathsf{snr}} h(\boldsymbol{r}_1, \ldots, \boldsymbol{r}_{n_c})), \tag{8.45}$$

where $\boldsymbol{w} = [w_1, \ldots, w_{n_c}]^{\mathrm{T}}$, $\boldsymbol{r}_q = [r_{q,1}, \ldots, r_{q,G}]^{\mathrm{T}}$ *with* $q = 1, \ldots, n_c$,

$$g(\boldsymbol{r}_1, \ldots, \boldsymbol{r}_{n_c}) \triangleq \prod_{q=1}^{n_c} \left[\binom{d_q}{\boldsymbol{r}_q} \prod_{g=1}^{G} \xi_{q,g}^{r_{q,g}} \right], \tag{8.46}$$

$$h(\boldsymbol{r}_1, \ldots, \boldsymbol{r}_{n_c}) \triangleq \frac{2\Delta}{\sqrt{2d}} \sum_{q=1}^{n_c} \sum_{g=1}^{G} r_{q,g}(2g-1). \tag{8.47}$$

Proof: Similarly to (6.381), the qth self-convolution in (8.18) with $p_{\Lambda_q|C_q}(\lambda|0)$ given by (8.37) can be written as

$$\{p_{\Lambda_q|C_q}(\lambda|0)\}^{*d_q} = \sum_{\substack{\boldsymbol{r}_q \in \mathbb{N}^G \\ \|\boldsymbol{r}_q\|_1 = d_q}} \binom{d_q}{\boldsymbol{r}_q} \Psi\left(\lambda, -\sum_{g=1}^{G} r_{q,g}\mu_g, d_q\sigma_1^2\right) \prod_{g=1}^{G} \xi_{q,g}^{r_{q,g}}, \tag{8.48}$$

where because of the assumption of the ZcM we use $\sigma_g^2 = \sigma_1^2$. Using (8.48) in (8.18) we obtain

$$p_{\Lambda^\Sigma}(\lambda) = \sum_{\substack{\boldsymbol{r}_1 \in \mathbb{N}^G \\ \|\boldsymbol{r}_1\|_1 = d_1}} \binom{d_1}{\boldsymbol{r}_1} \Psi\left(\lambda, -\sum_{g=1}^{G} r_{1,g}\mu_g, d_1\sigma_1^2\right) \prod_{g=1}^{G} \xi_{1,g}^{r_{1,g}} * \cdots *$$

$$\sum_{\substack{\boldsymbol{r}_{n_c} \in \mathbb{N}^G \\ \|\boldsymbol{r}_{n_c}\|_1 = d_{n_c}}} \binom{d_{n_c}}{\boldsymbol{r}_{n_c}} \Psi\left(\lambda, -\sum_{g=1}^{G} r_{n_c,g}\mu_g, d_{n_c}\sigma_1^2\right) \prod_{g=1}^{G} \xi_{n_c,g}^{r_{n_c,g}} \tag{8.49}$$

$$= \sum_{\substack{\boldsymbol{r}_1 \in \mathbb{N}^G \\ \|\boldsymbol{r}_1\|_1 = d_1}} \cdots \sum_{\substack{\boldsymbol{r}_{n_c} \in \mathbb{N}^G \\ \|\boldsymbol{r}_1\|_1 = d_{n_c}}} g(\boldsymbol{r}_1, \ldots, \boldsymbol{r}_{n_c}) \Psi\left(\lambda, -\sum_{q=1}^{n_c}\sum_{g=1}^{G} r_{q,g}\mu_g, \sigma_1^2 \sum_{q=1}^{n_c} d_q\right), \tag{8.50}$$

where $g(\boldsymbol{r}_1, \ldots, \boldsymbol{r}_{n_c})$ is given by (8.47). The proof is completed by using (8.50) in (8.20) together with $\sigma_1^2 = 8\Delta^2$ and $\mu_g = 4\Delta^2(2g-1)$ in (6.387). □

Example 8.9 (BICM-M with 4PAM **and** $R_c = 1/2$**)** *Consider a 4PAM constellation and a rate* $R_c = 1/2$ *turbo encoder (TENC) or CENC* ($n_c = 2$ *and* $m = 2$), *both giving a spectral efficiency of* $R = 1$ *bit/symbol. For the CENC we use the optimal distance spectrum (ODS) CENC* $\mathbf{G} = [23, 35]$ ($\nu = 4$) *from Table 2.1 where the qth generator polynomial is associated with the qth encoder's output. For the TENC, we use the* $R_c = 1/2$ *TENC from Example 2.31. While the TENC defined in Fig. 2.21 has nominally three outputs (systematic bits, first and second code bits), we make no distinction between the parity bits, and we consider them to belong to the same output class* $\underline{c}_2$, *while the systematic bits are gathered in* $\underline{c}_1$.

For $n_c = m = 2$ *we see from (8.9) that there is only one degree of freedom when selecting* $\mathbf{U}$, *i.e.,* $u_{1,1}$. *The two simplest BICM-M interleavers that can be constructed in this case are* $u_{1,1} \in \{0, 1\}$ *while* BICM-S *is obtained using* $u_{1,1} = 1/2$. *The corresponding multiplexer matrices* $\mathbf{U}$ *are then given by*

$$\mathbf{U}_{\mathrm{M},1} = \begin{bmatrix} 1 & 0 \\ 0 & 1 \end{bmatrix}, \quad \mathbf{U}_{\mathrm{M},2} = \begin{bmatrix} 0 & 1 \\ 1 & 0 \end{bmatrix}, \quad \mathbf{U}_{\mathrm{S}} = \begin{bmatrix} 1/2 & 1/2 \\ 1/2 & 1/2 \end{bmatrix}. \tag{8.51}$$

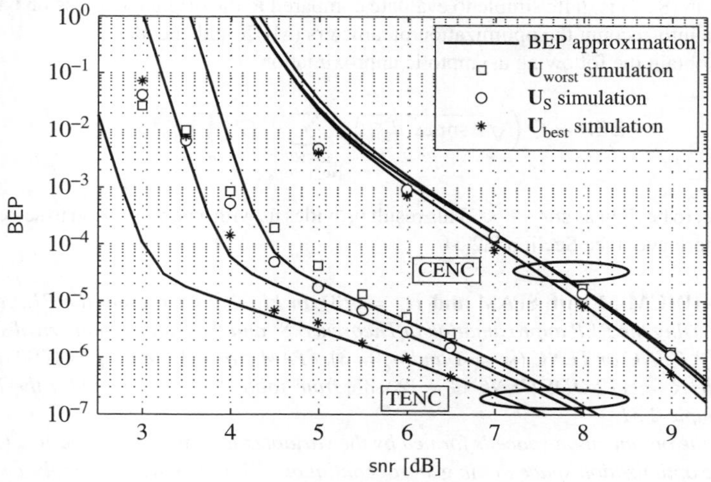

Figure 8.8 BEP approximation in (8.22) based on Theorem 8.8 (lines) and simulated BEP (markers) for BICM-M, $R_c = 1/2$, 4PAM ($n_c = m = 2$) and the three interleavers defined in (8.51). We used the ODS CENC $\mathbf{G} = [23, 35]$ from Table 2.1 ($\nu = 4$) and the TENC in Example 2.31. The interleaver size of the TENC Π_{tc} is $N_b = 1000$ and 10 iterations are performed by the turbo decoder

In Fig. 8.8 the BEP approximations from (8.22) *based on Theorem 8.8 are compared against simulation results.*[4] *The approximations and simulation match quite well for small values of the BEP in the case of the CENC and, for the TENC, they match in the "error floor" region.*

For the CENC in this example, the best interleaver design, which we denote by $\mathbf{U}_{\text{best}}$, *is obtained with* $\mathbf{U} = \mathbf{U}_{\text{M},1}$, *i.e., when the bits coming from the first encoder's output (generator polynomial* 23*) are more protected by the channel than the second encoder's output. The worst interleaver design—denoted by* $\mathbf{U}_{\text{worst}}$*—is obtained setting* $\mathbf{U} = \mathbf{U}_{\text{M},2}$, *while the S-interleaver gives a performance between* $\mathbf{U}_{\text{best}}$ *and* $\mathbf{U}_{\text{worst}}$. *The performance gains offered by* $\mathbf{U}_{\text{best}}$ *are relatively small* (0.3 dB *at* $\text{BEP} = 10^{-6}$*); however, we will later see that for other encoders or code rates, the gains can be much larger.*

For the TENC the optimum interleaver is $\mathbf{U}_{\text{best}} = \mathbf{U}_{\text{M},2}$, *i.e., setting* $u_{1,1} = 0$, *which means that the parity bits should be more protected than the systematic bits (i.e., they are transmitted using the first position of the mapper). This in fact contradicts a somewhat general belief that systematic bits should always be transmitted in the more reliable positions. We emphasize that for this TENC, and for a target BEP of* 10^{-6}, *the difference between* $\mathbf{U}_{\text{best}}$ *and* $\mathbf{U}_{\text{worst}}$ *is* 1 dB, *which is obtained without complexity increase but only by properly assigning the code bits to the bit positions in the* 4PAM *symbol. The results in Fig. 8.8 also show that the optimum assignment depends on the encoder used (defined by its GIWD).*

We note that all combinations in (8.49) are in general tedious to evaluate (specially for large values of n_c and/or m), and thus, we propose further approximations. One straightforward simplification is to consider, for each q, only the Gaussian density with the smallest mean-to-standard deviation ratio (e.g., $g = 1$). The intuition behind this approximation is that the error terms related to other Gaussian functions decrease quickly with snr. This approximation yields the following approximation on the BEP:

$$\text{BEP} \approx \sum_{d=d_{\text{free}}}^{d_{\max}} Q(\sqrt{2d\,\text{snr}\,\Delta^2}) \sum_{\substack{\boldsymbol{w}\in\mathbb{N}^{n_c}\\ \|\boldsymbol{w}\|_1=d}} C_{\boldsymbol{w}}^{\mathcal{C}} \prod_{q=1}^{n_c} \xi_{q,1}^{d_q}. \tag{8.52}$$

[4] We used $d_{\max} = 100$ for the TENC and $d_{\max} = 50$ for the CENC.

The expression in (8.52) is quite simple to evaluate compared to the original expression based on (8.45), and it still takes into account the optimization parameters (MUX and encoder).

We can also obtain the following asymptotic approximation

$$\mathsf{BEP} \approx Q\left(\sqrt{2\,\mathsf{snr}\,\Delta^2 d_{\text{free}}}\right) \sum_{\substack{\boldsymbol{w}\in\mathbb{N}^{n_c}\\ \|\boldsymbol{w}\|_1=d_{\text{free}}}} C_{\boldsymbol{w}}^{\mathcal{C}} \prod_{q=1}^{n_c} \xi_{q,1}^{d_q}, \tag{8.53}$$

which we expect to be tight as $\mathsf{snr} \to \infty$. This result provides us with a new criterion to select the optimum encoder and interleaver (see Section 8.2.3).

Example 8.10 (BICM-M with 8PAM **and** $R_c = 1/3$**)** *In Fig. 8.9 we present the BEP approximation given by* (8.22) *using the PEP approximation in Theorem 8.8*[5] *and the result of numerical simulations for a rate* $R_c = 1/3$ *CENC or TENC* ($k_c = 1$ *and* $n_c = 3$), *8PAM* ($m = 3$) *and* $N_b = 1500$. *For the CENC we used the rate* $R_c = 1/2$ *ODS CENC* $\mathbf{G} = [23, 35]$ *from Table 2.1* ($\nu = 4$) *and for the TENC we used the one in Example 2.31.*

In this case, the optimization space is formed by the variables $u_{1,1}$, $u_{1,2}$, $u_{2,1}$, *and* $u_{2,2}$ *(see* (8.9)*). The variables of the optimization space are in general continuous. However, we only analyze six simple configurations, where* $u_{q,k} \in \{0, 1\}$ *as well as* BICM-S, *i.e.,* $u_{q,k} = 1/3$. *We present in Fig. 8.9 the results obtained for the best and worst MUX as well as for the S-interleaver. The best (or worst) BICM-M was found by selecting the matrix* $\mathbf{U}$ *that minimizes (maximizes) the BEP approximations at a target BEP of* $\mathsf{BEP} = 10^{-6}$*: however, we noted that varying* BEP *to any value between* 10^{-7} *and* 10^{-4} *does not change the conclusions about the best (or worst) BICM-M. For this particular encoder, the matrices found are*

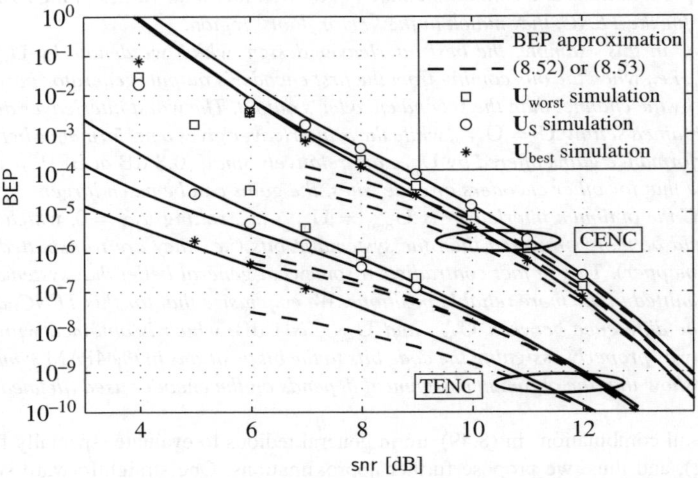

Figure 8.9 BEP approximation in (8.22) based on Theorem 8.8 (lines) and simulated BEP (markers) for BICM-M for the rate $R_c = 1/2$ ODS CENC $\mathbf{G} = [23, 35]$ from Table 2.1 ($\nu = 4$) and for the TENC in Example 2.31 for $R_c = 1/3$ and 8PAM ($k_c = 1$, $n_c = m = 3$). The interleaver size of the TENC Π_{tc} is $N_b = 1500$ and 10 iterations are performed by the turbo decoder. The asymptotic bounds based on (8.52) for the TENC and on (8.53) for the CENC are also shown

[5] We used $d_{\max} = 50$ for the TENC and $d_{\max} = 25$ for the CENC.

$$\mathbf{U}_{\text{best}} = \begin{bmatrix} 0 & 0 & 1 \\ 0 & 1 & 0 \\ 1 & 0 & 0 \end{bmatrix}, \quad \mathbf{U}_{\text{worst}} = \begin{bmatrix} 1 & 0 & 0 \\ 0 & 1 & 0 \\ 0 & 0 & 1 \end{bmatrix}. \tag{8.54}$$

The results in Fig. 8.9 again show that the BEP approximations match the simulation results and that for the TENC and a target BEP of 10^{-6} *there is difference of approximately* 2 dB *between* $\mathbf{U}_{\text{worst}}$ *and* $\mathbf{U}_{\text{best}}$.

In Fig. 8.9 we also present the BEP approximations in (8.52) *and* (8.53). *For the TENC we used* (8.52) *with* $d_{\max} = 50$ *and for the CENC we used* (8.53). *The computations for these bounds are very simple compared with* (8.45), *and yet they predict the asymptotic performance of the system as shown in Fig. 8.9.*

From the results presented in Figs. 8.8 and 8.9, we can draw some conclusions:

- *For a given target BEP of* 10^{-6}, *the SNR gains between the best and the worst interleaver configuration are between some tenths of* dB *and up to* 2 dB *(see the TENC in Fig. 8.9).*
- *The BEP approximation in* (8.22) *based on* (8.45) *is tight for BEP values less than* 10^{-3} *for the CENC and for the error floor region of the TENC, while* (8.52) *and* (8.53) *can be used to predict the asymptotic performance of a TENC and a CENC, respectively.*
- *For the analyzed cases, an optimized M-interleaver is always better than the S-interleaver and an improperly designed M-interleaver (*$\mathbf{U}_{\text{worst}}$*) can degrade the system performance compared to* $\mathbf{U}_{\text{S}}$. *Therefore, when using BICM-M, optimization of the MUX becomes a mandatory step.*
- *Since the notion of "best" and "worst" refers only to the constrained set of matrices with integer entries (*$u_{q,k} \in \{0, 1\}$*),* $\mathbf{U}_{\text{S}}$ *having fractional entries can be worse than* $\mathbf{U}_{\text{worst}}$ *(see, e.g., the CENC in Fig. 8.9). Therefore, S-interleavers cannot be considered, in general, a "conservative" solution between* $\mathbf{U}_{\text{best}}$ *and* $\mathbf{U}_{\text{worst}}$.

Analyzing the approximation of the BEP expression (8.22) together with (8.45), we can appreciate three terms: $C^{\mathcal{C}}_{\boldsymbol{w}}$, which depends only on the encoder; $Q(\sqrt{\mathsf{snr}} h(\boldsymbol{r}_1, \ldots, \boldsymbol{r}_{n_c}))$, which depends only on the channel (see (8.46)); and $g(\boldsymbol{r}_1, \ldots, \boldsymbol{r}_{n_c})$, which depends on the interleaver (see (8.36)). Assuming that the constellation and the labeling are fixed, the optimum performance of the system will be achieved by a joint design of the interleaver *and* the encoder. We study this in the following section.

8.2.3 Joint Encoder and Interleaver Design

It is well known that ODS CENCs in Table 2.1 are the optimum CENCs for binary transmission. However, according to (8.53), when UEP is introduced by the channel, the optimization criterion is different, i.e., a joint optimization of the MUX (i.e., how the code bits are assigned to different bit positions in the constellation) and the CENC should be done. More specifically, the expression in (8.53) shows that we first need to optimize the *free Hamming distance* (FHD) of the encoder and then we need to optimize the GIWD of the encoder and the MUX matrix $\mathbf{U}$. To formally define the optimum design for a given constraint length ν and code rate R_c, we define $\mathcal{G}^{\text{MFHD}}_{k_c,n_c,\nu}$ as the set of encoders in $\mathcal{G}_{k_c,n_c,\nu}$ that give MFHD, where $\mathcal{G}_{k_c,n_c,\nu}$ is the CENC universe defined in Section 2.6.1.

Definition 8.11 (Optimal BICM-M) *A CENC* $\mathbf{G}^* \in \mathcal{G}^{\text{MFHD}}_{k_c,n_c,\nu}$ *and a MUX* $\mathbf{U}^*$ *are said to be optimum if they produce an asymptotic BEP that is a minimum compared to the values that any other encoder and MUX combination can generate, i.e.,*

$$[\mathbf{G}^*, \mathbf{U}^*] \triangleq \underset{\substack{\mathbf{G}\in\mathcal{G}^{\text{MFHD}}_{k_c,n_c,\nu} \\ \mathbf{U}\in\mathbb{R}^{(n_c-1)\times(m-1)}}}{\operatorname{argmin}} \left\{ \sum_{\substack{\boldsymbol{w}\in\mathbb{N}^{n_c} \\ \|\boldsymbol{w}\|_1 = d_{\text{free}}}} G^{\mathcal{C}}_{\boldsymbol{w}} \prod_{q=1}^{n_c} \xi_{q,1}^{d_q} \right\}. \tag{8.55}$$

Note that for given values of n_c and m, the problem of selecting the optimum interleaver configuration (selection of $\mathbf{U}$) is a continuous multidimensional optimization problem. For simplicity, however,

Table 8.1 Optimum CENC and *multiplexers* (MUXs) for $R_c = 1/2$ wit 4PAM, 8PAM, and 16PAM. New encoders found, better than the ODS CENCs in Table 2.1, are highlighted

		4PAM ($m=2$)		8PAM ($m=3$)			16PAM ($m=4$)			
ν	d_{free}	$\mathbf{G}^*$	$u^*_{1,1}$	$\mathbf{G}^*$	$u^*_{1,1}$	$u^*_{1,2}$	$\mathbf{G}$	$u^*_{1,1}$	$u^*_{1,2}$	$u^*_{1,3}$
2	5	$[5,7]$	0	$[5,7]$	0	1/3	$[5,7]$	1/2	1/2	0
3	6	$[15,17]$	1	$[15,17]$	2/3	1/3	$[15,17]$	1/2	1/2	0
4	7	$[23,35]$	1	$[27,31]$	0	1/3	$[23,35]$	1/2	1/2	0
5	8	$[53,75]$	0	$[53,75]$	0	1/3	$[53,75]$	1/2	1/2	0
6	10	$[133,171]$	1	$[\mathbf{135,147}]$	0	1/3	$[\mathbf{135,147}]$	0	0	1/2
7	10	$[247,371]$	1	$[\mathbf{225,373}]$	0	1/3	$[247,371]$	1/2	1/2	0
8	12	$[\mathbf{515,677}]$	0	$[\mathbf{557,751}]$	0	1/3	$[\mathbf{457,755}]$	1/2	1/2	0
9	12	$[1151,1753]$	0	$[1151,1753]$	0	1/3	$[1151,1753]$	1/2	1/2	0

the optimization is performed over only a limited number of points. Using Definition 8.11, an exhaustive search for pairs $[\mathbf{G}^*, \mathbf{U}^*]$ with constraint length up to $\nu = 9$ was performed. Three different configurations were tested: code rate $R_c = 1/2$ ($k_c = 1$, $n_c = 2$) with 4PAM ($m = 2$), 8PAM ($m = 3$), or 16PAM ($m = 4$). The (simplified) optimization space for $\mathbf{U}$ in these cases was $u_{1,1} \in \{0, 1/2, 1\}$ for $m = 2$, $u_{1,1}, u_{1,2} \in \{0, 1/3, 2/3\}$ for $m = 3$, and $u_{1,1}, u_{1,2}, u_{1,3} \in \{0, 1/2, 1\}$ for $m = 4$. The results are presented in Table 8.1, where we highlight CENCs found that are different (in terms of their IWD) from the ODS CENCs in Table 2.1. Among the 24 combinations studied, 6 resulted in new optimal encoders.

The cost function in (8.55) is shown in Fig. 8.10 as a function of the interleaver parameter $u_{1,1}$ for $R_c = 1/2$, 4PAM, and $\nu = 8$. The ODS CENC [561, 753] is shown with a thick black line. Analyzing this curve, it is clear that the performance of this encoder can be optimized by setting $u_{1,1} = 1$, and that the curve has a maximum for $u_{1,1} = 0.4$, which will result in the worst interleaver design for this

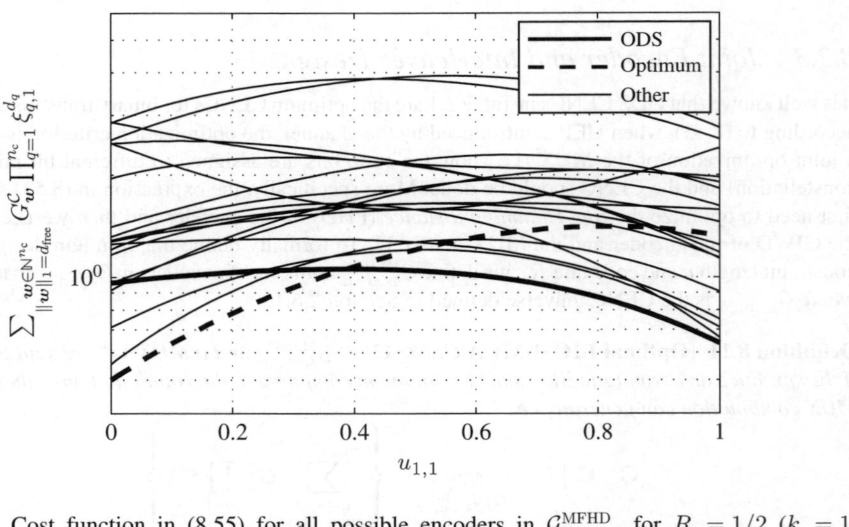

Figure 8.10 Cost function in (8.55) for all possible encoders in $\mathcal{G}^{\text{MFHD}}_{k_c,n_c,\nu}$ for $R_c = 1/2$ ($k_c = 1$, $n_c = 2$), 4PAM, and $\nu = 8$ as a function of the MUX parameter $u_{1,1}$. The thick solid line represents the ODS CENC $\mathbf{G} = [561, 753]$, and the thick dashed line the optimum BICM-M design based on $\mathbf{G}^* = [515, 677]$

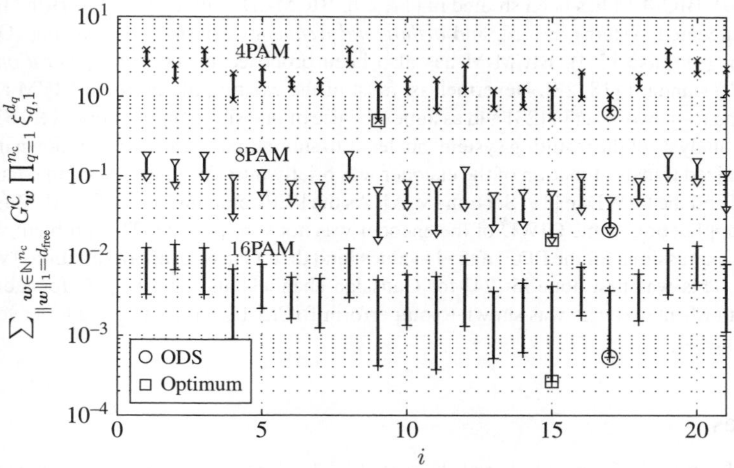

Figure 8.11 Cost function in (8.55) for all the 21 possible encoders in $\mathcal{G}_{1,2,8}^{\text{MFHD}}$, for the best and worst interleaver design: $m = 2$ ('×'), $m = 3$ ('∇'), and $m = 4$ ('+'). The lines represent the range of variation between the best and the worst interleaver design

particular encoder. The cost function obtained for the encoder $\mathbf{G}^* = [515, 677]$ (thick dashed line) attains the smallest value among all other encoders (including the ODS one). Consequently, if the MUX is adequately designed setting $u_{1,1} = 0$ (best BICM-M), this encoder is the optimal encoder for this particular configuration. However, if the interleaver is not optimized, e.g., setting $u_{1,1} = 1/2$ (BICM-S), the new encoder is not optimal anymore.

We conclude by showing in Fig. 8.11 the performance of the optimum design $[\mathbf{G}^*, \mathbf{U}^*]$ compared with all encoders with $\nu = 8$ and $d_{\text{free}} = 12$ (enumerated using the variable i) using the best and the worst interleaver design ($\mathbf{U}_{\text{best}}$ and $\mathbf{U}_{\text{worst}}$). The lines represent the range of variation between the best and the worst interleaver design, i.e., any other interleaver configuration will have a coefficient between the corresponding pair of markers. We note that the optimum design may significantly outperform other encoders, e.g., 16PAM and $i = 15$ in Fig. 8.11. The improvement with respect to ODS CENCs is less evident but clear. Thus, the results presented in this section indicate that optimizing the interleaver and encoder should be a mandatory step in the design of BICM-M.

8.3 Bibliographical Notes

UEP in terms of BEP for uncoded transmission with MPSK and MQAM constellations labeled by the BRGC has been studied in [1] and for 16QAM and 64QAM in [2, Figs. 5.2 and 5.4]. Unequal power allocation for systematic/parity bits to impose UEP is an idea first used for turbo-encoded BICM in [3] and later analyzed in [4–8]. UEP for turbo-encoded schemes has been studied in [3, 4, 6, 8–10]. The influence of the block length and code rate for optimal power allocation was analyzed in [4, 11] and interleaver design aiming to assign the code bits to different bit positions for high-order modulation schemes was studied in [12].

The BICM transceivers with M-interleaver we analyzed in this chapter in fact correspond to the original model introduced by Zehavi in [13] for BICM (and also for *BICM with iterative demapping* (BICM-ID) in [14]), where the application of parallel interleavers was postulated. Over the years, different names have been given to these interleavers: e.g., "in-line" 15, "intralevel" [16], "M" [17], "dual" [18], or "modular" [19]. BICM-M has been studied in [17, 20], BICM-ID with M-interleavers in [14], BICM for serially concatenated systems in [15], and *orthogonal frequency-division multiplexing* (OFDM)-based BICM in [16] (see also [21]). BICM-M has also been proposed in the *third-generation partnership project* (3GPP) standard [18, 22]. Nevertheless, most of the existing literature on BICM and BICM-ID still follows the framework set in [23] and assumes the use of one singe-input interleaver (BICM-S). This simplifies the analysis of the resulting system, but leads to suboptimality, as we have shown in this chapter.

The generalized transfer function of the encoder was briefly introduced in the original paper of Zehavi [13, eq. (4.8)] and the GIWD in [19, Section IV-A]. As the PEP computation for BICM-M is not straightforward, the application of the GIWD of the encoder was not very popular. This problem was solved by the analytical expressions for the PDF of the L-values (and their Gaussian approximations) we introduced in Chapter 5. This was first done in [24] for Gaussian channels and later in [25] for fading channels. BICM-ID with M-interleavers was shown to outperform BICM-ID with a single interleaver in [26] (see also [27, Chapter 4]).

References

[1] Lassing, J., Ström, E. G., Agrell, E., and Ottosson, T. (2003) Unequal bit-error protection in coherent M-ary PSK. IEEE Vehicular Technology Conference (VTC-Fall), October 2003, Orlando, FL.

[2] Hanzo, L., Webb, W., and Keller, T. (2000) *Single- and Multi-Carrier Quadrature Amplitude Modulation*, John Wiley & Sons.

[3] Hokfelt, J. and Maseng, T. (1996) Optimizing the energy of different bitstreams of turbo code. International Turbo Coding Seminar, August 1996, Lund, Sweden.

[4] Salah, M. M., Raines, R. A., Temple, M. A., and Bailey, T. G. (2000) Energy allocation strategies for turbo codes with short frames. The International Conference on Information Technology: Coding and Computing (ITCC), March 2000, Las Vegas, NV.

[5] Zhang, W. and Wang, X. (2004) Optimal energy allocations for turbo codes based on distributions of low weight codewords. *Electron. Lett.*, **20** (19), 1205–1206.

[6] Duman, T. and Salehi, M. (1997) On optimal power allocation for turbo codes. International Symposium on Information Theory (ISIT), June 1997, Ulm, Germany.

[7] Mohammadi, A. H. S. and Khandani, A. K. (1997) Unequal error protection on the turbo-encoder output bits. IEEE International Conference on Communications (ICC), June 1997, Montreal, QC, Canada.

[8] Mohammadi, A. H. S. and Khandani, A. K. (1999) Unequal power allocation to the turbo-encoder output bits with applications to CDMA systems. *IEEE Trans. Commun.*, **47** (11), 1609–1610.

[9] Aydinlik, M. and Salehi, M. (2008) Turbo coded modulation for unequal error protection. *IEEE Trans. Commun.*, **56** (4), 555–564.

[10] Kousa, M. A. and Mugaibel, A. H. (2002) Puncturing effects on turbo codes. *Proc. IEE*, **149** (3), 132–138.

[11] Choi, Y. M. and Lee, P. J. (1999) Analysis of turbo codes with symmetric modulation. *Electron. Lett.*, **35** (1), 35–36.

[12] Rosnes, E. and Ytrehus, O. (2006) On the design of bit-interleaved turbo-coded modulation with low error floors. *IEEE Trans. Commun.*, **54** (9), 1563–1573.

[13] Zehavi, E. (1992) 8-PSK trellis codes for a Rayleigh channel. *IEEE Trans. Commun.*, **40** (3), 873–884.

[14] Li, X. and Ritcey, J. A. (1997) Bit-interleaved coded modulation with iterative decoding. *IEEE Commun. Lett.*, **1** (6), 169–171.

[15] Nilsson, A. and Aulin, T. M. (2005) On in-line bit interleaving for serially concatenated systems. IEEE International Conference on Communications (ICC), May 2005, Seoul, Korea.

[16] Stierstorfer, C. and Fischer, R. F. H. (2007) Intralevel interleaving for BICM in OFDM scenarios. 12th International OFDM Workshop, August 2007, Hamburg, Germany.

[17] Abramovici, I. and Shamai, S. (1999) On turbo encoded BICM. *Ann. Telecommun.*, **54** (3–4), 225–234.

[18] Dahlman, E., Parkvall, S., Sköld, J., and Beming, P. (2008) *3G Evolution: HSPA and LTE for Mobile Broadband*, 2nd edn, Academic Press.

[19] Li, X., Chindapol, A., and Ritcey, J. A. (2002) Bit-interlaved coded modulation with iterative decoding and 8PSK signaling. *IEEE Trans. Commun.*, **50** (6), 1250–1257.

[20] Hansson, U. and Aulin, T. (1996) Channel symbol expansion diversity—improved coded modulation for the Rayleigh fading channel. IEEE International Conference on Communications (ICC), June 1996, Dallas, TX.

[21] Stierstorfer, C. (2009) A bit-level-based approach to coded multicarrier transmission. PhD dissertation, Friedrich-Alexander-Universität Erlangen-Nürnberg, Erlangen, Germany.

[22] 3GPP (2009) Universal mobile telecommunications system (UMTS); multiplexing and channel coding (FDD). Technical Report TS 25.212, V8.5.0 Release 8, 3GPP.

[23] Caire, G., Taricco, G., and Biglieri, E. (1998) Bit-interleaved coded modulation. *IEEE Trans. Inf. Theory*, **44** (3), 927–946.

[24] Alvarado, A., Agrell, E., Szczecinski, L., and Svensson, A. (2010) Exploiting UEP in QAM-based BICM: Interleaver and code design. *IEEE Trans. Commun.*, **58** (2), 500–510.

[25] Hossain, Md. J., Alvarado, A., and Szczecinski, L. (2011) Towards fully optimized BICM transceivers. *IEEE Trans. Commun.*, **59** (11), 3027–3039.

[26] Alvarado, A., Szczecinski, L., Agrell, E., and Svensson, A. (2010) On BICM-ID with multiple interleavers. *IEEE Commun. Lett.*, **14** (9), 785–787.

[27] Alvarado, A. (2010) Towards fully optimized BICM transmissions. PhD dissertation, Chalmers University of Technology, Göteborg, Sweden.

9

BICM Receivers for Trellis Codes

In this chapter we study a trellis encoder and a *bit-interleaved coded modulation* (BICM) receiver, i.e., the *coded modulation* (CM) encoder is a serial concatenation of a *convolutional encoder* (CENC) and a memoryless mapper, and the bitwise interleaver is not present. This in fact corresponds to the model in Fig. 2.4, where the *maximum likelihood* (ML) decoder is replaced by a BICM decoder. The study of this somehow unusual combination is motivated by numerical results that show that for the *additive white Gaussian noise* (AWGN) channel and *convolutional encoders* (CENCs), removing the interleaver in a BICM transceiver results in improved performance.

This chapter is organized as follows. In Section 9.1 we introduce the idea of BICM with "trivial interleavers" (i.e., no interleaving) and in Section 9.2 we study the optimal selection of CENCs for this configuration. Motivated by the results in Section 9.2, we study binary labelings for trellis encoders in Section 9.3.

9.1 BICM with Trivial Interleavers

The design philosophies behind *trellis-coded modulation* (TCM) and BICM for the AWGN channel are quite different. On one hand, TCM is constructed coupling together a CENC and a constellation using a *set-partitioning* (SP) labeling. On the other hand, BICM is typically a concatenation of a binary encoder, a bit-level interleaver, and a mapper with constellation labeled by the *binary reflected Gray code* (BRGC). The BRGC is often used in BICM because it maximizes the *bit-interleaved coded modulation generalized mutual information* (BICM-GMI) for medium and high signal-to-noise ratios, as we showed in Section 4.3. In TCM, the selection of the CENC is done so that the *minimum Euclidean distance* (MED) of the resulting trellis code is maximized, while in BICM the CENCs are designed for binary transmission, i.e., to have *maximum free Hamming distance* (MFHD) or—more generally—to have *optimal distance spectrum* (ODS).

In this chapter we consider the BICM transceiver where the interleaver Π in Fig. 2.7 (see also Fig. 8.2) is not present. Equivalently, we might say that the interleaving vector defined in Section 2.7 is "trivial," i.e., $\boldsymbol{\pi} = [1, 2, \ldots, N_{\mathrm{c}}]$. Owing to this interpretation, we will refer to this transceiver as *BICM with trivial interleavers* (BICM-T). For clarity of presentation, we consider the simplest 1D case, i.e., a rate $R_{\mathrm{c}} = 1/2$ ($k_{\mathrm{c}} = 1$ and $n_{\mathrm{c}} = 2$) CENC and a (4PAM) ($m = 2$) constellation labeled by the BRGC over the AWGN channel. The resulting transceiver is shown in Fig. 9.1. Furthermore, we do not consider a multiplexer block (as we did in Chapter 8), as in this case the two most relevant cases can be obtained by swapping the order of the generator polynomials of the encoder.

Bit-Interleaved Coded Modulation: Fundamentals, Analysis, and Design, First Edition.
Leszek Szczecinski and Alex Alvarado.

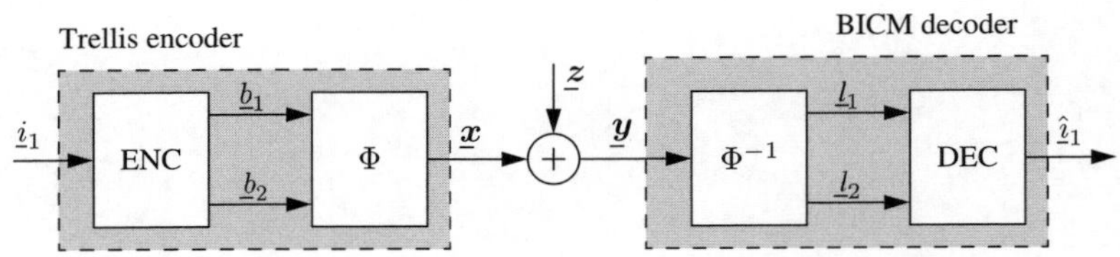

Figure 9.1 BICM-T system under consideration: a trellis encoder and a BICM decoder

A quick examination of Fig. 9.1 reveals that the structure of the transmitter of BICM-T corresponds to the trellis encoder shown in Fig. 2.4. This type of encoders are typically used with a decoder that implements the ML detection rule in Definition 3.2. In BICM-T, however, the receiver in Fig. 9.1 is composed of a demapper and a binary decoder, i.e., it is a suboptimal decoder that implements the BICM decoding rule in Definition 3.4.

When the (nontrivial) interleaver is present (e.g., in BICM-S or BICM-M), the use of a BICM decoder is mandatory as there is no efficient way of implementing the ML rule. Nevertheless, the receiver in Fig. 9.1 uses the BICM decoding rule, even though the interleaver is not present anymore. As we will see later, this in fact gives an advantage over BICM-S. Therefore, the receiver of BICM-T in Fig. 9.1 corresponds to a conventional BICM receiver, where L-values for each bit are computed and then fed to a *binary* decoder. Here we consider a CENC and a binary decoder based on the Viterbi algorithm.

9.1.1 PDF of the L-values in BICM-T

At each time instant n, two bits $b_1[n]$ and $b_2[n]$ are transmitted using the same symbol $\boldsymbol{x}[n]$. The two corresponding L-values $l_1[n]$ and $l_2[n]$ are then obtained from the same observation $\boldsymbol{y}[n]$, and thus, L_1 and L_2 are dependent. In principle, we might design a decoder, which is aware of the fact that the L-values are not independent in BICM-T. However, this would go against the BICM principle of using a binary decoder operating "blindly" on the L-values and we thus continue to use the BICM decoder, which applies the decision rules defined in Section 6.2.4.

In particular, the decoding metric (6.183) for BICM-T is expressed as

$$\mathrm{M}(\underline{\boldsymbol{b}}, \underline{\boldsymbol{l}}) = \sum_{n=1}^{N_s} \sum_{k=1}^{2} b_k[n] l_k[n], \tag{9.1}$$

where because of the absence of the interleaver $\underline{\boldsymbol{l}} = \underline{\boldsymbol{\lambda}}$ and $\underline{\boldsymbol{b}} = \underline{\boldsymbol{c}}$.

The *pairwise-error probability* (PEP) in (6.188) for the metric in (9.1) is then given by

$$\mathrm{PEP}(\underline{\boldsymbol{b}}, \hat{\underline{\boldsymbol{b}}}) = \Pr\left\{ \sum_{n=1}^{N_s} L[n] \geq 0 | \underline{\boldsymbol{B}} = \underline{\boldsymbol{b}} \right\}, \tag{9.2}$$

where

$$L[n] \triangleq \sum_{k=1}^{2} (-1)^{b_k[n]} e_k[n] L_k[n], \tag{9.3}$$

and

$$e_k[n] = \hat{b}_k[n] \oplus b_k[n] \tag{9.4}$$

indicates that the L-value $L_k[n]$ contributes to the PEP only for the values of n where $\hat{b}_k[n] \neq b_k[n]$. Because the channel is memoryless, the metrics in (9.3) for different time instants n are independent, and thus, without loss of generality, from now on we drop the time index n.

Table 9.1 Possible values of the metric L in (9.3)

	$\hat{\boldsymbol{b}} = [0,0]^{\mathrm{T}}$	$\hat{\boldsymbol{b}} = [0,1]^{\mathrm{T}}$	$\hat{\boldsymbol{b}} = [1,1]^{\mathrm{T}}$	$\hat{\boldsymbol{b}} = [1,0]^{\mathrm{T}}$
$\boldsymbol{b} = [0,0]^{\mathrm{T}}$	0	L_2	$L_1 + L_2$	L_1
$\boldsymbol{b} = [0,1]^{\mathrm{T}}$	$-L_2$	0	L_1	$L_1 - L_2$
$\boldsymbol{b} = [1,1]^{\mathrm{T}}$	$-L_1 - L_2$	$-L_1$	0	$-L_2$
$\boldsymbol{b} = [1,0]^{\mathrm{T}}$	$-L_1$	$-L_1 + L_2$	L_2	0

Depending on values of $\boldsymbol{b}$ and $\hat{\boldsymbol{b}}$, the metric L in (9.3) is calculated as specified in Table 9.1. We will now compute the *probability density function* (PDF) of L in (9.3) for the 12 cases in which $e_1 = 1$ or $e_2 = 1$.

We start with the simplest case, i.e., $\pm L_1$ and $\pm L_2$. We use the *zero-crossing model* (ZcM) from Section 5.5.2 to approximate the PDF of the L-values, and thus, we can use the results in Example 5.19 for the cases $\boldsymbol{e} = [0,1]^{\mathrm{T}}$ and $\boldsymbol{e} = [1,0]^{\mathrm{T}}$. In particular, from (5.198), we obtain

$$p_{L_2|\boldsymbol{X}}(l|-3\Delta) = p_{L_2|\boldsymbol{X}}(-l|-\Delta) = p_{L_2|\boldsymbol{X}}(-l|\Delta) = p_{L_2|\boldsymbol{X}}(l|3\Delta) = \Psi(l, -4\gamma, 8\gamma), \tag{9.5}$$

and from (5.197) we obtain

$$p_{L_1|\boldsymbol{X}}(l|-3\Delta) = p_{L_1|\boldsymbol{X}}(-l|3\Delta) = \Psi(l, -12\gamma, 8\gamma), \tag{9.6}$$

$$p_{L_1|\boldsymbol{X}}(l|-\Delta) = p_{L_1|\boldsymbol{X}}(-l|\Delta) = \Psi(l, -4\gamma, 8\gamma), \tag{9.7}$$

where $\gamma = \mathsf{snr}\,\Delta^2$ (see (5.7)).

The expressions (9.5)–(9.7) give the parameters of 8 out of the 12 PDFs. These parameters are shown in Table 9.2 (not highlighted). In what follows, we consider the problem of computing the remaining four PDFs in Table 9.1, i.e., the PDFs $p_{L_1+L_2|\boldsymbol{X}}(l|-3\Delta)$, $p_{L_1-L_2|\boldsymbol{X}}(l|-\Delta)$, $p_{-L_1-L_2|\boldsymbol{X}}(l|\Delta)$, and $p_{-L_1+L_2|\boldsymbol{X}}(l|3\Delta)$.

As shown in Section 5.1.2, the max-log L-values are piece-wise linear functions of $\boldsymbol{Y}$, and thus, the metrics L in (9.3), being a linear combinations of L_1 and L_2, are also piece-wise linear functions of $\boldsymbol{Y}$. In Fig. 9.2, we show L in (9.3) for the four possible cases when $\boldsymbol{e} = [1,1]^{\mathrm{T}}$. These functions are obtained by linearly combining the piece-wise functions in Fig. 5.10 (see also (5.53) and (5.54)).

For a given transmitted symbol $\boldsymbol{X} = \boldsymbol{x}$, the received signal $\boldsymbol{Y}$ is a Gaussian random variable with mean $\boldsymbol{x}$ and variance $\mathsf{N}_0/2$. Therefore, the PDF of L in (9.3) is a sum of piece-wise truncated Gaussian functions. In order to obtain expressions that are easy to work with, we use again the ZcM, which replaces all the Gaussian pieces by a single Gaussian function. To clarify this, we show in Fig. 9.2 the linear

Table 9.2 Mean values and variances of the Gaussian approximation for L in Table 9.1. Eight out of the 12 relevant cases follow directly from Example 5.19. The remaining four (highlighted) are given by (9.9)–(9.10). To indicate that the Gaussian distributions correspond to the same $\boldsymbol{e} = \boldsymbol{b} \oplus \hat{\boldsymbol{b}}$, we use circles (for $\boldsymbol{e} = [0,1]^{\mathrm{T}}$), stars (for $\boldsymbol{e} = [1,1]^{\mathrm{T}}$), and diamonds (for $\boldsymbol{e} = [1,0]^{\mathrm{T}}$)

	$\hat{\boldsymbol{b}} = [0,0]^{\mathrm{T}}$	$\hat{\boldsymbol{b}} = [0,1]^{\mathrm{T}}$	$\hat{\boldsymbol{b}} = [1,1]^{\mathrm{T}}$	$\hat{\boldsymbol{b}} = [1,0]^{\mathrm{T}}$
$\boldsymbol{b} = [0,0]^{\mathrm{T}}$	—	$(-4\gamma, 8\gamma)^{\circ}$	$(\boldsymbol{-16\gamma, 32\gamma})^{*}$	$(-12\gamma, 8\gamma)^{\diamond}$
$\boldsymbol{b} = [0,1]^{\mathrm{T}}$	$(-4\gamma, 8\gamma)^{\circ}$	—	$(-4\gamma, 8\gamma)^{\diamond}$	$(\boldsymbol{-16\gamma, 32\gamma})^{*}$
$\boldsymbol{b} = [1,1]^{\mathrm{T}}$	$(\boldsymbol{-16\gamma, 32\gamma})^{*}$	$(-4\gamma, 8\gamma)^{\diamond}$	—	$(-4\gamma, 8\gamma)^{\circ}$
$\boldsymbol{b} = [1,0]^{\mathrm{T}}$	$(-12\gamma, 8\gamma)^{\diamond}$	$(\boldsymbol{-16\gamma, 32\gamma})^{*}$	$(-4\gamma, 8\gamma)^{\circ}$	—

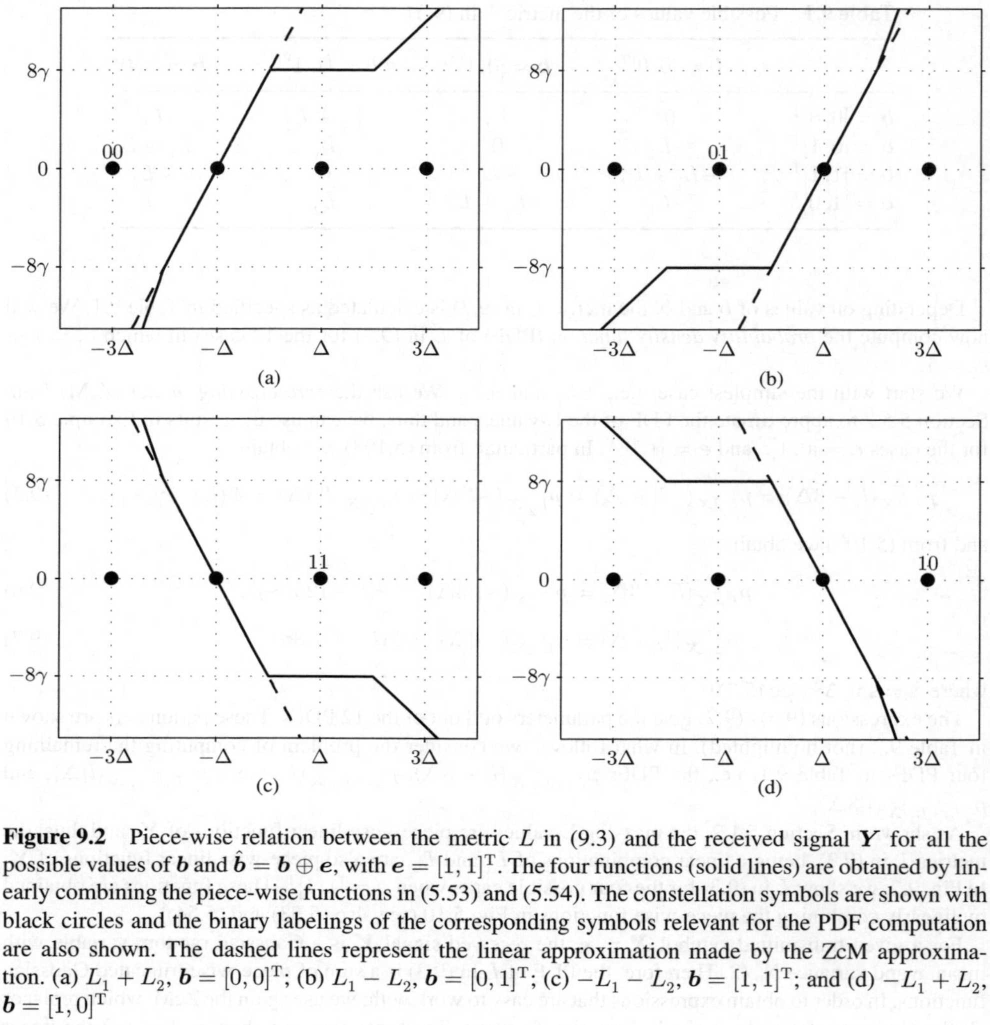

Figure 9.2 Piece-wise relation between the metric L in (9.3) and the received signal $\boldsymbol{Y}$ for all the possible values of $\boldsymbol{b}$ and $\hat{\boldsymbol{b}} = \boldsymbol{b} \oplus \boldsymbol{e}$, with $\boldsymbol{e} = [1, 1]^{\mathrm{T}}$. The four functions (solid lines) are obtained by linearly combining the piece-wise functions in (5.53) and (5.54). The constellation symbols are shown with black circles and the binary labelings of the corresponding symbols relevant for the PDF computation are also shown. The dashed lines represent the linear approximation made by the ZcM approximation. (a) $L_1 + L_2$, $\boldsymbol{b} = [0, 0]^{\mathrm{T}}$; (b) $L_1 - L_2$, $\boldsymbol{b} = [0, 1]^{\mathrm{T}}$; (c) $-L_1 - L_2$, $\boldsymbol{b} = [1, 1]^{\mathrm{T}}$; and (d) $-L_1 + L_2$, $\boldsymbol{b} = [1, 0]^{\mathrm{T}}$

approximation the ZcM does. For example, for $L_1 + L_2$ (i.e., $\boldsymbol{b} = [0, 0]^{\mathrm{T}}$, and $\hat{\boldsymbol{b}} = [1, 1]^{\mathrm{T}}$), we have

$$L_1 + L_2 \approx \frac{8\gamma}{\Delta}\boldsymbol{Y} + 8\gamma, \tag{9.8}$$

and because $\boldsymbol{Y} \sim \mathcal{N}(\boldsymbol{x}, \mathsf{N}_0/2)$, we obtain

$$p_{L_1+L_2|\boldsymbol{B}}(l|[0, 0]^{\mathrm{T}}) \approx \Psi(l, -16\gamma, 32\gamma). \tag{9.9}$$

A similar analysis can be done for the remaining three cases, yielding the same result, i.e.,

$$p_{L_1-L_2|\boldsymbol{B}}(l|[0, 1]^{\mathrm{T}}) = p_{-L_1-L_2|\boldsymbol{B}}(l|[1, 1]^{\mathrm{T}}) = p_{-L_1+L_2|\boldsymbol{B}}(l|[1, 0]^{\mathrm{T}}) \approx \Psi(l, -16\gamma, 32\gamma). \tag{9.10}$$

The parameters of the Gaussian PDFs in (9.9) and (9.10) are the ones we highlight in Table 9.2. This completes the task of finding the 12 densities for the metrics in Table 9.1.

9.1.2 *Performance Analysis*

Now that we have analytical approximations for the PDF of L in (9.3), we return to the problem of computing the PEP in (9.2). First, from the results in Table 9.2 we see that there are three Gaussian PDFs with different parameters, where the number of Gaussian functions needed to evaluate the PEP in (9.2) depends on the transmitted codeword $\underline{\boldsymbol{b}}$ and the competing codeword $\hat{\underline{\boldsymbol{b}}}$. In principle, this is not a problem, and the PEP can be computed for each possible pair $(\underline{\boldsymbol{b}}, \hat{\underline{\boldsymbol{b}}})$. However, the disadvantage of such an approach is that the final PEP expression will have to consider a distance spectrum of the CENC that takes into account all possible pairs of coded sequences that leave a trellis state and remerge after an arbitrary number of trellis stages. In other words, in such a case, the *generalized input-dependent weight distribution* (GIWD) we considered in Section 8.1.4, which assumes the all-zero codeword was transmitted, cannot be used. This is not surprising as the trellis encoder we consider now is not linear.

To overcome the problem described above, we assume the pseudorandom scrambler in Fig. 3.3 is used. At each time instant n, four scrambler outcomes are possible: $\boldsymbol{s} = [0,0]^{\mathrm{T}}$, $\boldsymbol{s} = [0,1]^{\mathrm{T}}$, $\boldsymbol{s} = [1,0]^{\mathrm{T}}$, and $\boldsymbol{s} = [1,1]^{\mathrm{T}}$. For any given pair $(\boldsymbol{b}, \hat{\boldsymbol{b}})$, the distribution of the nth metric contribution is given in Table 9.2. It can be shown that for each scrambler outcome, the distribution of the nth metric contribution $L[n]$ changes over the four distributions shown with the same marker in Table 9.2. For example, for $\boldsymbol{b} = [0,1]^{\mathrm{T}}$ and $\hat{\boldsymbol{b}} = [0,0]^{\mathrm{T}}$, we see that the nth metric contribution is distributed as $\Psi(l, -4\gamma, 8\gamma)$, which is marked with a circle. For $\boldsymbol{s} = [0,0]^{\mathrm{T}}$ no additional analysis is necessary (the bits are not scrambled), but if $\boldsymbol{s} = [0,1]^{\mathrm{T}}$, then $\boldsymbol{b} = [0,0]^{\mathrm{T}}$ and $\hat{\boldsymbol{b}} = [0,1]^{\mathrm{T}}$, and thus, the nth metric contribution is given by the entry $(1,2)$ in the table (marked with a circle), i.e., $\Psi(l, -4\gamma, 8\gamma)$. If $\boldsymbol{s} = [1,0]^{\mathrm{T}}$ the relevant entry in the table is $(3,4)$ and if $\boldsymbol{s} = [1,1]^{\mathrm{T}}$ the entry is $(4,3)$. The key observation here is that all these four densities (marked with a circle) correspond to the same binary difference $\boldsymbol{e}$. In the case we just described, $\boldsymbol{e} = [0,1]^{\mathrm{T}}$.

A similar analysis can be done for other cases. For $\boldsymbol{e} = [1,1]^{\mathrm{T}}$, all the four distributions marked with a star are covered by changing the scrambler outcome. In this case, the distribution is always given by $\Psi(l, -16\gamma, 32\gamma)$.

When $\boldsymbol{e} = [1,0]^{\mathrm{T}}$ is considered, we have $L[n] = \pm L_1[n]$, and then, *unequal error protection* (UEP) effect is apparent: in two out of the four cases the distribution $\Psi(l, -4\gamma, 8\gamma)$ is obtained (low protection), while in the other two cases we obtain the distribution $\Psi(l, -12\gamma, 8\gamma)$ (high protection). This means that while scrambler does not affect the distribution for the cases $\boldsymbol{e} = [0,1]^{\mathrm{T}}$ and $\boldsymbol{e} = [1,1]^{\mathrm{T}}$, it does affect the case $\boldsymbol{e} = [1,0]^{\mathrm{T}}$.

We can summarize the previous discussion by expressing the PDF of the metric in (9.3) (after marginalization over the scrambler outcomes) conditioned on the random variable $\boldsymbol{E}$ modeling the error $\boldsymbol{e} = \boldsymbol{b} \oplus \hat{\boldsymbol{b}}$, i.e.,

$$p_{L|\boldsymbol{E}}(l|\boldsymbol{e}) = \begin{cases} p_{L_1|B_1}(l|0), & \text{if } \boldsymbol{e} = [1,0]^{\mathrm{T}} \\ p_{L_1|B_2}(l|0), & \text{if } \boldsymbol{e} = [0,1]^{\mathrm{T}} \\ p_{L_1+L_2|\boldsymbol{B}}(l|[0,0]^{\mathrm{T}}), & \text{if } \boldsymbol{e} = [1,1]^{\mathrm{T}} \end{cases}, \tag{9.11}$$

where

$$p_{L_1|B_1}(l|0) = \frac{1}{2}(\Psi(l, -12\gamma, 8\gamma) + \Psi(l, -4\gamma, 8\gamma)), \tag{9.12}$$

$$p_{L_2|B_2}(l|0) = \Psi(l, -4\gamma, 8\gamma), \tag{9.13}$$

$$p_{L_1+L_2|\boldsymbol{B}}(l|[0,0]^{\mathrm{T}}) = \Psi(l, -16\gamma, 32\gamma). \tag{9.14}$$

For any linear code, the PEP can then be expressed as

$$\mathrm{PEP}(\underline{\boldsymbol{b}}, \hat{\underline{\boldsymbol{b}}}) = \mathrm{PEP}(\underline{\boldsymbol{0}}, \underline{\boldsymbol{b}} \oplus \hat{\underline{\boldsymbol{b}}}) = \mathrm{PEP}(\underline{\boldsymbol{0}}, \underline{\boldsymbol{e}}), \tag{9.15}$$

which means that we can assume the all-zero codeword was transmitted and use the PDFs in (9.11)–(9.14) to evaluate the PEP.

To take into account the contribution of the errors $\boldsymbol{e}[n]$ in the PEP, we make a "super-generalization" of the *Hamming weight* (HW) of $\underline{\boldsymbol{e}}$ via the vector $\tilde{\boldsymbol{w}} = [\tilde{w}_1, \tilde{w}_2, \tilde{w}_{1,2}]^{\mathrm{T}}$, where

$$\tilde{w}_1 = \sum_{n=1}^{N_s} \mathbb{I}_{[\boldsymbol{e}[n]=[1,0]^{\mathrm{T}}]}, \tag{9.16}$$

$$\tilde{w}_2 = \sum_{n=1}^{N_s} \mathbb{I}_{[\boldsymbol{e}[n]=[0,1]^{\mathrm{T}}]}, \tag{9.17}$$

$$\tilde{w}_{1,2} = \sum_{n=1}^{N_s} \mathbb{I}_{[\boldsymbol{e}[n]=[1,1]^{\mathrm{T}}]}. \tag{9.18}$$

Then, the PEP in (9.15) can be expressed as

$$\mathsf{PEP}(\underline{\boldsymbol{0}}, \underline{\boldsymbol{e}}) = \mathsf{PEP}(\tilde{\boldsymbol{w}}) \tag{9.19}$$

$$\approx \int_0^\infty \{p_{L_1|B_1}(l|0)\}^{*\tilde{w}_1} * \{p_{L_2|B_2}(l|0)\}^{*\tilde{w}_2} * \{p_{L_1+L_2|\boldsymbol{B}}(l|[0,0]^{\mathrm{T}})\}^{*\tilde{w}_{1,2}}\, \mathrm{d}l. \tag{9.20}$$

Using the PDF of the L-values in (9.12)–(9.14), we can now calculate the PEP in (9.20). The elements in the integral in (9.20) can be expressed as

$$\{p_{L_1|B_1}(l|0)\}^{*\tilde{w}_1} = \left(\frac{1}{2}\right)^{\tilde{w}_1} \sum_{r_1=0}^{\tilde{w}_1} \binom{\tilde{w}_1}{r_1} \Psi\left(l, -4\gamma(2r_1 + \tilde{w}_1), 8\gamma\tilde{w}_1\right), \tag{9.21}$$

$$\{p_{L_2|B_2}(l|0)\}^{*\tilde{w}_2} = \Psi\left(l, -4\gamma\tilde{w}_2, 8\gamma\tilde{w}_2\right), \tag{9.22}$$

$$\{p_{L_1+L_2|\boldsymbol{B}}(l|[0,0]^{\mathrm{T}})\}^{*\tilde{w}_{1,2}} = \Psi\left(l, -16\gamma\tilde{w}_{1,2}, 32\gamma\tilde{w}_{1,2}\right), \tag{9.23}$$

where we used $\Psi(l, \mu_1, \sigma_1^2) * \ldots * \Psi(l, \mu_J, \sigma_J^2) = \Psi(l, \sum_{j=1}^J \mu_j, \sum_{j=1}^J \sigma_j^2)$ and where the binomial coefficient in (9.21) appears because of the convolution of the sum of the two Gaussian functions in (9.12).

Using the relationships (9.21)–(9.23) in (9.20) yields

$$\mathsf{PEP}(\tilde{\boldsymbol{w}}) \approx \int_0^\infty \left(\frac{1}{2}\right)^{\tilde{w}_1} \sum_{r_1=0}^{\tilde{w}_1} \binom{\tilde{w}_1}{r_1} \Psi\left(l, 4\gamma\mu_{\mathrm{tot},r_1}, 8\gamma\sigma^2_{\mathrm{tot}}\right)\, \mathrm{d}l, \tag{9.24}$$

where

$$\mu_{\mathrm{tot},r_1} = -(\tilde{w}_1 + \tilde{w}_2 + 4\tilde{w}_{1,2} + 2r_1), \tag{9.25}$$

$$\sigma^2_{\mathrm{tot}} = (\tilde{w}_1 + \tilde{w}_2 + 4\tilde{w}_{1,2}). \tag{9.26}$$

By using $\gamma = \mathsf{snr}\,\Delta^2 = \mathsf{snr}/5$ (see (2.47)) and (9.25) and (9.26) in (9.24), we obtain

$$\mathsf{PEP}(\tilde{\boldsymbol{w}}) \approx \left(\frac{1}{2}\right)^{\tilde{w}_1} \sum_{r_1=0}^{\tilde{w}_1} \binom{\tilde{w}_1}{r_1} Q\left(\sqrt{\frac{\mu^2_{\mathrm{tot},r_1}}{\sigma^2_{\mathrm{tot}}} \frac{2\,\mathsf{snr}}{5}}\right). \tag{9.27}$$

In analogy to (8.22), the *bit-error probability* (BEP) for BICM-T can be approximated as

$$\mathsf{BEP} \approx \sum_{d=d_{\mathrm{free}}}^{d_{\mathrm{max}}} \sum_{\substack{\tilde{\boldsymbol{w}} \in \mathbb{N}^3 \\ \|\tilde{\boldsymbol{w}}\|_1 = d}} G^{\mathcal{C}}_{\tilde{\boldsymbol{w}}} \mathsf{PEP}(\tilde{\boldsymbol{w}}), \tag{9.28}$$

which used with (9.27) gives the final PEP approximation for BICM-T:

$$\mathsf{BEP} \approx \sum_{d=d_{\text{free}}}^{d_{\max}} \sum_{\substack{\tilde{\boldsymbol{w}}\in\mathbb{N}^3 \\ \|\tilde{\boldsymbol{w}}\|_1=d}} \left(\frac{1}{2}\right)^{\tilde{w}_1} G^{\mathcal{C}}_{\tilde{\boldsymbol{w}}} \sum_{r_1=0}^{\tilde{w}_1} \binom{\tilde{w}_1}{r_1} Q\left(\sqrt{\frac{\mu^2_{\text{tot},r_1}}{\sigma^2_{\text{tot}}} \frac{2\,\mathsf{snr}}{5}}\right). \tag{9.29}$$

Similarly, an expression for the *word-error probability* (WEP) for BICM-T can be found, namely,

$$\mathsf{BEP} \approx \sum_{d=d_{\text{free}}}^{d_{\max}} \sum_{\substack{\tilde{\boldsymbol{w}}\in\mathbb{N}^3 \\ \|\tilde{\boldsymbol{w}}\|_1=d}} \left(\frac{1}{2}\right)^{\tilde{w}_1} N_{\text{q}} F^{\mathcal{C}}_{\tilde{\boldsymbol{w}}} \sum_{r_1=0}^{\tilde{w}_1} \binom{\tilde{w}_1}{r_1} Q\left(\sqrt{\frac{\mu^2_{\text{tot},r_1}}{\sigma^2_{\text{tot}}} \frac{2\,\mathsf{snr}}{5}}\right). \tag{9.30}$$

In (9.29) and (9.30), we use $G^{\mathcal{C}}_{\tilde{\boldsymbol{w}}}$ and $F^{\mathcal{C}}_{\tilde{\boldsymbol{w}}}$, which are the GIWD and *generalized weight distribution* (GWD) of the CENC. These GIWD and GWD for BICM-T should consider neither the total HW d (needed for BICM-S) nor the *generalized Hamming weight* (GHW) $\boldsymbol{w} = [w_1, w_2]^{\text{T}}$ (needed for BICM-M), but a GHW $\tilde{\boldsymbol{w}} = [\tilde{w}_1, \tilde{w}_2, \tilde{w}_{1,2}]^{\text{T}}$. In this super-generalization of the HW, $\tilde{w}_1$ counts the number of columns of the divergent sequence in the trellis where $\underline{c}_1 = 1$ and $\underline{c}_2 = 0$, $\tilde{w}_2$—the number of columns where $\underline{c}_1 = 0$ and $\underline{c}_2 = 1$, and $\tilde{w}_{1,2}$—the number of columns where $\underline{c}_1 = \underline{c}_2 = 1$. In other words, it generalizes the GIWD used in BICM-M in the sense that it considers the case $c_1 = c_2 = 1$ as a special case. This new GIWD of the encoder is such that for a divergent codeword with HW d,

$$d = \tilde{w}_1 + \tilde{w}_2 + 2\tilde{w}_{1,2}. \tag{9.31}$$

The next example shows how to calculate the new, super-generalized $G^{\mathcal{C}}_{\tilde{\boldsymbol{w}}}$ and $F^{\mathcal{C}}_{\tilde{\boldsymbol{w}}}$.

Example 9.1 (GWD and GIWD of the Encoder $\mathbf{G} = [5, 7]$ for BICM-T) *For the CENC $\mathbf{G} = [5, 7]$, and following a procedure analogous to the one in Examples 6.23 and 8.6, we obtain the generalized transfer function*

$$T(V, \tilde{W}_1, \tilde{W}_2, \tilde{W}_{1,2}) = \frac{V\tilde{W}^2_{1,2}(V(\tilde{W}^2_1 - \tilde{W}^2_2) + \tilde{W}_2)}{1 - V(V(\tilde{W}^2_1 - \tilde{W}^2_2) + 2\tilde{W}_2)}. \tag{9.32}$$

In Fig. 9.3 we show the modified state machine, where the occurrences of $\tilde{W}_1\tilde{W}_2$ in Fig. 8.7 are replaced by the dummy variable $\tilde{W}_{1,2}$. In analogy to (8.31) and (8.32), the GWD and GIWD of the encoder can

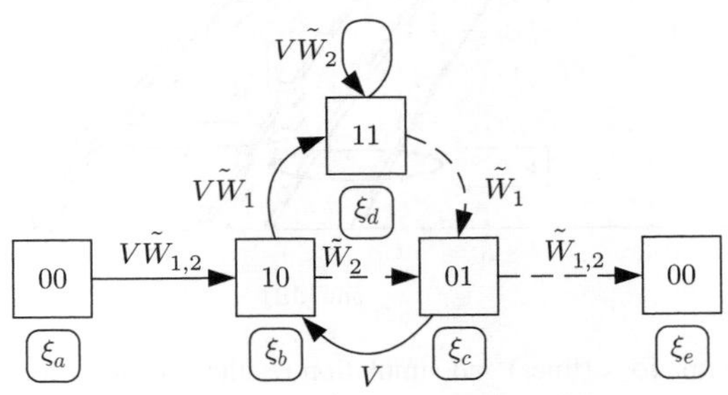

Figure 9.3 Generalized state machine for $\mathbf{G} = [5, 7]$ used in the analysis of BICM-T. The path that generates the FHD event is $\xi_a \to \xi_b \to \xi_c \to \xi_e$

be calculated as

$$F^{\mathcal{C}}_{\tilde{\boldsymbol{w}}} = \frac{1}{\tilde{w}_1!\tilde{w}_2!\tilde{w}_{1,2}!} \frac{\partial^{\tilde{w}_1}}{\partial \tilde{W}_1^{\tilde{w}_1}} \frac{\partial^{\tilde{w}_2}}{\partial \tilde{W}_2^{\tilde{w}_2}} \frac{\partial^{\tilde{w}_{1,2}}}{\partial \tilde{W}_{1,2}^{\tilde{w}_{1,2}}} T(1, \tilde{W}_1, \tilde{W}_2, \tilde{W}_{1,2}) \Bigg|_{\tilde{W}_1=0,\tilde{W}_2=0,\tilde{W}_{1,2}=0}, \tag{9.33}$$

$$G^{\mathcal{C}}_{\tilde{\boldsymbol{w}}} = \frac{1}{\tilde{w}_1!\tilde{w}_2!\tilde{w}_{1,2}!} \frac{\partial^{\tilde{w}_1}}{\partial \tilde{W}_1^{\tilde{w}_1}} \frac{\partial^{\tilde{w}_2}}{\partial \tilde{W}_2^{\tilde{w}_2}} \frac{\partial^{\tilde{w}_{1,2}}}{\partial \tilde{W}_{1,2}^{\tilde{w}_{1,2}}} \frac{\partial}{\partial V} T(V, \tilde{W}_1, \tilde{W}_2, \tilde{W}_{1,2}) \Bigg|_{\tilde{W}_1=0,\tilde{W}_2=0,\tilde{W}_{1,2}=0,V=1}. \tag{9.34}$$

The free Hamming distance (FHD) of this encoder is $d_{\text{free}} = 5$ *and* $G^{\mathcal{C}}_5 = 1$*, i.e., there is one divergent path at Hamming distance (HD) five from the all-zero codeword, and the nonzero elements of the corresponding codeword are given by*

$$[\boldsymbol{e}[t_0], \boldsymbol{e}[t_0+1], \boldsymbol{e}[t_0+2]] = \begin{bmatrix} 1 & 0 & 1 \\ 1 & 1 & 1 \end{bmatrix}, \tag{9.35}$$

where t_0 *is the time instant for which the input sequence has a unique nonzero element, i.e.,* $i[t_0] = 1$ *and* $i[n] = 0, n \neq t_0$*. For this particular error codeword, we have* $w_{\text{H}}(\underline{\boldsymbol{e}}) = 5$*,* $\tilde{w}_1 = 0$*,* $\tilde{w}_2 = 1$*, and* $\tilde{w}_{1,2} = 2$ *(while in BICM-M we would have* $w_1 = 2$ *and* $w_2 = 3$*).*

Example 9.2 (BICM-T with $R_c = 1/2$ Encoder) *In Fig. 9.4, simulation results for BICM-T using the ODS CENC* $\mathbf{G} = [5, 7]$ *(*$\nu = 2$*) and* $\mathbf{G} = [247, 371]$ *(*$\nu = 7$*) from Table 2.1 are shown. We also consider swapping the generator polynomials, i.e., we consider the encoders* $\mathbf{G} = [7, 5]$ *and* $\mathbf{G} = [371, 247]$*. We distinguish between the best and worst configurations, where the best configuration is when all the bits generated by the polynomial* 7 *or* 371 *are sent over* $k = 1$ *and all the bits generated by the polynomial* 5 *or* 247 *are sent over* $k = 2$*. In Fig. 9.4 we also show the simulation results for* BICM-S *and for* BICM-M*. For the latter, we consider the optimal values of* $u^*_{1,1}$ *in Table 8.1 (*$m = 2$*).*

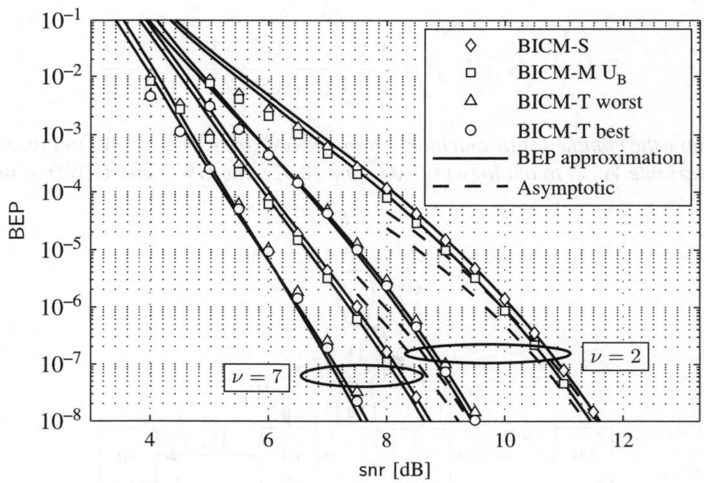

Figure 9.4 BEP approximations (lines) and simulation results (markers) for BICM-T, BICM-M, and BICM-S, 4PAM labeled by the BRGC, and for the ODS CENC in Table 2.1: $\mathbf{G} = [5, 7]$ ($\nu = 2$) and $\mathbf{G} = [247, 371]$ ($\nu = 7$). The BEP approximations are given by (9.29) for BICM-T, by (8.52) for BICM-M, and (6.398) for BICM-S. The asymptotic bounds given by (9.36) are shown with dashed lines

In Fig. 9.4 we also show the BEP approximation for BICM-T *in* (9.29) *as well as the BEP for* BICM-M *in (8.52) and for* BICM-S *in (6.398).*[1] *The results in Fig. 9.4 show that the developed BEP approximations predict well the simulation results and that the gains by using* BICM-T *instead of BICM-S for a BEP target of* 10^{-7} *are approximately* 2 dB *and* 1 dB *for* $\nu = 2$ *and* $\nu = 7$*, respectively. What is maybe most surprising is that these gains are obtained by decreasing the complexity of the system, i.e., without interleaving/deinterleaving.*

We conclude this section by comparing the results for BICM-T with those obtained for BICM-M and BICM-S. First of all, we note that unlike BICM-M where for $R_c = 1/2$ only two variables are needed (w_1 and w_2, see Section 8.1.4), in BICM-T three variables are needed ($\tilde{w}_1$, $\tilde{w}_2$, and $\tilde{w}_{1,2}$). If we now consider BICM-S, owing to the interleaver, the case $\boldsymbol{e} = [1, 1]^{\mathrm{T}}$ in (9.11) is ignored because it does not occur frequently (see Fig. 6.13 (a)). In such a case, only circles and diamonds in Table 9.2 should be considered, and thus, only two Gaussian distributions appear in the final density, namely, $\Psi(l, -12\gamma, 8\gamma)$ and $\Psi(l, -4\gamma, 8\gamma)$, with probabilities $1/4$ and $3/4$, respectively. This can also be seen from (9.12) and (9.13) (the PDF in (9.14) is not considered). As expected, this corresponds to the model in Example 6.34, where the probabilities $1/4$ and $3/4$ correspond to the values of w_1 and w_2 studied in that example.

9.2 Code Design for BICM-T

The results from the previous section show relatively large gains offered by BICM-T over BICM-S. In this section, we quantify these gains and we analyze the performance of BICM-T for asymptotically high *signal-to-noise ratio* (SNR). Asymptotically optimum encoders are defined and BICM-T is also compared with Ungerboeck's TCM.

9.2.1 Asymptotic Bounds

The BEP approximation in (9.29) is a sum of weighted Q-functions, and thus, as $\mathsf{snr} \to \infty$, the BEP is dominated by the Q-functions with the smallest argument. For each $\tilde{\boldsymbol{w}} = [\tilde{w}_1, \tilde{w}_2, \tilde{w}_{1,2}]^{\mathrm{T}}$, the dominant Q-function in (9.28) is obtained for $r_1 = 0$, and thus, the BEP of BICM-T can be (asymptotically) approximated as

$$\mathsf{BEP} \approx MQ\left(\sqrt{\frac{2\,\mathsf{snr}\,A}{5}}\right), \tag{9.36}$$

where

$$A \triangleq \min_{\substack{\tilde{\boldsymbol{w}} \in \mathbb{N}^3 \\ G^{\mathcal{C}}_{\tilde{\boldsymbol{w}}} \neq 0}} \{\tilde{w}_1 + \tilde{w}_2 + 4\tilde{w}_{1,2}\}, \tag{9.37}$$

$$M \triangleq \sum_{\substack{\tilde{\boldsymbol{w}} \in \mathbb{N}^3 \\ \tilde{w}_1 + \tilde{w}_2 + 4\tilde{w}_{1,2} = A}} \left(\frac{1}{2}\right)^{\tilde{w}_1} G^{\mathcal{C}}_{\tilde{\boldsymbol{w}}}. \tag{9.38}$$

In Fig. 9.4, we show asymptotic approximations for $\nu = 2$. For BICM-T we used (9.36), for BICM-S we used (6.399), and for BICM-M we used (8.53). All of them are shown to follow well the simulation results. Similar results can be obtained for the encoder with $\nu = 7$; however we do not show them so that the figure is not overcrowded.

The *asymptotic gain* (AG) provided by using BICM-T instead of BICM-S, denoted by $\mathrm{AG}_{\mathrm{S}\to\mathrm{T}}$, is obtained directly from (9.36) and (6.399), as stated in the following corollary.

[1] All the BEP approximations were obtained with $d_{\max} = 30$.

Corollary 9.3 *The AG provided by* BICM-T *with respect to* BICM-S *for any CENC with GIWD $G^{\mathcal{C}}_{\tilde{\boldsymbol{w}}}$ is*

$$\mathrm{AG}_{\mathrm{S}\rightarrow\mathrm{T}} = 10\log_{10}\left(\frac{A}{d_{\mathrm{free}}}\right), \tag{9.39}$$

where d_{free} is the FHD of the convolutional code (CC) and A is given by (9.37).

Corollary 9.3 holds for any CENC; however, because ODS CENCs are the asymptotically optimal encoders for BICM-S, studying their loss is of particular interest. To clarify this, consider the following example.

Example 9.4 (AG for the Encoder $\mathrm{G} = [5,7]$**)** *For the particular encoder $\mathbf{G} = [5,7]$, it is possible to prove that the solution of (9.37) corresponds to the event at FHD,*[2] *i.e., $d_{\mathrm{free}} = 5$, $\tilde{w}_1 = 0$, $\tilde{w}_2 = 1$, $\tilde{w}_{1,2} = 2$ (see Example 9.1), and therefore, $A = 9$. This results in an AG in (9.39) equal to*

$$\mathrm{AG}_{\mathrm{S}\rightarrow\mathrm{T}} = 10\log_{10}\left(\frac{9}{5}\right) \approx 2.55\ \mathrm{dB}. \tag{9.40}$$

Moreover, because the input sequence that generates the codeword at FHD has HW equal to one ($G^{\mathcal{C}}_{\tilde{\boldsymbol{w}}} = 1$ for $\tilde{\boldsymbol{w}} = [0,1,2]^{\mathrm{T}}$), we obtain $M = 1$, i.e., as $\mathsf{snr} \rightarrow \infty$,

$$\mathsf{BEP} \approx Q\left(\sqrt{\frac{18\,\mathsf{snr}}{5}}\right). \tag{9.41}$$

If the generator polynomials are swapped, i.e., if we consider the CENC $\mathbf{G} = [7,5]$, we obtain $\tilde{w}_1 = 1$, $\tilde{w}_2 = 0$, $\tilde{w}_{1,2} = 2$ and the same A (because A does not depend on the order of the polynomials). However, in this case $M = 1/2$, which gives

$$\mathsf{BEP} \approx \frac{1}{2}Q\left(\sqrt{\frac{18\,\mathsf{snr}}{5}}\right). \tag{9.42}$$

The asymptotic bounds in (9.41) and (9.42) are shown in Fig. 9.4, where the influence of the coefficient M ($M = 1$ for $\mathbf{G} = [5,7]$ and $M = 1/2$ for $\mathbf{G} = [7,5]$) is also visible in the simulation results.

Although the gains offered by BICM-T over BICM-S are quite large, they are bounded, as shown in the following theorem.

Theorem 9.5 *For any CENC, the AG provided by* BICM-T *with respect to* BICM-S *is upper bounded as*

$$\mathrm{AG}_{\mathrm{S}\rightarrow\mathrm{T}} \leq 3\ \mathrm{dB}. \tag{9.43}$$

Proof: For any CENC, we have

$$\begin{aligned} A &= \min_{\substack{\tilde{\boldsymbol{w}}\in\mathbb{N}^3\\ G^{\mathcal{C}}_{\tilde{\boldsymbol{w}}}\neq 0}} \{\tilde{w}_1 + \tilde{w}_2 + 4\tilde{w}_{1,2}\} \\ &\leq \tilde{w}_1^{\mathrm{free}} + \tilde{w}_2^{\mathrm{free}} + 4\tilde{w}_{1,2}^{\mathrm{free}} && (9.44)\\ &= 2d_{\mathrm{free}} - \tilde{w}_1^{\mathrm{free}} - \tilde{w}_2^{\mathrm{free}} && (9.45)\\ &\leq 2d_{\mathrm{free}}, && (9.46) \end{aligned}$$

where we use $[\tilde{w}_1^{\mathrm{free}}, \tilde{w}_2^{\mathrm{free}}, \tilde{w}_{1,2}^{\mathrm{free}}]^{\mathrm{T}}$ to denote a "super-generalized" weight of the event(s) at FHD and d_{free} is the MFHD of the CC. The inequality in (9.44) holds because the event(s) at FHD belong to the

[2] However, this is not always true for other CENCs.

elements in the minimization. The equality in (9.45) follows by recognizing $2\tilde{w}_1^{\text{free}} + 2\tilde{w}_2^{\text{free}} + 4\tilde{w}_{1,2}^{\text{free}}$ as twice the FHD of the CC (see (9.31)) The inequality in (9.46) holds because $\tilde{w}_1^{\text{free}} + \tilde{w}_2^{\text{free}} \geq 0$. The use of $A \leq 2d_{\text{free}}$ in (9.39) completes the proof. □

The results in Theorem 9.5 are valid for any CENC. In particular, they also hold for the ODS CENCs. As expected, the AG obtained in (9.40) is consistent with (9.43). The AG offered by BICM-T over BICM-S for all the ODS CENCs $\mathbf{G}$ (up to $\nu = 7$) are shown in the eighth column of Table 9.3. The values obtained are around 2 dB, which is consistent with Theorem 9.5.

9.2.2 *Optimal Encoders for BICM-T*

ODS CENCs are the asymptotically optimal CENCs for binary transmission and for BICM-S, as shown in (6.385). For BICM-M, we showed in Section 8.2.3 that ODS CENCs are not optimal anymore; the optimal joint interleaver and encoder design for BICM-M is given in Definition 8.11. If BICM-T is considered, and as a direct consequence of (9.36), optimal encoders for BICM-T can be defined too. These encoders are asymptotically optimal for BICM-T and guarantee gains for high SNR; however, we will also show that gains appear for low and moderate SNR.

Definition 9.6 (Optimal Encoders for BICM-T) *A CENC $\mathbf{G}^* \in \mathcal{G}_{k_c,n_c,\nu}$ is said to be optimal for* BICM-T *if it produces an asymptotic BEP, which is a minimum compared to those obtained using any other encoder with the same R_c and ν, i.e., it is an encoder with the lowest multiplicity M among the encoders with the highest A.*

We performed an exhaustive numerical search based on Definition 9.6 and found the optimal encoders for BICM-T. The results are shown in Table 9.3, where for comparison we also show the respective coefficients A and M of the ODS CENCs. The search is done in lexicographic order and considering a truncated spectrum $\tilde{w}_1 + \tilde{w}_2 + 4\tilde{w}_{1,2} \leq d_{\text{free}} + 8$, where d_{free} is the FHD of the ODS CENC for that memory ν. If there is more than one optimal encoder for a given ν, we present the first one in the list. As shown in Table 9.3, new encoders (highlighted) are found in four out of six analyzed cases.

We emphasize that, during the search, the optimal encoders for BICM-T are not assumed to be encoders with MFHD, i.e., $\mathbf{G}^* \in \mathcal{G}_{k_c,n_c,\nu}$ (and not $\mathbf{G}^* \in \mathcal{G}_{k_c,n_c,\nu}^{\text{MFHD}}$). The results in Table 9.3 confirm this by showing that in general the FHD of the encoder is not the adequate criterion in BICM-T, i.e., encoders with small FHD could perform better than the ODS CENCs (which are obtained by maximizing the FHD, see Section 2.6.1). For example, for $\nu = 5$ the FHD of the optimal encoder is $d_{\text{free}} = 7$ while for the

Table 9.3 Optimal encoders for BICM-T, their FHD, as well as their coefficients A and M. New encoders found, better than the ODS CENCs in Table 2.1, are highlighted. The asymptotic gains are also shown

ν	Optimal				ODS		AG [dB]		
	$\mathbf{G}^*$	d_{free}	A	M	A	M	$\text{AG}_{\text{S}\to\text{T}}$	$\text{AG}_{\text{UC}\to\text{T}}$	$\text{AG}_{\text{UC}\to\text{S}}$
2	$[7, 5]$	5	9	0.50	9	0.50	2.55	2.55	0
3	$[13, 17]$	6	10	0.50	10	0.50	2.22	3.01	0.79
4	$\mathbf{[23, 33]}$	7	11	0.38	11	0.88	1.96	3.42	1.46
5	$\mathbf{[45, 55]}$	7	13	1.62	12	0.50	2.11	4.15	2.04
6	$\mathbf{[107, 135]}$	9	14	0.50	14	3.09	1.46	4.47	3.01
7	$\mathbf{[313, 235]}$	10	16	8.02	15	0.61	2.04	5.05	3.01

ODS CENC the FHD is larger ($d_{\text{free}} = 8$). The same happens for $\nu = 6$. In fact, only for $\nu = 2$ the ODS CENC is also optimum for BICM-T.[3] The results in Table 9.3 also show that the gains offered by the optimal encoders in four out of six cases come from a reduced multiplicity M, i.e., both the optimal and the ODS CENC have the same A, and thus, the same asymptotic behavior. On the other hand, for $\nu = 5$ and $\nu = 7$, there is a difference in A that will result in nonzero asymptotic gains. Namely, for $\nu = 5$

$$\text{AG}_{\text{T}^{\text{ODS}}\to\text{T}^*} = 10\log_{10}\left(\frac{13}{12}\right) \approx 0.35 \text{ dB}, \tag{9.47}$$

and for $\nu = 7$

$$\text{AG}_{\text{T}^{\text{ODS}}\to\text{T}^*} = 10\log_{10}\left(\frac{16}{15}\right) \approx 0.28 \text{ dB}, \tag{9.48}$$

where we used the notation $\text{AG}_{\text{T}^{\text{ODS}}\to\text{T}^*}$ to show the asymptotic gain of BICM-T using the optimal encoders versus BICM-T using ODS CENCs.

9.2.3 *BICM-T versus TCM*

As explained before, the difference between the system in Fig. 9.1 and TCM is the receiver used. In what follows we compare the asymptotic performance of BICM-T and the optimum ML decoder.

We have previously defined in (9.39) the AG of BICM-T over BICM-S. It is also possible to define the AG of BICM-T compared to uncoded transmission with the same spectral efficiency (uncoded 2PAM). This AG is

$$\begin{aligned} \text{AG}_{\text{UC}\to\text{T}} &= 10\log_{10}\left(\frac{2\ \mathsf{snr} A/5}{2\ \mathsf{snr}}\right) \\ &= 10\log_{10}\left(\frac{A}{5}\right), \end{aligned} \tag{9.49}$$

which follows from (9.36) and (6.33).

The AG in (9.49) is shown in Table 9.3. For $\nu = 2$, $\text{AG}_{\text{UC}\to\text{T}}$ is equal to 2.55 dB, which is the same as $\text{AG}_{\text{S}\to\text{T}}$. This is because BICM-S with $\nu = 2$ does not offer any AG compared to uncoded 2PAM.

For comparison, consider the TCM design proposed by Ungerboeck. This is shown in Table 9.4, where the CENCs with their corresponding FHD for different values of ν are shown. This table also includes the

Table 9.4 Ungerboeck's trellis encoders for $R_c = 1/2$ and 4PAM labeled by the *natural binary code* (NBC) for different ν

ν	$\mathbf{G}$	d_{free}	$\text{AG}_{\text{UC}\to\text{TCM}}$ [dB]
2	$[5, 2]$	3	2.55
3	$[13, 4]$	4	3.01
4	$[23, 4]$	4	3.42
5	$[45, 10]$	4	4.15
6	$[103, 24]$	5	4.47
7	$[235, 126]$	8	5.05

[3] For $\nu = 3$ the optimal encoder $\mathbf{G} = [13, 17]$ has the same spectrum $G^{\mathcal{C}}_{\mathbf{w}}$ as the ODS CENC $\mathbf{G} = [15, 17]$. The optimal encoder appears in the list because of the lexicographic order search.

AG offered by TCM over uncoded transmission with the same spectral efficiency. Analyzing the values of $\mathrm{AG}_{\mathrm{UC}\rightarrow\mathrm{T}}$ in Table 9.3, we find that they are the same as those in Table 9.4. These results show that if BICM-T is used with the correct CENC, it performs asymptotically as well as TCM, and therefore, it should be considered as a good alternative for CM in nonfading channels.[4] This is, however, not the case if BICM-S is used, or if BICM-T is used with the ODS CENC. For example, BICM-T for $\nu = 5$ and an ODS CENC gives an AG

$$\mathrm{AG}_{\mathrm{UC}\rightarrow\mathrm{T}} = 10\log_{10}\left(\frac{12}{5}\right) \approx 3.80 \text{ dB}, \tag{9.50}$$

which is less than the AG of 4.15 dB offered by BICM-T with an optimal encoder (or by TCM).

For completeness, in Table 9.3, we also show the AG offered by BICM-S over uncoded transmission, defined as $\mathrm{AG}_{\mathrm{UC}\rightarrow\mathrm{S}} = \mathrm{AG}_{\mathrm{UC}\rightarrow\mathrm{T}} - \mathrm{AG}_{\mathrm{S}\rightarrow\mathrm{T}}$. The performance of uncoded transmission does not depend on ν, but increasing ν changes the performance of BICM-T and BICM-S. This is reflected in an increase of the *asymptotic gains* (AGs) $\mathrm{AG}_{\mathrm{UC}\rightarrow\mathrm{T}}$ and $\mathrm{AG}_{\mathrm{UC}\rightarrow\mathrm{S}}$ in Table 9.3 when ν increases. Finally, we also note this is not the case for $\nu = 6, 7$, where the use of BICM-S with the ODS CENC results in both cases in an asymptotic gain of 3.01 dB compared to uncoded transmission (see $\mathrm{AG}_{\mathrm{UC}\rightarrow\mathrm{S}}$ in Table 9.3). This can be explained from the fact that the ODS CENCs for $\nu = 6, 7$ have the same $d_{\mathrm{free}} = 10$.

9.3 Equivalent Labelings for Trellis Encoders

Conventionally, CM designs based on CENCs either optimize the encoder for a constellation labeled by a SP labeling (TCM), or simply connect an encoder designed for binary transmission with the BRGC (BICM). Of course, none of these approaches is optimal. Instead, a joint optimization over the labeling universe $\mathcal{Q}_m$ (see Section 2.5.2) *and* the CENC universe $\mathcal{G}_{k_{\mathrm{c}},n_{\mathrm{c}},\nu}$ (see Section 2.6.1) should be performed. This problem is very difficult, even for the simplest cases. As an example, we consider the design of a trellis encoder based on an $R_{\mathrm{c}} = 1/3$ CENC with $\nu = 5$ and an 8-ary constellation. In this case, the labeling universe includes $8! = 40320$ different binary labelings and the CENC universe (discarding catastrophic encoders) includes about 160000 encoders.[5] An exhaustive search for this quite simple configuration would require checking about 6 billion combinations of encoders and labelings.

In this section we address the problem of the joint design of the CENC and the labeling for a trellis encoder, which, based on the analysis in the previous sections, can be used in combination with a BICM decoder. The analysis is based on a classification of binary labelings, which allows us to formally prove that in any trellis encoder the NBC can be replaced by many other labelings (including the BRGC) without causing any performance degradation, provided that the encoder is properly selected. This is also used to explain the asymptotic equivalence between BICM-T and TCM presented in Section 9.2.3.

9.3.1 Equivalent Trellis Encoders

We have seen in Fig. 9.4 that if the interleaver is removed in BICM, its performance over the AWGN channel is improved. Moreover, we showed that in terms of MED, a properly designed BICM-T performs asymptotically as well as TCM (see Tables 9.3 and 9.4). The transmitters for TCM and BICM-T for the 8-state (overall constraint length $\nu = 3$) CENC are shown in Fig. 9.5 (a) and c, respectively. A careful examination of the results in Table 9.3 reveals that the optimal trellis encoder found when analyzing BICM-T, in fact, is *equivalent* to the one proposed by Ungerboeck 30 years ago. For a 4PAM constellation, Ungerboeck's SP labeling (NBC) can be generated using the BRGC plus one binary addition (which we call *transform*) applied to its inputs, as shown in Fig. 9.5 (b) (see also Example 2.12). If the transform

[4] This asymptotic equivalence between BICM-T and TCM will be studied in more detail in Section 9.3.

[5] Which are found through an exhaustive enumeration of all the elements in $\mathcal{G}_{k_{\mathrm{c}},n_{\mathrm{c}},\nu}$.

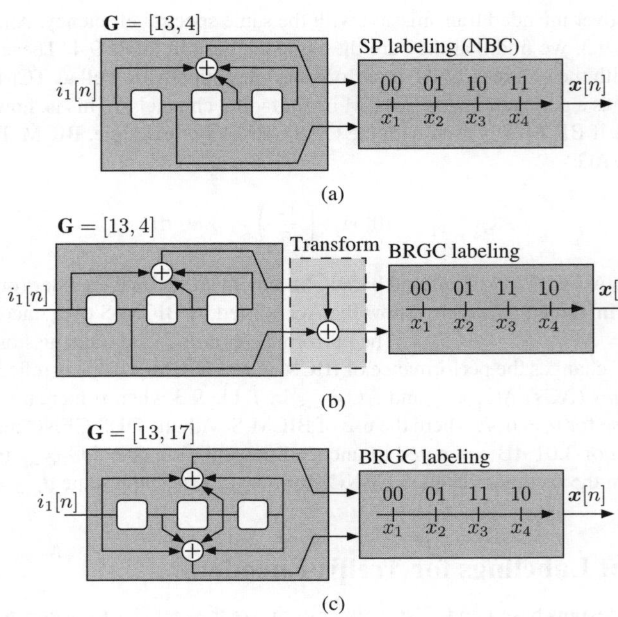

Figure 9.5 Three equivalent trellis encoders: (a) encoder for TCM, (b) equivalent encoder, and (c) encoder for BICM; (a) Ungerboeck's CENC $\mathbf{G} = [13, 4]$ from Table 9.4 and the NBC; (c) Optimal CENC $\mathbf{G} = [13, 17]$ Table 9.3 and the BRGC. The trellis encoder in (b) shows how a transformation based on a binary addition can be included in the mapper (to go from (b) to (a)) or in the code (to go from (b) to (c))

is included in the mapper, the encoder in Fig. 9.5 (a) is obtained, while if it is included in the CENC, the trellis encoder in Fig. 9.5 (c) is obtained. This equivalence also applies to encoders with larger number of states[6] and simply reveals that for 4PAM a TCM transceiver based on the BRGC will have identical performance to Ungerboeck's TCM if the encoder is properly modified, where the modification is the application of a simple transform.

To formally study this equivalence, we consider the trellis encoder in Fig. 2.4, where for simplicity we assume that $n_c = m$. For a given constellation $\mathbf{X}$ and overall constraint length ν, a *CM encoder* is fully defined by the CENC matrix $\mathbf{G}$ and the labeling of the constellation $\mathbf{Q}$, and thus, a trellis encoder is defined by the pair $\Xi = [\mathbf{G}, \mathbf{Q}]$. In this section we show how a joint optimization over all $\mathbf{G} \in \mathcal{G}_{k_c,m,\nu}$ (i.e., over the CENC universe) and $\mathbf{Q} \in \mathcal{Q}_m$ (i.e., over the labeling universe) can be restricted, without loss of generality, to a joint optimization over all $\mathbf{G} \in \mathcal{G}_{k_c,m,\nu}$ and over a subset of $\mathcal{Q}_m$.

The output of a given trellis encoder $\Xi = [\mathbf{G}, \mathbf{Q}]$ at time n depends only on $\nu + k_c$ information bits. Using (2.90), the transmitted symbol at time n can then be expressed as

$$\boldsymbol{x}[n] = \Phi_{\mathbf{Q}}(\boldsymbol{b}[n]) = \Phi_{\mathbf{Q}}(\mathbf{G}^{\mathrm{T}}\boldsymbol{j}[n]), \tag{9.51}$$

where $\boldsymbol{j}[n] = [\underline{i}'_1, \underline{i}'_2, \ldots, \underline{i}'_{k_c}]^{\mathrm{T}}$, $\underline{i}'_p = [i_p[n], i_p[n-1], \ldots, i_p[n-\nu_p]]$ and $p = 1, 2, \ldots, k_c$, i.e., $\boldsymbol{j}[n] = \boldsymbol{u}[n]^{\mathrm{T}}$, where $\boldsymbol{u}[n]$ was used in (2.90). From now on, we use the notation $\Phi_{\mathbf{Q}}(\cdot)$ to explicitly show the dependence of the mapper on the binary labeling $\mathbf{Q}$.

[6] This equivalence does not always directly hold because Table 9.3 lists the encoders in lexicographic order and because for some values of ν there are more than one encoder with identical performance.

The following definition gives a condition for the trellis encoders to be equivalent, where this equivalence implies equivalence in terms of BEP and WEP.

Definition 9.7 *Two trellis encoders* $\Xi = [\mathbf{G}, \mathbf{Q}]$ *and* $\tilde{\Xi} = [\tilde{\mathbf{G}}, \tilde{\mathbf{Q}}]$ *are said to be* equivalent *if they produce the same output symbol for the same information-bit sequence, i.e., if* $\Phi_{\mathbf{Q}}(\mathbf{G}^{\mathrm{T}}\boldsymbol{j}) = \Phi_{\tilde{\mathbf{Q}}}(\tilde{\mathbf{G}}^{\mathrm{T}}\boldsymbol{j})$ *for any* $\boldsymbol{j} \in \mathbb{B}^{\nu+k_c}$.

In the following, we use $\mathcal{T}_m$ to denote the set of all binary invertible $m \times m$ matrices and $\mathcal{R}_m$ to denote the set of all reduced column echelon binary matrices (see Definition 2.14) of size $m \times M$.

Lemma 9.8 $\Phi_{\mathbf{Q}}(\boldsymbol{q}) = \Phi_{\tilde{\mathbf{Q}}}(\mathbf{T}\boldsymbol{q})$*, where* $\tilde{\mathbf{Q}} = \mathbf{T}\mathbf{Q}$*, for any two mappers* $\Phi_{\mathbf{Q}}$ *and* $\Phi_{\tilde{\mathbf{Q}}}$ *that use the same constellation* $\mathbf{X}$*, any* $\mathbf{T} \in \mathcal{T}_m$*, and any* $\boldsymbol{q} \in \mathbb{B}^m$.

Proof: Let $\boldsymbol{u}_i \triangleq [0, \ldots, 0, 1, 0, \ldots, 0]^{\mathrm{T}}$ be a vector of length M, where the one is in position i. Thus, $\boldsymbol{q}_i = \mathbf{Q}\boldsymbol{u}_i$ for $i = 1, \ldots, M$. The mapping $\Phi_{\mathbf{Q}}$ satisfies by definition $\Phi_{\mathbf{Q}}(\boldsymbol{q}_i) = \boldsymbol{x}_i$ for $i = 1, \ldots, M$, or, making the dependency on $\mathbf{Q}$ explicit,

$$\Phi_{\mathbf{Q}}(\boldsymbol{q}) = \boldsymbol{x}_i, \quad \text{if } \boldsymbol{q} = \mathbf{Q}\boldsymbol{u}_i, \tag{9.52}$$

for any $\boldsymbol{q} \in \mathcal{B}^m$. Similarly, for any $\boldsymbol{q} \in \mathbb{B}^m$,

$$\begin{aligned}\Phi_{\tilde{\mathbf{Q}}}(\mathbf{T}\boldsymbol{q}) &= \boldsymbol{x}_i, \quad \text{if } \mathbf{T}\boldsymbol{q} = \tilde{\mathbf{Q}}\boldsymbol{u}_i \\ &= \boldsymbol{x}_i, \quad \text{if } \boldsymbol{q} = \mathbf{Q}\boldsymbol{u}_i,\end{aligned} \tag{9.53}$$

where the last step follows because $\mathbf{Q} = \mathbf{T}^{-1}\tilde{\mathbf{Q}}$. As the right-hand sides of (9.52) and (9.53) are equal, $\Phi_{\tilde{\mathbf{Q}}}(\mathrm{T}\boldsymbol{q}) = \Phi_{\mathbf{Q}}(\boldsymbol{q})$ for all $\boldsymbol{q} \in \mathbb{B}^m$. □

Theorem 9.9 *For any* $\mathbf{G} \in \mathcal{G}_{k_c, m, \nu}$*,* $\mathbf{Q} \in \mathcal{Q}_m$*, and* $\mathbf{T} \in \mathcal{T}_m$*, the two trellis encoders* $\Xi = [\mathbf{G}, \mathbf{Q}]$ *and* $\tilde{\Xi} = [\tilde{\mathbf{G}}, \tilde{\mathbf{Q}}]$ *are equivalent, where* $\tilde{\mathbf{Q}} = \mathbf{T}\mathbf{Q}$ *and* $\tilde{\mathbf{G}} = \mathbf{G}\mathbf{T}^{\mathrm{T}}$.

Proof: For any $\boldsymbol{j} \in \mathbb{B}^{\nu+k_c}$, $\Phi_{\tilde{\mathbf{Q}}}(\tilde{\mathbf{G}}^{\mathrm{T}}\boldsymbol{j}) = \Phi_{\tilde{\mathbf{Q}}}(\mathrm{T}\mathbf{G}^{\mathrm{T}}\boldsymbol{j}) = \Phi_{\mathbf{Q}}(\mathbf{G}^{\mathrm{T}}\boldsymbol{j})$, where the last equality follows by Lemma 9.8. The theorem now follows using Definition 9.7. □

Theorem 9.9 shows that a full search over $\mathcal{G}_{k_c, m, \nu}$ and $\mathcal{Q}_m$ will include many pairs of equivalent trellis encoders. Therefore, an optimal trellis encoder with given parameters can be found by searching over a subset of $\mathcal{G}_{k_c, m, \nu}$ and the whole set $\mathcal{Q}_m$ or vice versa. Here we choose the latter approach, searching over a subset of $\mathcal{Q}_m$.

The following theorem will be used to develop an efficient search algorithm we introduce in the next section. For a proof of the following theorem, we refer the reader to the references we provide in Section 9.4.

Theorem 9.10 *Any binary labeling* $\mathbf{Q} \in \mathcal{Q}_m$ *can be uniquely factorized as*

$$\mathbf{Q} = \mathbf{T}\mathbf{Q}_{\mathrm{R}}, \tag{9.54}$$

where $\mathbf{T} \in \mathcal{T}_m$ *and* $\mathbf{Q}_{\mathrm{R}} \in \mathcal{R}_m$.

Theorem 9.10 shows that all binary matrices $\mathbf{Q}$ can be uniquely generated by finding all the invertible matrices $\mathbf{T}$ (the set $\mathcal{T}_m$) and all the different reduced column echelon matrices $\mathbf{Q}_{\mathrm{R}}$ (the set $\mathcal{R}_m$). In particular, we have

$$M_{\mathrm{T}} \triangleq |\mathcal{T}_m| = \prod_{k=1}^{m}(M - 2^{k-1}), \tag{9.55}$$

$$M_{\mathrm{R}} \triangleq |\mathcal{R}_m| = \frac{M!}{\prod\limits_{k=1}^{m}(M - 2^{k-1})}. \tag{9.56}$$

Table 9.5 Number of modified Hadamard classes ($M_{\mathrm{R}} = |\mathcal{R}_m|$), their cardinality ($M_{\mathrm{T}} = |\mathcal{T}_m|$), and the total number of labelings ($M!$) for different values of m

m	1	2	3	4	5	6
M_{R}	2	4	240	$1.038 \cdot 10^9$	$2.632 \cdot 10^{28}$	$6.2943 \cdot 10^{78}$
M_{T}	1	6	168	20 160	$9.999 \cdot 10^6$	$2.016 \cdot 10^{10}$
$M!$	2	24	40 320	$2.092 \cdot 10^{13}$	$2.631 \cdot 10^{35}$	$1.2689 \cdot 10^{89}$

In Table 9.5, the values for M_{R} and M_{T} for $1 \leq m \leq 6$ are shown. In this table we also show the number of binary labelings ($|\mathcal{Q}_m| = M! = M_{\mathrm{R}} M_{\mathrm{T}}$), i.e., the number of matrices $\mathbf{Q}$ in the labeling universe.

A *modified Hadamard class* is defined as the set of matrices $\mathbf{Q}$ that can be generated via (9.54) using the same reduced column echelon matrix $\mathbf{Q}_{\mathrm{R}}$. There are thus M_{R} modified Hadamard classes, each with cardinality M_{T}.

As a consequence of Theorems 9.9 and 9.10, the two trellis encoders $[\mathbf{G}, \mathbf{Q}]$ and $[\mathbf{G}(\mathbf{T}^{-1})^{\mathrm{T}}, \mathbf{Q}_{\mathrm{R}}]$ are equivalent for any $\mathbf{G} \in \mathcal{G}_{k_c,m,\nu}$ and $\mathbf{Q} \in \mathcal{Q}_m$, where $\mathbf{Q}_{\mathrm{R}}$ and $\mathbf{T}$ are given by the factorization (9.54). In other words, all nonequivalent trellis encoders can be generated using one member of each modified Hadamard class only, and thus, a joint optimization over all $\mathbf{G} \in \mathcal{G}_{k_c,m,\nu}$ and $\mathbf{Q} \in \mathcal{Q}_m$ can be reduced to an optimization over all $\mathbf{G} \in \mathcal{G}_{k_c,m,\nu}$ and $\mathbf{Q} \in \mathcal{R}_m$ with no loss in performance. This means that the search space is reduced by at least a factor of $M_{\mathrm{T}} = M!/M_{\mathrm{R}}$. For example, for 8-ary constellations ($m = 3$), the total number of different binary labelings that must be tested is reduced from $8! = 40320$ to 240. Moreover, as shown in Section 9.3.4, this can be reduced even further if the constellation $\mathbf{X}$ possesses certain symmetries.

9.3.2 *Modified Full Linear Search Algorithm*

The problem of finding the set $\mathcal{R}_m$ of reduced column echelon matrices for a given m can be solved by using a modified version of the so-called *full linear search algorithm* (FLSA). Such a *modified full linear search algorithm* (MFLSA) generates one member of each modified Hadamard class, namely, the labeling that corresponds to a reduced column echelon matrix $\mathbf{Q}_{\mathrm{R}}$. Its pseudocode is shown in Algorithm 1. In this algorithm, the (row) vector $\boldsymbol{r} = [r_1, \ldots, r_M]$ denotes the decimal representation of the columns of the matrix $\mathbf{Q}_{\mathrm{R}}$. The first labeling generated (line 1) is always the NBC. Then the algorithm proceeds by generating all permutations thereof, under the condition that no power of two $(1, 2, 4, \ldots)$ is preceded by a larger value. By Definition 2.14, this simple condition assures that only reduced column echelon matrices are generated.

Example 9.11 (MFLSA for $m = 2$**)** *For* $m = 2$, *the MFLSA returns the following reduced column echelon matrices:*

$$\mathcal{R}_2 = \left\{ \begin{bmatrix} 0 & 0 & \mathbf{1} & 1 \\ 0 & \mathbf{1} & 0 & 1 \end{bmatrix}, \begin{bmatrix} 0 & 0 & \mathbf{1} & 1 \\ \mathbf{1} & 0 & 0 & 1 \end{bmatrix}, \begin{bmatrix} 0 & \mathbf{1} & 0 & 1 \\ \mathbf{1} & 0 & 0 & 1 \end{bmatrix}, \begin{bmatrix} 0 & \mathbf{1} & 1 & 0 \\ \mathbf{1} & 0 & 1 & 0 \end{bmatrix} \right\}, \tag{9.57}$$

where the first element in $\mathcal{R}_2$ *is the NBC in Definition 2.11 and we highlighted the pivots of the matrices. Alternatively, the set* $\mathcal{R}_2$ *in decimal notation (as the MFLSA generates it) is*

$$\mathcal{R}_2 = \{[0, 1, 2, 3], [1, 0, 2, 3], [1, 2, 0, 3], [1, 2, 3, 0]\}. \tag{9.58}$$

The six binary invertible matrices for $m = 2$ are

$$\mathcal{T}_2 = \left\{ \begin{bmatrix} 0 & 1 \\ 1 & 0 \end{bmatrix}, \begin{bmatrix} 0 & 1 \\ 1 & 1 \end{bmatrix}, \begin{bmatrix} 1 & 0 \\ 0 & 1 \end{bmatrix}, \begin{bmatrix} 1 & 0 \\ 1 & 1 \end{bmatrix}, \begin{bmatrix} 1 & 1 \\ 0 & 1 \end{bmatrix}, \begin{bmatrix} 1 & 1 \\ 1 & 0 \end{bmatrix} \right\}. \tag{9.59}$$

Using Theorem 9.10, all the 24 binary labelings in $\mathcal{Q}_2$ (see Table 9.5) can be generated by multiplying the matrices in $\mathcal{R}_2$ and in $\mathcal{T}_2$.

Algorithm 1 Modified full linear search algorithm (MFLSA)

Input: The order m

Output: Print the M_R different reduced column echelon vectors $\boldsymbol{r}$

```
r ← [0, 1, ..., M−1]
loop
    print r
    index ← 0
    while r_M = index do
        Circularly shift [r_index + 1, ..., r_M] one step to the right
        index ← index + 1
        while index is a power of 2 do
            index ← index + 1
        end while
        if index = M−1 then
            Quit
        end if
    end while
    Find pointer such that r_pointer = index
    Swap r_pointer and r_pointer + 1
end loop
```

Example 9.12 (MFLSA for $m = 3$) *For $m = 3$, the MFLSA generates the reduced column echelon matrices (given in decimal notation) shown in Table 9.6 (read row by row). The first column in the table corresponds to the output of the FLSA. Columns two to seven show the additional matrices generated by the MFLSA, which differ from the first column only in the position of the all-zero codeword.*

Example 9.13 (Classification of 8PSK Labelings) *Analyzing the labelings in Example 2.20, we find that the labeling in Fig. 2.17 (b) belongs to the first modified Hadamard class ($\mathbf{Q}_{\mathrm{R}} = \mathbf{Q}_{\mathrm{NBC}}$) and that the labeling in Fig. 2.17 (c) belongs to a different class, i.e.,*

$$\mathbf{Q}_{\mathrm{SSP/M8}} = \begin{bmatrix} 1 & 0 & 0 \\ 0 & 1 & 0 \\ 1 & 0 & 1 \end{bmatrix} \mathbf{Q}_{\mathrm{NBC}}, \quad \mathbf{Q}_{\mathrm{MSP}} = \begin{bmatrix} 1 & 1 & 1 \\ 0 & 1 & 0 \\ 0 & 0 & 1 \end{bmatrix} \mathbf{Q}_{\mathrm{R}}, \tag{9.60}$$

where $\mathbf{Q}_{\mathrm{R}} = [0, 1, 2, 4, 7, 6, 5, 3]$ (in decimal). This shows that in an exhaustive search, the NBC does not span all the SP labelings.

9.3.3 NBC and BRGC

Another way of interpreting the result in Theorem 9.9 is that for any trellis encoder $\tilde{\Xi} = [\tilde{\mathbf{G}}, \tilde{\mathbf{Q}}]$, a new equivalent trellis encoder can be generated using an encoder $\mathbf{G} = \tilde{\mathbf{G}}(\mathbf{T}^{-1})^{\mathrm{T}}$ and a labeling $\mathbf{Q} = \mathbf{T}^{-1}\tilde{\mathbf{Q}}$ that belongs to the same modified Hadamard class as the original labeling $\tilde{\mathbf{Q}}$. One direct consequence of

Table 9.6 Output of the MFLSA for $m = 3$ (read row by row)

01234567	10234567	12034567	12304567	12340567	12345067	12345607	12345670
01243567	10243567	12043567	12403567	12430567	12435067	12435607	12435670
01245367	10245367	12045367	12405367	12450367	12453067	12453607	12453670
01245637	10245637	12045637	12405637	12450637	12456037	12456307	12456370
01245673	10245673	12045673	12405673	12450673	12456073	12456703	12456730
01234657	10234657	12034657	12304657	12340657	12346057	12346507	12346570
01243657	10243657	12043657	12403657	12430657	12436057	12436507	12436570
01246357	10246357	12046357	12406357	12460357	12463057	12463507	12463570
01246537	10246537	12046537	12406537	12460537	12465037	12465307	12465370
01246573	10246573	12046573	12406573	12460573	12465073	12465703	12465730
01234675	10234675	12034675	12304675	12340675	12346075	12346705	12346750
01243675	10243675	12043675	12403675	12430675	12436075	12436705	12436750
01246375	10246375	12046375	12406375	12460375	12463075	12463705	12463750
01246735	10246735	12046735	12406735	12460735	12467035	12467305	12467350
01246753	10246753	12046753	12406753	12460753	12467053	12467503	12467530
01234576	10234576	12034576	12304576	12340576	12345076	12345706	12345760
01243576	10243576	12043576	12403576	12430576	12435076	12435706	12435760
01245376	10245376	12045376	12405376	12450376	12453076	12453706	12453760
01245736	10245736	12045736	12405736	12450736	12457036	12457306	12457360
01245763	10245763	12045763	12405763	12450763	12457063	12457603	12457630
01234756	10234756	12034756	12304756	12340756	12347056	12347506	12347560
01243756	10243756	12043756	12403756	12430756	12437056	12437506	12437560
01247356	10247356	12047356	12407356	12470356	12473056	12473506	12473560
01247536	10247536	12047536	12407536	12470536	12475036	12475306	12475360
01247563	10247563	12047563	12407563	12470563	12475063	12475603	12475630
01234765	10234765	12034765	12304765	12340765	12347065	12347605	12347650
01243765	10243765	12043765	12403765	12430765	12437065	12437605	12437650
01247365	10247365	12047365	12407365	12470365	12473065	12473605	12473650
01247635	10247635	12047635	12407635	12470635	12476035	12476305	12476350
01247653	10247653	12047653	12407653	12470653	12476053	12476503	12476530

this result is that any trellis encoder using the NBC labeling can be constructed using the BRGC and an appropriately selected encoder. To clarify this, consider the following example.

Example 9.14 *The BRGC and the NBC of order $m = 2$ belong to the same modified Hadamard class, which follows directly from Example 2.12. Because of this, CENCs can be chosen to make the two resulting trellis encoders equivalent. This was illustrated in Fig. 9.5, where the transform block corresponds to the transform matrix $\mathbf{T} = [[1,1]^{\mathrm{T}}, [0,1]^{\mathrm{T}}] = \mathbf{T}^{-1}$. As $\mathbf{Q}_{\mathrm{NBC}} = \mathbf{T}^{-1}\mathbf{Q}_{\mathrm{BRGC}}$, the trellis encoders $[\mathbf{G}, \mathbf{Q}_{\mathrm{BRGC}}]$ and $[\mathbf{G}', \mathbf{Q}_{\mathrm{NBC}}]$ are equivalent, where*

$$\mathbf{G} = \begin{bmatrix} 1 & 1 \\ 0 & 1 \\ 1 & 1 \\ 1 & 1 \end{bmatrix} = \mathbf{G}'(\mathbf{T}^{-1})^{\mathrm{T}} = \begin{bmatrix} 1 & 0 \\ 0 & 1 \\ 1 & 0 \\ 1 & 0 \end{bmatrix} \begin{bmatrix} 1 & 1 \\ 0 & 1 \end{bmatrix}. \tag{9.61}$$

The above relation between the NBC and the BRGC is generalized to an arbitrary order m in the following theorem.

Theorem 9.15 *The BRGC and the NBC of any order m belong to the same modified Hadamard class.*

Proof: The BRGC and NBC are related via $\mathbf{Q}_{\text{BRGC}} = \mathbf{T}\mathbf{Q}_{\text{NBC}}$, with $\mathbf{T}$ given by (2.56). The theorem now follows from Theorem 9.10 and the definition of a modified Hadamard class. □

Theorem 9.15 can be understood as follows. Any trellis encoder using the NBC $\mathbf{Q}_{\text{NBC}}$ and a CENC $\mathbf{G}$ is equivalent to a trellis encoder using the BRGC $\mathbf{Q}_{\text{BRGC}}$ and a CENC $\mathbf{G}\mathbf{T}^{\text{T}}$ with $\mathbf{T}$ given by (2.56). Example 9.14 and Theorem 9.15 explain, in part, the results obtained in Section 9.2.3, where it is shown that BICM-T with the optimal encoders and the BRGC perform asymptotically as well as TCM. The "in part" comes from the fact that in BICM-T a (suboptimal) BICM receiver is used while in TCM an ML decoder is used.

9.3.4 *M*PAM *and M*PSK *Constellations*

While all the previous results are fully general in the sense that they are valid for any constellation, any memoryless channel, and any receiver, in this section we study well-structured one- and two-dimensional constellations, i.e., MPAM and (MPSK) constellations defined in Section 2.5.

MPAM constellations are symmetric around zero. Because of this, two trellis encoders based on an MPAM constellation, the first one using the labeling $\mathbf{Q} = [\boldsymbol{q}_1, \boldsymbol{q}_2, \ldots, \boldsymbol{q}_{M-1}, \boldsymbol{q}_M]$ and the second one using a "reverse" labeling

$$\mathbf{Q}' = [\boldsymbol{q}'_1, \boldsymbol{q}'_2, \ldots, \boldsymbol{q}'_{M-1}, \boldsymbol{q}'_M] = [\boldsymbol{q}_M, \boldsymbol{q}_{M-1}, \ldots, \boldsymbol{q}_2, \boldsymbol{q}_1], \tag{9.62}$$

are equivalent for any M. This result implies that the number of binary labelings that give nonequivalent trellis encoders is $M_{\text{R}}/2$. To generate only the $M_{\text{R}}/2$ nonequivalent labelings for MPAM, the MFLSA in Algorithm 1 can be modified as follows. Replace r_M on lines 5 and 6 with $r_{e(index)}$, where the integer function $e(q)$ is defined as $M/2$ if $q = 0$ and M otherwise. This has the effect of only generating labelings, in which the all-zero label is among the first $M/2$ positions.

Example 9.16 (Equivalent Labelings for 4PAM) *For* 4PAM *($m = 2$), only* 2 *labelings need to be tested, instead of the* 24 *in an exhaustive search (see Table 9.5). These two labelings are the NBC (first labeling in* (9.57)*) and the second labeling in* (9.57)*. This labeling in fact corresponds to the BRGC, where the second bit position is negated. This shows that a negation of a bit position, which is an operation that has no impact on the BEP of uncoded transmission or on the BICM-GMI, does make a difference in this classification of labelings.*

Example 9.17 (Equivalent Labelings for 8PAM) *For* 8PAM *($m = 3$),* 120 *labelings need to be evaluated, instead of* 40320 *in an exhaustive search; see Table 9.5. These* 120 *labelings are listed in the first four columns of Table 9.6.*

A trellis encoder based on an MPSK constellation is not affected by a circular rotation of its labeling, i.e., without loss of generality it can be assumed that the all-zero label is assigned to the constellation point $\boldsymbol{x}_1 = [1, 0]$. The consequence of this is that for MPSK constellations, the number of reduced column echelon matrices that give nonequivalent trellis encoders is reduced further by a factor of M. For $m \geq 3$ the labelings can be obtained from the MFLSA by setting $index \leftarrow 3$ in line 4, which gives the FLSA.

Example 9.18 (Equivalent Labelings for 4PSK) *In view of the results in Table 9.5, for* 4PSK*, there is only one labeling that needs to be tested, e.g., the NBC, and thus, only a search over the encoders needs to be performed. Moreover, without loss of generality, we can use the BRGC instead (because it is in the same Hadamard class of the NBC) and search over encoders for this labeling. Since* 4PSK *labeled* *by the BRGC can be considered as two independent* 2PAM *constellations (one in each dimension), the design of trellis encoders in this case boils down to selecting CENCs with optimal spectrum. This implies that asymptotically optimal trellis encoders (in terms of BEP) can be constructed by coupling the ODS CENCs in Table 2.1 and a* 4PSK *constellation labeled by the BRGC.*

Example 9.19 (Equivalent Labelings for 8PSK**)** *For 8PSK ($m = 3$), there are* 30 *labelings that need to be tested, which correspond to the first column of Table 9.6.*

9.4 Bibliographical Notes

BICM-T was introduced in [1] and later analyzed in detail in [2] for a rate $R_c = 1/2$ encoder and 4PAM, where the performance analysis developed in this chapter as well as the optimal encoders were also introduced. Very recently, it was shown in [3] that for BICM-T, the asymptotic loss in terms of PEP introduced by the suboptimal BICM receiver is bounded by 1.25 dB. Furthermore, [3] also shows that this loss is in fact zero for a wide range of linear codes, including all rate $R_c = 1/2$ CENCs we studied in this chapter.

TCM designs based on SP are considered heuristic [4, pp. 525, 531], [5, Section 3.4], and thus, they do not necessarily lead to an optimal design [6, p. 680]. Indeed, the results in [7, Tables 2–3], [8, Chapter 6; 9] show the suboptimality of using an SP labeling. For a good explanation about this, we refer the reader to [10, Section I]. The problem of using non-SP labelings for TCM has been studied in the literature. In [6, Section 13.2.1], a trellis encoder that uses a rate $1/2$ ODS CENC and a 4PSK constellation labeled by the BRGC was presented (see also [11, Example 18.2]). The same configuration was presented in [6, Problem 13–11], where the problem of selecting an appropriate labeling for a given encoder was analyzed. Another example is the so-called pragmatic TCM [12, Section 8.6, 13] (see also [14, 15, Section 9.2.4]), where a non-SP labeling was used together with an off-the-shelf MFHD/ODS CENC. For 4PSK, this labeling corresponds to the BRGC, but for larger constellations it is not a Gray labeling (although it has a particular structure). A similar labeling was also used in [16]. A binary labeling for BICM-T was heuristically proposed in [1] for MPAM constellations for $M = 4, 8, 16$.

The idea of using a Gray labeling and an MFHD CENC in fact dates back to [17, 18], which correspond to the first pragmatic TCM approaches. Trellis encoders using a constellation labeled by the BRGC were designed in [7] where a search over CENCs maximizing the MED was performed. In [8, Chapter 6; 9], and in the same spirit of the ODS CENCs, Zhang used a non-Gray non-SP labeling and tabulated trellis encoders with optimal spectrum. Gray-coded TCM codes for fading channels were studied in [19–21].

The classification of labelings is inspired by the one in [22], where the so-called Hadamard classes were used to solve a related search problem in source coding and where the FLSA was also introduced. The modified Hadamard classes introduced in this chapter are narrower than the regular Hadamard classes defined in [22], each including M reduced column echelon matrices. For a proof of Theorem 9.10, we refer the reader to [23, p. 187, Corollary 1].

The idea of applying a linear transformation to the labeling/encoder was first introduced in [8, Fig. 6.5], (see also [9, 24, 25, Chapter 2]). The equivalence between TCM encoders and encoders optimized for the BRGC and the NBC as well as the relationship between the encoders in [2, 26] were first pointed out to us by R. F. H. Fischer [27]. In the same way the encoders in [26, Table I], (which we reproduced in Table 9.4) and [2, Table III] are equivalent, it was shown in [25, Chapter 2], that the encoders found in [8, Chapter 6], using a non-Gray labeling are equivalent to those found with a Gray-labeled constellation in [7]. The formalism of using modified Hadamard classes to study the equivalence of trellis encoders was introduced in [28, 29], where the MFLSA was also introduced and trellis encoders with optimum distance spectrum were tabulated.

In a related work, Wesel *et al.* introduced in [10] the concept of *edge profile* (EP) of a labeling, and argued that in most of the cases, the EP can be used to find equivalent trellis encoders in terms of MED. The EP is also argued to be a good indication of the quality of a labeling for TCM in [10, Section I]; however, its optimality is not proven. Nevertheless, all the labelings found in [28] have optimal EP. This makes us conjecture that good trellis encoders can be found by using the EP of [10] on top of the Hadamard-based classification. This approach would indeed reduce the search space; however, the optimality would not be guaranteed.

References

[1] Stierstorfer, C., Fischer, R. F. H., and Huber, J. B. (2010) Optimizing BICM with convolutional codes for transmission over the AWGN channel. International Zurich Seminar on Communications, March 2010, Zurich, Switzerland.

[2] Alvarado, A., Szczecinski, L., and Agrell, E. (2011) On BICM receivers for TCM transmission. *IEEE Trans. Commun.*, **59** (10), 2692–2702.

[3] Ivanov, M., Alvarado, A., Brännström, F., and Agrell, E. (2014) On the asymptotic performance of bit-wise decoders for coded modulation. *IEEE Trans. Inf. Theory*, **60** (5), 2796–2804.

[4] Proakis, J. G. (2000) *Digital Communications*, 4th edn, McGraw-Hill.

[5] Schlegel, C. B. and Perez, L. C. (2004) *Trellis and Turbo Coding*, 1st edn, John Wiley & Sons.

[6] Barry, J. B., Lee, E. A., and Messerschmitt, D. G. (2004) *Digital Communication*, 3rd edn, Springer.

[7] Du, J. and Kasahara, M. (1989) Improvements of the information-bit error rate of trellis code modulation systems. *Trans. IEICE*, **E 72** (5), 609–614.

[8] Zhang, W. (1996) Finite state systems in mobile communications. PhD dissertation, University of South Australia, Adelaide, Australia.

[9] Zhang, W., Schlegel, C., and Alexander, P. (1994) The bit error rate reduction for systematic 8PSK trellis codes by a Gray scrambler. IEEE International Conference on Universal Wireless Access, April 1994, Melbourne, Australia.

[10] Wesel, R. D., Liu, X., Cioffi, J. M., and Komninakis, C. (2001) Constellation labeling for linear encoders. *IEEE Trans. Inf. Theory*, **47** (6), 2417–2431.

[11] Lin, S. and Costello, D. J. Jr. (2004) *Error Control Coding*, 2nd edn, Prentice Hall, Englewood Cliffs, NJ.

[12] Clark, G. C. Jr. and Cain, J. B. (1981) *Error-Correction Coding for Digital Communications*, 2nd edn, Plenum Press.

[13] Viterbi, A. J., Wolf, J. K., Zehavi, E., and Padovani, R. (1989) A pragmatic approach to trellis-coded modulation. *IEEE Commun. Mag.*, **27** (7), 11–19.

[14] Wolf, J. K. and Zehavi, E. (1995) p^2 codes: pragmatic trellis codes utilizing punctured convolutional codes. *IEEE Commun. Mag.*, **33** (2), 94–99.

[15] Morelos-Zaragoza, R. H. (2002) *The Art of Error Correcting Coding*, 2nd edn, John Wiley & Sons.

[16] Hagenauer, J. and Sundberg, C.- E. (1988) On the performance evaluation of trellis-coded 8-PSK systems with carrier phase offset. *Arch. Elektron. Uebertragungstechnik*, **5** (42), 274–284.

[17] Odenwalder, J. P., Viterbi, A. J., Jacobs, I. M., and Heller, J. A. (1973) Study of information transfer optimization for communication satellites. Technical Report NASA-CR-114561, Linkabit Corporation, San Diego, CA.

[18] Digeon, A. (1977) On improving bit error probability of QPSK and 4-level amplitude modulation systems by convolutional coding. *IEEE Trans. Commun.*, **COM-25** (10), 1238–1239.

[19] Du, J. and Vucetic, B. (1990) New M-PSK trellis codes for fading channels. *Electron. Lett.*, **26** (16), 1267–1269.

[20] Du, J. and Vucetic, B. (1991) New 16-QAM trellis codes for fading channels. *Electron. Lett.*, **27** (12), 1009–1010.

[21] Du, J. and Vucetic, B. (1995) Construction of new MPSK trellis codes for fading channels. *IEEE Trans. Commun.*, **43** (2/3/4), 776–478.

[22] Knagenhjelm, P. and Agrell, E. (1996) The Hadamard transform—a tool for index assignment. *IEEE Trans. Inf. Theory*, **42** (4), 1139–1151.

[23] Birkhoff, G. and Mac Lane, S. (1977) *A Survey of Modern Algebra*, 4th edn, Macmillan, New York.

[24] Gray, P. K. and Rasmussen, L. K. (1995) Bit error rate reduction of TCM systems using linear scramblers. IEEE International Symposium on Information Theory (ISIT), September 1995, Whistler, BC, Canada.

[25] Gray, P. K. (1999) Serially concatenated trellis coded modulation. PhD dissertation, University of South Australia, Adelaide, Australia.

[26] Ungerboeck, G. (1982) Channel coding with multilevel/phase signals. *IEEE Trans. Inf. Theory*, **28** (1), 55–67.

[27] Fischer, R. F. H. Private communication, Jan. 2011.

[28] Alvarado, A., Graell i Amat, A., Brännström, F., and Agrell, E. (2012) On the equivalence of TCM encoders. IEEE International Symposium on Information Theory (ISIT), July 2012, Cambridge, MA.

[29] Alvarado, A., Graell i Amat, A., Brännström, F., and Agrell, E. (2013) On optimal TCM encoders. *IEEE Trans. Commun.*, **61** (6), 2178–2189.

Index